2014

INTERNATIONAL DESIGN EXCHANGE PROJECT

The 5th Hidden Space Project

RE-ENERGISING SPACE

01

The Cass London Metropolitan University

The Sir John Cass Faculty of Art, Architecture and Design

Interior Design BA (Hons)

UK

02

Hunan University

Dept. of Industrial Design, in school of Design

CHN

03

Dongyang Mirae University

Dept. of Interior Design, in school of Design

Rep. of KOREA

The 5th Hidden Space Project

RE-ENERGISING SPACE

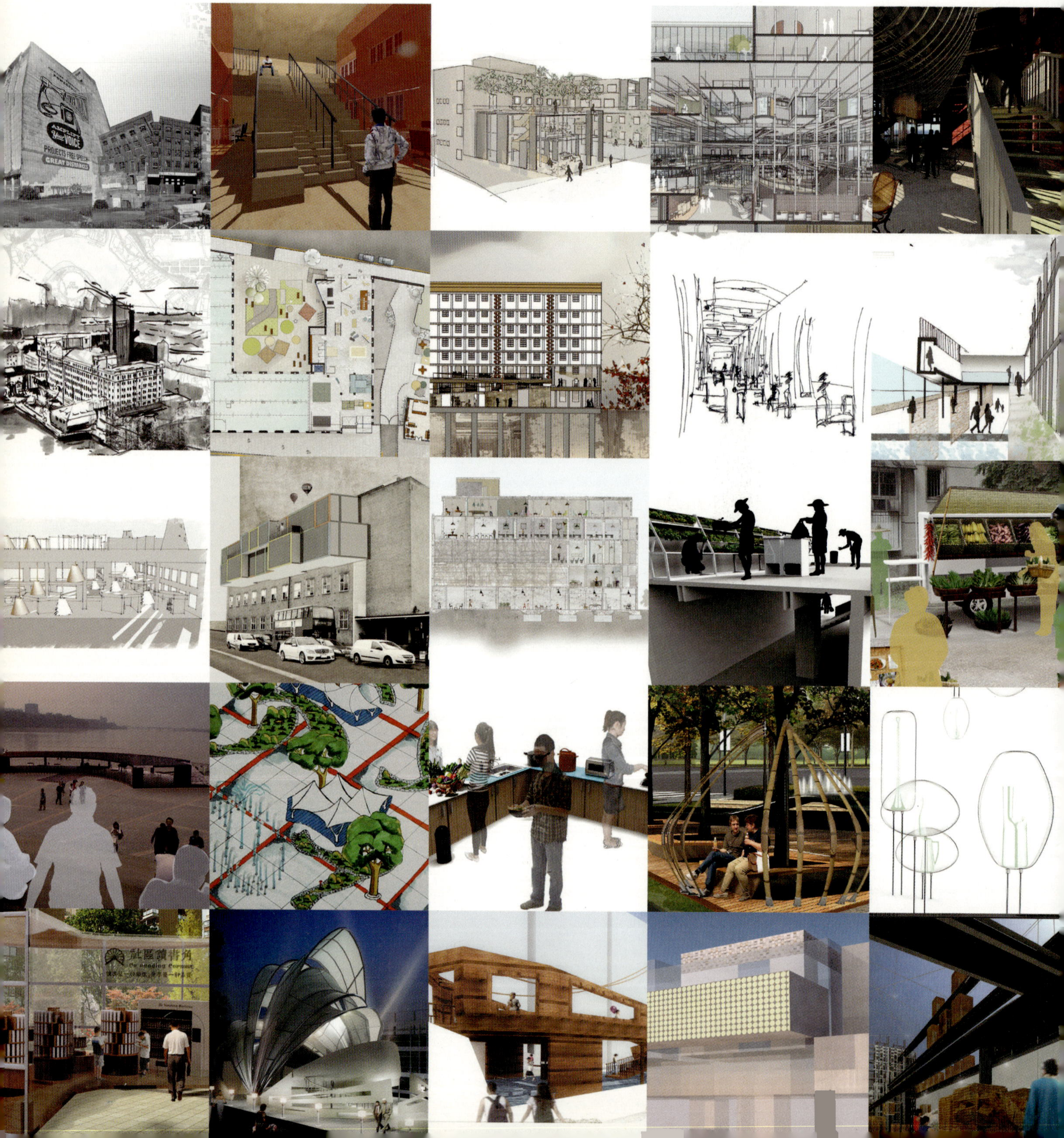
PROJECTS FREE SPEECH
GREAT DISTANCES

_ 49 Projects

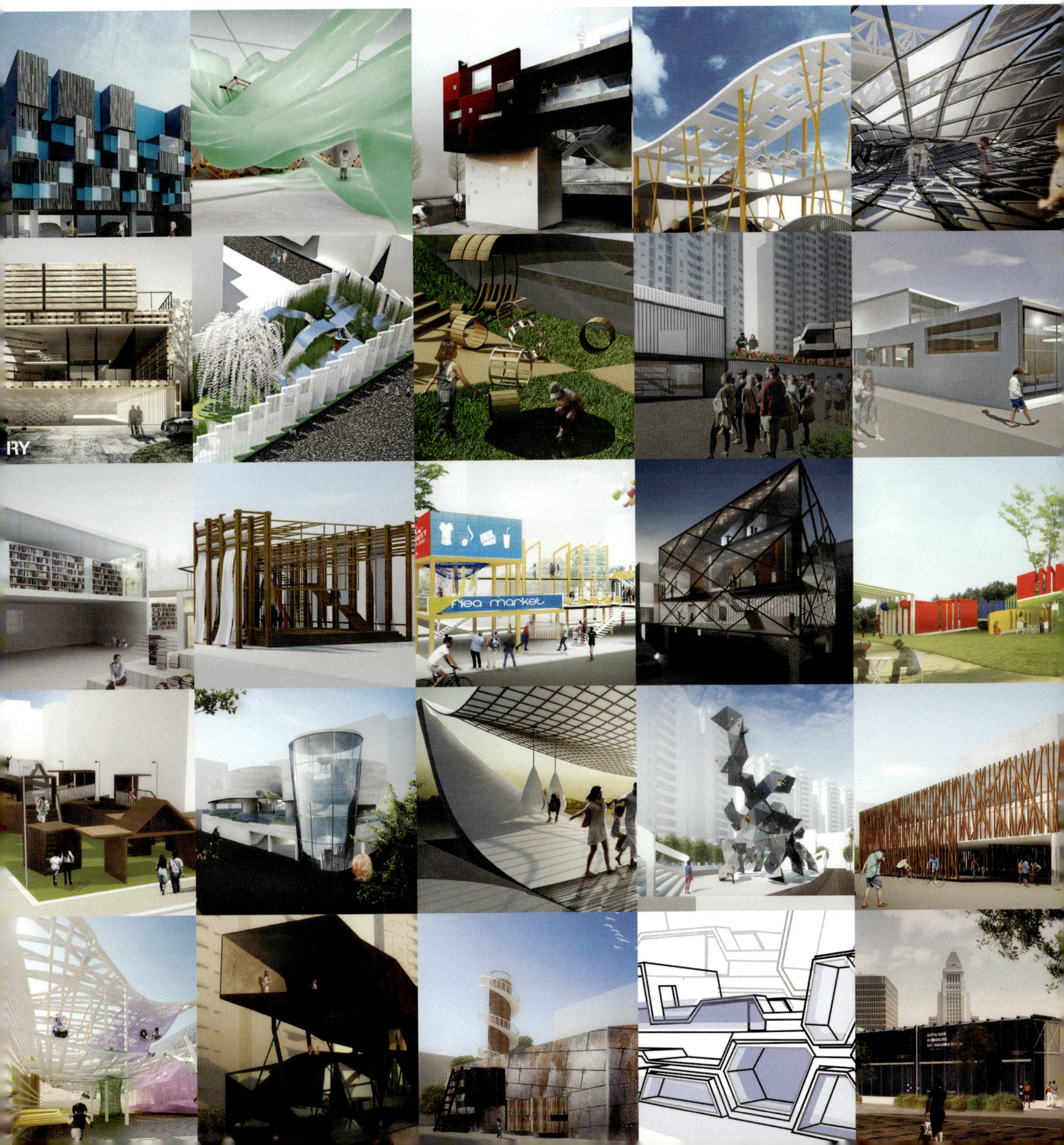

contents

01 The Cass London Metropolitan University _UK

INTRODUCTION

PROJECTS

02 Hunan University _CHN

INTRODUCTION

PROJECTS

03 Dongyang Mirae University _Rep. of KOREA

INTRODUCTION

PROJECTS

1

The Cass London Metropolitan University
The Sir John Cass Faculty of Art, Architecture and Design
Interior Design BA (Hons)

The Sir John Cass Faculty of Art, Architecture and Design
41-47 Commercial Road
Lonodon E1 1LA
Switchboard 020 7423 0000
www.thecass.com

Kaye Newman

Course Organizer
Interior Design BA (Hons)
The Cass Faculty of Art, Architecture and Design, London Metropolitan Univerity

Re-Energising Space

This is now the 5th Book in the Hidden Space series. It has historically been a project between the Interior design students of London Metropolitan University, Faculty of Art, Architecture and Design and now called the CASS and the students of Dong Yang University, Seoul, South Korea and at points we have also collaborated with Kyung Won University also in Seoul.
This year we have invited Hunan University of Changsha in China to join with us, so we welcome our new collaborators and look forward to the exchange of ideas and the cultural interchange that is both fascinating but helps us to understand and reflect with different methodologies.

The CASS Faculty of Art, Architecture and Design now bases its projects within a studio, running an atelier system. Each studio is led by a practitioner or an expert with in the specific discipline which not only allows the students to develop skills and understanding it shows precise insights of practice.

The studios have allowed us to broaden our range of projects, as we ran two projects rather than one this year that looked at Re-energising space.

Studio 1, Unspoken Revolution has looked at a large and complex set of derelict warehouse factory buildings situated in the 'out of town' South East of London where the river edges into the estuary. The Millennium Mills sits opposite London City airport abandoned and dilapidated. But it's this quality of dilapidation and the memory of older practices that takes the imagination. The studio have endeavoured to keep the romantic materiality but suggest interventions that make this location a place for its community to enjoy.

Studio 2, The Shoreditch Vibe, has looked at a semi vacant space situated in the East End, where it shares space with the local infrastructure of the telephone exchange. Being originally built by the government owned GPO in 1956, the cost of refurbishment is now too expensive to contemplate when considering the upheaval of the telephone exchange. However this part of the East End has many buildings whose architectural character has remained as until recently the developers overlooked this area in favour of the lucrative City and West End so the whole area offers an eclectic mix of authenticity, unpolished and makeshift spatial creativity.

Kaye Newman
Course Organizer - 코스 기획자

이 책은 숨겨진 공간(Hidden Space) 시리즈 중 5번째의 책이다. 이 책은 역사적으로 런던 메트로폴리탄 대학의 인테리어 디자인 학생들과 지금은 CASS라 부르는 예술, 건축 및 디자인 교수진, 그리고 서울 동양미래대학의 학생들 간의 프로젝트였으며, 때로 우리는 서울 경원(현재의 가천)대학교와도 협력을 했었다.
올해 우리는 중국 창사의 Hunan University에게 우리와 함께 해줄 것을 요청했으며, 따라서 우리는 새로운 협력자들을 환영하고, 우리가 다른 방법들을 이해하고 생각해보는데 도움이 될 문화 교류 및 아이디어의 교류가 이루어질 것을 기대하고 있다.

CASS 예술, 건축 및 디자인 교수진들은 이제 이 프로젝트를 아틀리에 시스템으로 운영되는 스튜디오 내에서 진행하고 있다. 각 스튜디오는 특정 분야에서 활동하고 있는 사람이나 전문가가 이끌고 있으며, 이는 학생의 능력 개발을 촉진할 뿐 아니라 현장에 대한 정확한 인사이트를 가지게 하는데 도움이 될 것이다.

우리는 공간에 다시 활력을 불어넣기 위해 올해에는 1개의 프로젝트가 아니라 2개의 프로젝트를 운영하고 있기 때문에 이 스튜디오들을 통해 프로젝트의 범위를 확대할 수 있었다.

스튜디오 1은 알려지지 않은 혁명(Unspoken Revolution)으로서, 강 유역에 있는 런던 남동부의 "오래된 마을"에 위치한 크고 복잡한 버려진 창고 공장 건물을 대상으로 하고 있다. 이 밀레니엄 밀은 런던 시티 공항 반대편에 위치해 있으며, 버려진 채 다 허물어져 가고 있는 건물들이다. 하지만, 상상력을 자극하는 것이 바로 그러한 버려졌다는 점과 오래 전의 일에 대한 기억들이다. 스튜디오 1은 낭만적 유형성을 유지하기 위해 노력해왔지만 이 장소를 지역사회가 즐길 수 있는 장소로 개조할 것을 제안하고 있다.

스튜디오 2는 Shoreditch Vibe로서, East End에 위치한 절반은 비어 있는 공간을 대상으로 하고 있으며, 이 공간은 전신교환국 인프라와 공간을 공유하고 있다. 원래는 정부가 소유한 GPO가 1956년에 지은 건물로서, 이 공간을 개조하는 비용은 전신교환국의 변화를 고려할 때 너무 많은 비용이 든다. 하지만 최근까지도 개발업체들이 수익성이 좋은 West End쪽을 선호해 이 지역을 간과해왔기 때문에 이 East End 부지에는 아직도 건축적 특성을 유지하고 있는 건물들이 많이 남아있다. 따라서 이 지역 전체에는 순수함과 손대지 않은 임기응변적인 창의력을 결합시켜 볼 수 있는 여지가 남아 있다.

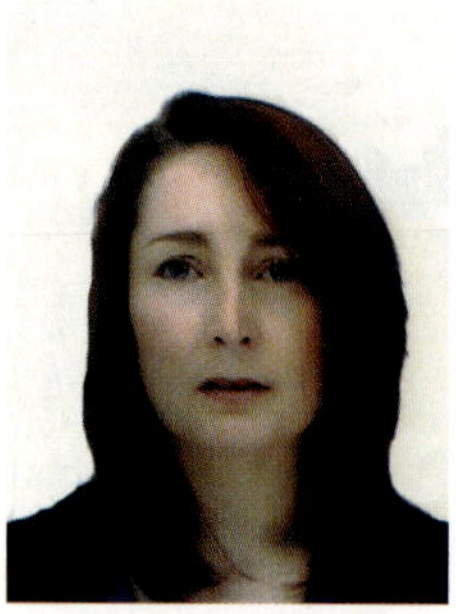

Janette Harris MALTHE, SEDA

Senior Lecturer
Module Leader
Senior Lecturer
Interior Design BA (Hons)
The Cass Faculty of Art, Architecture and Design, London Metropolitan Univerity
Studio Leader of Rise of the Creative District

Pedagogical Approaches

Studio system

- Year long modules
- Atelier style learning and teaching with a mixture of Year 2 and 3 in the same studio
- Studio Leader plus supporting Lecturers
- Average 28 students to a studio

Our approach is essentially constructivist, encouraging connections of learning, while offering opportunities for experimentation, testing and innovation, whether through, personal or group activities. The goal is to support students to develop existing skills to work in harmony with new digital technologies working towards developing resources and mechanisms to fulfil the needs of the 21st Century learner. Critical thinking is at the heart of our philosophy. We engage students in immersive research and reflective action, working towards an authentic learning experience, valued by industry and professional practice. Projects are developed with the aim of building skills for the future knowledge economy, and inspire life-long learning. Face-to-face sessions supported by online environments encourage a shift from tutor centred teaching to student centred learning. Innovative live projects employ themes that act as real world experiences supported by professional practice and serve to provide meaningful contextual provision for students.

All learning partners (staff and students) under our guidance work as a community of practice by sharing, gifting and developing skills through proximity. Cooperation, communication, social responsibility, coordination and cohesive approaches enable students and staff to become independent learners, with the support of a learning community. A framework supporting learners can be adapted to the individual and group needs to shift emphasis if appropriate.

The Projects

The projects are concerned with the exploration and realisation of structure and materials within a space in order to re-energise the existing. Workshop activities encourage a wide range of skills, developing reflection leading to action. The idea of expressing concepts through theoretical narratives, developed into 3D outcomes asks the students to develop, form, structure, texture and materiality from a framework of research. Visual communication is key to start the process of expression, the heart of the project. Diverse learning styles are taken into consideration, with strengths being supported and building on skills that enable a successful outcome.

- Learning styles are adapted and developed through active learning within a community within a blended learning environment
- Face-to-face sessions are active and shift students' domains through reflection and action
- Students are prompted (in early stages of a project) to experiment, testing outcomes to their limits, allowing a choice of approaches
- Materials choices and experimentation are key to the experience of the outcome. Students test, but always try to relate back to their concept.
- Students recognise when to take a step back, and not seeing this as failure, rather a positive step to recognising a need to shift emphasis
- Formative assessment is scheduled at two key stages during the module, where issues can be resolved and reviewed from the student and tutor's perspective
- The outcomes demonstrate key skills for employability, including graphic communication for publishing

Conclusion

The most successful learning and teaching approach this year was from Shoreditch Vibe ran by Kaye Newman.

- Started the Major project from day 1
- Punctuated the year with a series of Critiques with Professional practice and clients on the live brief, which kept the students on track and allowed for reflection and action
- The development of ideas through modelling, testing, experimenting and reflecting
- Evidencing decisions through theoretical underpinning
- Evidencing the experiential materiality of the spatial journey
- The development of personal presentation responses through hand drawing and digital technologies
  The tutors are very keen that each person has their own style, developed through the understanding of themselves, their actions and reactions. The result is unique portfolio's that the industry takes notice of.

Janette Harris
Senior Lecturer
Module Leader – 교과목 지도교수

교육 접근방식

스튜디오 시스템

- 연간 지속되는 모듈
- 동일 스튜디오에 2년제와 3년제가 혼합되어 있는 아틀리에 스타일의 학습 및 수업
- 스튜디오 리더 + 지원 강사진
- 한 스튜디오 당 평균 28명의 학생

우리의 접근방식은 본질적으로 구성주의적 접근방식으로서, 학습의 연관성을 강조하면서 동시에 개인 활동 또는 그룹 활동을 통해 실험, 테스트 및 혁신의 기회를 제공해주고 있다. 교육 목표는 학생들이 갖고 있는 기존의 능력들을 21세기 학습자들의 니즈를 충족시킬 수 있는 자원과 메커니즘을 개발할 수 있는 새로운 디지털 기술과 조화를 이룰 수 있게 발전시켜나가는데 있다. 우리 철학의 핵심에 있는 것이 비판적 사고이다. 우리는 학생들이 많은 연구활동과 사색적 활동을 하는 것을 권장하며, 산업 현장 및 전문 직업 현장에서 인정받을 수 있는 참된 학습 경험을 해보기를 격려하고 있다. 프로젝트들은 미래의 지식 산업을 위한 능력 계발과 평생 학습이라는 목표 하에 이루어지고 있다. 면대면 수업 시간과 이를 보완한 온라인 환경을 통해 교사 중심의 수업에서 학생 중심의 학습으로의 전환을 꾀하고 있다. 혁신적인 생생한 프로젝트들에서는 현장에서 실제로 경험할 수 있는 주제들을 채택하고 있으며, 전문 직업 경험으로 이를 보완하고 있고, 이 프로젝트들을 통해 학생들에게 의미 있는 경험을 제공해주려고 한다.

우리의 지도 하에 있는 모든 학습 파트너들(교직원과 학생)은 근접성을 통해 능력을 공유하고, 기부하며, 개발하는 과정을 통해 일종의 훈련 커뮤니티로서의 역할을 담당하고 있다. 협력, 의사소통, 사회적 책임, 조율 및 응집력 있는 접근방식 등을 통해 학생과 교직원들이 학습 커뮤니티의 지원을 통해 독립적 학습자가 될 수 있게 하고 있다. 필요한 경우 학습 중점을 변화시켜야 하는 개인의 니즈나 그룹 니즈에 맞게 학습자들을 지원하는 기본틀을 변경시킬 수도 있다.

프로젝트

프로젝트들은 기존의 것을 리엔지니어링 하기 위해 공간 내 구조 및 재료를 탐색하고 현실화 시키는 것에 중점을 두고 있다. 작업실 활동들은 다양한 능력과 행동으로 이어질 수 있는 사색을 촉진시키고 있다. 이론적 내러티브를 통해 컨셉을 표현하고 3D 결과물로 발전시킨다는 아이디어는 학생들에게 연구 틀로부터 구조, 텍스처, 유형성을 개발하고 구성할 것을 요구하고 있다. 시각적 커뮤니케이션이 프로젝트의 핵심인 표현 과정을 시작하는 중요한 열쇠라 할 수 있다. 다양한 학습 스타일들을 고려하며, 장점을 강화시키고, 성공적 결과물을 낼 수 있는 능력을 계발시킨다.

- 혼합적인 학습 환경 내 커뮤니티 안에서 적극적인 학습을 통해 학습 스타일을 조절하고 개발시킨다.
- 면대면 수업시간은 적극적이며 사색과 행동을 통해 학생들의 관심 영역의 변화가 이루어질 수 있다.
- 학생들에게는 자신의 한계를 시험해볼 것을 권장하며, 다양한 접근방식들을 선택할 수 있게 해준다.
- 결과물을 경험하는데 있어 소재 선택과 실험과정이 핵심적인 부분이다. 학생들은 자신이 선택한 소재들을 테스트 해보지만 항상 자신의 컨셉과 어떤 관련성이 있는지를 모색해봐야 한다.
- 학생들은 언제 뒤로 물러나야 하는지를 알아야 하며, 그것을 실패로 보는 것이 아니라 중점을 바꾸기 위해 필요한 긍정적 단계로 인식해야 한다.
- 이 모듈에서는 중요한 2단계에서 공식 평가가 계획되어 있는데, 이 평가에서는 학생들과 교사의 관점에서 이슈들을 해결하고 검토해볼 수 있다.
- 결과는 출판을 위한 그래픽 커뮤니케이션을 포함해 고용 가능성을 위한 중요 능력들을 입증하게 된다.

결론

올해 가장 성공적인 학습 및 교수 방식은 Kaye Newman이 운영하는 Shoreditch Vibe의 접근방식이었다.

- 1일차부터 주요 프로젝트 시작
- 학생들을 지속적으로 교육시키고 학생들의 사색과 행동을 촉진하는 전문 직업적 훈련에 대한 비판적 평가와 고객에 대한 현장 브리핑으로 1년을 마감
- 모델링, 테스트, 실험 및 사색을 통한 아이디어 개발
- 이론적 기초를 통한 증거에 입각한 결정
- 공간적 여행의 실험적 유형성의 증거 제시
- 손으로 직접 그린 도면과 디지털 기술을 통해 개인적인 프레젠테이션 능력의 개발

교사들은 개인 각자가 그 나름의 스타일을 가지고, 자기 자신과 자신의 행동과 반응을 이해하게 되기를 간절히 바라고 있다. 그 결과는 산업체가 주목할 수 있는 독특한 포트폴리오로 나타나게 될 것이다.

Studio 1 Unspoken Revolution
Studio Leader Janette Harris
Elizabeth Freemont

Intro

The 'Unspoken Revolution' explores the narrative of the 'Derelict', spanning sensory physical spaces, communities and those who sit on the fringe. The translation of the research aims, through creative process, to develop social awareness and community practice. The understanding of personalities, of community and materiality will be explored and expressed through dynamic exercises for architectural and interior interventions.

How can we rebuild sustainable public areas, tapping into nostalgia that encourages physical meaningful interaction? How can we develop interior topography and spatial language that reflects and accommodate a diverse cultural contex, developing relaxed, approachable spaces that the local community can embrace? The location for the projects will be within the East of London, including the Lea Valley.

Who are the Community?

Kant refers to space being public, while time is our own personal event. Yet as time shrinks in this 21st Century digital era, the notion of space has shifted from reality to virtual environments, changing our social and cultural contexts, creating inward facing dynamics. The need for public environments appears to be shrinking. Can we reverse this trend by providing meaningful opportunities to personal needs of the local communities? Can the environment evoke a sense of belonging through the revealing the buildings personality, reaffirming the communities connection to it and those who once inhabited its confines?

The Scope

The Studio will be working with professional practice and clients that will enable the idea of the 'Unspoken Revolution' to be explored through a number of mechanisms. Data collection will be undertaken through video, drawing/illustration, and conceptual material exploration through sculpture, modelling and orthographic detailing, supported by critical theoretical approach. The aim is to challenge the process thereby telling a story through unconventional representation developing a questioning relationship with the viewer. The use of materiality to enhance the experience may enable a relationship to develop through direct contact with appropriate medium.

The process

- Engaging
- Experimentation
- Evaluation
- Expression

The projects range from:

- Conceptual Instillation Year 2 and 3
- Interior and exterior Interventions, systems and services Year 2 and 3
- Competitions Year 3
- Self-determined projects Year 3

The project scope will encourage a range of outcomes suitable for your communities. This may include:

- Community Spaces
- Gathering
- Performance
- work space
- Outlets
- Lost and future skills
- Archive
- Event
- Other

The Framework

The urban master planners and the local communities have to consider a more responsible composition of ideas to our environmental needs. Sustainable energy use and material choice are to be rewarded but how does this connect with the individual and the surroundings? The studio will investigate responsible approaches to materials and how new emerging technologies can have a positive impact on those who inhabit your spaces.

Learning and Teaching:

- Investigation into community, environment and materials of the site, with an emphasis on building a strong narrative
  to support the design process
- Development through extensive model making and fine art approaches, leading to a professional output.
- Experimentation and discovery through traditional craft skills and digital media to demonstrate and further your ideas
- Workshops and competitions
- Mentoring and critiques with professional practice
- Producing a collaborative book for publication
- Undertaking a series of exhibitions
- Opportunities to win internships with leading practices

스튜디오 1 알려지지 않은 혁명(Unspoken Revolution)
스튜디오 리더 Janette Harris
Elizabeth Freemont

서론

알려지지 않은 혁명(Unspoken Revolution) 스튜디오는 감각적인 물리적 공간, 커뮤니티와 그 가장자리에 있는 사람들을 포괄하는 "버려진" 이란 개념을 탐색한다. 연구 목적은 창의적 과정을 통해 사회적 인식을 일깨우고 커뮤니티의 활용을 유도하는 것이다. 역동적인 건축적 개입과 인테리어 개입을 통해 성격과 커뮤니티, 유형성의 이해를 모색하고 이를 표현하게 될 것이다.

물리적으로 의미 있는 상호교류를 촉진할 수 있는 향수를 자극하는 지속 가능한 공공 공간을 어떻게 재건할 수 있을까? 다양한 문화적 맥락을 반영하고 포용할 수 있는 인테리어와 공간 언어를 어떻게 만들어낼 수 있을까? 지역사회가 수용할 수 있는 편안하고 접근 가능한 공간을 어떻게 연출해낼 수 있을까? 프로젝트의 위치는 Lea Valley를 포함한 영국 동부 내에 위치한 장소가 될 것이다.

커뮤니티는 누구인가?

칸트는 공간을 공적인 것인 반면 시간은 사적인 것이라 말했었다. 하지만 21세기의 디지털 시대에서 공간에 대한 생각은 현실에서 가상 환경으로 바뀌어졌으며, 우리의 사회적, 문화적 맥락을 바꾸어 역동적인 것들에 직면한 내면을 창조해야 하는 시대가 되었다. 공공 환경에 대한 필요는 줄어들고 있는 것으로 보인다. 우리가 지역사회의 니즈에 의미 있는 기회를 제공함으로써 이러한 트렌드를 거스를 수 있을까? 그러한 환경이 건물의 성격을 드러내고 그러한 건물의 성격과 지역사회 및 과거 그곳에서 살았던 사람들과의 관련성을 확인시켜 줌으로써 소속감을 자극할 수 있을까?

범위

알려지지 않은 혁명 스튜디오는 여러 메커니즘을 통해 "알려지지 않은 혁명"이라는 아이디어를 탐색할 수 있는 전문가들 및 고객들과 함께 작업할 것이다. 비디오, 그림/일러스트레이션, 조각, 모델링 등을 통한 개념적 소재 탐색을 이용해 데이터 수집을 할 것이며, 이를 보완해 비판적인 이론적 접근방식도 취하게 될 것이다. 목표는 보는 사람이 의문을 제기할 수 있는 관계를 설정함으로써 비전통적인 재현(representation)을 통해 스토리를 말해준다는 것이다. 경험을 촉진하기 위한 물질적 요소들(materiality)을 사용함으로써 적절한 매체와의 직접적인 접촉을 통한 관계 형성이 가능할 수 있을 것이다.

과정은

- 관여 과정
- 실험 과정
- 평가 과정
- 표현 과정

프로젝트의 범위는:

- 2년 차와 3년 차 –개념적 주입
- 2년 차 및 3년 차 인테리어 및 익스테리어 개입, 시스템 및 서비스
- 3년 차 대회
- 3년 차 스스로 결정한 프로젝트

프로젝트의 적용 범위는 자신의 커뮤니티에 적합한 다양한 결과들을 촉진하게 될 것이다. 여기에는 다음과 같은 사항들이 포함될 수 있다.

- 커뮤니티 공간
- 모임
- 공연
- 작업 공간
- 아울렛
- 과거 잃어버린 능력 및 미래의 능력
- 기록
- 이벤트
- 기타

프레임워크

도시 마스터 플래너 및 지역 커뮤니티들은 환경적 니즈에 보다 책임감 있는 아이디어를 검토해보아야 한다. 지속 가능한 에너지 사용 및 소재 선택은 보상을 받아야 하지만, 이것이 개인 및 주변환경에 어떻게 연관될 수 있는가? 스튜디오는 소재에 대한 책임감 있는 접근방식을 모색할 것이며, 새로 부상하고 있는 기술들이 그 공간에 살고 있는 사람들에게 어떻게 긍정적 영향을 미칠 수 있는지를 탐색해볼 것이다.

학습 및 교수(Teaching):

- 디자인 과정을 지원할 수 있는 강력한 내러티브를 만드는 것에 중점을 두고 커뮤니티, 부지의 환경 및 소재를 탐색
- 전문적인 결과물을 산출할 수 있는 방대한 모델 제작 및 예술적 접근 방식을 통한 개발
- 자신의 아이디어를 보여주고 더욱 발전시키기 위해 전통적인 작업 기술과 디지털 매체를 통한 실험 및 발견
- 워크샵 및 대회
- 전문가들로부터의 멘토링 및 비판
- 출판을 위한 협동 서적 제작
- 일련의 전시 실시
- 선도적인 업체에서의 인턴십 기회 획득

CASS Faculty of Art, Architecture and Design
Interior Design BA(Hons)
The Shoreditch Vibe

Studio Position: 2

Shoreditch, London E1 is buzzing with innovative, ground breaking small businesses that have been created by pioneering entrepreneurs. British Telecom has allowed us to use The Shoreditch Exchange in Shoreditch. It was built by the GPO, General Post Office in 1956 by the government, so is a solid building envelope of basic details. The building still maintains the telephone exchange for the East End but 3000sqm is now redundant and looking for to associate itself with another enterprise..

The Shoreditch location offers space that has a long history and many stories of former trades and well known household names. It also offers a lively diverse cross section of people who live and work in the area and has now become extremely popular with those who want to make a difference and seek an original and authentic locale to develop their businesses ideas.

The studio has asked the students to explore a large, redundant, almost empty, brick and concrete space situated in Shoreditch to create a Co-worker hub, for new business start-ups, a place where many companies can co-exist together, small or large. It will be space where not only the facilities are shared but more importantly it will encourage communication and collaboration and inter-company connections developing a spatial co-operative.

The organisation of space needs to encompass the flexibility required to accommodate change and growth, looking at kinetic and dynamic structures that provide options of movement allowing for adaptability and variable spatial solutions making this working space as exciting and resourceful as possible.

'Fujimoto himself speaks of architecture as a hazy dream a zone of blurry areas. Everything is a subtle play of difference in space and time,'

Essay: Gleiter.J, The most extreme aesthetic: Sou Fujimoto Futurospective Architecture, Verlag der Buchhandlung Walther Konig, 2012 p329.

The journey throughout should consider the spontaneous, the chance and serendipitous encounters which will provide opportunities to cultivate a business richness. The studio has looked at the possibilities of altering the structure to design a vertical language as well considering the open space, so all should enjoy being part of the whole building.

The environmental research will reveal the opportunities to develop public and private space, working creatively with sound and light. It will look at the individual identity of each occupier and how these are settled within the BT Exchange building and its new brand. We will look at the ideas of movement and being kinetic in terms of the scope within the space and also as an extension of the space within the community. It will look at the idea of temporary both occupation and building, collapsible and demountable. The concepts should add an energy that inspires, that allows for think tanks and innovation where the space instigates and is part of the business activation.

The studio has asked some important questions. What is creative space? How does space support initiative and fresh thinking? How do people work together in a multi-occupancy space? How do we consider the whole person in terms of the physical space throughout the day?

Adaptability and Flexibility

This studio is researching into ideas of adaptability and flexibility within a Co-worker or Start Up environment. The needs of the 'workspace' have changed as new technology and wireless connections have allowed the office inhabitant to be far more mobile, nomadic and autonomous, where they have the ability to work in many different scenarios, in or out of the office environment.

The office space is now less about territory and more about shared facilities, networking opportunities and inspiring its inhabitant but it's also about the immediate, fast response and information at hand.

Through research carried out looking at Co-worker hubs and new Business Start-Ups these new environments champion the vibe and the positive engagement. The lack of a corporate identity removes both hierarchy and management control allowing the worker to define their own practice, their own pace and the choice of space.

However many of these new co-workers spaces rely on tiny budgets, the overall comfort and creative space is in many cases is compromised. Is this environment tolerable because of shortened working timespans in the office or could these simple design solutions support this new environment more comprehensively?

CASS 예술, 건축 및 디자인
인테리어 디자인 BA(Hons)
Shoreditch Vibe

스튜디오 2

Shoreditch, London E1는 개척적인 기업가들이 세운 혁신적이고 혁명적인 소기업들로 붐비고 있다. 브리티시 텔레콤(British Telecom)은 우리가 Shoreditch 에 있는 Shoreditch Exchange를 사용할 수 있게 허락해 주었다. 이 건물은 1956년 정부 소유 우체국(GPO)이 지은 건물로서, 기본적인 세부구조들은 매우 튼튼한 건물이다. 이 건물에는 아직도 East End를 위한 전신 교환국이 있지만, 현재 3천 평방미터가 비어있어 이를 다른 기업에 양도할 기회를 모색하고 있다.

Shoreditch 지역은 오랜 역사를 지니고 있으며, 과거 무역과 관련된 많은 얘기들이 전해져 오고 있으며, 이곳 출신의 유명한 가문들도 많이 있다. 이 지역은 또한 이곳에 살고 있는 사람과 이 지역에서 일하고 있는 사람들이 만날 수 있는 다양한 장소들을 제공하고 있으며, 현재는 이곳을 바꾸어 자신들의 사업적 아이디어를 펼칠 수 있는 장소로 만들기 원하는 사람들에게는 매우 잘 알려져 있는 장소이기도 하다.

Shoreditch Vibe 스튜디오는 학생들에게 여러 크고 작은 회사들이 공존하면서 새로운 사업 시작을 위해 동료들의 허브 역할을 해줄 수 있는 공간을 창출하기 위해 Shoreditch에 위치한 크고, 거의 비어 있는 벽돌 콘크리트 공간을 탐색해보게 하고 있다. 이 공간은 시설만 공유하는 것이 아니라 그 보다 더 중요한 커뮤니케이션과 협력, 회사 간 연결을 촉진해 공간적 협력을 유도할 수 있는 공간이 될 것이다.

공간의 구성에는 변화와 성장을 담아낼 수 있으며, 이 작업 공간을 최대한 흥미롭고 유용하게 만들 수 있는 적응성과 다양한 공간적 솔루션이 가능한 역동적인 구조로 만들기 위해 필요한 융통성이 있어야 한다.

'후지모토는 건축을 경계가 희미한 꿈 같은 분야라고 말하고 있다. 모든 것은 시간과 공간의 미묘한 차이들이 만들어내는 것이다."

에세이: Gleiter,J, The most extreme aesthetic: Sou Fujimoto Futurospective Architecture,Verlag der Buchhandlung Walther Konig, 2012 p.329

이러한 과정은 사업을 풍부하게 만드는 자발성, 기회, 우연한 발견 등을 고려해야 한다. Shoreditch Vibe 스튜디오는 모든 부분들이 건물의 일부로서 제 역할을 할 수 있게 하기 위해 수직적 언어를 디자인할 수 있게 구조를 변경할 수 있는 가능성을 모색해 왔으며, 또한 개방 공간을 검토해 오고 있다.

환경 연구를 통해 공공 공간과 사적 공간을 개발할 수 있는 기회가 드러날 것이며, 소리와 빛을 창조적으로 활용할 수 있는 방법이 드러날 수 있을 것이다. 이 건물을 사용하고 있는 사람들의 개인적 정체성이 드러나고, 그러한 개인적 정체성이 BT 교환국 건물과 새로운 형태의 BT 건물과 조화를 이룰 수 있는 방법을 모색하게 될 것이다. 우리는 공간 내의 범위라는 측면에서 움직임과 동적인 것이라는 아이디어에 중점을 두고 있고, 또한 커뮤니티 내에서의 공간 확장에도 중점을 두고 있다. 또한 건물의 사용과 건물, 붕괴와 해체라는 아이디어에도 중점을 두고 있다. 이러한 컨셉들은 공간이 사업 활성화를 자극하고 그 일부가 될 수 있는 씽크 탱크와 혁신을 자극하고 이를 허용할 수 있는 에너지를 불어넣을 것이다.

Shoreditch Vibe 스튜디오는 몇 가지 중요한 질문을 제기하고 있다.
즉, 창의적 공간이란 무엇인가?
공간이 어떻게 혁신적이고 신선한 생각들에 도움이 될 수 있는가?
사람들이 복수의 사람들이 이용하고 있는 공간에서 어떻게 함께 작업할 수 있는가?
물리적 공간이라는 측면에서 우리는 사람을 어떻게 생각해야 하는가?

적응성 및 융통성

Shoreditch Vibe 스튜디오는 협업자들 또는 가동 환경 내에서의 적응성과 융통성이라는 개념을 연구하고 있다. 새로운 기술과 무선 연결을 통해 사무실에서 일하는 사람들이 훨씬 더 먼 거리를 이동할 수 있게 해줌에 따라 "작업공간" 에 대한 니즈는 변화하고 있으며, 이로 인해 사무실에서 일하는 사람들은 여러 다른 시나리오 하에서 일할 수 있게 되었다.

사무 공간은 이제 어떤 물리적 구역의 문제가 아니라 시설 공유, 네트워킹, 일하는 사람들을 고취시키는 것이 되었으며, 또한 즉각적이고 빠른 반응과 바로 사용할 수 있는 정보의 공간이 되었다.

협업자들의 허브와 새 비즈니스의 시작을 염두에 두고 실시한 연구들을 통해 이러한 새로운 환경이 Vibe 스튜디오의 관심을 끌었으며 우리 스튜디오의 긍정적 관여를 이끌어내었다. 기업 ID의 부족은 위계 및 경영 통제를 없애 일하는 사람들이 스스로의 업무와 업무 속도를 정하고 스스로 공간을 선택하게 하였다.

하지만 이러한 새로운 협업 공간 중 많은 부분이 작은 예산에 의존하고 있으며, 전체적인 편의와 창조적 공간은 많은 경우 훼손되고 있다. 이러한 상황이 근무시간을 줄인다고 허용될 수 있는가? 아니면 이러한 단순한 디자인 솔루션들이 이 새로운 환경을 보다 종합적으로 지원해줄 수 있는가?

Julie Berdou
julie.berdou@gmail.com
www.behance.net/julieberdou

Adaptive WorkspaceHUB

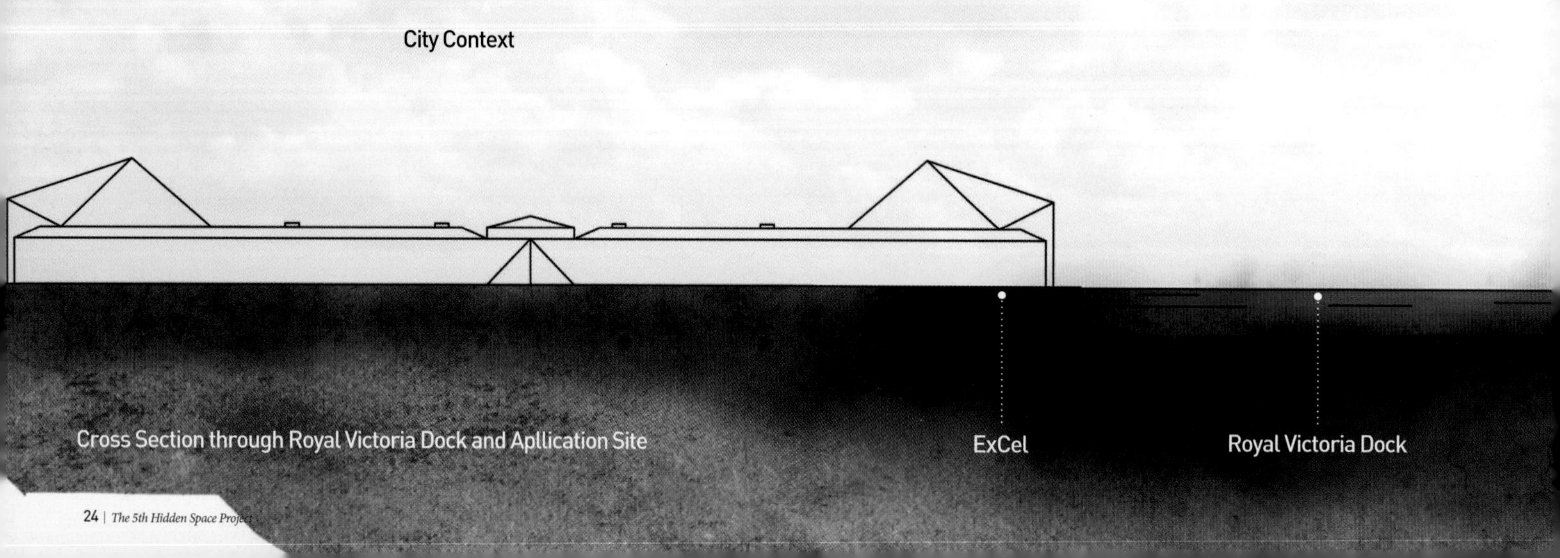

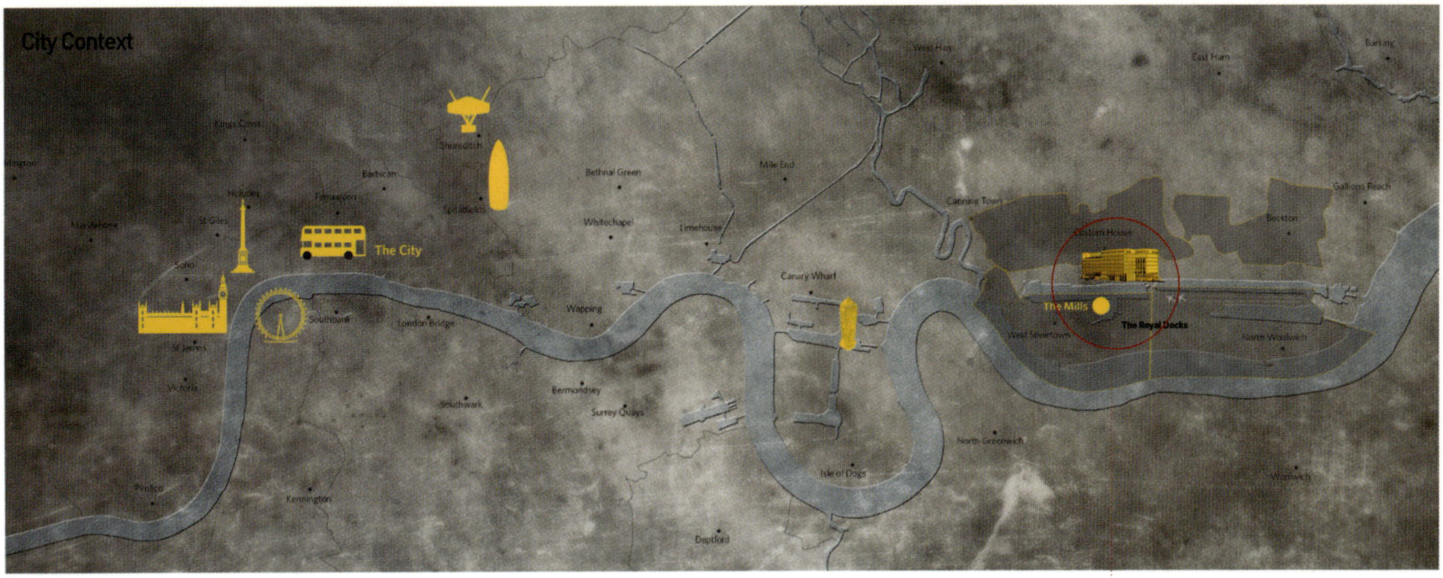

The Site

The site comprises the derelict 'Spillers' Millennium Mills building and its Grade 2 listed Silo D, a former flower mill factory, situated on the southern side of the Royal Victoria Dock in East London.
Millennium Mills sits at the south end of London's fast-evolving Lea Valley. Spanning six of London's boroughs, from the M25 down to Leamouth, Lea Valley is packed with urban diversity and challenges aplenty. There is good and bad to react to, and there is big potential for real-world collaborations.

사이트는 "Spillers" 밀레니엄 밀스 건물과 이 건물의 2등급 사일로 D(이전의 화초 공장)으로 구성되어 있으며, 런던 동부 왕립 빅토리아 선창의 남쪽에 위치해 있다.
밀레니엄 밀은 런던에서 빠르게 성장하고 있는 지역인 Lea Valley의 남쪽 끝단에 위치해 있다. M25에서부터 Leamouth까지 6개의 자치구에 걸쳐 있는 Lea Valley는 도시의 다양성과 다양한 문제들이 존재하고 있는 곳이다. 좋은 것도 있고 나쁜 것도 있으며, 실제 세계에서의 협력을 이끌어낼 수 있는 가능성도 큰 곳이다.

The Aim

The transition from post-industrial decay to an era of revitalisation and liveability is proposed. The remake of London's industrial legacy Millennium Mills as a contemporary place to live and work and its embedment as a distinctive element in the social, economic and physical infrastructure of the urban frame of London.

산업화 이후의 쇠퇴에서 새로운 활력과 활성화의 시대로의 변화를 제안한다. 런던의 산업적 레전드였던 밀레니엄 밀스를 현대적인 거주 및 업무 공간으로 재구성하며, 이곳을 런던의 사회적, 경제적, 물리적 인프라에 있어 특별한 장소로 만든다는 제안이다.

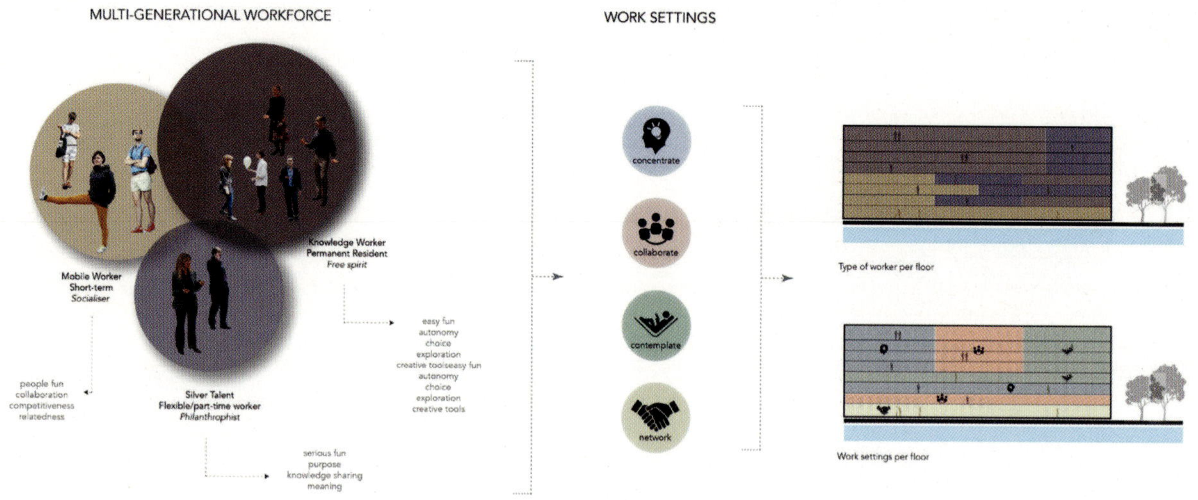

The Concept

To provide a workplace that is inclusively designed for the needs of a multi-generational workforce (knowledge workers, Generation Y, Silver Talent) by offering a variety of breakout, meeting and social areas with a high segmentation and choice concept, as opposed to one-size fits all.

The aim is to design a physical office environment for three main work settings, that:
1. aid concentration
2. facilitate collaboration and
3. provide escape in form of contemplative spaces.

단일 크기의 공간 보다는 분획이 잘 되어 있고 선택이 가능한 다양한 휴식 공간, 모임 공간 및 사회적 공간을 제공함으로써 다양한 세대로 구성된 근로자(지식 근로자, Y 세대 근로자, 실버 경력자)들의 니즈에 맞게 디자인된 업무 공간을 제공한다.

목표는 3개의 주요 업무 공간을 위한 물리적 사무 공간을 설계하는 것이다.
1. 지원이 집중되어 있는 업무 공간
2. 협력을 촉진할 수 있는 업무 공간
3. 사색을 할 수 있는 형태의 업무 휴식 공간

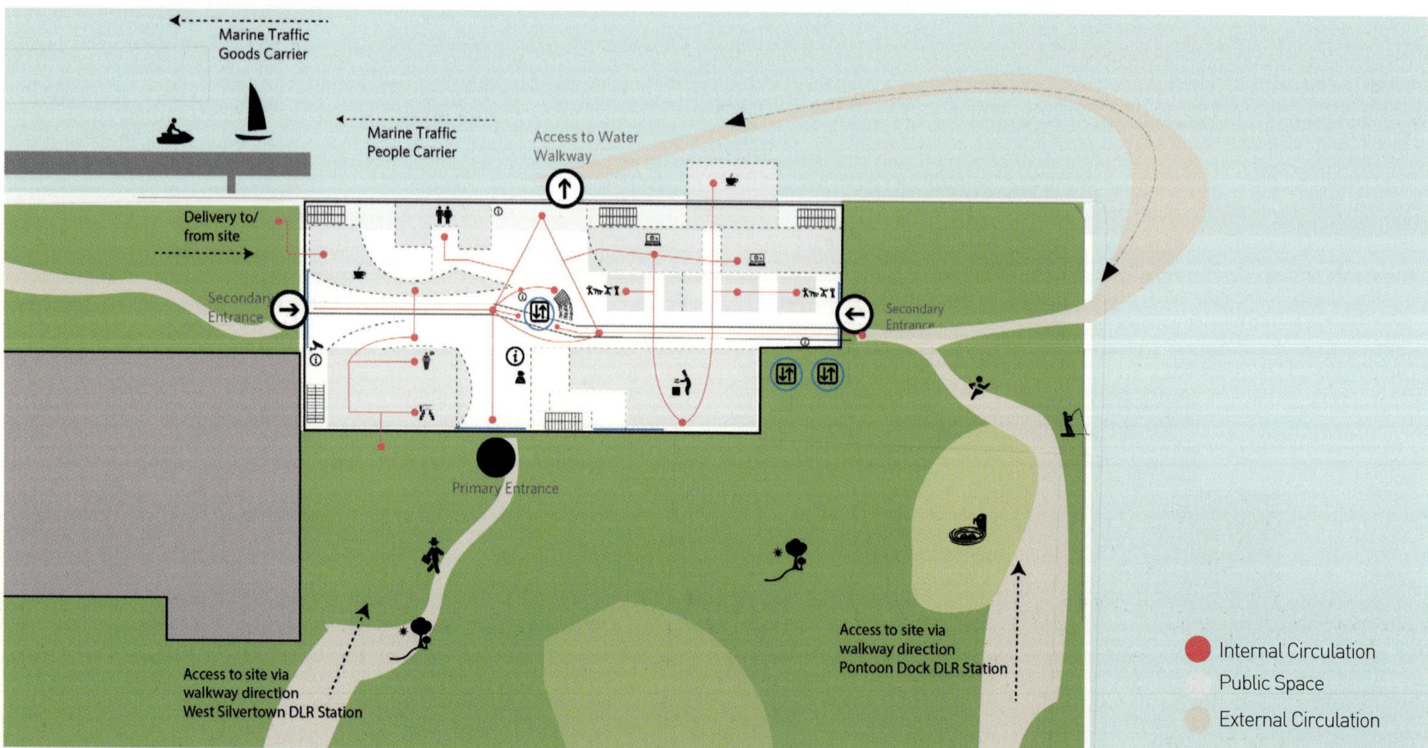

Master plan: revitalising an obsolete site that was once a flourishing trade gateway with a dynamic horizontal and vertical circulation of goods and people to and from the site.

Central to all three work settings is the individual's opportunity to personalise their work style and to work in a number of different ergonomic positions (seated, standing, reclining) that allow for a healthier work style.

To make design interventions that target acoustics, lighting, information technology, furniture, ergonomics and ambience that impact productivity and comfort of all generations of workers and that allow for a positive relationship between habitat and behaviour. In a work environment that is becoming less independent on place, more mobile, diffuse and multi-sited, Millennium Mills act as an anchor point for people to congregate in a highly effective, productive and imaginative work environment, appropriate to the individual worker's needs.
The development of an interior topography and spatial language that reflects and accommodates diverse human movement, developing relaxed, approachable spaces that the local community can embrace.

이러한 3가지의 업무 공간 모두에 핵심적인 것이 업무 스타일의 특성을 살릴 수 있고 보다 건강한 업무 스타일이 가능한 다양한 인체공학적 자세(앉은 자세, 서 있는 자세, 기댈 수 있는 자세)로 업무를 볼 수 있게 할 수 있는 기회를 제공해주는 것이다.

모든 세대의 근로자들에게 생산성을 촉진시킬 수 있고 편안함을 줄 수 있으며 거주와 행동 간 긍정적 관계 형성이 가능한 음향, 조명, 정보 기술, 가구, 인체공학 및 분위기를 목표로 하는 디자인을 만드는 것. 점점 더 장소에 덜 제한을 받고 이동성이 크며, 여러 장소에서 업무를 봐야 하는 업무 환경으로 변화하고 있는 상황에서 밀레니엄 밀은 개별 근로자들의 니즈에 적합한 효과적이고 생산적이며 창의적인 업무 환경에 사람들이 모일 수 있는 기준이 될 것이다.
다양한 인간 활동을 반영하고 수용할 수 있으며, 지역사회가 수용할 수 있는 편안하고 접근 가능한 공간을 만들 수 있는 인테리어 디자인과 공간 언어를 개발한다.

Adaptive WorkspaceHUB
Cross Section

1:200 @ A3

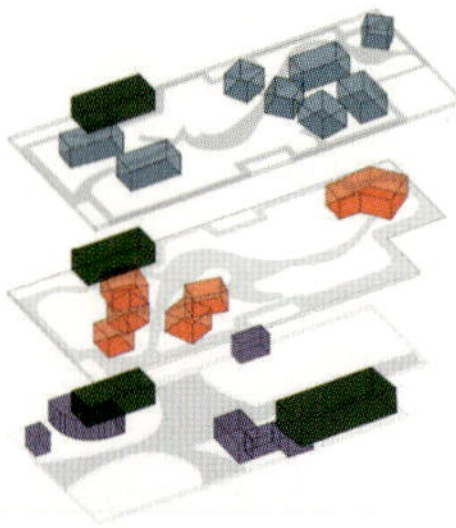

Program Blocks

Undercroft, new Splitlevel
and Level 1

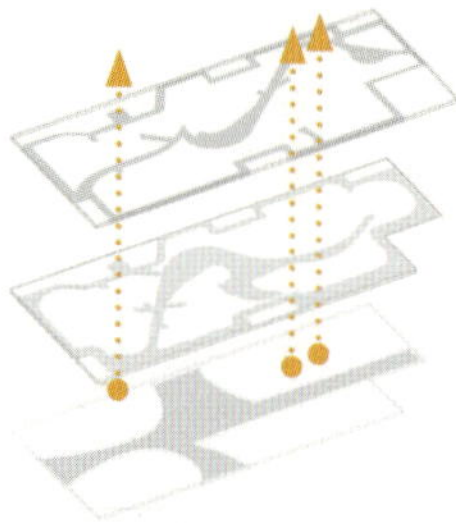

Circulation

Vertical and horizontal

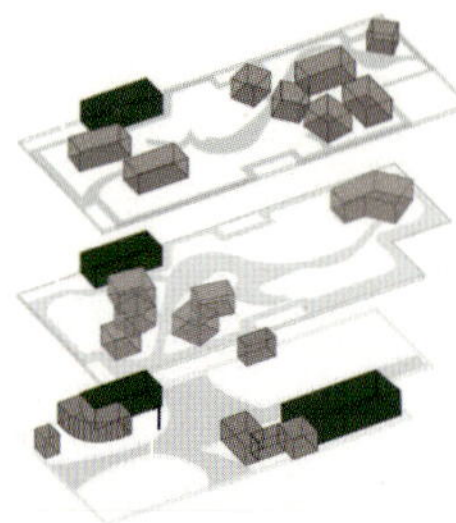

Public Spaces

1. Toilets
2. Restaurant

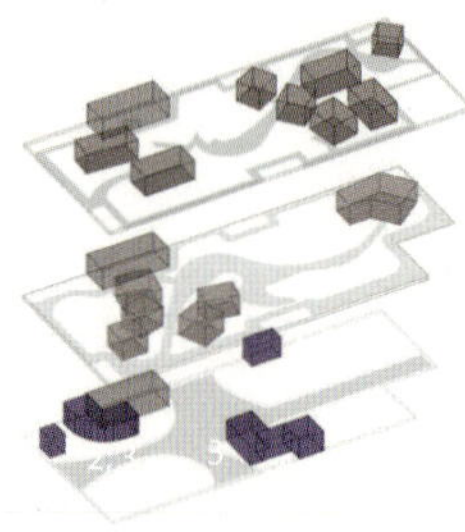

Service Spaces

1. Service Lift
2. Storage and Mechanical
3. Administrative
4. Restaurant Kitchen

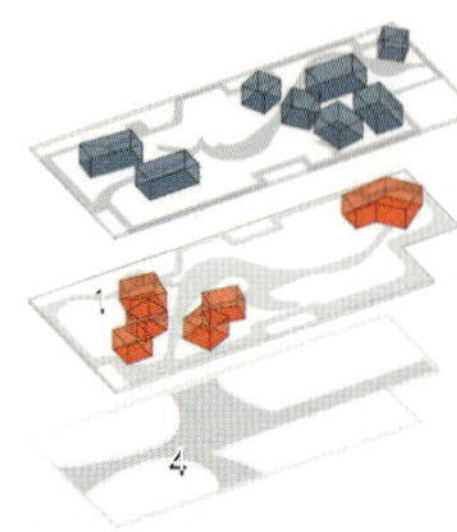

Office Spaces

1. Collaborate
2. Concentrate
3. Contemplate
4. Network

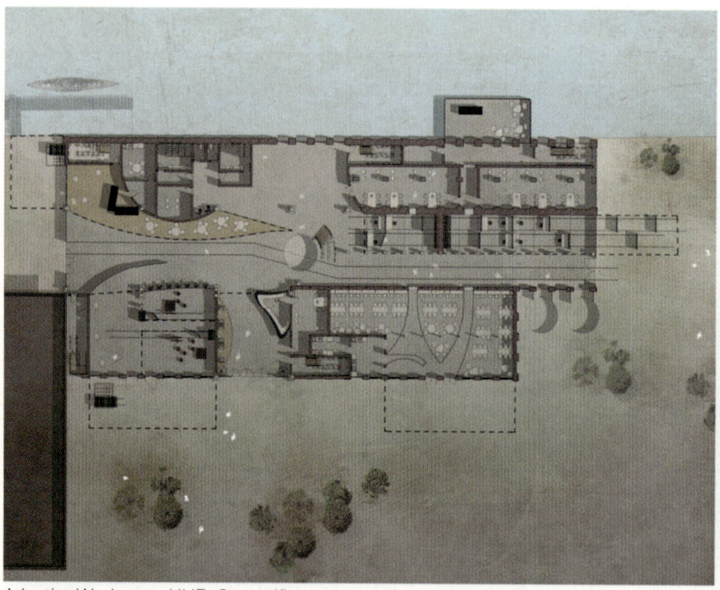

Adaptive WorkspaceHUB, Groundfloor - *network*

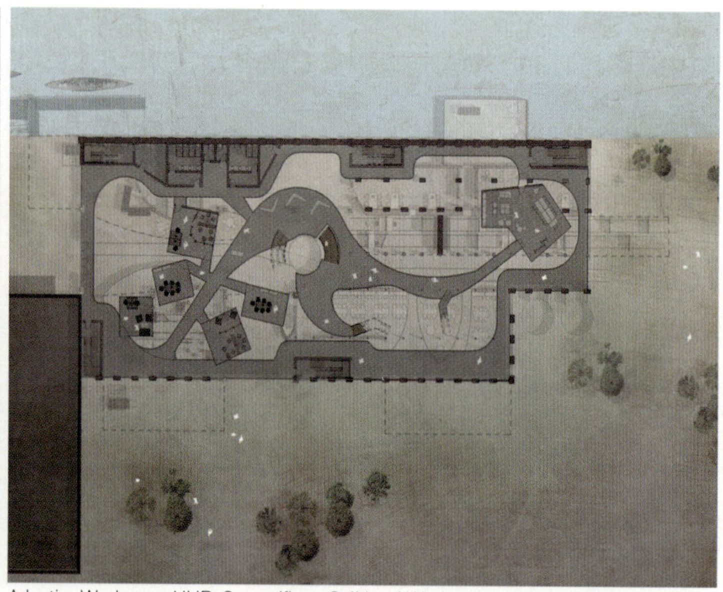

Adaptive WorkspaceHUB, Groundfloor, Splitlevel Workpods - *collaboration*

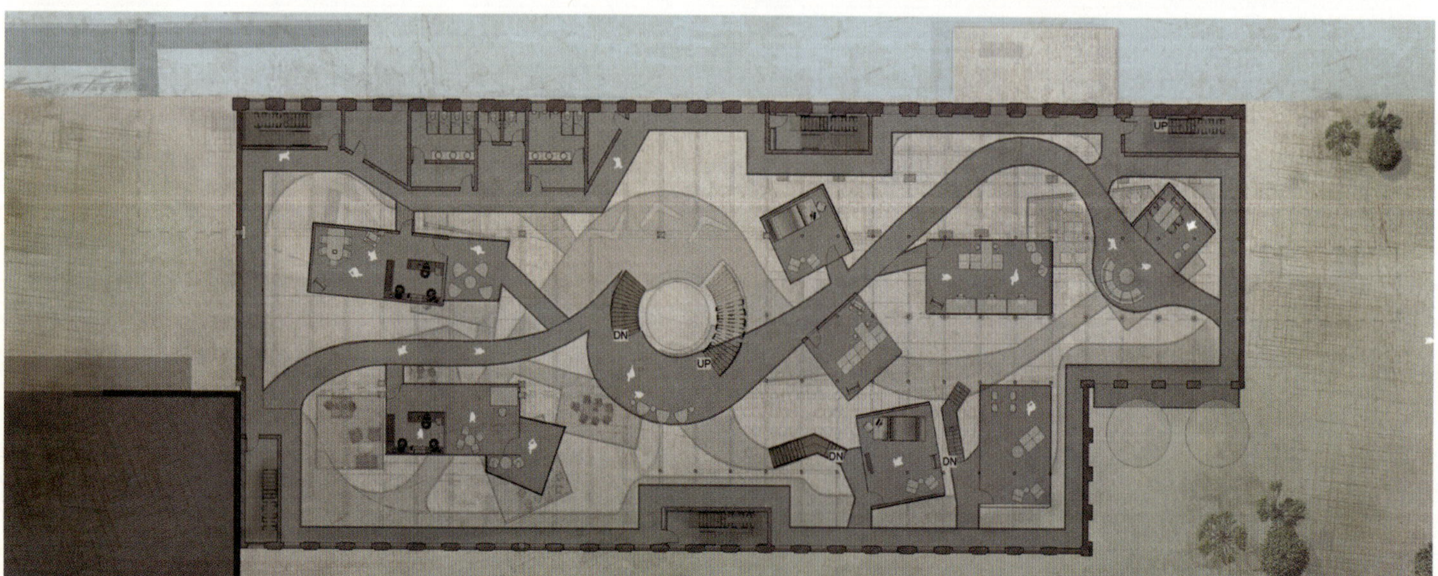

1st Floor Plan, Workpods - *concentration*

Adaptive WorkspaceHUB, Groundfloor, Splitlevel Workpods - *collaboration*

Adaptive WorkspaceHUB, 1st floor Workpods - *concentration*

Adaptive WorkspaceHUB, Splitlevel looking down on entrance and Workpods - *collaboration*

Adaptive WorkspaceHUB Sectional Perspective

Implication

A dynamic new set of circulation and work spaces working in respectful harmony but offering a refreshing and reviving new element, contrastive to the rigid and logical aspect of the building's inherent rhythm of columns and supporting steel beams. Voids allow the building and user to breathe, and create a dynamic composition of vistas and views, horizontally and vertically.

The realisation of an environment that fosters ideas, creates growth and overturns assumptions that get in the way of imagining new possibilities and visionary conceptions.

다이내믹한 새로운 순환체계 및 업무 공간들은 조화를 이루고 있지만, 활기를 불어넣을 수 있는 새로운 요소들을 제공해주고 있으며, 건물에 내재해 있는 기둥의 리듬과 지지 강철 빔의 엄격하고 논리적 측면과는 대조를 보이고 있다. 빈 공간들은 건물과 사용자가 호흡하며, 다이내믹한 경치와 전망을 수직, 수평적으로 연출해내게 한다.

아이디어를 자극하는 환경의 실현으로 성장을 이끌어내며, 새로운 가능성을 상상해내고 비전을 세우는데 방해가 되는 가정들을 없애는 역할을 해준다.

Vaida Kvedaraite
vaida.kvedaraite@gmail.com

The Nest

The Brief

The aim of our project is to explore a large, redundant, almost empty, concrete space situated in Shoreditch and create a Co-worker hub, (New Business Start-Ups) a place where many companies can co-exist together, small or large.
It will be place where not only the facilities are shared but more importantly it will encourage communication and collaboration and inter-company connections developing a spatial co-operative. The initial research through Shoreditch and throughout the Hoxton area; created ideas to reenergize The BT Exchange building for its potential with respect to its location and purpose. A project based on creating a creative multifunctional building in a rapidly evolving and vibrant environment - with a great focus on sustainability and a functioning future.

우리 프로젝트의 목표는 Shoreditch에 위치한 대형 규모의 비어 있는 콘크리트 공간을 탐색해서 작든 크든 많은 회사들이 공존할 수 있는 장소인 협업 허브(새 비즈니스 시작)를 만드는 것이다.

이 장소는 시설만 공유하는 곳이 아니라 보다 중요하게는 커뮤니케이션과 협력, 회사 간 연계를 촉진해 공간적 협력을 창출해내는 장소가 될 것이다. Shoreditch와 Hoxton 지역에 대한 초기 연구를 통해 BT 교환국 건물의 위치와 목적을 존중하면서 이 건물이 가지고 있는 잠재력에 활력을 불어 넣을 수 있는 아이디어를 생성해내었다. 이 프로젝트는 급속히 변화하고 있는 활력에 넘치는 환경 속에서 창의적인 다기능 건물을 만드는 것에 중점을 두고 있으며, 아울러 지속가능성과 미래의 기능에도 초점을 두고 있다.

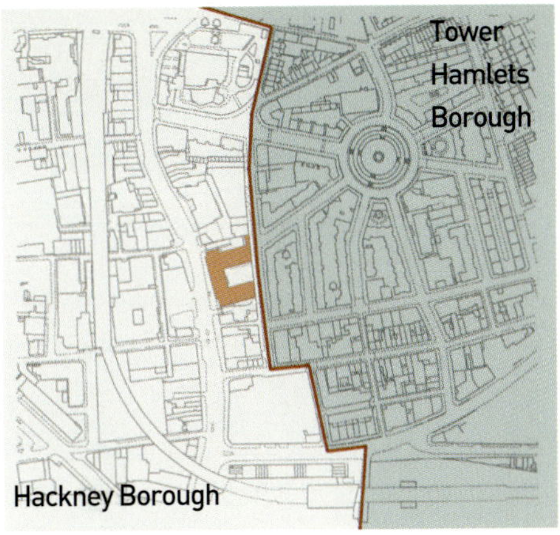

Location

The 3D map showing the location of the BT building in context with other buildings around

The BT Building

BT Exchange Building was erected by GPO in 1956. It has the construction of pale red bricks, concrete and steel. Building is U shape with the entrance from Shoreditch High Street.

BT 교환국 건물은 1956년에 GPO가 세운 건물이다. 이 건물은 붉은 적벽돌과 콘크리트, 강철로 건축되어 있다. 이 건물은 U자 형의 형태로 Shoreditch High Street에 입구를 두고 있다.

The Concept

The concept "Nest' comes from idea to bring back artist and design community back to Shoreditch which left it because they could not afford increasing rent prices. The Start Up "Nest" office will be a hub to start their business, grow up, get strong in competitive market and leave the place for new comers. "Nest" will provide with creative and inspiring working and recreational spaces. Places for group work, a lot of communal spaces which will encourage communication, collaboration and networking and inclosed small cellular offices for meetings and individual space to work.

"둥지" 라는 컨셉은 예술가들과 디자인 커뮤니티가 이들이 증가하는 임대비용을 감당할 수 없어 떠났던 Shoreditch로 다시 돌아오게 만든다는 아이디어에서 비롯된 것이었다. "둥지" 사무실은 이들이 사업을 다시 시작하고, 경쟁 시장에서 더욱 강력해지며, 이 장소를 새로운 곳으로 만드는 허브가 될 것이다. 둥지는 이들에게 창의적이고 영감을 자극하는 작업 및 오락 공간을 제공해줄 것이다. 그룹 작업을 위한 장소, 커뮤니케이션과 협력, 네트워킹을 촉진하는 의사소통의 장소, 회의와 개별 작업 공간을 위한 소규모 셀 형식의 사무실 공간 등을 제공하게 될 것이다.

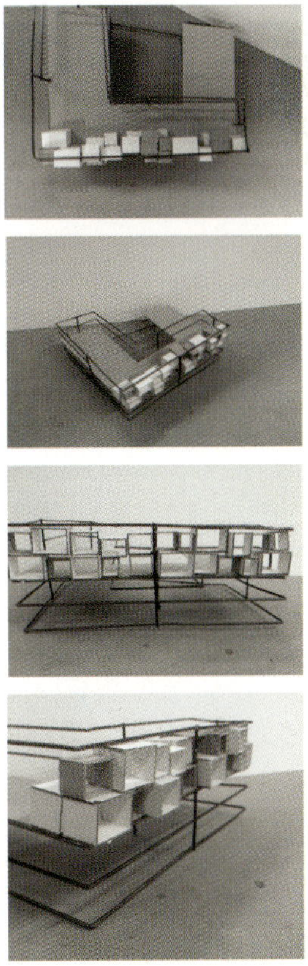

Plans

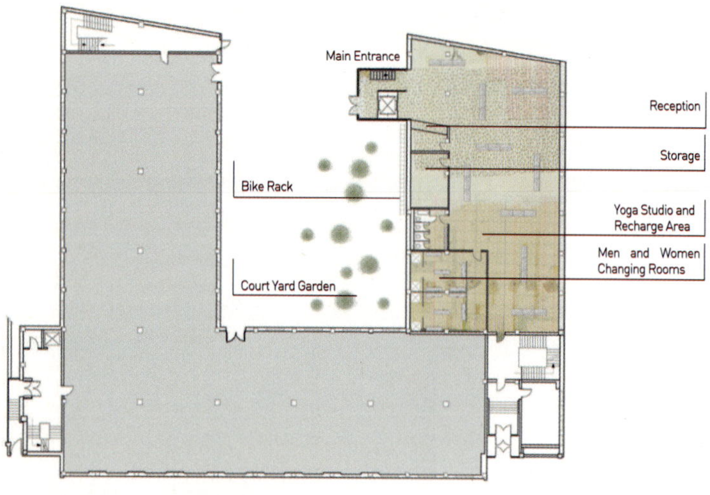

Ground Floor

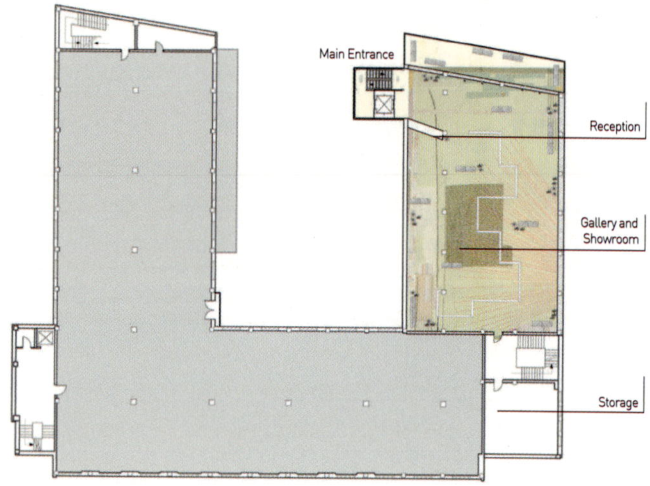

First Floor

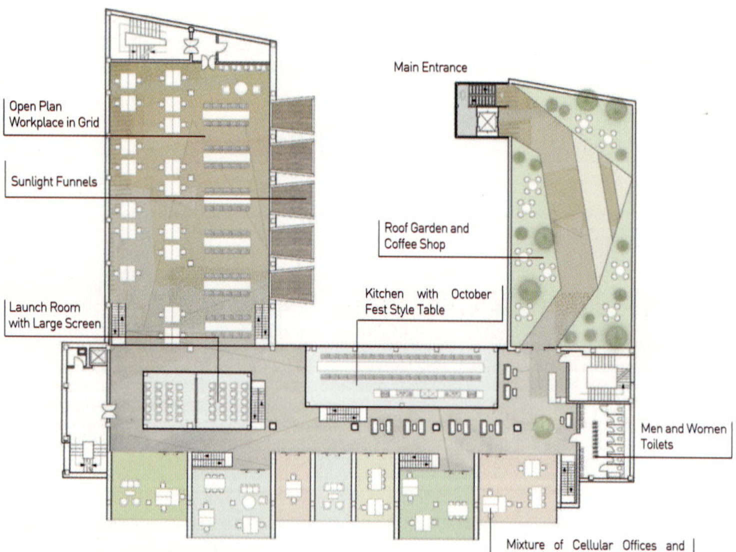

Second Floor

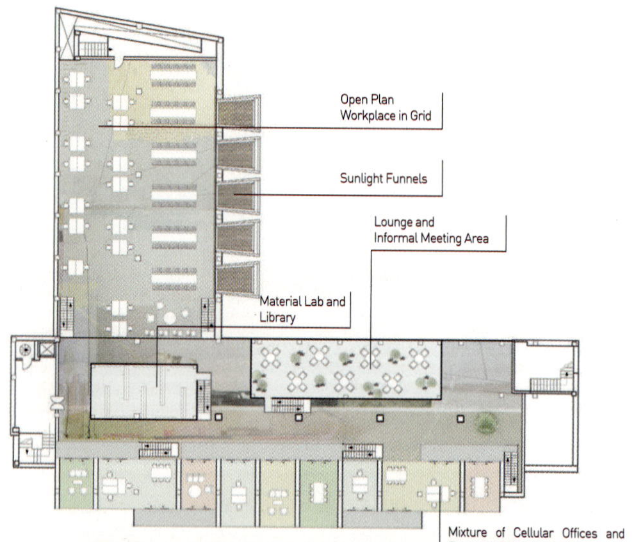

Third Floor

Materials

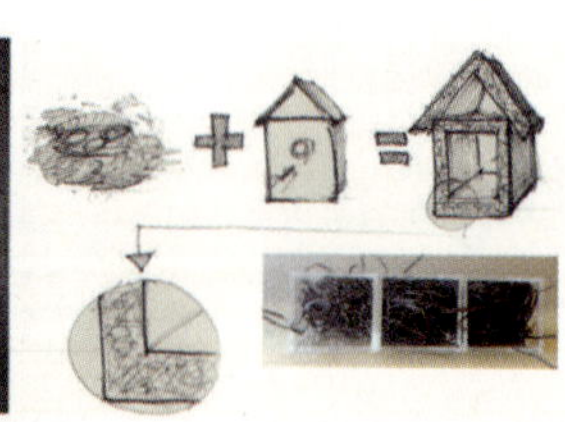

Straw Bale Construction

Traditional straw bale construction advantages:
- Sustainable;
- Cheap (1 straw bale about £3)
- Noise insulation;
- Temperature insulation;
- Lightweight.

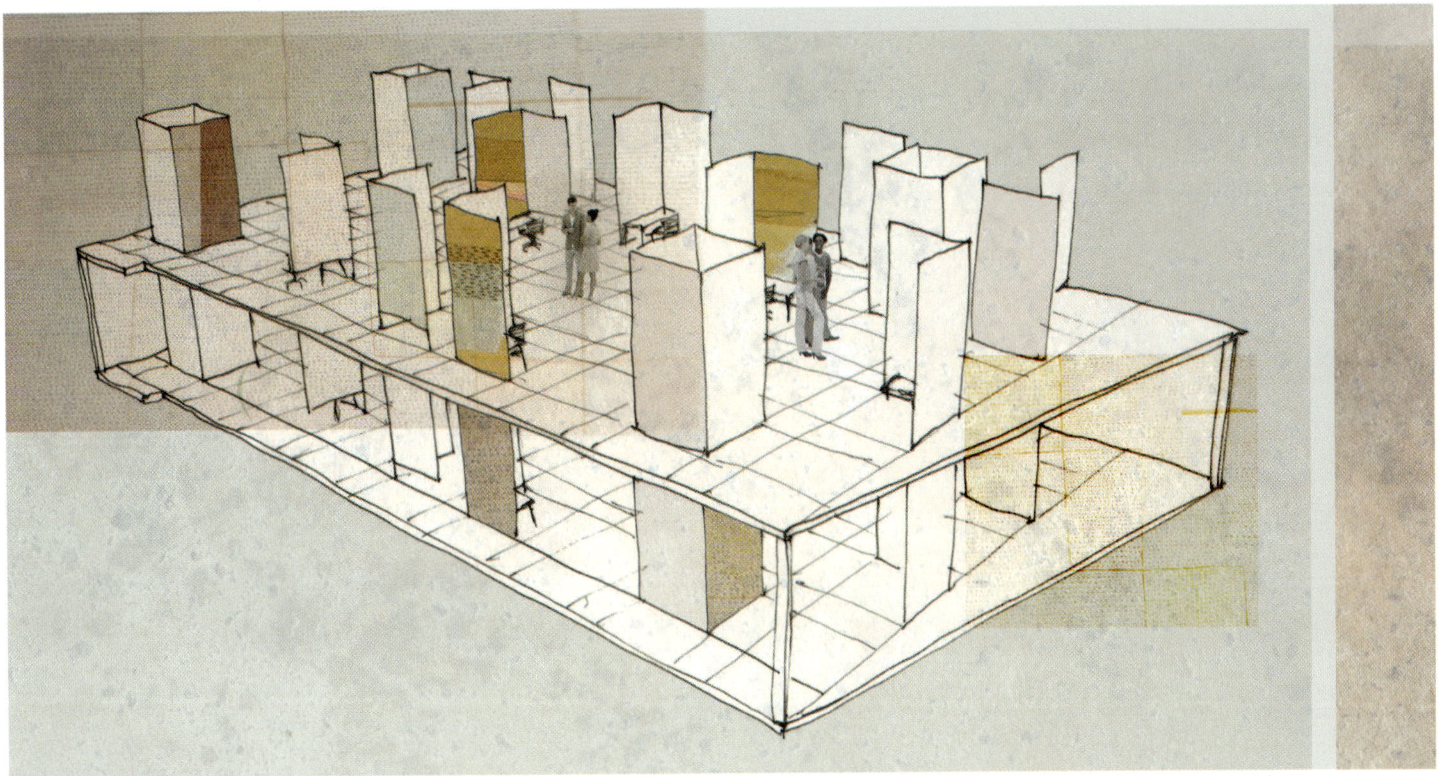

Open plan grid work space

APP Logo

Day at the Nest Office

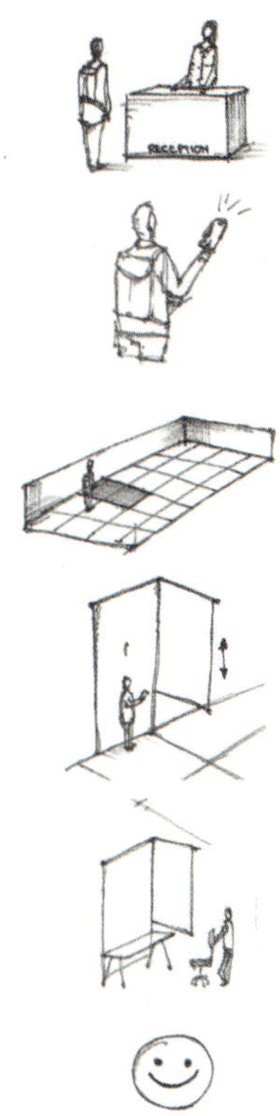

1. At the Reception the person books a space for himself or/and his colleagues. Earlier you come better place you get.
2. He get's a notification on his iPhone about booked space.
3. Following the coordinates on App he finds a booked space.
4. Then he pulls down the curtains as many and as low as he wants creating more private space.
5. Finally he takes a simple, light, wheelie furniture from the storage room and places it where needed.
6. Enjoys the day.

Visual of grid open plan work space with pulled down curtain dividers

Visual of double height ceiling corridor - space for communication

Visual of light funnels for enjoying the sunshine during the day.

Grid Workplace

The concept of the Grid Office is based on the nomadic working type, where people are working remotely, from home or any other public spaces. In my concept a working person does not have fixed desk at the office. Nest office's part of open plan work space has a simple grid design and adjustable projection screen curtains, working as a dividers of the space and screens for projection of companies logo or computer's interface. Through the VO App the worker can find and book a space. It will allow a person to search for the available space in the Nest office for him and/or his colleagues. When worker arrives to the office he finds the allocated space. He pulls down the screen curtains. On the projection screens, which also works as a space dividers, worker can project the logo of the company he works. As well as he can project his own files, images or programs he works with.

그리드 사무실이라는 컨셉은 노마딕 업무 형태에 기초한 것으로서, 그리드 사무실에서는 사람들이 원거리, 자택 또는 다른 공용 공간에서 일한다. 내가 제시한 개념에서는 일하는 사람이 반드시 사무실에 책상을 둘 필요가 없다. 둥지 사무실의 업무 공간은 단순한 그리드 디자인과 조정 가능한 프로젝션 스크린 커튼이 있어 이 커튼이 공간을 구분하고 회사 로고를 투사하는 스크린이나 컴퓨터 인터페이스 스크린의 역할을 한다. VO App를 통해 근로자들은 공간을 찾아 예약할 수 있다. 이는 자신과 자신의 동료를 위해 둥지 오피스에서 이용 가능한 공간을 찾을 수 있게 해준다. 근로자가 사무실에 도착하면, 그는 할당된 공간을 찾게 된다. 그리고 난 뒤 스크린 커튼을 내린다. 프로젝션 스크린 상에(이는 또한 공간을 분할해주는 역할도 함) 근로자는 자신이 일하고 있는 회사 로고를 투사한다. 그는 또한 자신이 작업하고 있는 파일, 이미지나 프로그램을 띄울 수도 있다.

Sections

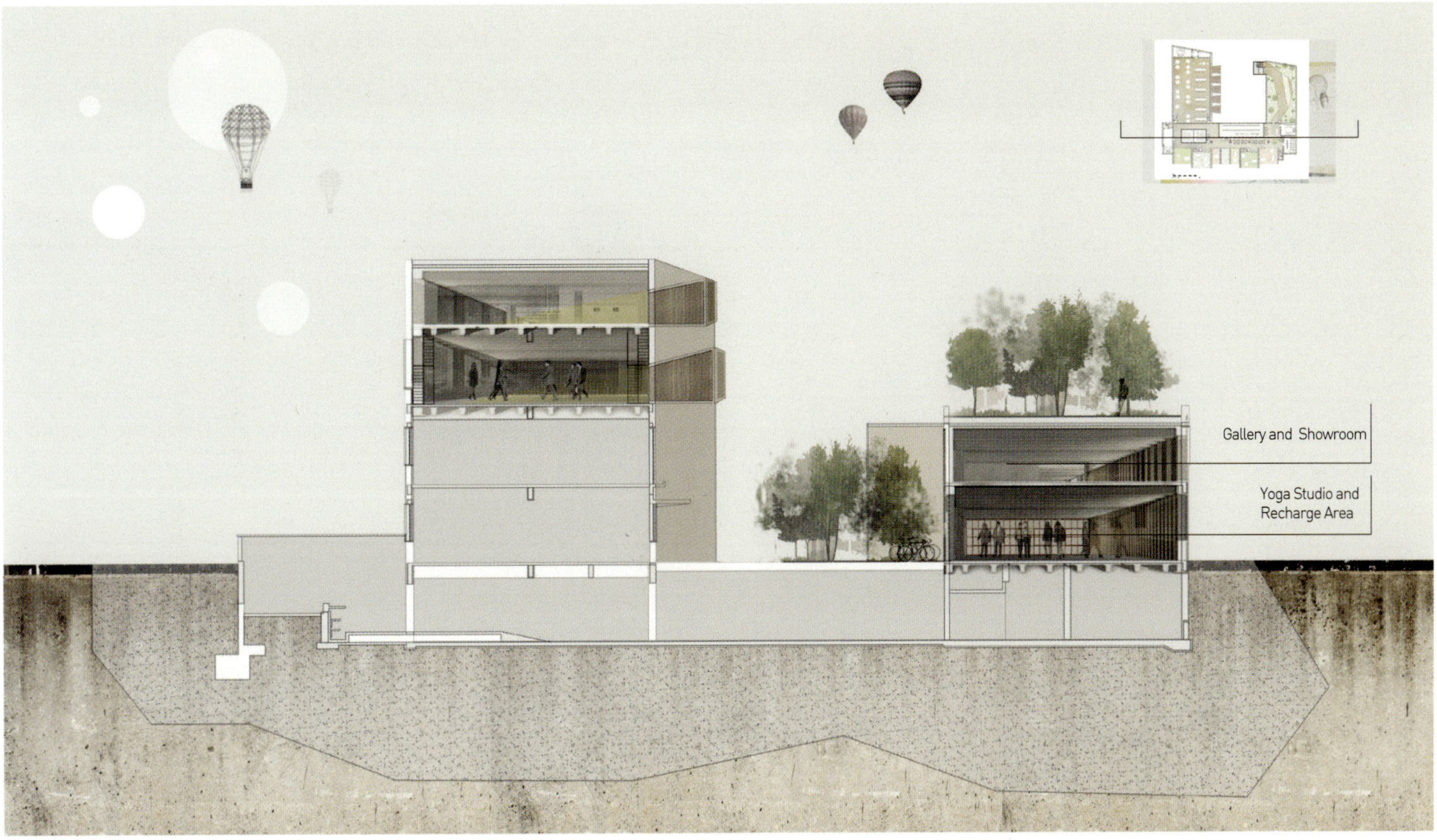
Gallery and Showroom
Yoga Studio and Recharge Area

Cellular Offices and Formal Meeting Rooms
Men and Women Toilets

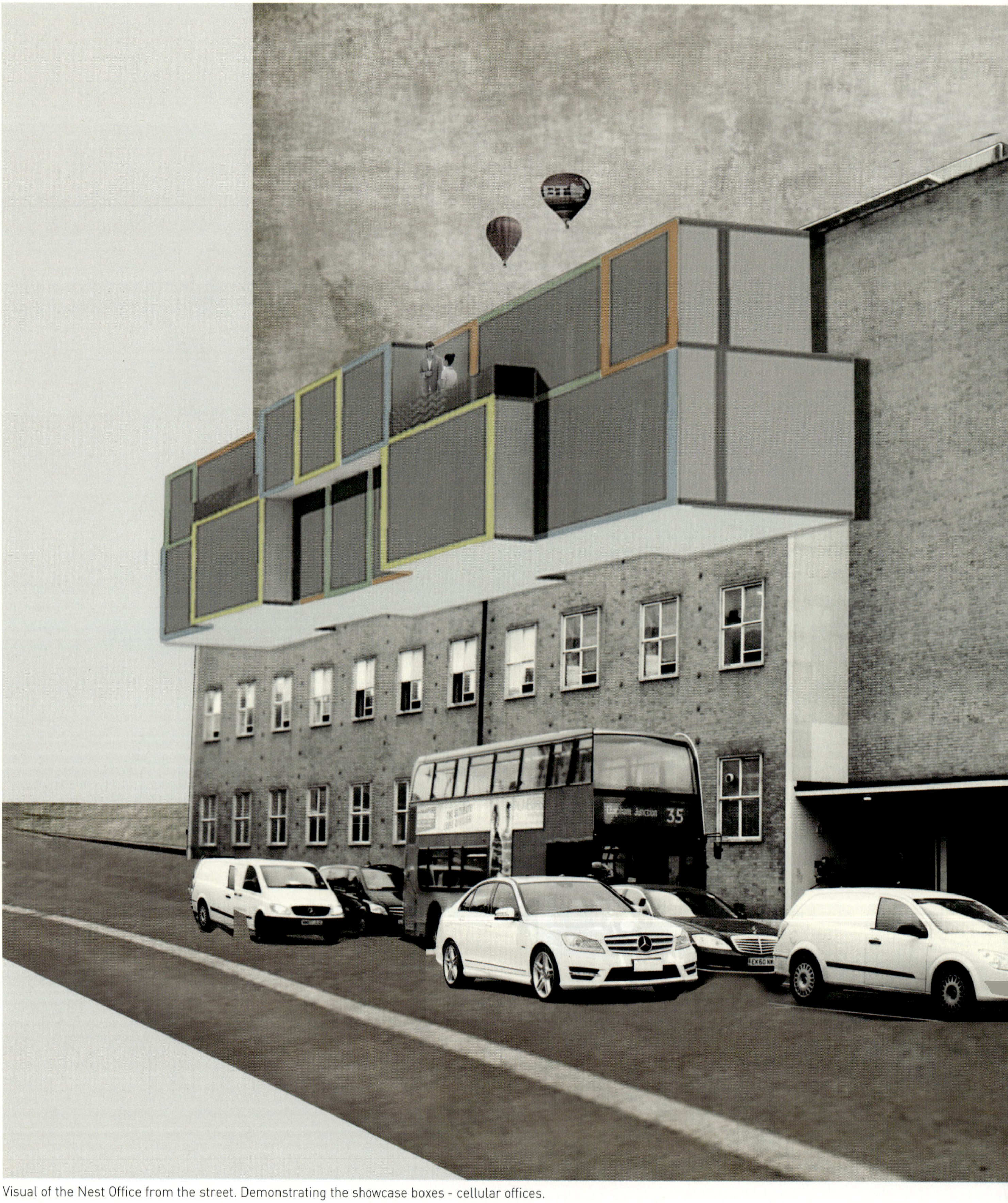

Visual of the Nest Office from the street. Demonstrating the showcase boxes - cellular offices.

International
UK Overnight
UK Sameday
MY
08000 14 14 79
info@mymessenger.co.uk
LOVE
ME

Arin Aurora S. Menes
arin.menes@gmail.com

the SHUFFLE building

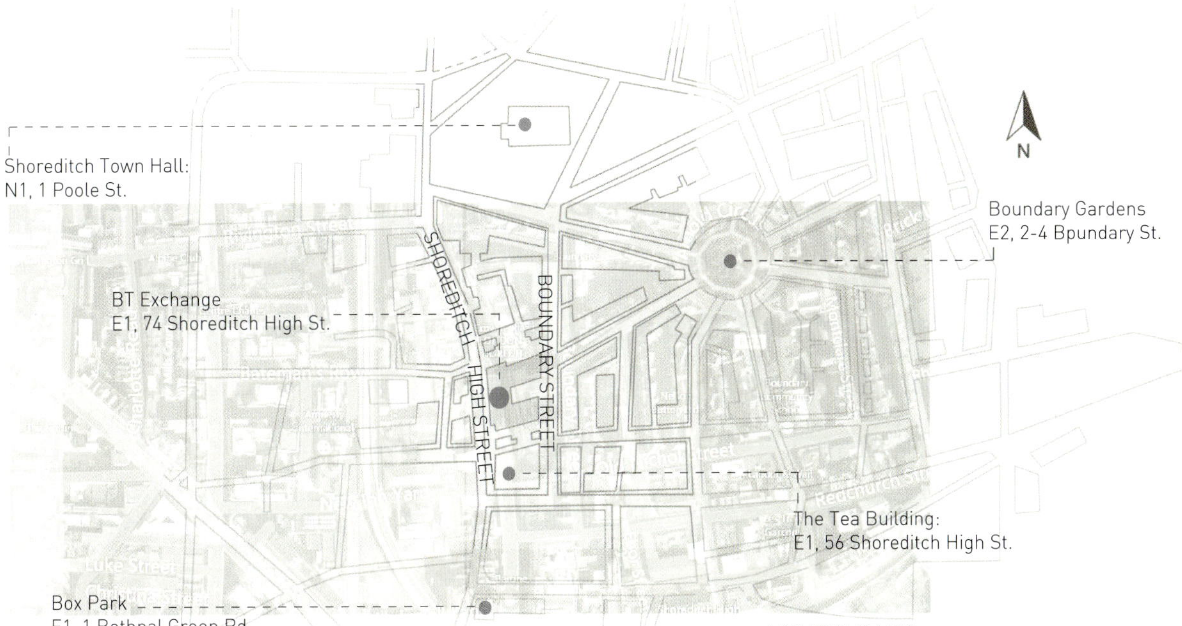

LOCATIONS MAP

The aim of this project is to re-energize BT`s redundant Telephone Exchange Building situated in the heart of Shoreditch. To reate a co-workers hub, for the private and the public, engaging the local community. A place were many companies can co-exist together, small or large, focusing on flexibility and adaptability to accommodate change and growth. It will be a place where not only the facilities are shared but more importantly it will encourage comminication and collabration and intercompany connect- ions developing a spatial co-operative. Reseaching the Shoreditch area and discover the Vibe, its heritage and the creative trades that exist in the area.

Explore through model making and sketching that will test public and private space, light and orientation, flexibility and adaptability of form and space. We were asket to look at different aspects around the topic: What is creative space? How does space support initiative and fresh thinking? How do people work with new technology? How do we consider the whole person and live/work issues?

이 프로젝트의 목표는 Shoreditch의 심장부에 위치한 비어 있는 BT 전신 교환국 건물을 부활시키는 것이다. 공공 또는 사적 협업 허브를 만들어 지역사회의 관여를 이끌어 내기 위한 것이다. 이 장소에는 크든 작든 많은 회사들이 공존하게 되며, 변화와 성장을 수용할 수 있는 융통성과 적응성에 초점을 둔다. 이 장소는 또한 시설만 공유하는 것이 아니라 보다 중요하게는 커뮤니케이션과 협력, 회사 간 연계를 촉진해 공간적 협력을 창출해낼 수 있는 장소가 될 것이다. Shoreditch를 연구해본 결과, Vibe와 그 유산, 이 지역에 존재했던 창의적인 무역 역사 등을 찾을 수 있었다.

공용 공간과 사적 공간, 조명과 방향, 형태 및 공간의 융통성과 적응성을 테스트할 모델과 스케치를 통해 탐색한다. 우리는 주제의 여러 측면을 살펴보라는 지시를 받았다. 창의적인 공간이란 무엇인가? 공간이 창의적이고 신선한 생각을 어떻게 도울 수 있는가? 사람들이 새로운 기술로 어떻게 작업할 수 있는가? 우리가 사람 전체와 삶/일의 문제를 어떻게 고려할 수 있는가?

The Concept
The Shuffle Building

Through the concept of a flexible and adaptable co-workers hub that shall engage communication both between the private and public users of the space within Shoreditch, comes the idea to design a building inspired by Shuffleboard. A table game that goes back over 500 years, is now yet again a popular pub activity amongst the younger generation. Two or more can *play* the game, where you *slide* a puck over the, up to 7 meters, long polished timber. A co-worker space that encourage collabration and creativity. A place you would want to visit for the futuristic surroundings and architectural experience amongst others not just to do your office work.

Shoreditch 내 공용 공간 사용자와 사적 공간 사용자 간 의사소통을 촉진할 수 있는 융통성 있고 적응력 있는 협업 허브라는 컨셉을 통해 셔플보드에서 영감을 받은 건물을 디자인하겠다는 생각을 할 수 있게 되었다. 500년도 더 된 이 테이블 게임은 아직도 젊은 세대에 인기를 얻고 있다. 2사람 이상이 게임을 할 수 있으며, 기다란 막대기로 7미터까지 퍽을 밀어낼 수 있다. 협력과 창의성을 촉진시키는 협업 공간이 될 것이다. 단지 업무 뿐 아니라 미래지향적인 주변 환경과 건축물들을 경험해보기 위해 다시 방문하고 싶은 공간이 될 것이다.

The Exchange still houses the telephone apparatus on the ground and first floor and therefore we are to design the unoccupied space and not compromise the security of the remaining BT space.

교환국 건물에는 아직도 지상층과 1층에 전화 시설들이 남아 있다. 따라서 우리는 남은 BT 공간의 안전을 훼손하지 않은 사람이 살지 않는 공간을 디자인해야 한다.

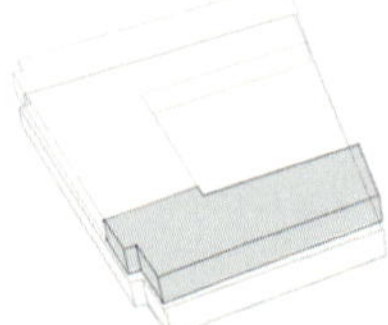

GROUND FLOOR AREA

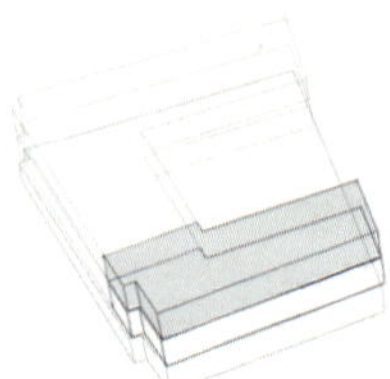

FIRST FLOOR AREA

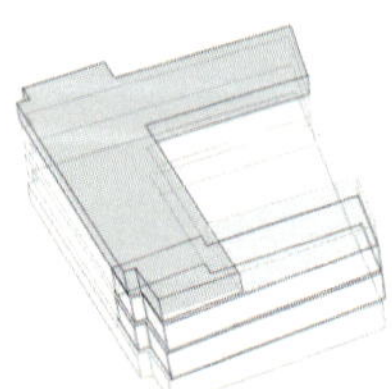

SECOND FLOOR AREA

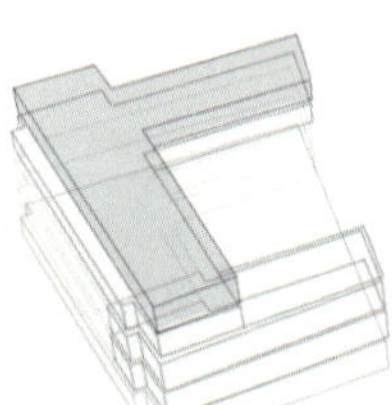

THIRD FLOOR AREA

The Exchange Building is approx. 6650 sqm (100%), BT with the basement occupies approxamately 3542 sqm (55%), that leaves 3108 sqm (45%) for the new Co-working space.

교환국 건물은 약 6650평방미터(100%)의 규모로서, 지하층이 있는 BT가 약 35432평방미터(55%)를 차지하고 있어, 3108평방미터(45%)를 새로운 협업 허브 공간으로 사용할 수 있다.

The building is between busy and noisy Shoreditch High Street and residential Boundary Street, therefore the noise levels around the building varies from practically silent on the East side (up to 54.9 dB), and to very loud (75+ dB) on the West side. For this reason it is very important to allocate office spaces correctly as well as use suitable sound insulation where necessary.
On the North/West part of the building there is a blue and the temperature there remains constant. This conditions is necessary for office and computer work. On the South/East side of the building is a hot and yellow light.

이 건물은 사람들로 붐비고 시끄러운 Shoreditch High Street와 거주구역 경계 도로 사이에 위치해 있기 때문에, 건물 주변의 소음 수준이 실질적으로 전무한 곳에서부터(동쪽 최대 54.9dB) 매우 시끄러운 곳(74dB 이상)(서쪽)에 이르기까지 다양하다. 이런 이유로, 사무실 공간을 정확하게 배분하고, 필요한 경우 적절한 방음을 하는 것이 매우 중요하다.
건물의 북/서쪽에는 그늘이 져 있어 온도가 일정하다. 이 조건은 사무실과 컴퓨터 작업에 필요한 조건이다. 건물 남서쪽은 온도가 높고 노란색 조명이 있다.

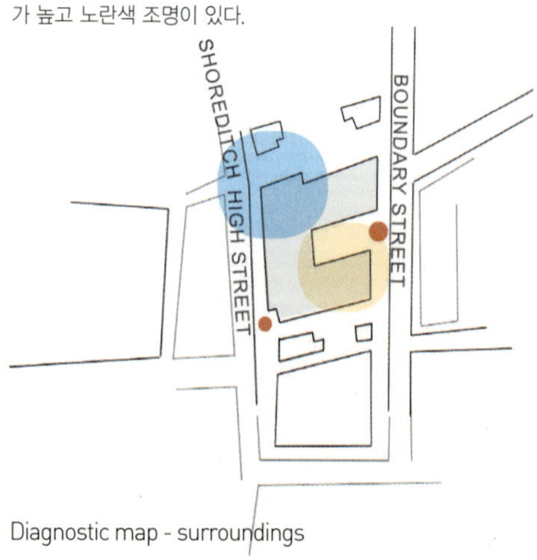

Diagnostic map - surroundings

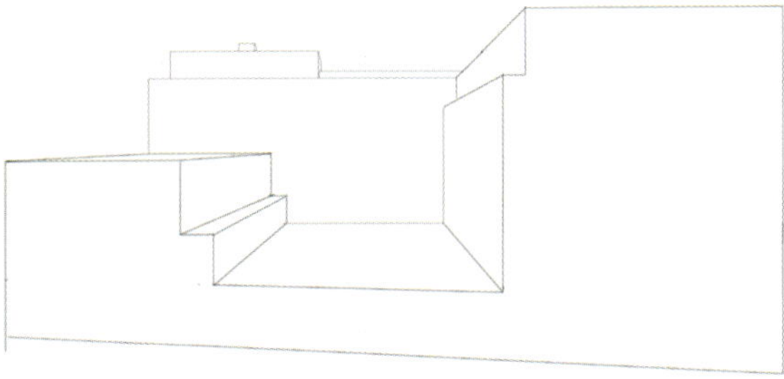

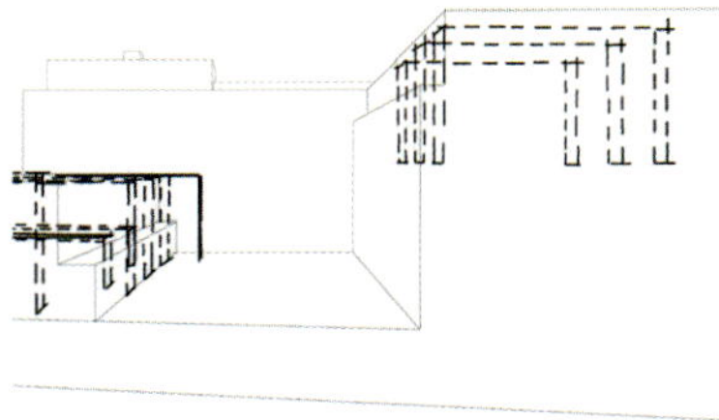

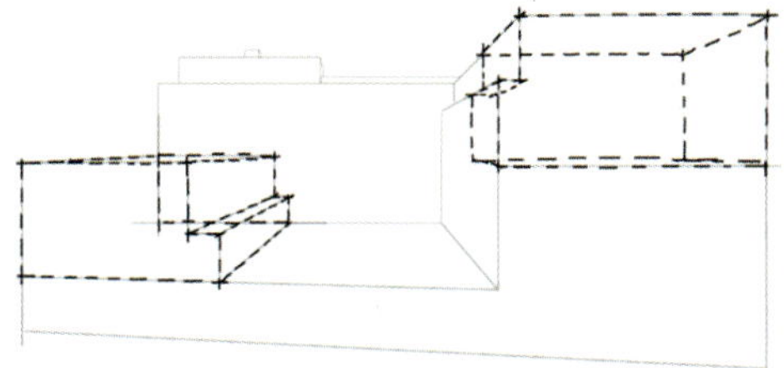

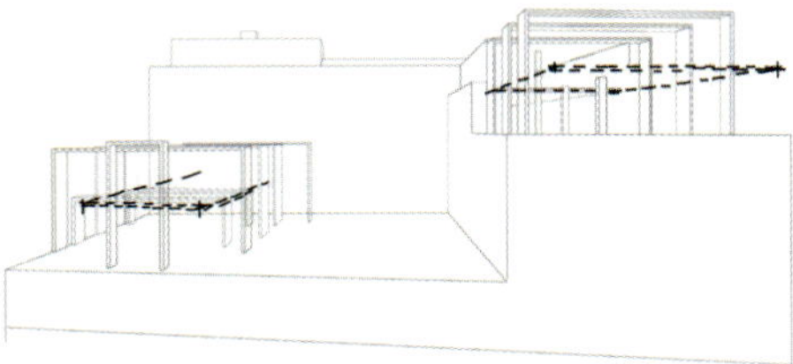

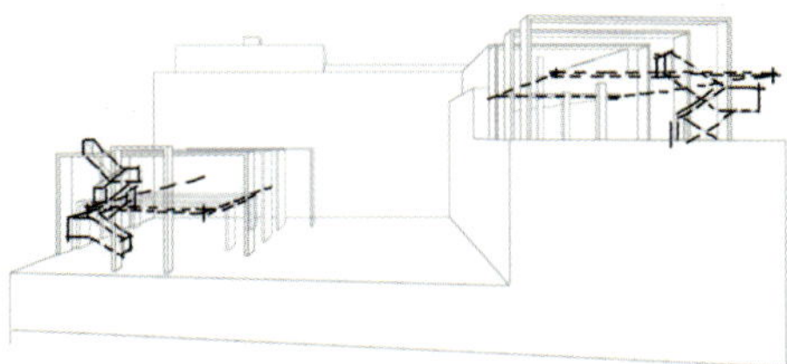

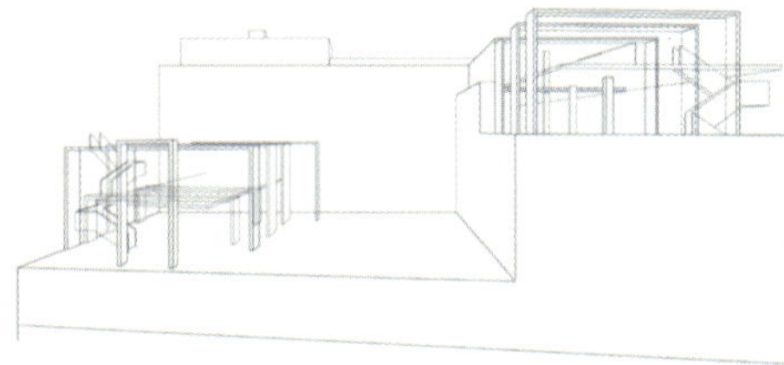

Bing map image of the building - viewed from boundary st.

View edited in photoshop to develope external changes.

Based on the buildings limitations I arrange my spaces, minding BT still occupying parts of it, experimenting how to co-exsist together as well as creating a public and private space within.
Re-creating the building (and excisting staircase) in different matersals; card, foamboard, paper, wire and acrylic helped me to destinguish the areas I will use within my proposed design and to see the volume of space.
I turned the building around so the main access is from residential Boundary Street instead of busy Shoreditch High Street. I then removed large areas going on to Boundary street from the excisting building to replace them with large glass constructions.
This immediatly includes the surroundings by inviting the outside inside, but most importantly brings the inside outside.

The new exterior additions is inspired by the dark and outstanding signs and stripes crossing the timer. The dark and massive constructual arches creates two outstanding and beautiful spaces within the building, connecting large glass facades, allowing the building maximum access of daylight, creating a exsiting view from both the inside and outside and also provides advertisement for the buildings own content.

건물의 제약조건에 기초해 나는 BT가 아직 일부를 차지하고 있다는 사실을 염두에 두고 회사들이 공존할 수 있는 방법과 그 안에 공용 공간과 사적 공간을 만들 수 있는 방법을 시험해보면서 공간을 배치했다.
다른 소재들을 이용해 건물을 재구성하였다. 예를 들어, 내가 제안한 디자인에서 사용한 카드 폼 보드, 종이, 와이어, 아크릴 등이 공간을 구분하는데 도움이 되었으며, 공간 볼륨을 나타내는데 도움이 되었다.

나는 사람들로 북적대는 Shoreditch High Street에 나 있는 입구 대신에 주거지역 경계 도로 쪽으로 입구를 냈다. 그런 다음 기존 건물로부터 경계도로까지 펼쳐져 있는 큰 공간을 대형 유리 구조물로 대치해 그 공간을 없앴다.
이는 외부를 안에서 볼 수 있는 것 뿐 아니라 내부를 외부에서 볼 수 있게 하는 효과가 있었다.

새로운 익스테리어를 추가한 것은 어둡고 눈에 띄는 간판들과 줄무늬들에 영감을 받은 것이었다. 어둡고 거대한 구조 아치는 건물 내에 눈에 잘 띄고 아름다운 2개의 공간을 창출해 대형 유리 입면과 연결시킴으로써 건물에 최대한 자연광이 들 수 있게 하였고, 내외부에서 볼 수 있게 하였고, 또한 건물 자체의 내용에 대한 광고효과도 누릴 수 있다.

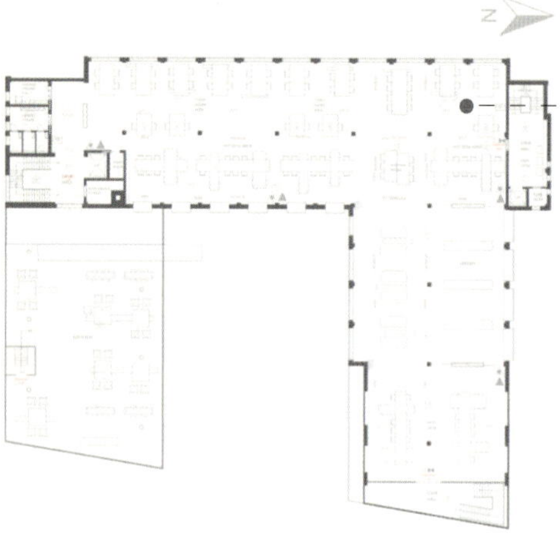

Concept Constructions
Sliding Wall System

Sliding wall and arches provides various opportunities and creates a spatial experience where the people working in the building can arrange their own sets very immediately.
How the wall and arches can become room dividers, acoustic walls, exhibtion walls or sunshade. Create a private room for a showcase, a presentation, a meetingroom or a preparation space for an event in the middle of the open work space.

슬라이딩 월과 아치는 다양한 기회를 제공하며, 건물에서 일하는 사람들이 자신의 세트들을 배열할 수 있는 공간 경험을 가능하게 한다.
벽과 아치는 방을 나누는 역할과 음양 벽, 전시 벽, 태양 가리개의 역할을 하며, 쇼 케이스, 프레젠테이션, 회의실 용도의 공간을 만들어 주며, 또는 개방형 업무 공간의 중앙에 이벤트를 준비할 수 있는 공간을 만들어주기도 한다.

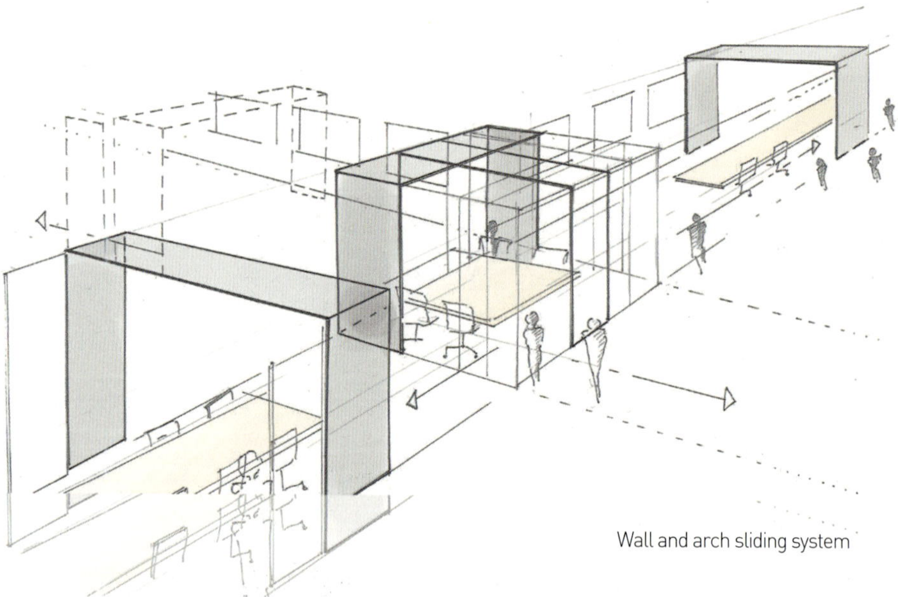

Wall and arch sliding system

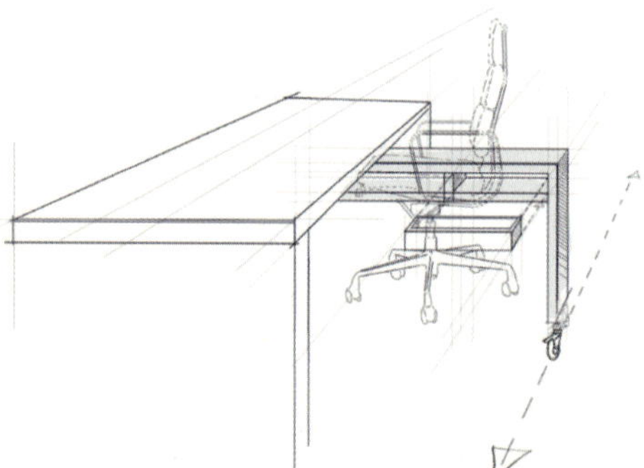

Work desk with moveable storage

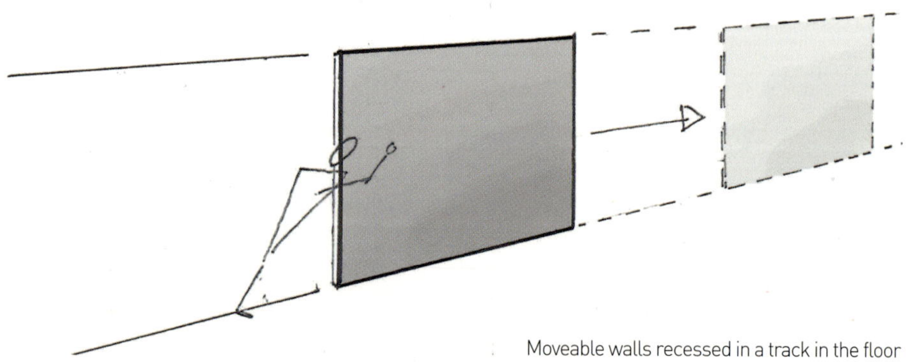

Moveable walls recessed in a track in the floor

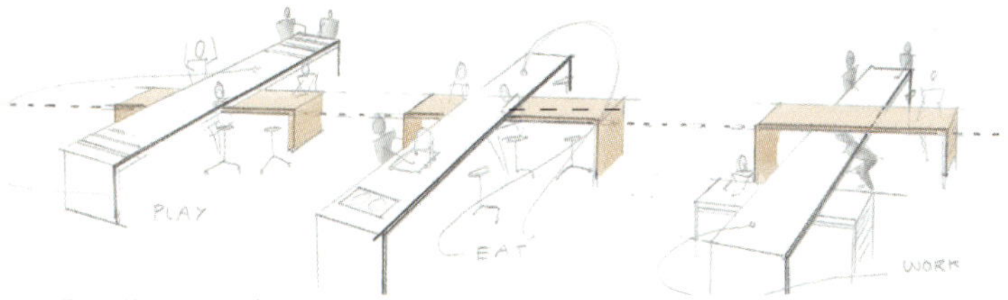

The different work areas

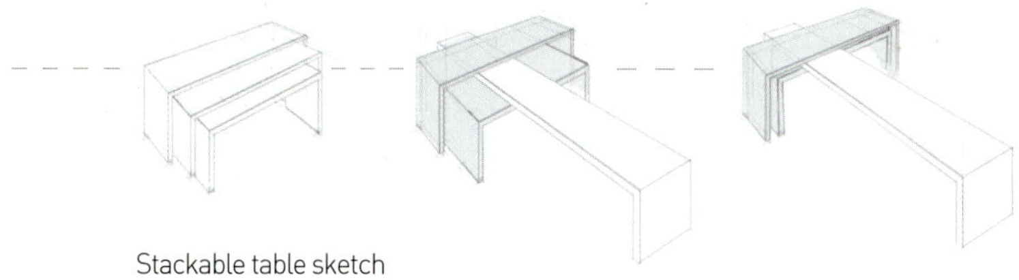

Stackable table sketch

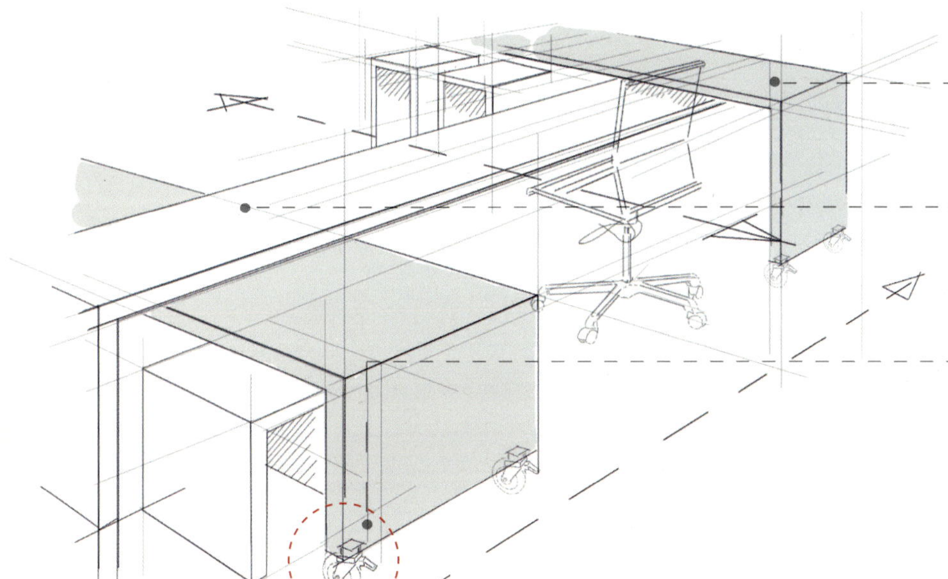

Construction sketch

Table arch structured with a light steel grid covered in fibre C dark grey concrete surface.

The long main table is the static core of the work group to keep some limitations within the working space. This table arch is built in Oak timber.

To make the table arches easily moveable they have steel wheels with breaks under each end of the arch.

Concept Constructions
Hot Desk Area

The hot desk area is also a set of moveable arches in various sizes so that they also are stackable. This allows the people working in the building to arrange their work settings; from a seperate and private situation to f.ex. one long workspace for team work or meetings, it also allows the idea of privacy although you are situated on the same work station.

핫 데스크 구역은 또한 다양한 크기로 이동할 수 있는 아치들의 세트로서, 쌓을 수도 있다. 이는 건물에서 일하는 사람들이 자신의 업무 환경을 배치할 수 있게 해준다. 즉, 별도의 개별 공간을 팀 작업이나 회의를 위한 긴 공간으로 전환시킬 수 있다. 이는 또한 동일한 워크스테이션에 있어도 프라이버시를 확보할 수 있게 해준다.

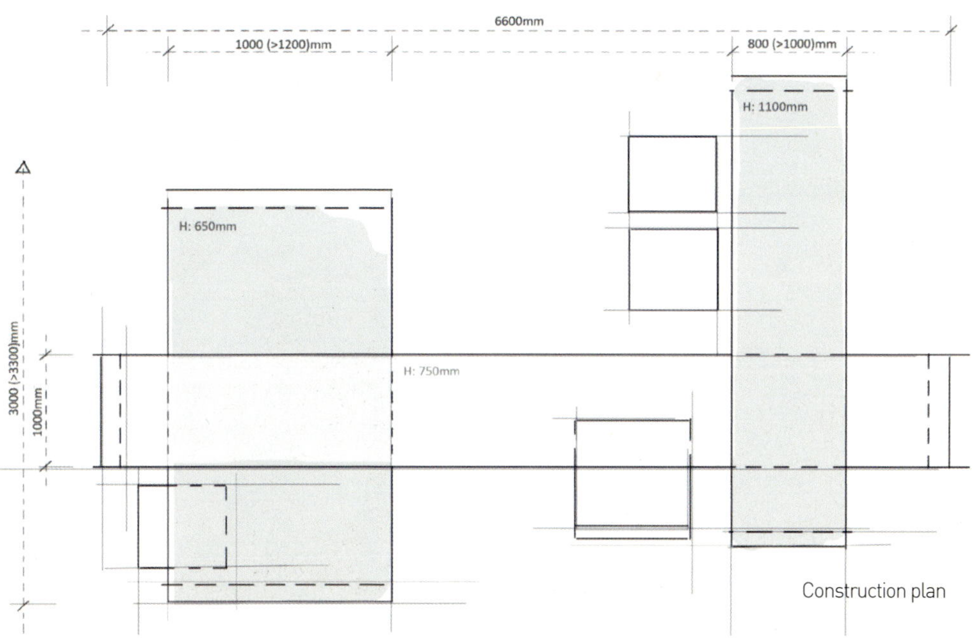

Construction plan

Wayfinding
The Staircase

The staircase play a important role in wayfinding and access through the building, for borth the people working there and the visitors. For the experience of the movement.
The massive floating staircase built in Oak timber breaks up the design in between concrete, glass and steel.
Oak is an outstanding material for building purpose, an Oak beam is as strong and as dependable as any steel joist. It is also durable and its beauty will increase as it ages. Wood has the lowest energy consumption and the lowest carbon dioide emission of any commoly used building material. It is also organic, non-toxic and naturally renewable material.

계단은 건물 내에서 일하는 사람이나 방문객 모두에게 있어 길 찾기나 출입구를 찾는데 중요한 역할을 담당한다. 참나무로 만든 거대한 부유식 계단이 콘크리트와 유리, 강철로 이루어진 디자인을 관통하고 있다.
참나무는 건축 용도로는 뛰어난 소재로서, 참나무 빔은 강철 호이스트만큼이나 튼튼하고 안정적이다. 참나무는 또한 내구성이 좋고, 시간이 지날수록 아름다움도 더해진다. 목재는 일반적으로 사용되는 다른 어떤 건축 소재 보다 에너지 소비가 낮고, 이산화탄소 배출량도 적다.

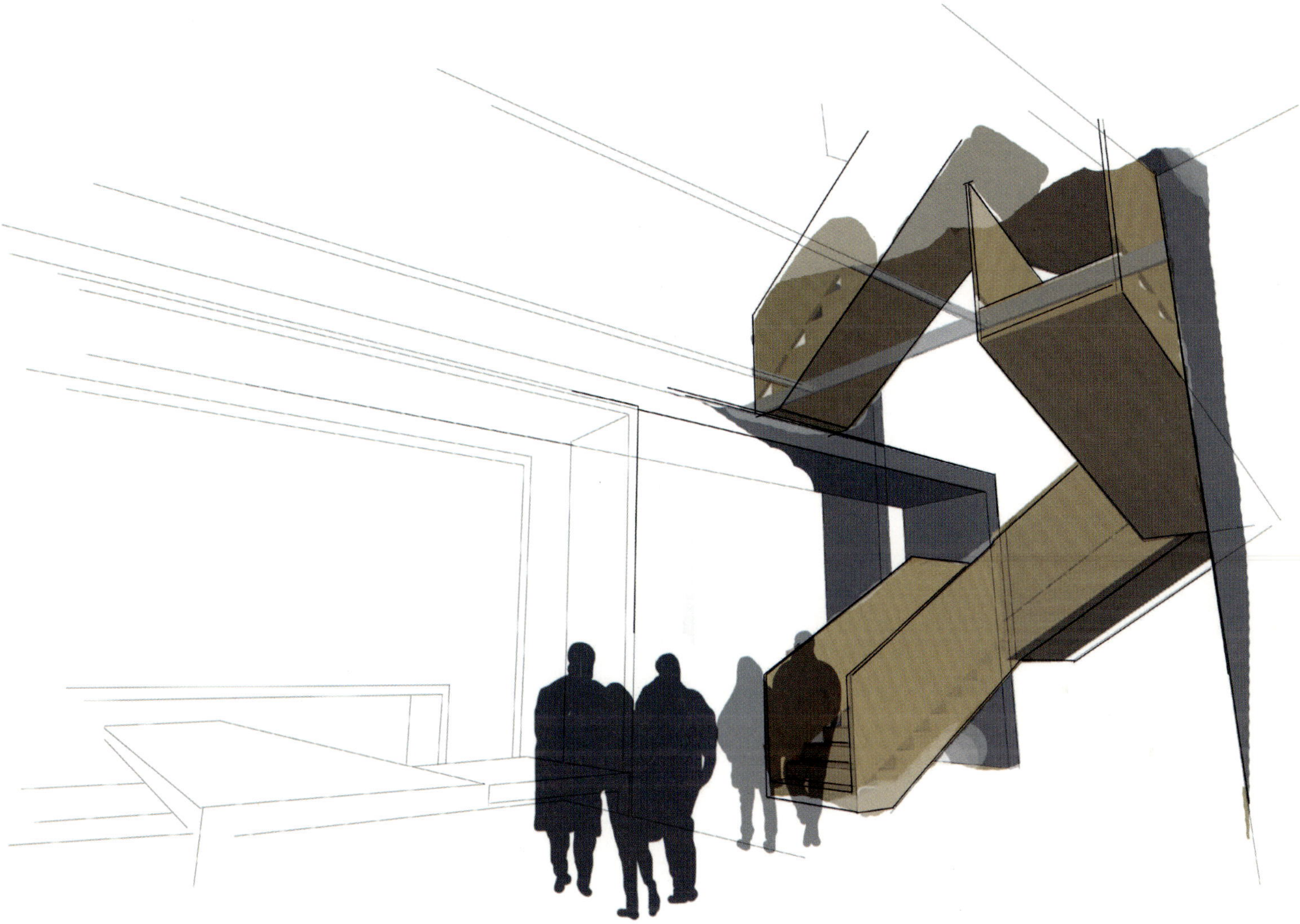

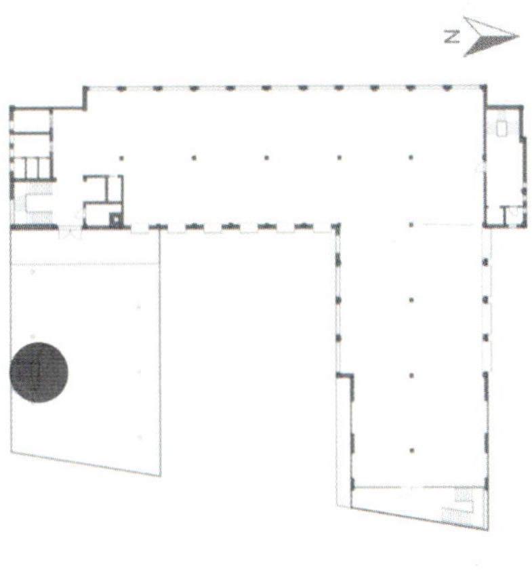

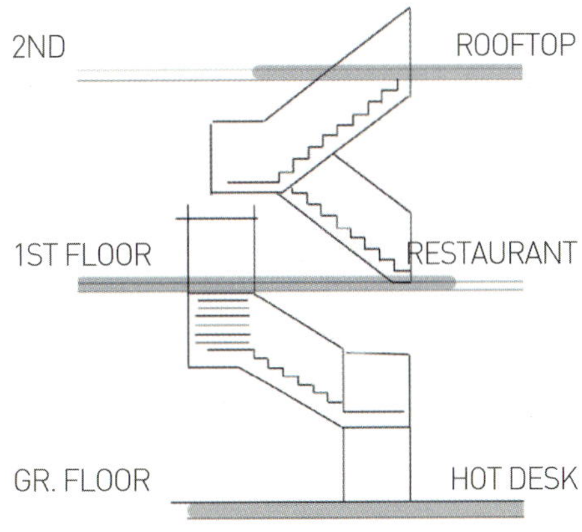
2ND
ROOFTOP
1ST FLOOR
RESTAURANT
GR. FLOOR
HOT DESK

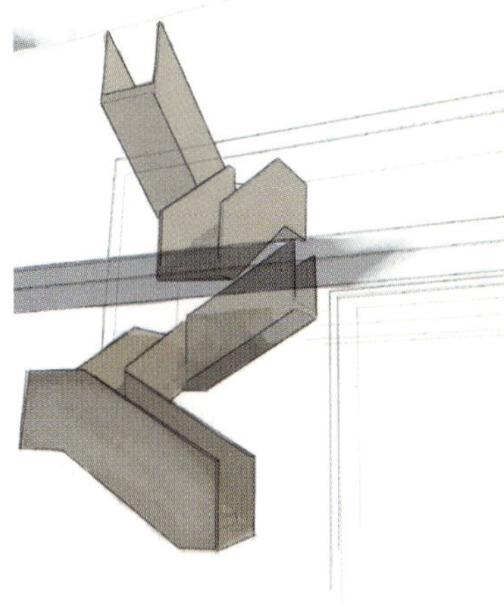

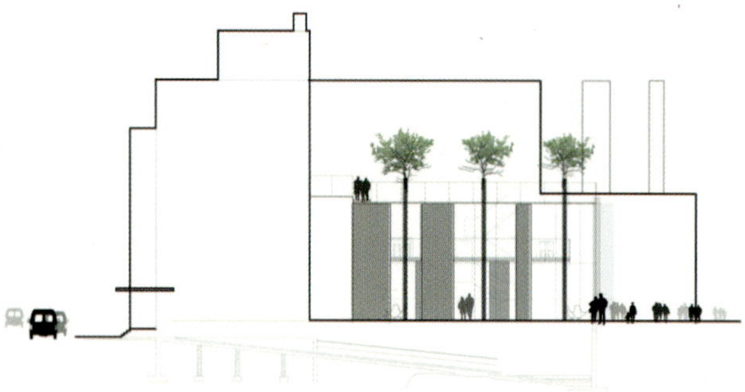

Northfacing facade drawings of the new extention

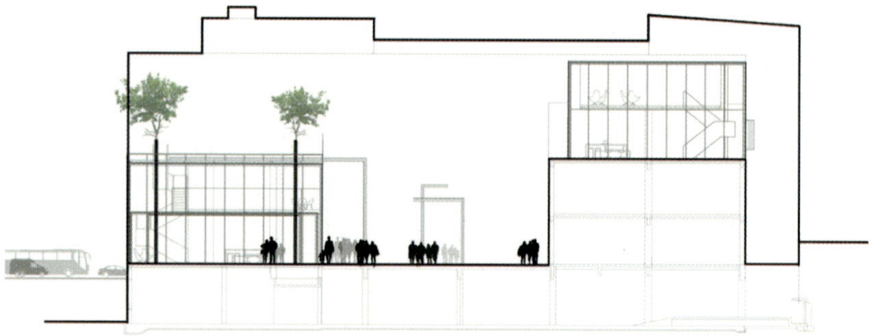

Westfacing facade drawings of the new extention

Final Design:
The Shuffle Building

The Shuffle is a new co worker hub and meeting place within the heart of Shoreditch. Here you can enjoy good food and drinks, a shufflegame or just our beatiful roof garden overlooking the quiet Boundary Street.
For our members there is full access to our public facilities but also including; bike parking, library, shuffleboard games in the office area, private meeting rooms, private desk in open office space, teamwork areas or hot desking both private and public. You have access to locker rooms with showers and private lockers, also a fully equipped kitchen, a opportunity to cook your own food, unless you want to eat in the building's own restaurant. The private lounge area on the third floor messanin offers amazing views overlooking the buildings roofgarden below and our bee nest garden.

셔플은 Shoreditch의 심장부에 위치한 새로운 동료 허브이며 만남의 장소이다. 이곳에서 음식과 음료를 즐길 수도 있고, 셔플 게임을 즐길 수도 있으며, 조용한 경계도로를 바라보며 아름다운 옥상 가든을 즐길 수도 있다.
회원들은 공공시설을 모두 이용할 수 있으며, 이 외에도 자전거 주차, 도서관, 사무실 구역에서의 셔플보드 게임, 개인 회의실, 개방형 사무실 공간에서의 개인 책상, 팀 작업 공간 또는 공공/개인 핫 데스크 등도 이용할 수 있다. 건물에 있는 식당을 이용하기를 원하지 않으면, 샤워기와 개인 라커, 주방이 구비되어 있는 라커룸도 이용할 수 있다. 3층의 개인 라운지 구역은 아래로 건물의 옥상 가든을 올려다보고 아래로는 꿀벌 둥지 가든을 내려다볼 수 있는 아름다운 조망을 제공한다.

Green intervention, bee nest garden for a different view

Public hot desk area within the new extention

SECTION

FACADE

Final Design:
The Shuffle Members

As a member all you need is to book your days in The Shuffle with one of our concierge; meeting rooms, banquet facilities or the roof garden for a fashionable summer party. If desired, we provide lockers in the reception area aswell as in the changingrooms; where you can store your belongings; Invite friends for a better meal or a game of Shuffleboard after work and you can be showered and ready for an evening out without the laptop or your gym equipment.

회원들은 회의실, 연회시설 또는 화려한 섬머 파티를 위한 옥상 가든 중 하나를 예약하기만 하면 된다. 필요한 경우, 리셉션 구역에 락커를 제공해줄 수도 있으며, 탈의실도 제공해 이곳에 소지품을 보관할 수도 있다. 업무 시간이 끝나고 나면 식사나 셔플 게임에 친구들을 초대할 수도 있고, 샤워를 한 뒤 노트북이나 체육관 장비 없이 외출을 즐길 수 있다.

Ewelina Ruminska
ewelina.ruminska@googlemail.com
www.behance.net/SolEwelina

The concept proposal strives to create a scheme which allows the community to reconnect to an iconic but forgotten building. One that will make space for a new vision, revitalize the local art scene, and bring back the lost sense of community to this diverse yet deprived area of London.
Seeking to embrace the rough beauty of the current derelict state, the proposal aims to bring the building back to life. The design of the New Mill Art Community Center will explore the divergence and convergence within the community.
Translating these elements into adjustable spaces, it connects and separates through light, texture and materials. The building transforms into a creative nest with endless possibilities for development by exploring the derelict and the role it plays. Empty floors are a platform for imagination.

이 컨셉에서는 한 때 상징적인 건물이었지만 지금은 잊혀진 건물과 지역사회가 다시 연계될 수 있는 장소를 만들고자 한다. 새로운 비전을 위한 공간을 만들게 될 이 컨셉에서는 지역 예술을 활성화시키고, 다양하지만 아직 활력이 부족한 런던의 이 지역에 지역사회라는 느낌을 되살리고자 한다.
현재 버려져 있는 상태가 지니고 있는 거친 아름다움을 포용하기 위해, 이 제안에서는 건물에 다시 활기를 불어넣고자 한다. 새로운 밀 아트 커뮤니티 센터의 디자인에서는 커뮤니티의 통합과 다양성을 모색하고 있다.
이러한 요소들을 조정 가능한 공간으로 전환시킴으로써, 이 디자인은 조명과 텍스쳐, 소재를 통해 연결과 분리를 시도한다. 버려진 장소를 탐색하고, 이 장소가 담당할 수 있는 역할들을 모색해봄으로써 건물은 무한한 가능성을 지닌 창의적인 둥지로 변환된다.

urban catalyst
{London Millenium Mills}

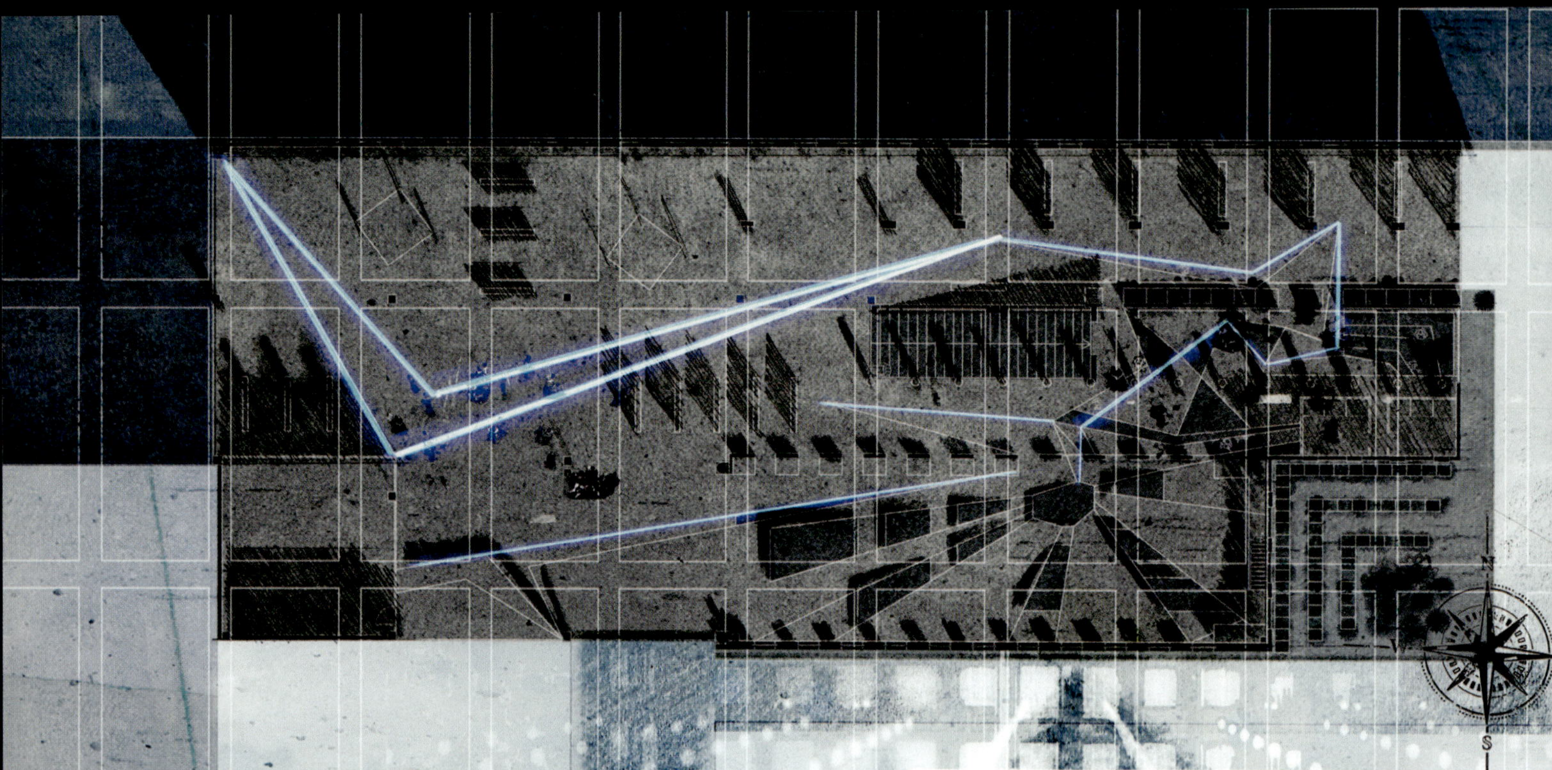

Ground floor plan

The ground floor serves as an event space, creating revenue, and the community plays an active part in maintaining the center, allowing them to learn, experience, and strive for success, thus boosting the confidence of the people.
In the new vision, the Mill retains its historical value while acting as a catalyst in creating opportunities for the surrounding community.

지상층은 이벤트 공간으로써, 수입을 창출하고, 커뮤니티가 센터를 유지하는데 있어 적극적인 역할을 담당함으로써 커뮤니티가 학습하고, 경험하며, 성공을 추구함으로써, 사람들의 자신감을 증진시키는 역할을 하게 된다.
이 새로운 비전에서 밀은 그 역사적 가치를 유지함과 동시에 주변 커뮤니티에 기회를 제공해줄 수 있는 촉매제 역할을 담당하게 된다.

Continuous desk

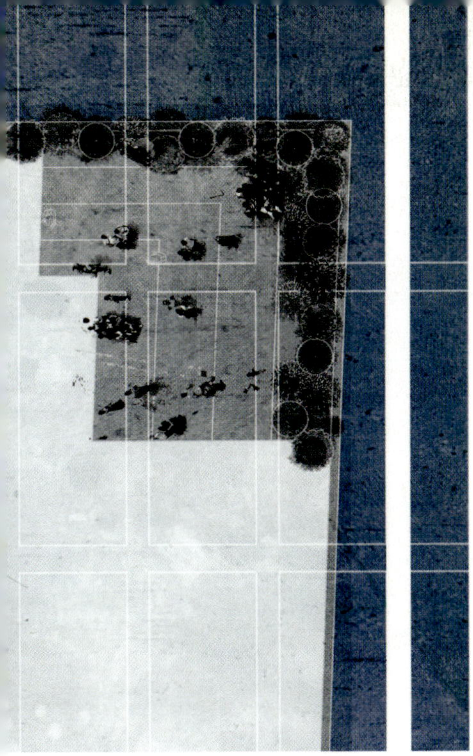

Art gallery

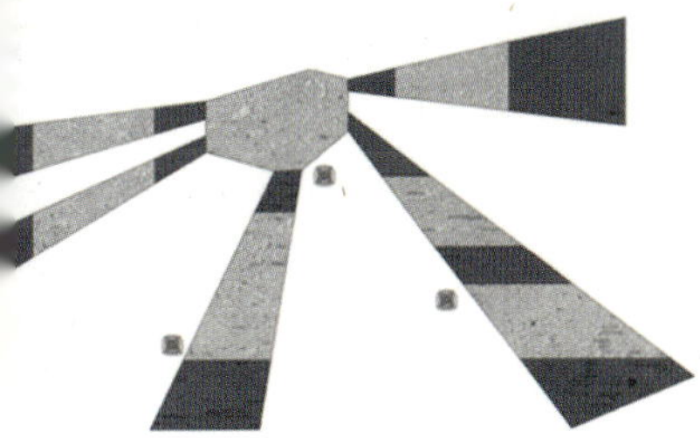

Foyer

art is either plagiarism
revolution

EVANGELIXALINA
Behance; Evangeline Louisa Cole
evangelicalina@gmail.com

Audi Gallery

An exhibition of performance, experience, history, and talent.

Front Elevation | South

Within the heart of West Silvertown stands the Millennium Flour Mills a defiant monument to a dynamic industrial past. It's life spanned nearly 80 years working 24 hours a day to provide flour to the nation. Since its closure in the late 1980's the Millennium Mills has been left for abandonment. Leaving the large empty rooms of concrete and steel to sit silently while the surrounding area and community carry on. There has only been one major project which was approved in 2007.
This would of have been Europe's largest Urban Regeneration project, creating 2000 jobs and offering a variety of different attributes including waterfront, apartments and an aquarium for the public but the funds could not be raised and the whole production fell through.

The issue being highlighted is the social aspect of Silver Town. The idea of the community is reflected in the current state of The Millennium Mills, abandoned and ignored.

웨스트 실버타운 심장부 내에 다이내믹한 과거 산업의 기념물인 밀레니엄 제분 공장이 서 있다. 이 제분공장의 수명은 매일 24시간씩 가동하면서 거의 80년 동안 국가에 밀가루를 공급해 왔었다. 1980년대 후반에 이 공장이 폐쇄된 이후 밀레니엄 밀은 버려진 채 방치되어 왔고, 주변 지역과 커뮤니티가 발전하는 동안 거대한 콘크리트 공장은 비어져 있었다. 이곳에는 유일하게 2007년에 승인된 프로젝트만이 진행되었었다.
이 프로젝트는 아마도 유럽에서 가장 큰 도시 재건 프로젝트로서, 200개의 일자리를 창출하고, 수변, 아파트, 공공 아쿠아리움 등을 포함한 다양한 특성들이 제안되었지만, 기금을 마련할 수 없었고, 결국 프로젝트는 좌절되었다. 중요하게 제기된 이슈는 실버타운의 사회적 측면이었다. 커뮤니티라는 아이디어가 현재 버려져 방치되어 있는 밀레니엄 밀의 현 상태에 반영되어 있다.

Concept

A space will be recreated to join the two sides of the Dock together, but not just the literal two sides, the social beings behind them which include Industrial, Corporate and Residential. These three main targets will hardly ever cross paths, the space will encourage the employees of the Excel, the vibrant new businesses and corporate events that are happening to be shared with the Industrial businesses and Residential in the area. For the Residential and Industrial areas to come from their typical surroundings and share the history and stories of Silver Town.

선창의 양쪽 면을 연결시키기 위해 공간을 재구성할 것이다. 이 양쪽 면은 단순히 문자 그대로의 두 측면이 아니라 산업, 기업과 거주를 포함한 그 이면에 있는 사회적 측면을 의미한다. 이러한 3개의 주요 측면들은 겹치는 부분이 거의 없기 때문에, 이 공간은 새로운 비즈니스와 기업 이벤트가 구역 내 산업적 비즈니스와 거주민들과 공유될 수 있게 할 것이다. 거주 구역 및 산업 구역이 전형적인 주변 환경에서 실버타운과 그 역사와 스토리를 공유하게 될 것이다.

Visual of the concrete stage with a corten steel backdrop/roof

Elevation and Section Visual | East

Sketches of the auditorium seating

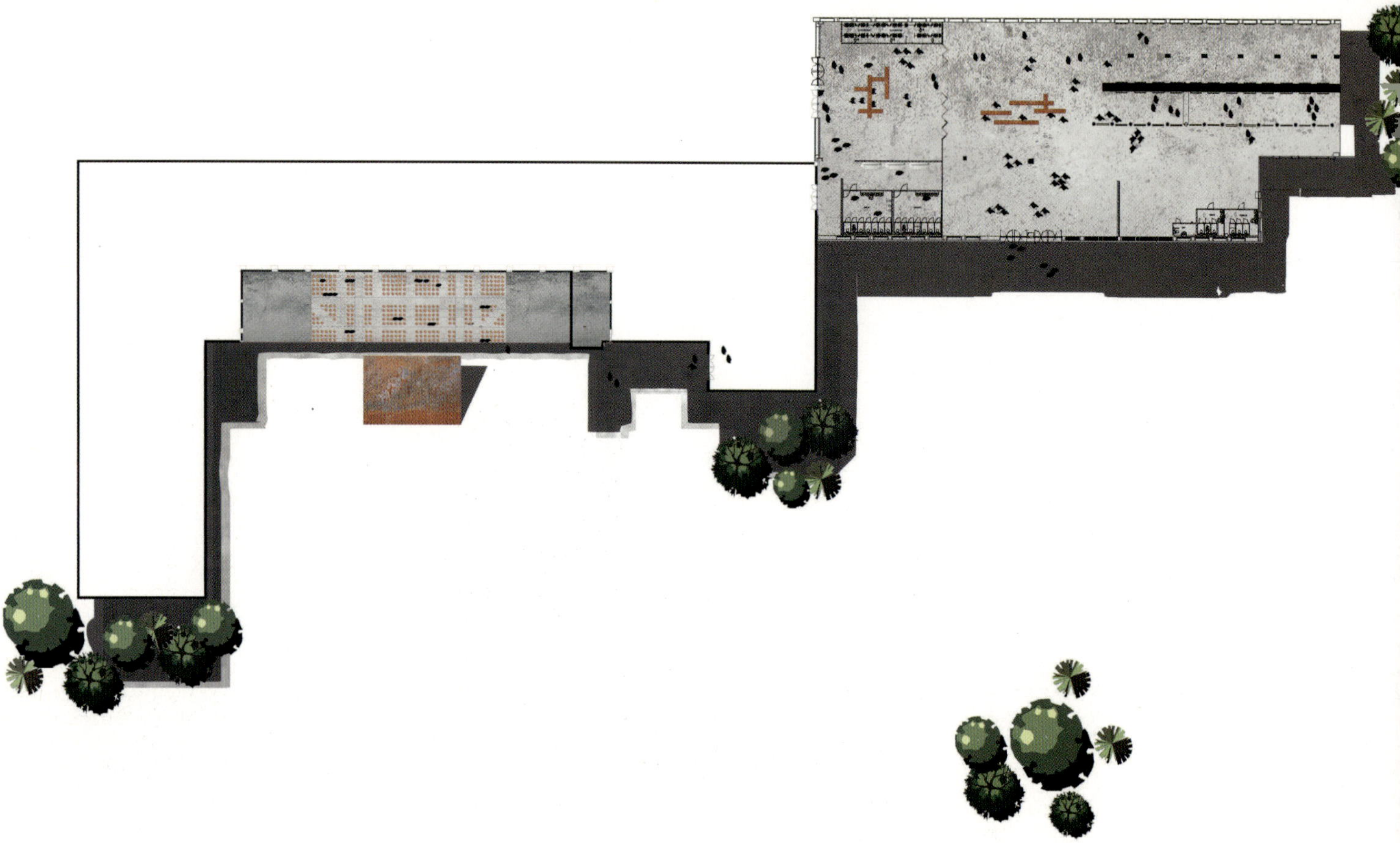

Elevation | East

The interior space of the Millennium Mills will be an inspirational auditorium. A place where local communities and those visiting can share knowledge, experience and histories. Utilising the generous personality of the interior spaces and embracing its rhythms and textures the aim is to develop and generate an archive of experiences, future proofing the communities lives. The materiality of the existing will be enhanced and emphasised by the punctuation of planes and apertures to create eventful spatial, inclusive journeys. The concoctions of movement, materials, form and structure will aim to reconnect a seemingly disparate society, through relational activities and events.

밀레니엄 밀의 내부 공간은 영감을 자극하는 강당이 될 것이다. 이곳에서는 지역 커뮤니티와 방문자들이 지식, 경험, 역사를 공유하는 장소가 될 것이다. 내부 공간의 넉넉함을 활용하고 그 리듬과 텍스처를 포용한 이 장소의 목표는 과거의 경험들을 기억하고, 앞으로의 커뮤니티의 삶을 증거하는 역할을 하는 것이다. 평면에 구멍을 내서 이벤트로 가득한 공간적 탐색을 가능하게 함으로써 기존의 유형성도 강화시키고 강조할 것이다. 움직임과 소재, 형태 및 구조를 혼합시킴으로써 겉으로 보기에는 분리된 사회를 관계 활동과 이벤트를 통해 재 연결시킬 것이다.

Section Visual | West

Enrique Santiago
kiquemenor@hotmail.com

dystopia utopia

1. A view from the DLR station

2. Thames Barrier Park

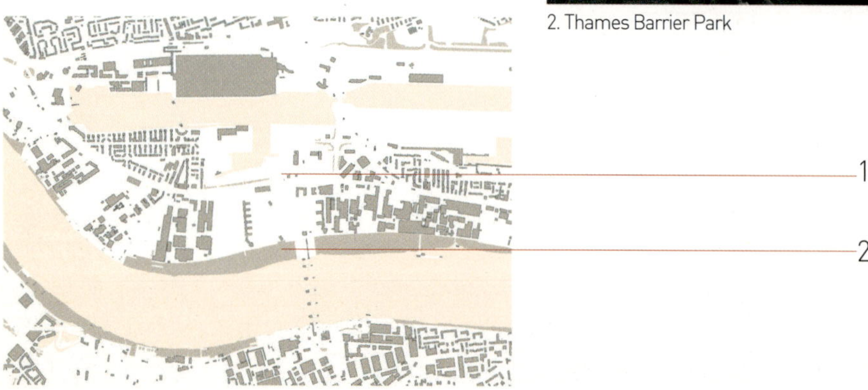

The Brief

The aim is to develop a thorough understanding of communities and materiality, in order to design meaningful spatial events.
The translation of the research aims, through a creative process, to develop social awareness and community practice. The understanding of personalities, of community and materiality will be explored and expressed through dynamic exercises of architectural and interior interventions.

목표는 의미있는 공간적 이벤트를 디자인하기 위해 커뮤니티와 유형성을 철저하게 이해하는 것이다.
연구의 목표는 창의적 과정을 통해 사회적 인식을 발전시키고 커뮤니티의 활동을 강화시키는 것이다. 다이내믹한 건축적 개입과 인테리어의 개입을 통해 커뮤니티의 특성과 유형성의 이해를 모색하고 표현할 것이다.

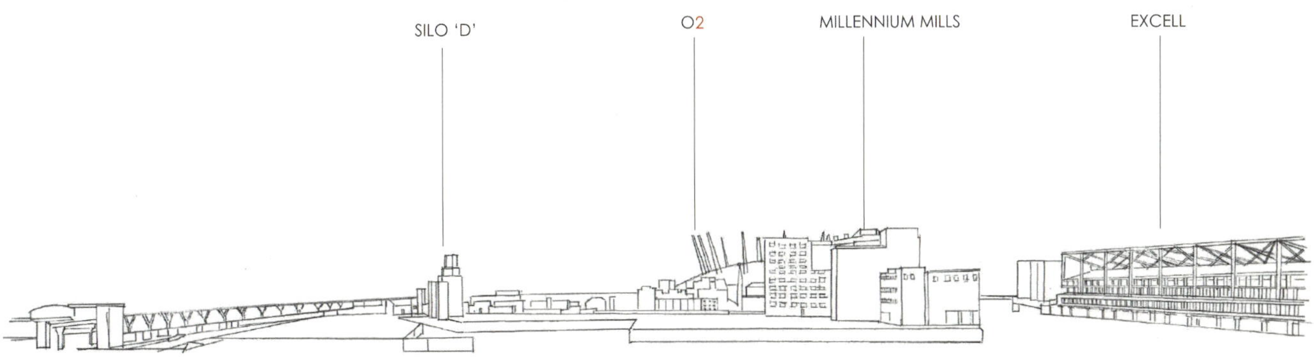

mood board
dystopia utopia

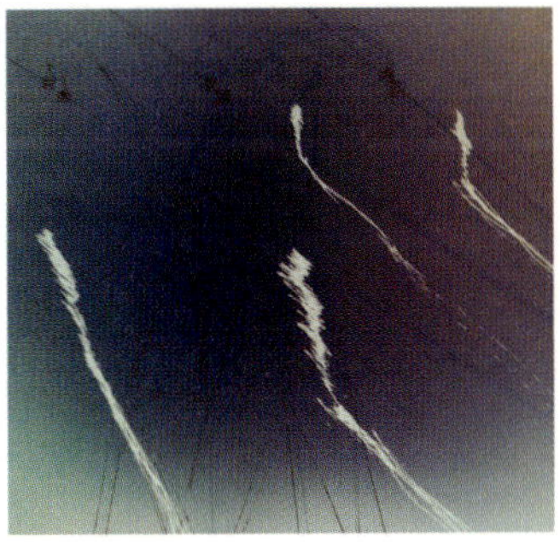

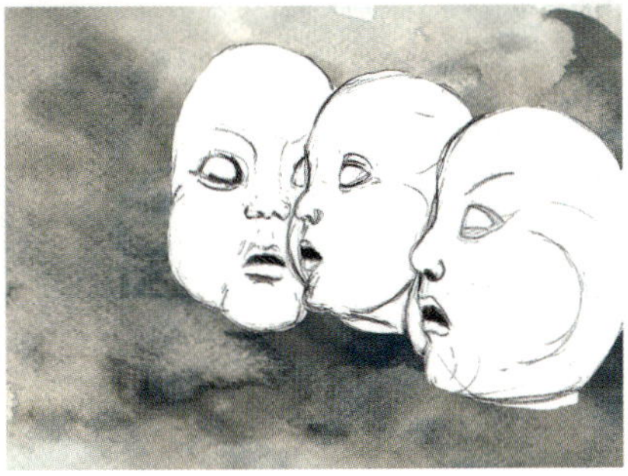

Intimate immensity

Immensity is a philosophical category of daydream, a sort of natural inclination, it contemplates the grandeur and this contemplation produces an attitude that is so special, an inner state that is so unlike any other, the daydreams transports the dreamer outside the immediate world to a world that bears the mark of infinity. In analysing images of immensity, we should realize within ourselves the pure being of pure imagination, it then becomes clear that works of art are by-products of this existentialism of the imagining being. In this direction of daydreams of immensity, the real product is consciousness of enlargement.
The space is the universal power enabling all the things to be connected.

광대무변은 백일몽의 철학적 범주로서, 일종의 자연적 경향으로서, 거대한 것을 추구하며, 그러한 추구로 인해 아주 특별하며, 다른 것과는 매우 다른 내적 상태를 산출해내고, 백일몽은 백일몽을 꾸는 사람을 현실에서 무한한 세계로 인도하는 역할을 한다. 광대무변의 이미지를 분석하는 것에 있어 우리는 우리 내부에 순수한 상상 속에서의 순수한 우리 존재를 인식해야 하고, 예술은 상상하는 존재의 실존에 따른 부산물이라는 사실을 인식해야 한다. 이러한 광대무변의 백일몽이라는 측면에서, 실제 산물은 확장을 의식하는 것이다.
이 공간은 모든 것이 연결될 수 있는 범우주적인 파워를 가진 곳이다.

The states of water

Water as a union of
past present and future

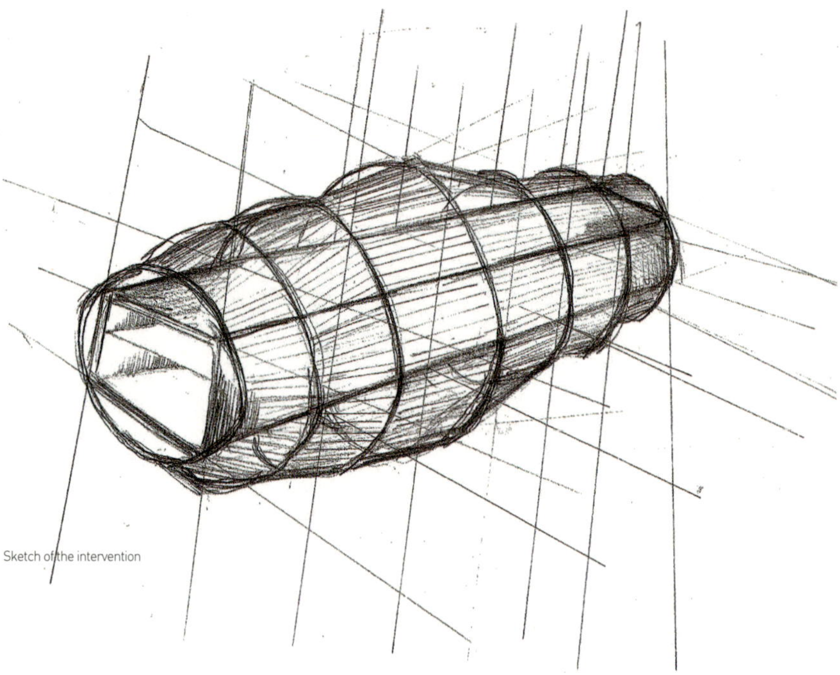

Sketch of the intervention

Ice

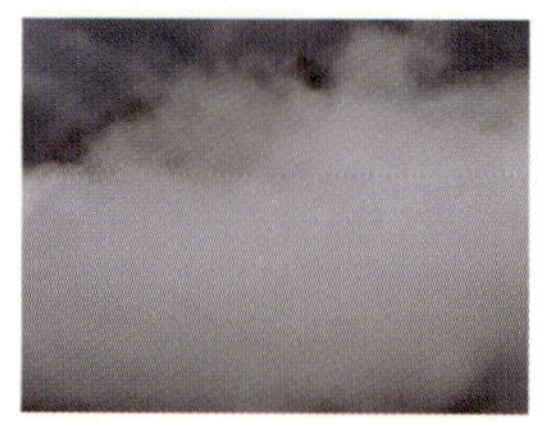

Steam

Liquid

The states of water

Community is constantly moving

View of the pond

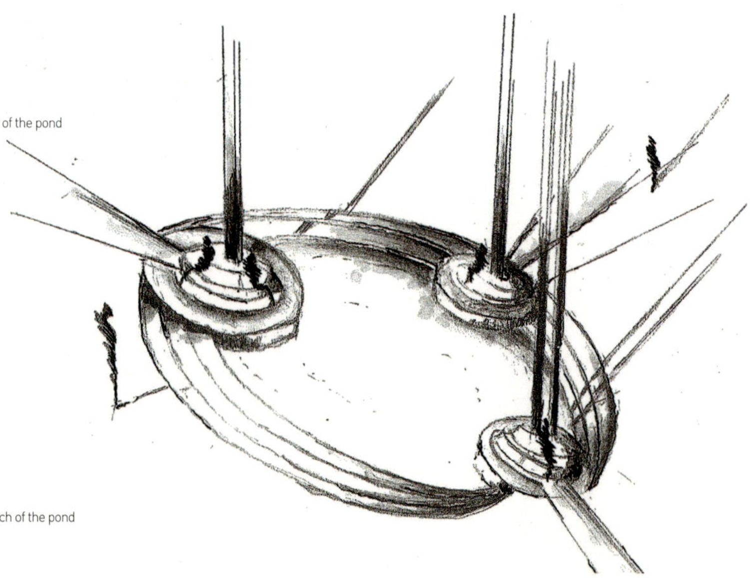

Sketch of the pond

View from the ground floor

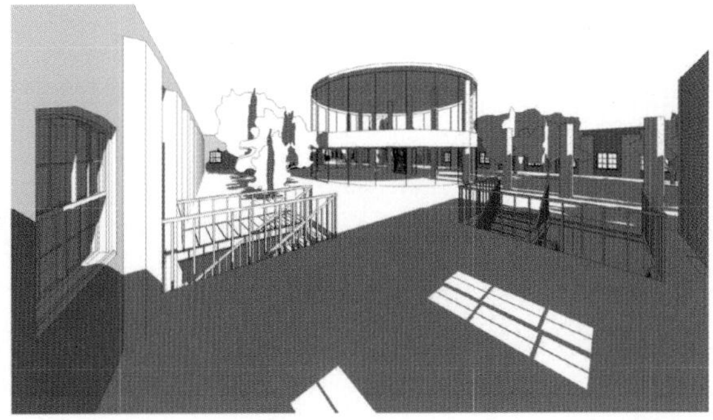

Image of the glass structure in the roof

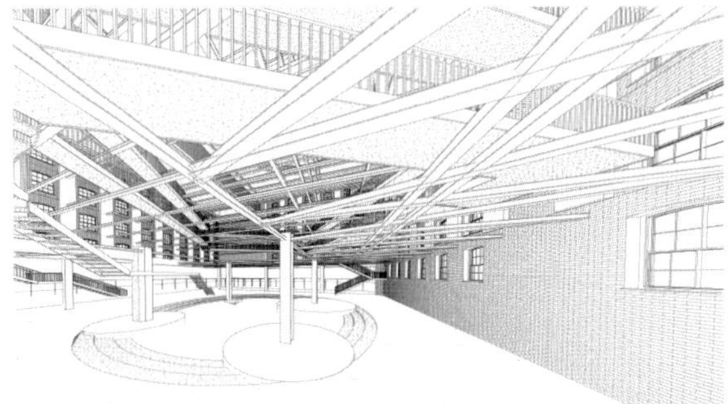

View of the pond

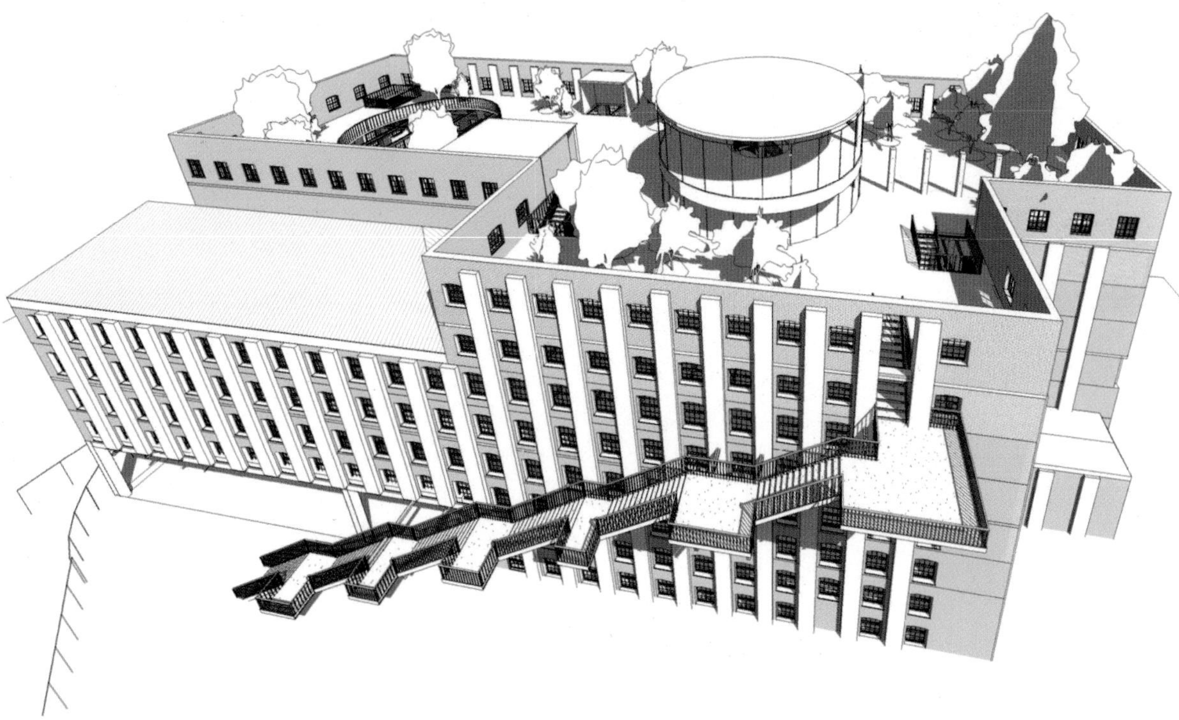

VIew of the south facade

Long section

Exhibition area

Phenomenology of the perception

M. Merleau-Ponty (1962) explains The essence of the building is reflected by the emotions of the subject rather than the object. People's relationship to a building is based upon the proximity of the building and the personal relationship to it. There are various factors that influence this connection. These factors maybe related to a historic bond created by a past community, such as employment, a sense of security, a place where human connections were made. The building can also engage peoples' interest by association to the nostalgia created by these factors.

M. Merleau-Ponty(1962)는 건물의 본질은 대상 보다 주체의 감성에 반영되어 있다고 주장하였다. 건물과 사람과의 관계는 건물의 근접성과 그에 대한 사람들의 관계에 기초한다. 이러한 연계에 영향을 미치는 다양한 요인들이 있다. 이러한 요인들은 아마도 고용이나 안정감과 같은 과거 커뮤니티에 의해 만들어진 역사적 유대 및 사람들 간의 연계가 이루어진 장소와 관련이 있다. 건물은 그러한 요인들에 의해 만들어진 향수와의 연관성을 통해 사람들의 관심을 자극할 수 있다.

Visual of one of the entrance and the pond.

Francesca Parpinel
https://www.behance.net/francescaparpinel

BINA

Before Industrial Now After Art Gallery

Urban catalyst project

The regeneration and re-energizing of Urban spaces Strive to embody a perceived community within its confines.
A redevelopment often gentrifies an area, changing it's Heart ans Soul.
The developers package a pre-ordinate way of working, living and playing, agenda's that do not always work within given Community.
Adding The Spirit of life and it's unpredictable behaviours and using this energy to drive a Design may embody a sense of belonging with those who interact with Space.

도심 공간의 재건 및 재활성화는 제약 조건 내에서 커뮤니티를 포용하고자 한다.
재개발이 그 구역의 격을 높이는 경우가 많고, 그 지역의 심장과 영혼을 변화시키는 경우가 많다.
개발업자들은 커뮤니티에 항상 효과적이지만은 않은 일과 생활, 여흥의 방식들을 결합시킨다.
삶의 정신과 예측할 수 없는 행동을 더하고 에너지를 사용해 디자인을 하는 것에는 공간과 상호작용하는 사람들의 소속감이 포함되어 있을 수 있다.

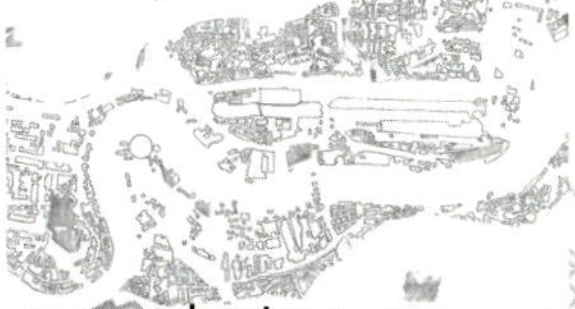
green and water

residential

industrial

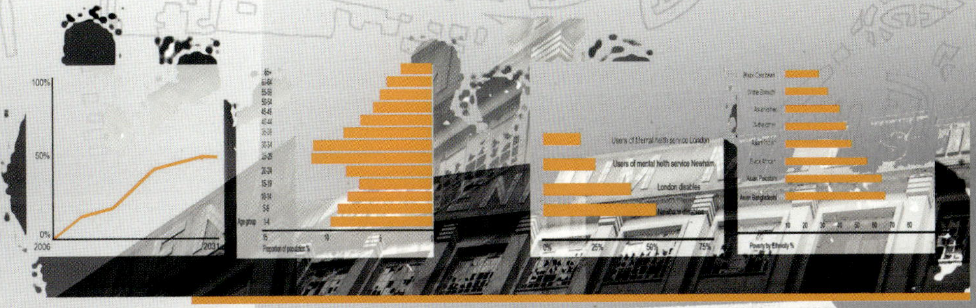
Newham demographic survey

Concept

Precedent Study have revealed that the area has been massively populated from people who use to work in the industrial sector which reflect a certain routine in life, the habit of doing the same action everyday.
Since the beginning of the century circumstances have radically changed , with the closure of the Docks as an active commercial market and productive area. Today the building is abandoned and what is left in the area is low -cost houses for rent relatively old, large abandoned spaces ; but also a new and growing side of the community.
Millennium Mills have changed it's main function, from a restless building is now become a silent but stately structure who is facing the reality of these days, very different from the past, now, the interest to explore the new era by telling other people what are the new ideas and concepts regarding culture is in the soul of the building.
After a period of quietness millennium mills need to wake it's soul, and with all the wisdom gained from the past, the process will be systematic and will follow the skeleton of the structure, as it will need to be respectful.
The passage from the past to the contemporary days will be demonstrated through the design and use of new building by the creative art centre, as new artists will be able to show and demonstrate through their installation and art works.

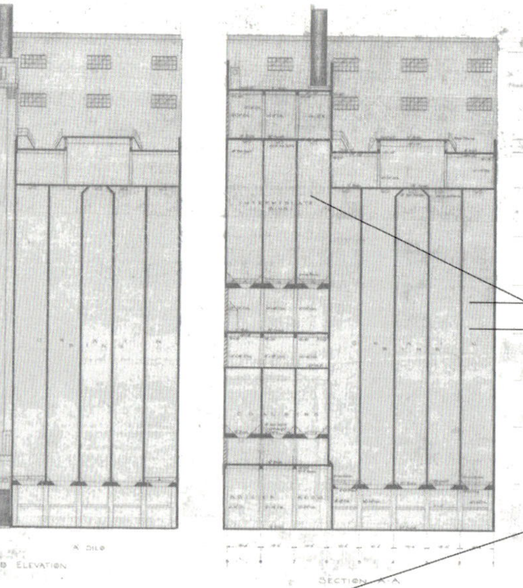

the history is in the building and the future is ready to go out

사전 연구를 통해 이 구역이 과거 산업 분야에서 일했던 사람들로 북적였다는 사실이 드러났고, 이는 매일 일정한 일상과 습관이 반복되었었다는 사실을 시사한다.
세기 초부터 상황이 급격하게 변해 왔고, 활발한 상업적 시장의 역할을 담당하던 선창가가 폐쇄되기도 하였다. 오늘날 이 건물은 버려져 있고, 이 구역에는 값싸고 상대적으로 오래된 임대 주택만이 남아 있고, 대부분은 버려져 있는 공간이 되었지만, 커뮤니티의 한 쪽은 새로 성장하고 있기도 하다.
밀레니엄 밀은 그 기능이 쉼 없이 돌아가던 건물에서 이제는 조용한 건물로 바뀌면서, 과거와는 매우 다른 현실을 직면하고 있으며, 다른 사람들에게 문화에 대한 새로운 아이디어와 컨셉에 대한 얘기를 함으로써 새로운 시대를 탐색해 보는 것이 이 건물의 정신이다.
조용한 시대를 거친 밀레니엄 밀의 영혼을 깨울 필요가 있고, 과거에 얻은 모든 지혜들을 이용해 이 과정은 체계적인 과정이 될 것이며, 구조물의 뼈대를 존중할 필요가 있기 때문에 그 뼈대를 이용하게 될 것이다.
과거로부터 현재로의 이행을 디자인과 창의적인 아트센터의 활용을 통해 입증할 것이며, 새 아티스트들은 자신의 설치작품과 예술작품을 통해 이를 보여줄 수 있을 것이다.

Working sketches

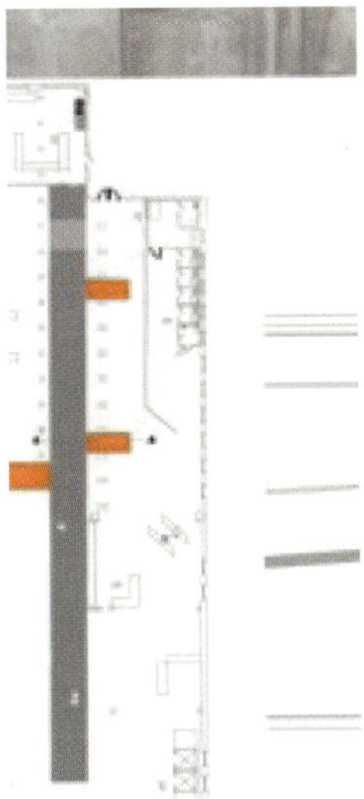

The library space

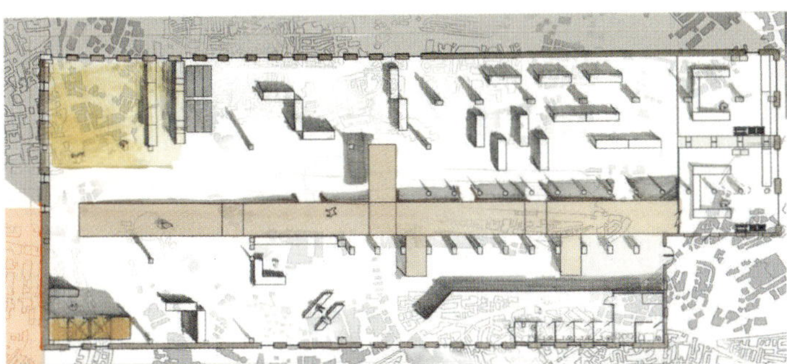

Gound floor plan

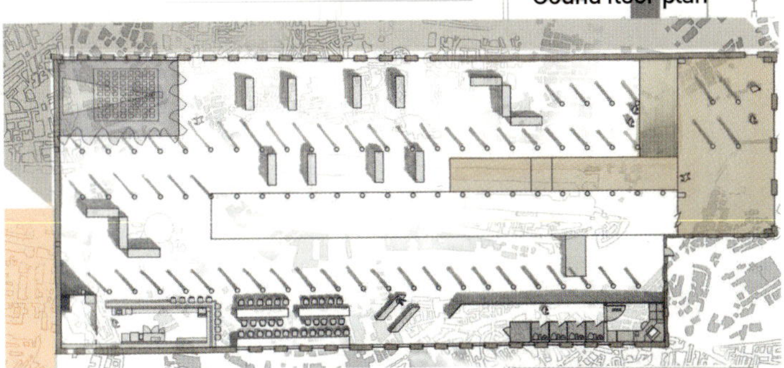

First floor Plan

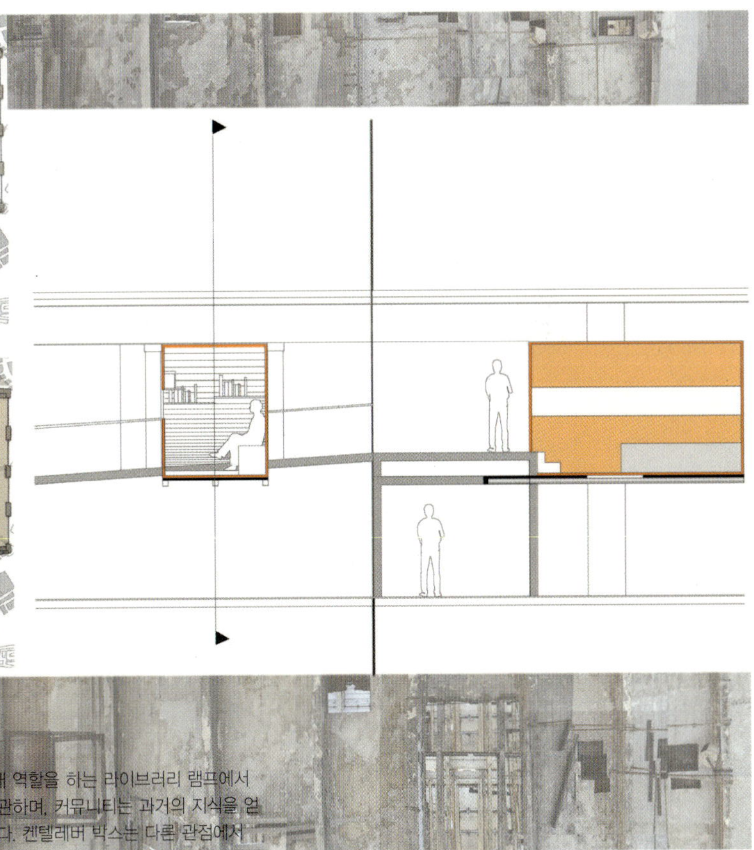

Plans showing the library ramp that acting as a new structural spine of the building; where the awareness of the past it could be stored and the community can have the advantage to gain knowledge of the past. The cantilever boxes are going to act as a private space where to read and enjoy the all space from a different point of view.

건물의 새로운 구조적 뼈대 역할을 하는 라이브러리 램프에서는 과거에 대한 인식을 보관하며, 커뮤니티는 과거의 지식을 얻는 혜택을 누리게 될 것이다. 켄텔레버 박스는 다른 관점에서 모든 공간을 읽고 향유하는 사적 공간이 될 것이다.

detail of cantilever structure

Sepinood Kakooei
sepinok@yahoo.com

The Shoreditch Vibe

Shoreditch Exchange located in the heart of Shoreditch within the heart of Hackney Borough. Boundary Street which is located on the east side of my building separates this site form Borough of Tower Hamlets. The building is facing the busy Shoreditch High Street from the west.

내 건물의 동쪽에 자리한 Hackney 자치구 경계도로 중심부 내 Shoreditch의 심장부에 위치한 Shoreditch 교환국이 이 부지를 Tower Hamlet 자치구와 구분시켜주고 있다. 이 건물은 사람들로 서쪽에 북적이는 Shoreditch High Street를 마주하고 있다.

Sunlight

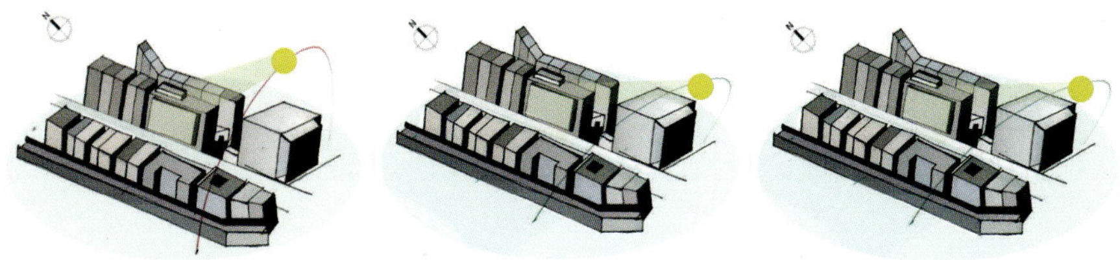

The brief is to renovate the interior space of the telephone exchange building, built by the GPO in 1956, into co-worker offices. The aim of the project is to encourage collaboration among the inhabiting companies. as well as defin and redefine public and private space. The creative use of light and sound, distinct depiction of each occupiers individual identity and ideas of movement bringing a kinetic dynamic was investigated within the space to allow both an eloquent and effective use of space.

1956년 GPO가 세운 전신 교환국 건물의 내부 공간을 협업 사무실로 리모델링하는 프로젝트이다. 이 프로젝트의 목표는 이곳에 입주한 회사들 간 협력을 촉진하고, 공공 공간 및 사적 공간을 규정하는 것이다. 조명, 음향, 각 입주자들의 개인 정체성을 창의적으로 활용하고, 역동적인 다이내믹을 유발하는 움직임의 아이디어를 창의적으로 이용하는 방안을 탐색해 공간을 우아하면서도 효율적으로 사용하고자 한다.

The concept behind the design is openness and visual stimulation. The aim is to create a dynamic and energetic feel in the workspace through natural light and industrial design. Shoreditch was once an area dense with warehouses. The concept aims to make a link between defining the Shoreditch characteristics, it's colour, personality and feel, whilst retaining it's architectural heritage is In addition, the aim is to create a sustainable environment within the building through use of existing structural properties and appropriate materials creating the correct level of acoustics. The focal point of the design focuses on dynamic structures with practical applications. The functional goal is to produce a light open office space in order to help the entrepreneurs to work in a more productive and healthier environment.

디자인 이면에 있는 컨셉은 개방성과 시각적 자극이다. 그 목표는 자연채광과 산업 디자인을 통해 업무 장소에 다이내믹하고 에너지 넘치는 분위기를 연출해내는 것이다. Shoreditch는 과거 창고들로 붐비던 곳이었다. 이 컨셉이 목표로 하는 것은 Shoreditchdml 특성들과 그 색상, 성격과 느낌 등을 연계시키고, 동시에 건축적 유산은 그대로 보존한다는 것이다. 또 다른 목표는 기존의 구조적 성격과 적절한 소재를 이용해 적절한 음향 수준을 만들어 냄으로써 건물 내 지속 가능한 환경을 창출해내는 것이다. 디자인이 중점을 두고 있는 점은 실용적 활용을 통해 다이내믹한 구조에 중점을 두는 것이다. 기능적 목표는 기업가들이 보다 생산적이고 건강한 환경에서 일할 수 있도록 개방형 사무실 공간을 만들어내는 것이다.

Concept of the Cafe- GroundFloor

Concept of theLounge- First Floor

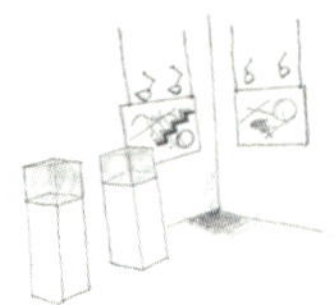

Concept of theGallery

Concept of the Cafe

Material- Mix Concrete

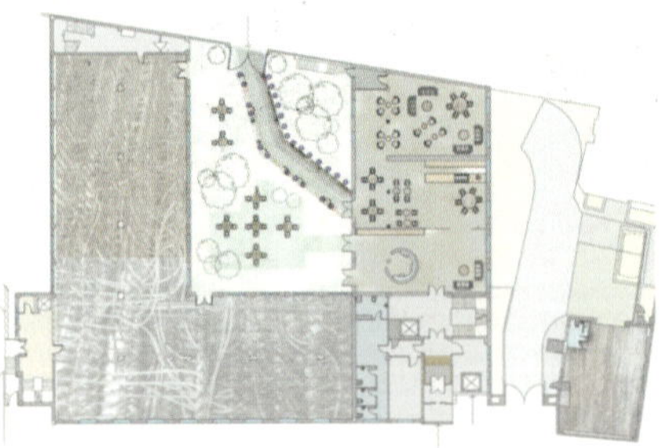

Ground Floor

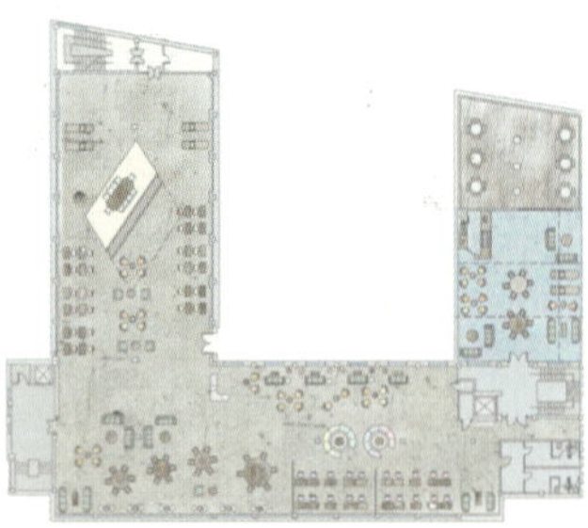

First Floor

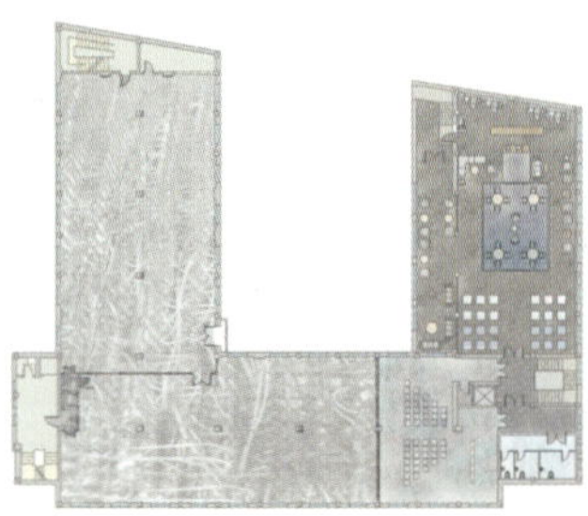

Second Floor

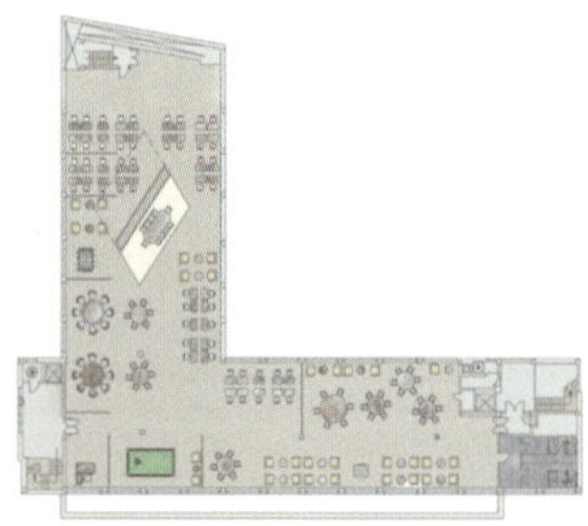

Third Floor

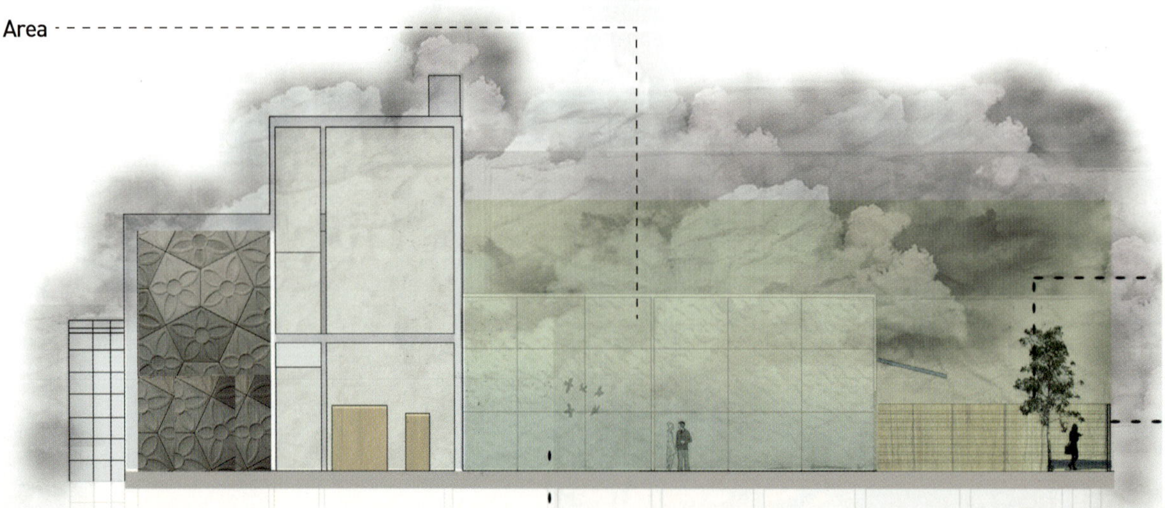

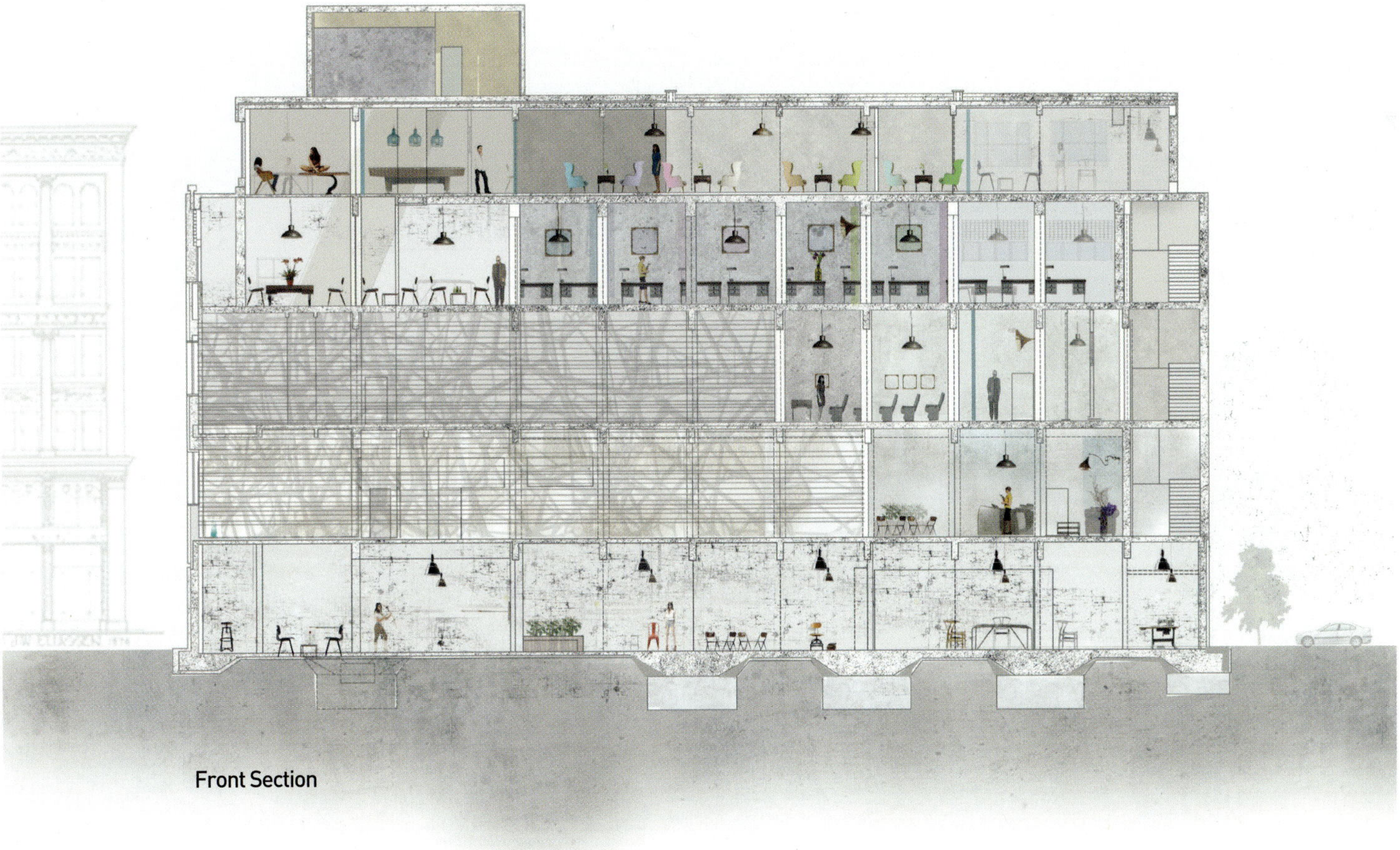

Front Section

Common Area where people can gather to relax, Play and work away from their quieter work areas. Frameless acoustics glass pannels reduce noise while retaining a sense of space.

사람들이 모여 휴식을 취할 수 있고, 놀이를 할 수 있고, 좀 더 조용한 업무 구역에서 벗어나 일할 수 있는 공용 공간. 프레임이 없는 음향 유리 패널은 소음을 줄여주는 동시에, 공간감은 그대로 유지시켜 준다.

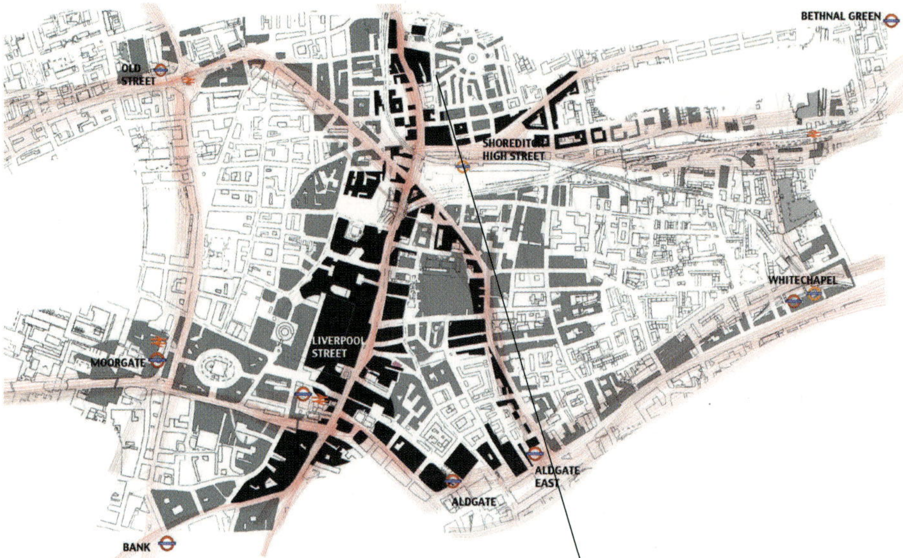

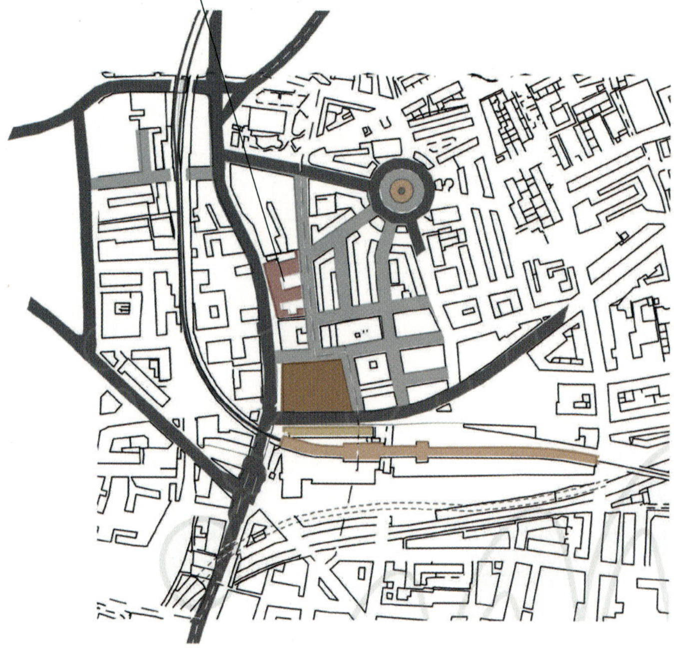

Delita Cole
delitac@hotmail.co.uk

The site

The Site is located in Shoreditch, East London. The area is on the border of Old Street which is well known for its Tech City and Silicon Roundabout. The location in which the space sits, is the ideal location for a start up space. Being right on the door step of old streets silicon valley, the area is on the rise for start up spaces.
Shoreditch famously has a lively diverse cross section of people who live and work within the area and has now become extremely popular with those who want to make a difference and seek inspiring and creative spaces to develop their businesses ideas.

사이트는 런던 동쪽의 Shoreditch에 위치해 있다. Tech 시티와 실리콘 로터리로 잘 알려져 있는 Old Street의 경계에 위치해 있다. 공간이 위치한 장소는 스타트 업 공간으로 이상적인 곳이다. Old Street 실리콘 밸리의 문간 바로 오른쪽에 위치해 있는 이 부지는 스타트 업 공간을 위한 언덕에 위치해 있다.
Shoreditch는 이 지역에서 거주하거나 직장을 가진 다양한 사람들이 모이는 장소로 유명하며, 이제는 자신의 사업 아이디어를 얻기 위해 영감을 자극하는 창의적 공간을 원하는 사람에게 매우 큰 인기를 얻고 있다.

My concept of interaction and collaboration devised from the idea of creating a big open working community in which knowledge and expertise is shared and given. With various companies and individuals occupying the space the idea is to connect the space, creating multiple meeting points and enabling chance encounters. When researching into office space it is apparent that the key factor to optimise productivity in workers is giving them a sense of control, through devices such as the thinking box and isolation slides users are in control over their environment and how isolated for included they are within the space.

상호교류 및 협력이라는 내 컨셉은 지식과 전문지식을 공유할 수 있는 커다란 개방형 업무 커뮤니티를 만든다는 아이디어에서 비롯되었다. 다양한 기업들과 개인들이 이 공간을 차지하게 될 아이디어는 공간을 연결해 다양한 모임의 장소를 만들어 내고, 사람들이 만날 수 있게 한다는 것이다. 사무실 공간에 대해 연구를 해보았을 때 근로자들의 생산성을 최적화할 수 있는 핵심 요소는 근로자들에게 생각 박스와 같은 장치를 통해 통제감을 주는 것이며, 고립 슬라이드 사용자들은 자신의 환경을 통제할 수 있고, 자신들이 공간 내에서 고립되는 방식을 통제할 수 있다.

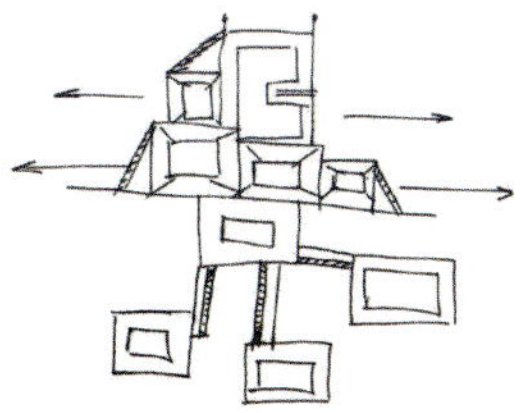

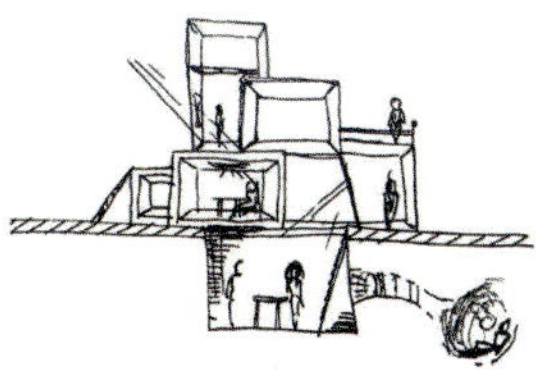

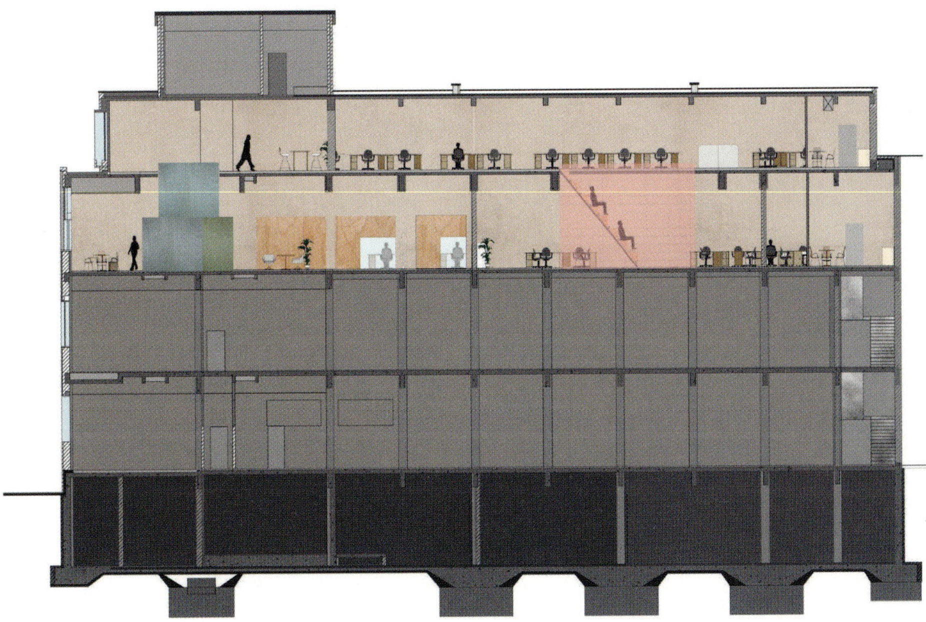

The Exchange will be a place in which contributes to the new and up coming businesses and trades of today. The space will house the leaders of the future, proving an efficient space for this is key. The ides of having all these different trades is something that drove the concept of collaboration. The idea of different trades and different business levels developed into the idea of accommodating for these trades on different levels - creating openings, spaces that are ever changing, like the trades over the years, influenced the design into these levels, into levels of space, privacy and interaction. Sink-holes, kinetic boxes and isolation slides are devices in which allow the user to shape their own environment allowing for which ever level of privacy meets there needs.

교환국은 새로운 비즈니스와 무역에 기여하는 장소가 될 것이다. 이 공간은 미래의 리더들을 유치하게 될 것이며, 그 핵심은 효율적인 공간을 제공하는 것이다. 이러한 서로 다른 사업 활동을 유치한다는 생각을 통해 협력이라는 컨셉을 생각해낼 수 있었다. 여러 다른 사업 활동과 사업 수준이라는 아이디어는 서로 다른 층에서 그러한 활동을 수용한다는 생각으로 발전되었고, 시간이 지남에 따라 변하는 사업과 같이 끊임없이 변화하는 열린 공간을 만들어낸다는 생각이 층들과 공간 층, 프라이버시 및 상호교류를 디자인하는 것에 영향을 미쳤다. 싱크 홀, 운동 박스 및 고립 슬라이드 등은 사용자들이의 프라이버시 니즈를 충족시킬 수 있게 자기 스스로 환경을 만들 수 있게 해주는 장치들이다.

Floor plans

GROUND FLOOR:

Court yard: outdoor seating, new main entrance, rental space for exhibitions and temporary food and drink stalls. Restaurant and cafe located on the ground, accessible to out side users, encourages people to explore space. Bike storage.

뜰: 야외 좌석, 새로운 주 출입구, 전시를 위한 임대 공간 및 임시 식음대, 외부 사용자들의 접근이 용이하며 사람들이 공간을 탐색할 수 있는 지상층의 레스토랑과 까페, 자전거 보관소 등이 있다.

FIRST FLOOR:

Reception with seating area next to mail and locker area. Private worker hubs sit next to the Reflective quite break out room, box enclosed devices, Isolation slides occupy the other side - space for a large group to work without disturbing the concentrated work in the opposite hubs.

1층: 우편 공간 및 락커 공간 바로 옆에 앉을 수 있는 공간이 있는 리셉션, 근로자 개인 허브가 조용하게 사색할 수 있는 휴게실 바로 옆에 위치해 있고, 고립 슬라이드(반대편 허브에 업무를 방해하지 않고 대규모 그룹이 일할 수 있는 공간)가 다른 쪽에 위치해 있다.

JOURNEY

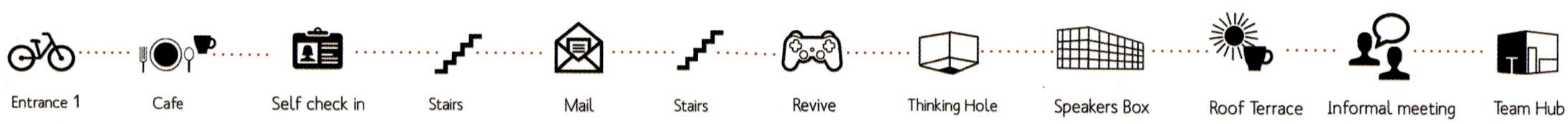

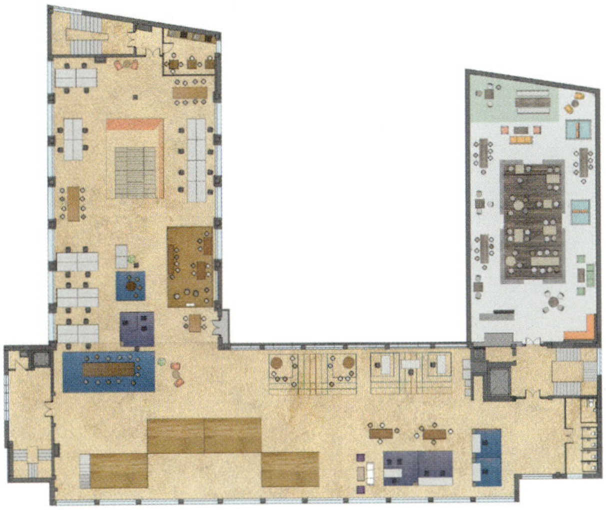

SECOND FLOOR:

Roof terrace; Outside seating area and cafe/ snack area on the second floor, as well as smoking area. Isolation slides located near enclosed spaces. The kinetic box moves and retracts, running from the second floor to third. Hot desks located near North and south for natural light exposure. Speakers box is central of the busy environment, bright bold colour stimulates users.

2층: 옥상 테라스: 야외의 앉는 구역과 카페/스낵 구역이 2층에 위치해 있으며, 흡연공간도 위치한다. 고립 슬라이드는 폐쇄형 공간 근처에 위치해 있다. 운동 박스는 2층에서 3층으로 움직일 수 있다. 북쪽과 남쪽 근처에 위치한 핫 데스크는 자연채광을 위한 것이다. 스피커 박스가 북적대는 환경의 핵심이고, 밝고 대담한 색상이 사용자들을 자극한다.

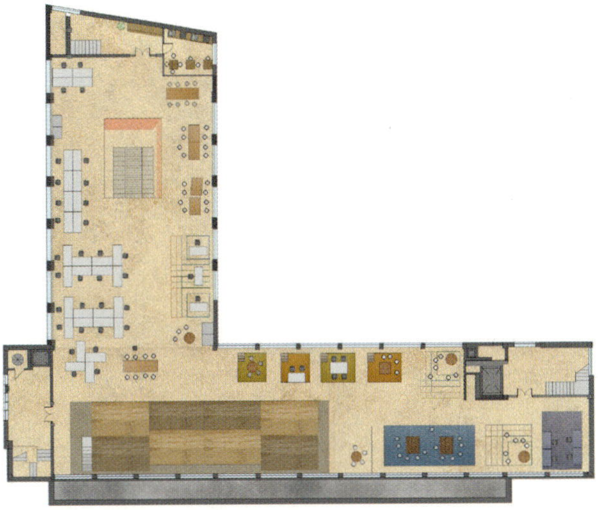

THIRD FLOOR:

Hot desks positioned on the North wing of the building, idea is with he connection of the space it will become a busy knowledge sharing environment, for all individual occupants. Thinking holes have been added on the third, to emphasis the idea of connecting levels, as cut outs will be visible underneath (second floor ceiling).

3층: 핫 데스크는 건물의 북쪽에 위치하며, 이 아이디어는 모든 개인 거주자들이 지식을 공유할 수 있는 공간이라는 아이디어와 연결되어 있다. (2층 바닥 천장) 아래에 자른 부위가 보일 것이기 때문에 생각 구멍(thinking hole)을 3층에 추가해 각 층을 연결한다는 아이디어를 강조하였다.

Cast in Concrete Floor and walls. White based low loading colour ' Putty'

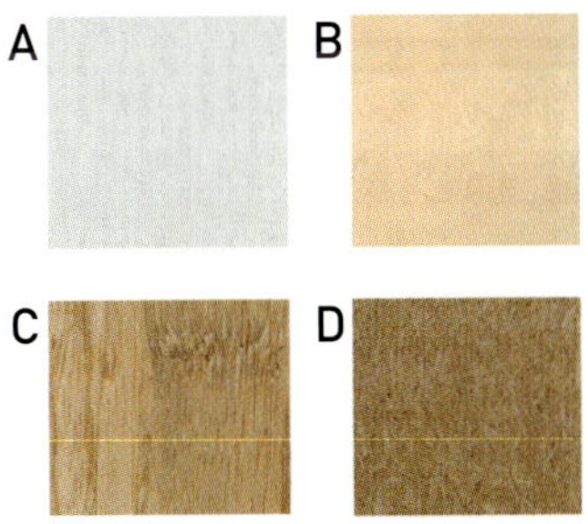

A) PLYWOOD
B) WHITE STAINED PLYWOOD
C) HONEY PLYWOOD
D) BIRCH PLYWOOD

Work area seating; plywood, light and flexible the simple seating areas can easily be moved, while the hot desk is a lot heavier and should be left stationary. Informal seating areas, vibrant colours used, comfortable and adds to sound absorption.

업무 구역 좌석; 플라이우드, 조명과 융통성 있는 단순한 좌석 구역을 쉽게 옮길 수 있는 동시에, 핫 데스크는 훨씬 더 무겁고, 안정적이어야 한다.
비공식 좌석 구역, 활기 넘치는 색상의 사용과 편안함과 소리 흡수 기능을 추가 하였다.

Various colours are painted on plywood to creative an exciting positive atmosphere within eh space. Also helps to indicate areas for users.

Glass will be used on the working hubs.
Panels of coloured glass will create a lecture/ speaking area in the space. The colours used are stimulating and positive to human emotion.

Devices for flexibility and adaptability

Flexibility and adaptability is a key theme running throughout the brief. With the rise of exciting and stimulating co worker spaces it is become more common for these spaces to accommodate to the users needs, much rather than the users working around the current design. Within the Exchange, the space has been designed with the end user in mind. As the space will be a co working office where individuals and groups will temporally rent a space for a short or long period of time, it is important to ensure that this space is flexible, exciting and adaptable to the users needs. Research on the workspace proves that employees are at there most productive when they are in a aesthetically appealing, flexible and stimulating environment. Through particular devices within my space, allows for users to be in control of there own environment, constructing and de-constructing temporary work places for themselves, the space is exciting with different levels of work space. Though the use of stimulating colours such as red and yellows in busy areas, in contrast to greens and blues in areas where concentration is required. It is though this the space psychologically and ergonomically appeals to the human senses.

융통성과 적응성이 개요 전체의 핵심 주제이다. 흥미롭고 자극적인 동료 공간과 더불어 이 공간들이 사용자의 니즈를 충족시키는 것이 훨씬 더 일반적이다. 교환국 내에서 이 공간은 최종 사용자를 염두에 두고 설계하였다. 이 공간은 개인과 집단이 일시적으로 단기 또는 장기간 이 공간을 임대할 공동 작업 사무실이 될 것이기 때문에, 이 공간이 융통성 있고, 흥미로우며, 사용자 니즈를 충족시킬 수 있는 공간이 될 수 있게 하는 것이 중요하다. 업무 공간에 대한 연구결과들은 직원들이 미적이며, 융통성 있고 자극적인 환경에 놓일 때 가장 생산적이라는 사실을 입증해주고 있다. 공간 내에 있는 특정 장치를 통해 사용자들이 자신의 환경을 통제하고, 자신을 위한 임시 공간을 구성했다가 해체할 수 있으며, 여러 업무 공간 층을 가진 흥미로운 공간이 되게 한다. 사람들이 북적대는 공간에 적색과 황색과 같은 자극적인 색상을 사용하고 이와는 대조적으로 집중이 필요한 곳에는 녹색과 청색을 사용한다. 이 공간은 심리적, 인체공학적으로 인간의 감각에 어필한다.

Different sized hubs accommodate for individuals and groups, allowing them to access a private area when required.

Spaces can be created according to a particulars users needs - two person job moving slide, encouraging iteration.

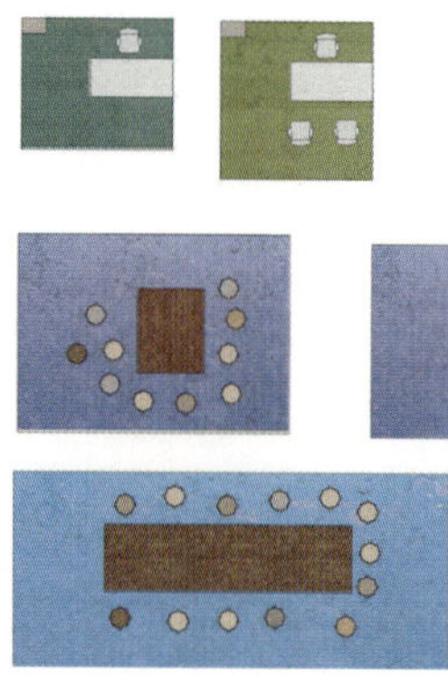

Different sized hubs, and different job tasks they accommodate.

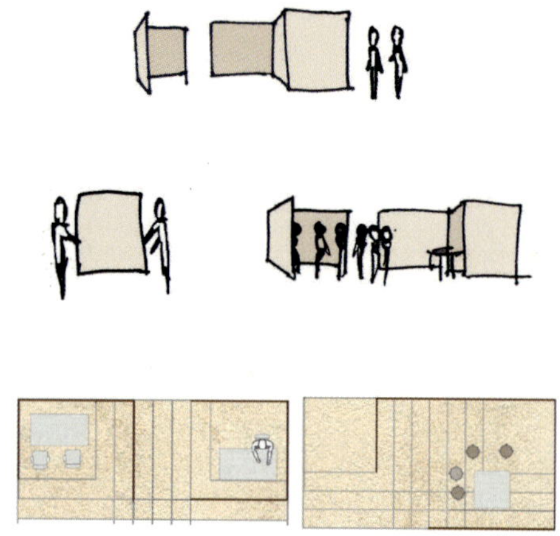

Isolation slides and different possible work areas.

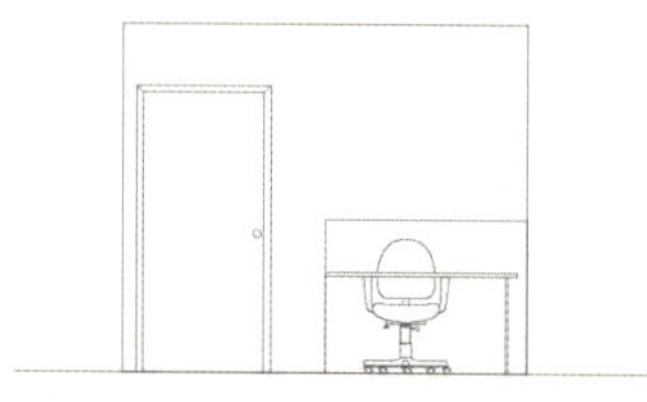

The private hubs, various sizes accommodate to every ones needs. The windows are 1100 mm high to allow for user privacy, avoiding total isolation. The hub allows for temporary privacy for confidential work or meetings, a short term work place for occupiers requiring complete concentration when finalising work. With the use of blinds darkness can be achieved for presentations and conference calls.

다양한 크기의 개인 허브는 모든 사람들의 니즈를 충족시킨다. 창문은 높이가 1100mm로서, 사용자들의 프라이버시를 보장하며, 완전한 고립을 피할 수 있다. 이 허브는 비밀을 유지해야 하는 업무나 회의에 대해 일시적으로 프라이버시를 보장하고, 업무를 마무리 할 때 집중이 필요한 근로자들에게 단기간의 업무 환경을 제공해준다. 블라인드를 사용해 어두운 환경이 필요한 프레젠테이션이나 컨퍼런스 콜을 할 때 어두운 환경을 조성할 수 있다.

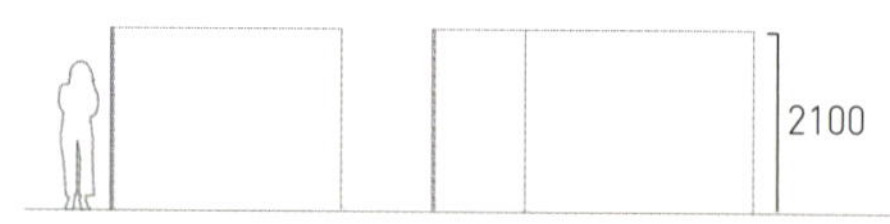

100 mm deep cut ins to the floor allows for 20mm thick plywood sheets of various sizes to fit creating multiple work areas. Slides include 2000mm, 2500mm, 3000mm, 3500mm. Various and multiple spaces can be created to accommodate for the work tasks of that day or certain individual(s).

100mm 깊이의 바닥 구멍은 다양한 크기의 20mm 두께의 플라이우드 시트를 이용해 복수의 작업 구역을 만들어낼 수 있게 할 수 있다. 슬라이드에는 2000mm, 2500mm, 3000mm 및 3500mm가 포함되어 있다. 다양한 복수의 공간들을 만들어 특정 날 또는 특정 개인의 업무를 수용할 수 있다.

The Thinking Hole will accommodate for various job tasks

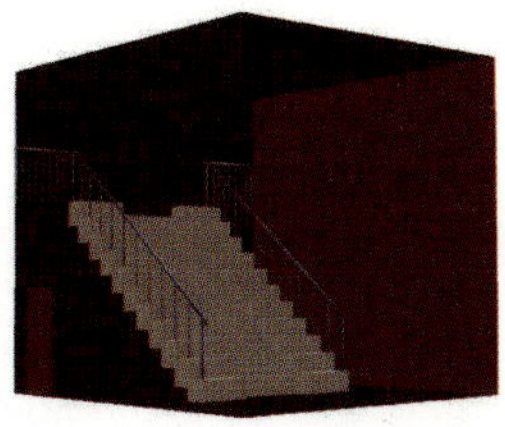

Multi functional glass box, vibrant colour chosen connecting space, inspiring users.

Fitted into a steel cage on runner tracks, each individual box is able to move forwards and backwards of 2000mm.

Think hole and different occupier and furniture choices.

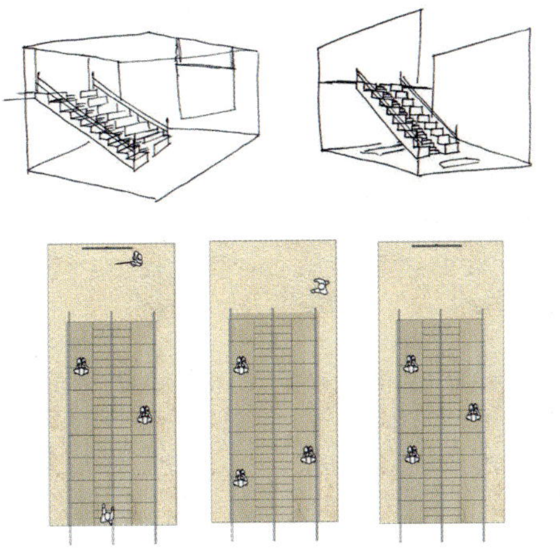

Different scenarios of space. Informal seating and talk/ cinema room.

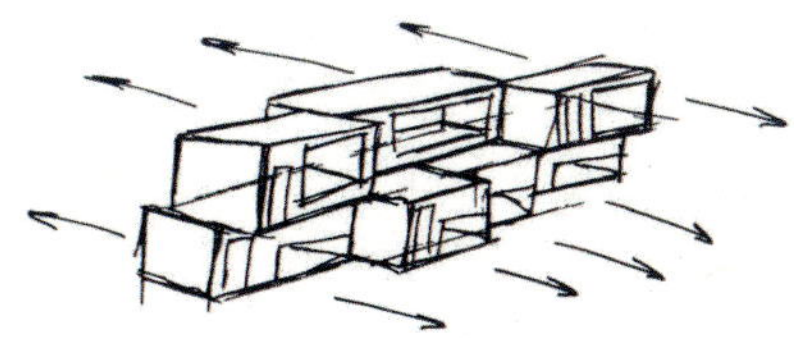

FUNCTIONS

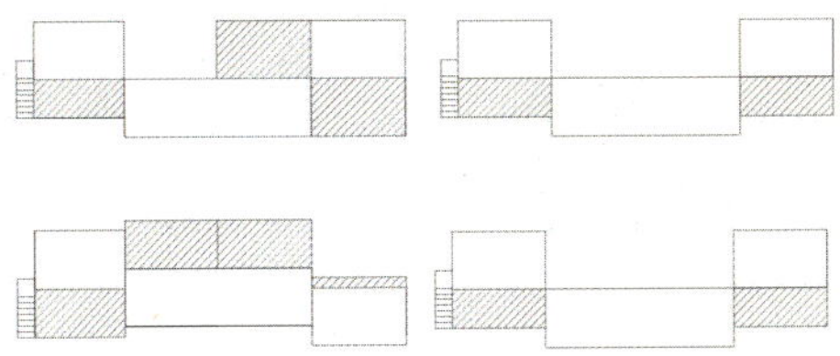

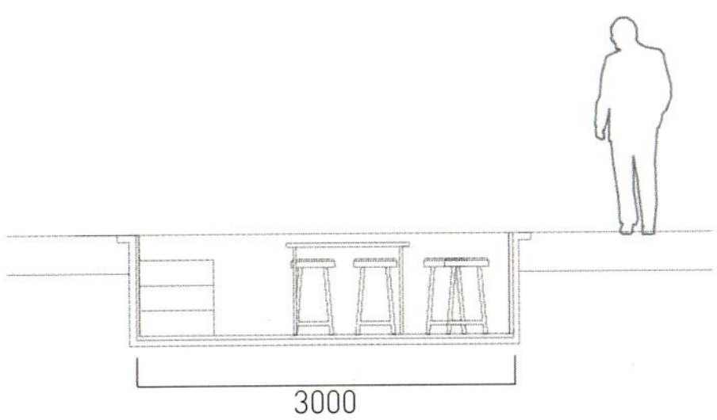

The 'Thinking Hole' is lowered 800 mm into the floor, suspended in a steal cage, finished with plywood and concrete, 100mm deep slots surround the box, allowing for 20mm thick plywood sheets to be inserted. Allows the user to be in control of there environment and how much privacy they have. User control is a key factor to optimising productivity within the workplace.

"사색 구멍" 은 바닥 안으로 800mm 아래로 뚫어 강철 케이지에 매달 수 있으며, 플라이우드와 콘크리트로 마감하며. 100mm 깊이의 슬롯이 박스를 둘러 싸 20mm 두께의 플라이우드 시트가 삽입될 수 있다. 사용자들이 자신의 환경을 통제할 수 있으며, 프라이버시의 정도를 통제할 수 있다. 사용자 통제가 업무 공간 내 생산성을 최적화하는데 있어 핵심적인 요인이다.

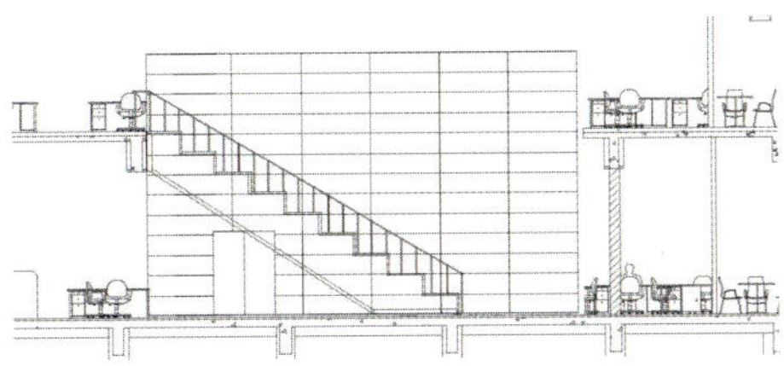

The Speakers box constructed of glass allows for visual openness in such a large enclose space, with a expanding screen situated at the front the space can be used for talks, lectures or networking events. The stair case has two sized steps, the outer, larger steps allow for users to sit to listen to a particular talk, or double as an informal seating/ meeting point. The inner smaller steps allow for general access.

유리로 만든 스피커 박스가 그러한 대형 규모의 공간에 시각적 개방성을 가져다주며, 공간 앞에 위치한 확장 스크린은 대화나 강의 또는 네트워킹에 사용될 수 있다. 계단은 2가지 크기의 계단이 있어 외부에 있는 공식 좌석/모임 장소의 두 배나 되는 큰 계단은 사람들이 앉아서 특정 연설을 들을 수 있다. 안쪽에 있는 작은 계단은 일반적인 접근이 가능하다.

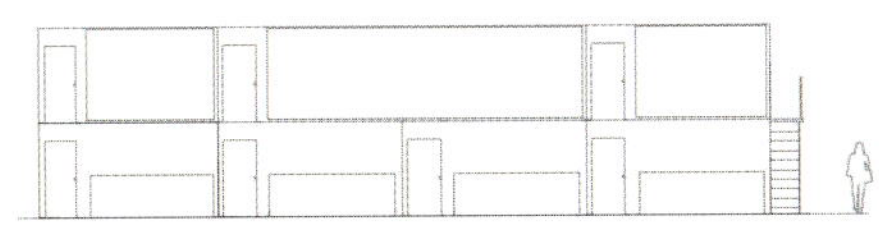

Lower boxes will be used for team work areas; private team working, brain storming. Shutter blinds allow for darkness within box for small team presentations. Upper boxes, have a large glass window on both the front and back of the box allowing for natural and artificial light to flood though, space will accommodate work areas for tailors, painters and dress/shoe makers. Can also be used as an exhibition space.

아래에 있는 박스는 팀 업무 공간, 사적인 팀 작업, 브레인스토밍으로 사용될 것이다. 셔터 블라인드는 소규모 팀 프레젠테이션을 위해 박스 안을 어둡게 할 수 있다. 상부의 박스는 박스 정면과 후면에 커다란 유리 창문이 있어, 자연 채광 및 인공조명이 들 수 있으며, 이 공간에는 재단사, 화가, 의상/신발 메이커들을 위한 작업 공간으로 사용할 수 있다. 이 공간은 또한 전시 공간으로도 사용될 수 있다.

1:50 sectional model showing devices and levels, and how they will effect the space from both levels.

Materials and colours are intended to be quite warm and welcoming, the use of white flexible furniture reinforces that of cleanliness through the space. Stimulating colours are used in various places, each colour has been carefully chosen for the area in order to get a certain user emotion.

재료와 색상은 따뜻함과 환영하는 분위기를 연출하기 위한 것이며, 흰색의 융통성 있는 가구들을 사용함으로써 공간을 통한 청결함을 강조하였다. 자극적인 색상은 다양한 공간에 사용하였으며, 각각의 색상은 특정의 사용자 정서를 위해 신중하게 선택된 것들이다.

Martina Terracciano
marti.t91@gmail.com

The Mill Market

Regeneration of the Millenium Flour Mill for the Silvertown Community

Site

The Millennium Mill is a derelict building that was built in 1905. It is the only surviving factory of the flour milling industry in London's docklands.
It is a symbol of the local Heritage as most of the people from the area used to work there. Since the factory closure in the 1980's, the building has been unused and no-one has worked or set foot in there ever since. Developers now plan to build 5000 new homes on this site and to convert the Millennium Mills into loft-style apartments.

밀레니엄 밀은 1905년에 세워진 버려진 건물이다. 이 건물은 런던 도크랜드에 있는 밀가루 제분 공장 중에 유일하게 남아 있는 공장이다. 이 건물은 이곳 출신의 대부분의 사람들이 이곳에서 일했었기 때문에 이 지역 유산의 상징이라 할 수 있다. 1980년대 이 공장이 폐쇄된 이후, 이 건물은 사용되지 않았었고, 이후로 이곳에 일하거나 발을 들여놓은 사람이 아무도 없었다. 개발업자들은 현재 이 부지에 5천 가구를 건설할 계획을 세우고 있으며, 밀레니엄 밀은 로프트 스타일의 아파트로 전환하고자 하고 있다.

Steel frame structure visible from the ground floor through a hall in the ceiling.

View from the West entrance.

Concept proposal

After an analysis of the Millennium Mill and the surrounding area of Silver town, a series of issues came to my attention.
In Silvertown there is a lack of a food stores that can be reached by walking. People need a car to go shopping and buy everyday essentials such as fresh meat and vegetables. Another issue that came to my mind is the lack of a community space where residents can go and get to know their neighbours. A concept proposal to these problems is to design a market hall that includes allotments space, a community garden and a cooking workshop.

밀레니엄 밀과 주변 실버타운을 분석한 이후 몇 가지 문제가 내 주의를 끌었다.
실버 타운에는 걸어서 갈 수 있는 식품점이 부족했다. 신선육이나 채소와 같은 일일 용품을 구입하기 위해 장을 보려면 차를 이용해야 했다. 또 다른 문제는 주민들이 모여 서로를 알 수 있는 커뮤니티 공간이 부족하다는 것이었다. 이러한 문제에 대한 컨셉 제안은 시민농원과 커뮤니티 가든 그리고 요리 작업실이 포함되어 있는 마캣 홀을 디자인하는 것이었다.

Concept sketches

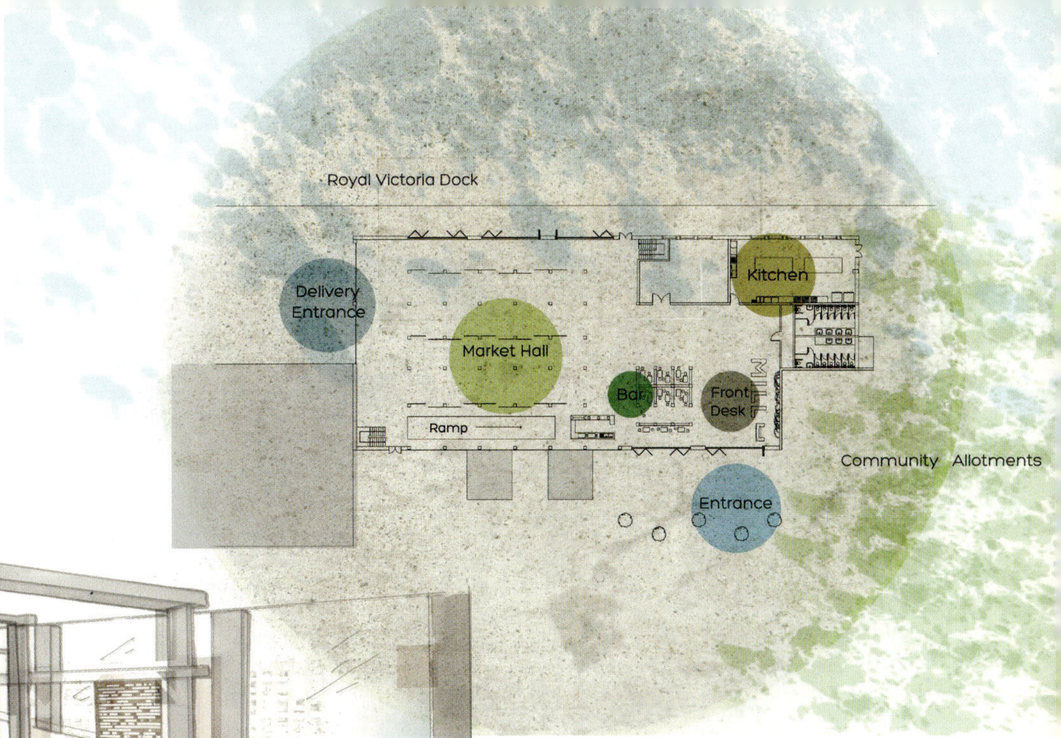

Gound floor plan

Movable screen in the Market Hall

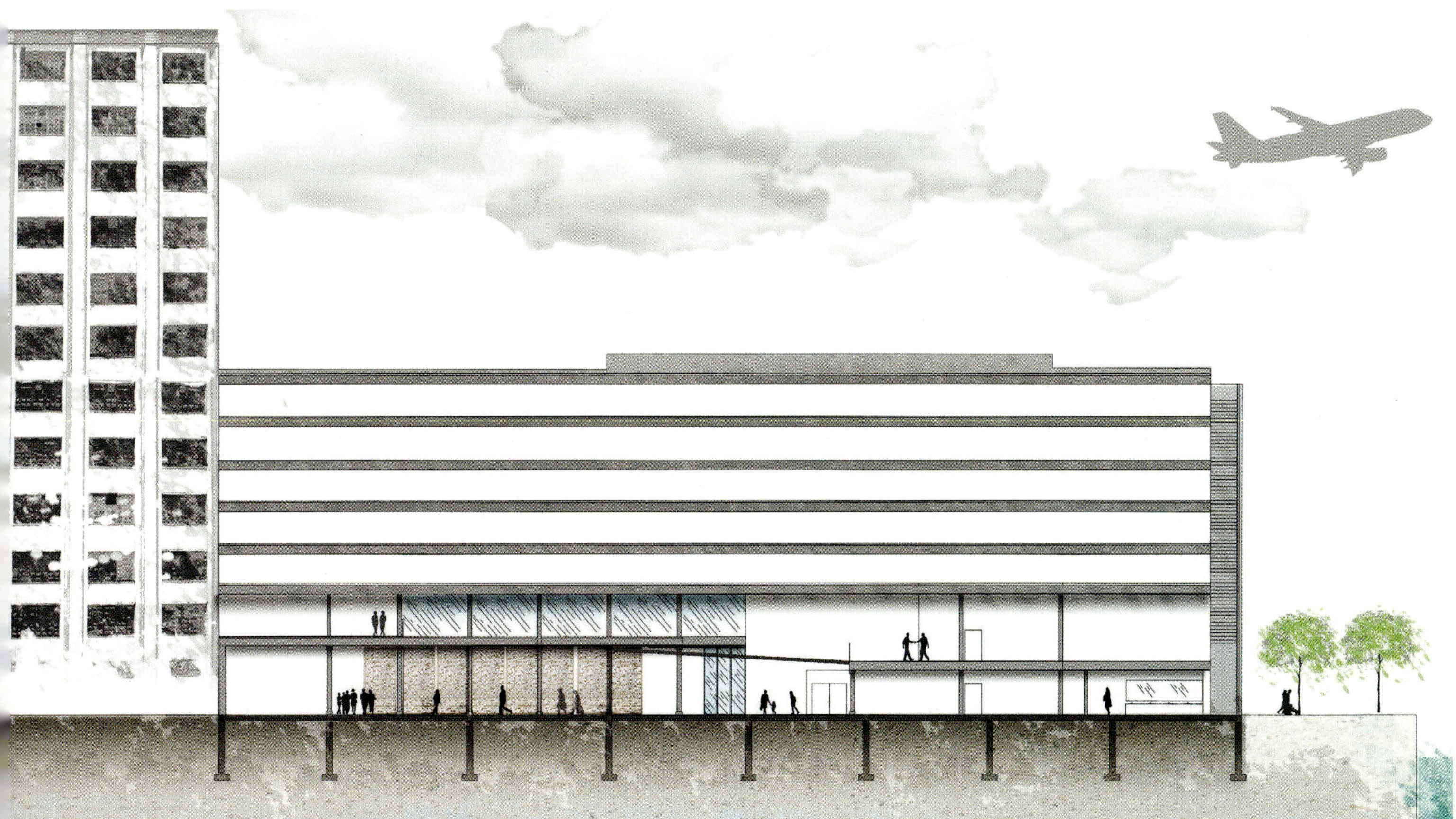

SOUTH SECTION VISUAL

The mill market

The market is integrated with allotments for the community to grow their own vegetables and a possibility to learn how to make food in the workshop/bakery. It is more than a place to buy or sell food, it is a place where people come to connect with the community, share their knowledge and learn useful skills.

이 시장은 주민들이 직접 스스로 채소를 기를 수 있고 조리실/빵집에서 음식을 만드는 법을 배울 수 있는 지역사회를 위한 시민농장에 통합되어 있다. 이 장소는 음식을 사고파는 장소 이상의 장소이며, 사람들이 모여 지역사회와 소통하고, 그들의 지식을 나누고 유용한 기술을 배울 수 있는 장소이다.

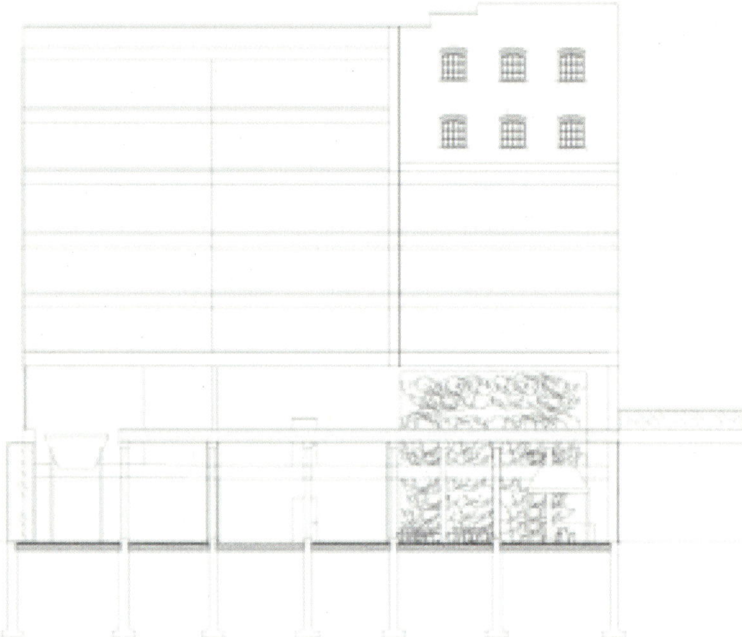

West Section

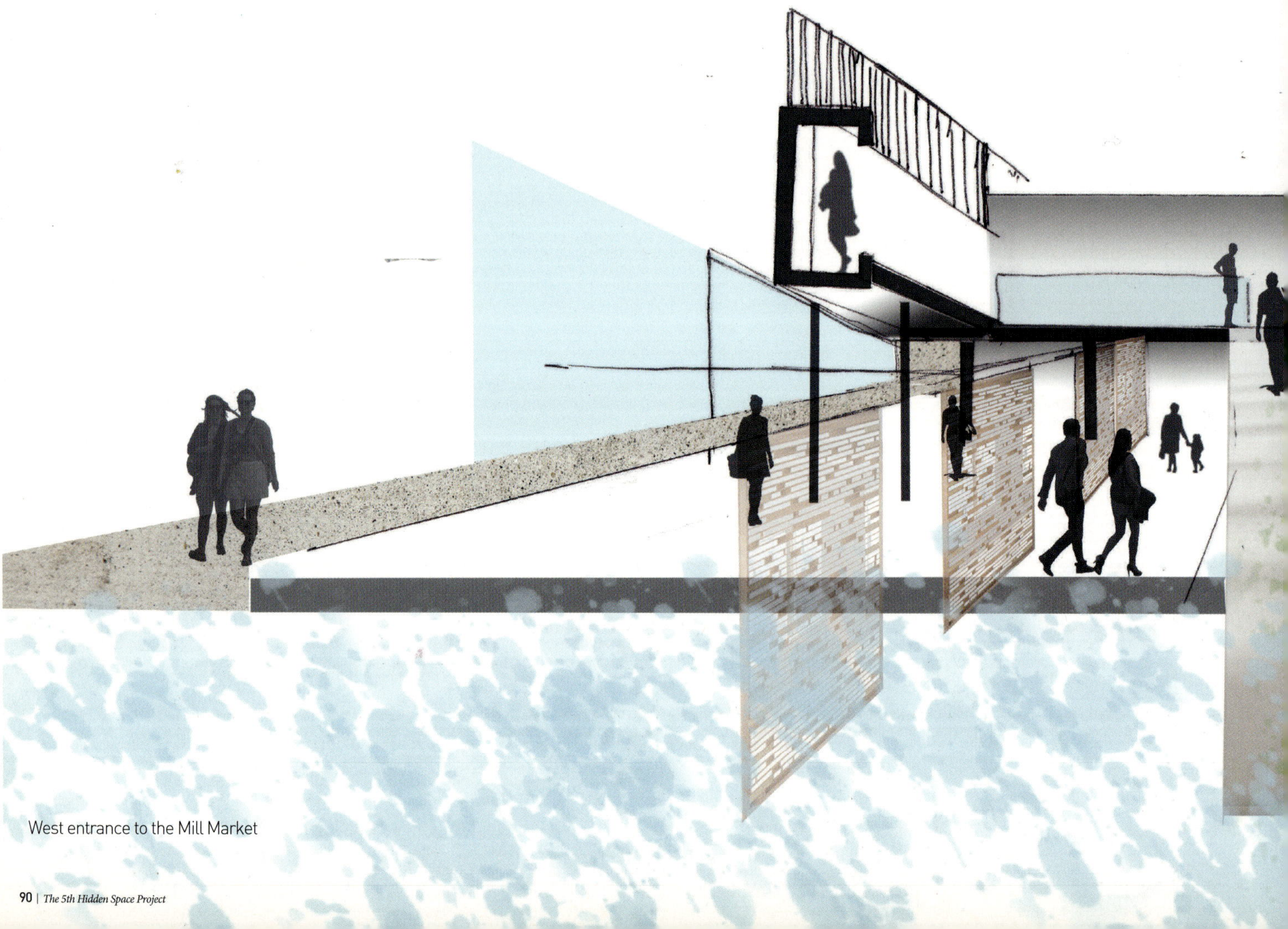

West entrance to the Mill Market

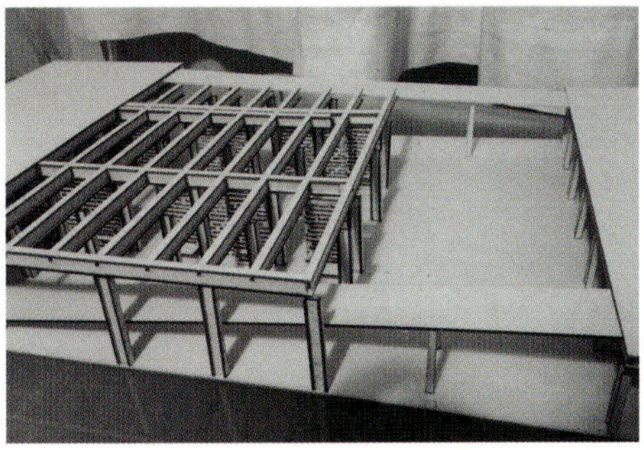

Model 1:100
Media: Laser machine
Finn board
PVA Glue

Selena Grasso
selenagrasso@yahoo.it

THE ANCHOR

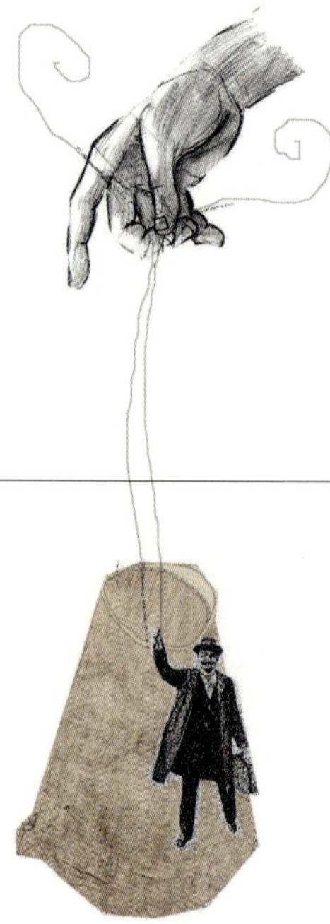

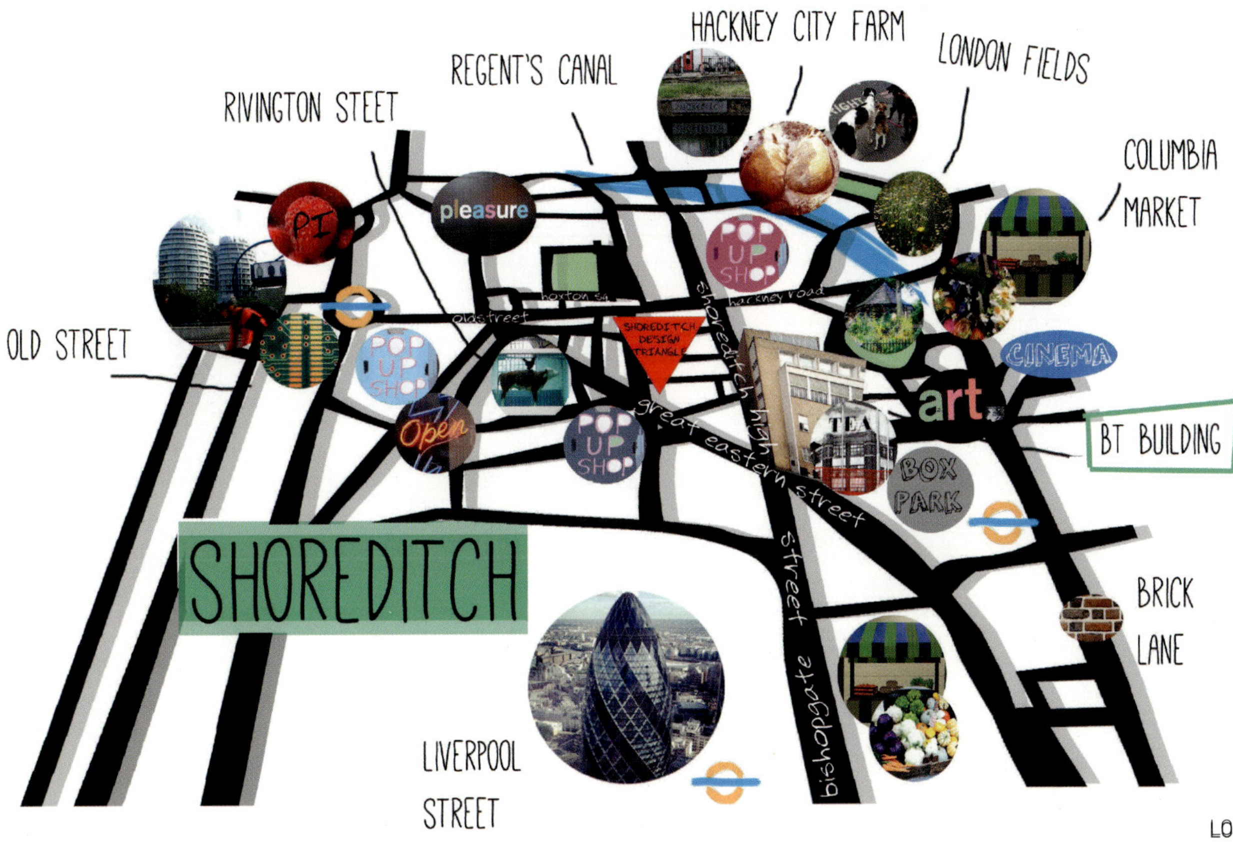

LOCATION MAP

The Brief

The brief asks to explore the redundant concrete and bricks bt building, situated in the heart of shoreditch within design a modern co-worker hub, looking at the new way of adaptability of upcoming start-up companies.
The aims is to create a place where many companies can coexist together, encourage communication and collaboration and inter-company connections developing a spatial co-operative inviting the community inside the building.

개요에서는 앞으로 새로 시작할 회사들이 적응할 수 있는 새로운 방법을 모색하는 현대적인 동료 허브를 디자인 하는 범위 내에서 Shoreditch의 심장부에 위치한 비어 있는 콘크리트 및 벽돌 BT 건물을 탐색하고자 한다. 목표는 많은 회사들이 공존할 수 있으며 커뮤니케이션과 협력, 회사 간 연계를 촉진해 공간적 협력을 창출해 낼 수 있는 공간을 만들어내고, 지역사회를 건물 안으로 끌어들이기 위한 것이다.

The shoreditch vibe

The studio encompass the new concept of flexibility , no long-term commitments in work and office space looking at the kinetic and dynamic structures that provide options of movement allowing for adaptability and variable spatial solutions making this working space as exciting and resourceful as possible.

Shoreditch Vibe 스튜디오는 융통성이라는 새로운 개념을 포용해, 업무 공간을 최대한 흥미롭고 유용한 곳으로 만들 수 있으며 적응성과 다양한 공간적 솔루션이 가능한 이동 옵션을 제공해줄 수 있는 역동적인 구조를 모색함으로써 업무와 사무 공간에 장기간 얽매이지 않게 하려 한다.

SHOREDITCH HIGHT STREET

Mood board

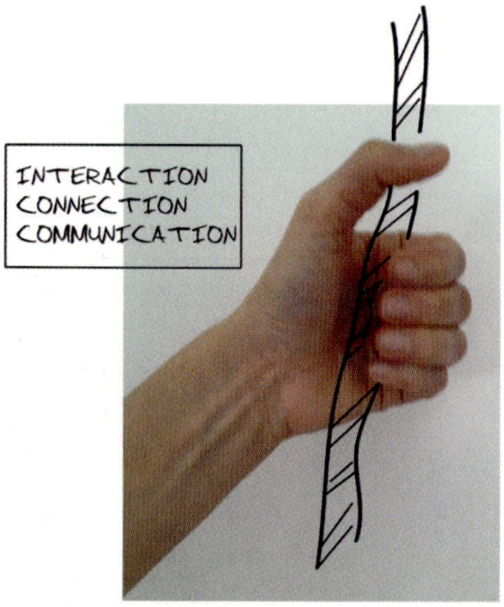

HTTP://INTERACTIVE.WXXI.ORG/

THE CONCEPT

The anchor concept comes from the inspiration of the acoustic experiment of the english architect robert hooke (1664) known as the very earliest phone device made up of two cans attached to a taut string or wire.
The concept gives the ability to choose and modify the components in the work space daily and or hourly, according to the need of use of the facilities by the quick re-programme of the plant system components.
The anchor is a flexible work place that means freedom to choose when and where to work, the start-up companies will cooperate and share the communal facilities enlarging the networks.
The design consists in a series of liftable cones by a manual pulling system, the cones are offices, meeting rooms, kitchen area, video rooms, re-charging area.
The cone system is a functional arrangement to allow privacy or either open plan to the workers, in a playful and no conventional way.
The conical shapes aims to protect the acoustic pollution of the busy high street providing games of light.

기본 개념은 줄이나 끈으로 두 개의 캔을 연결해서 만든 최초의 전화 장치를 발명한 것으로 알려져 있는 영국 건축가 Robert Hooke(1664)의 음향 실험에서 영감을 받은 것이다. 이 컨셉에서는 시설의 용도에 따라 플랜트 시스템 구성요소를 리프로그래밍 함으로써 업무 공간의 구성요소를 매일 선택해 변경할 수 있게 해준다.
그 기초는 융통성 있는 업무 공간으로서 언제 어디서 일할 것인가를 선택할 자유가 있고 새로 시작하는 회사들이 네트워크를 확장해 서로 협력하고 통신 시설을 공유할 수 있는 업무공간이라는 것이다.
이 디자인은 수동 당김 시스템을 이용해 들어 올릴 수 있는 원추들로 구성되어 있는데, 이 원추는 사무실, 회의실, 주방, 영상실, 재충전실 등등이다.
이 원추 시스템은 프라이버시를 확보하거나 보다 밝고 비전통적인 방식으로 근로자들에게 개방된 공간을 제공해줄 수 있는 기능 단위이다. 원추 형태가 목표하는 것은 북적거리는 Shoreditch High Street에서 발생하는 소음 공해를 줄이고, 조명 게임을 제공해주는 것이다.

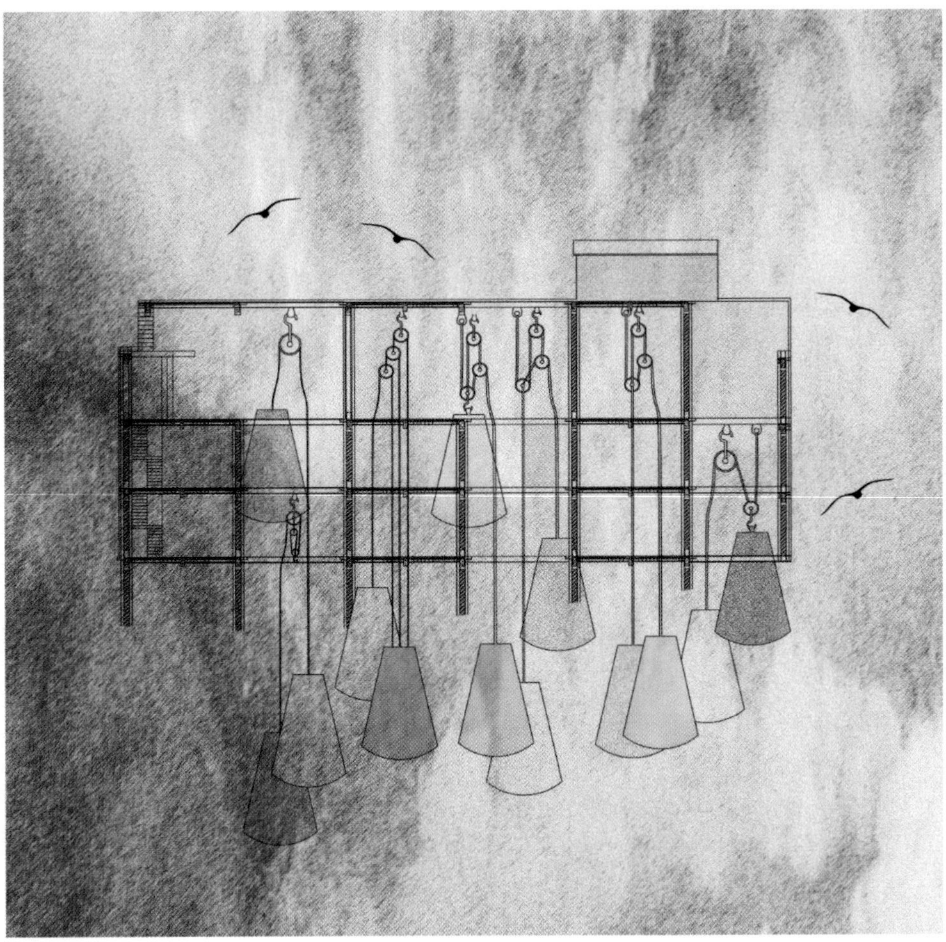

Tin can telephone & pulley system concept

Tin can telephone

The building

The bt exchange building's potential with respect to its location purpose is huge. It is build within the heart of shoreditch, now a prime area to locate a co-worker business as shoreditch is known for its creative industry and vibrant community.
The building was built on 1956 by the general post office, it is predominantly a concrete structure with a brick facade.
The site is overlooking the busy and noisy shoreditch hight street and the opposite side is facing the quite and green residential area of bounduary street.
The building has two main entrances: one from shoreditch high street, that will be used by the workers of the anchor to reach the offices in a fast and easy way, and the second entrance from boundary street, the access from the courtyard is for the private workers either for the public audience.
The courtyard open to the north wing with offices and east wing with restaurant, bar, material library & top roof garden.

BT 교환국 건물의 위치를 놓고 볼 때 이 건물이 지니고 있는 잠재력은 엄청나다. 이 건물은 Shoreditch의 심장부에 세워져 있으며, Shoreditch는 이제 창의적인 산업 커뮤니티와 활력이 넘치는 커뮤니티로 유명하기 때문에 동료 산업이 위치할 수 있는 핵심 지역이라 할 수 있다.
이 건물은 1956년에 GPO가 세운 건물로서, 주로 벽돌 입면에 콘크리트 구조물로 되어 있다. 부지는 북적대며 소란스러운 Shoreditch High Street를 내려다보고 있고, 그 반대편으로는 경계도로가 있는 조용한 녹지의 거주 지역을 마주하고 있다.
이 건물에는 주요 출입구가 2곳이 있다. 하나는 Shoreditch High Street 쪽에 있으며, 이곳은 근로자들이 빠르고 쉽게 접근할 수 있는 출입구이며, 다른 하나는 경계도로에 있으며, 뜰에서 접근할 수 있는 출입구이다. 뜰은 사무실이 있는 북쪽과 레스토랑, 바, 소재 도서관과 옥상 가든이 있는 동쪽으로 개방되어 있다.

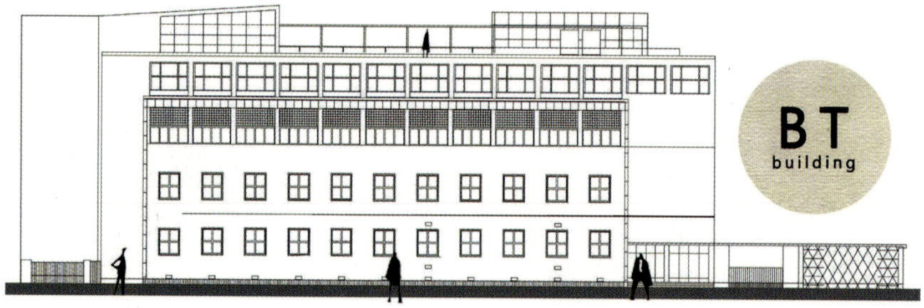

Front elevation

Environmental analysis

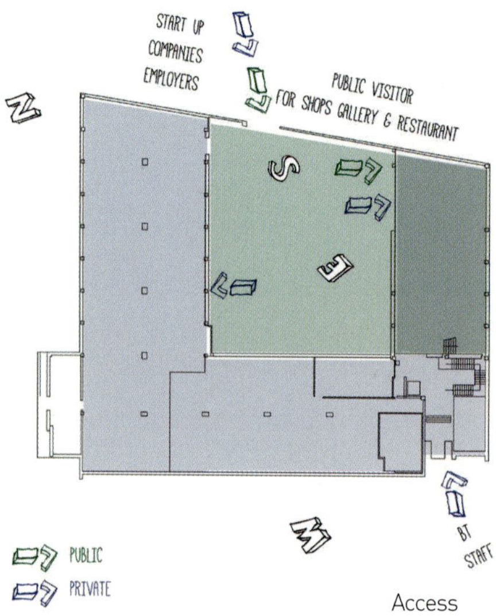

Access

Light

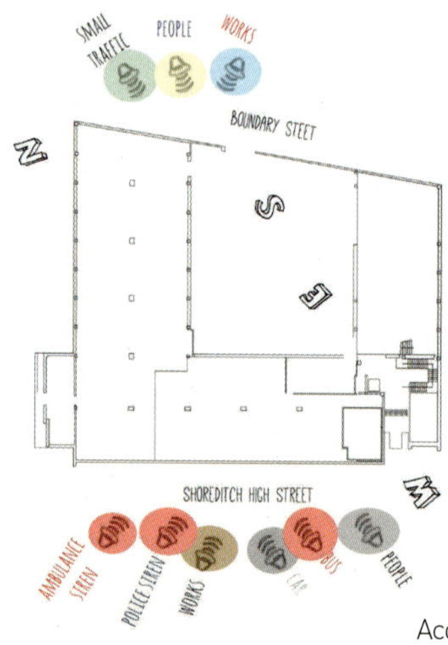

Acoustic

Sketch model

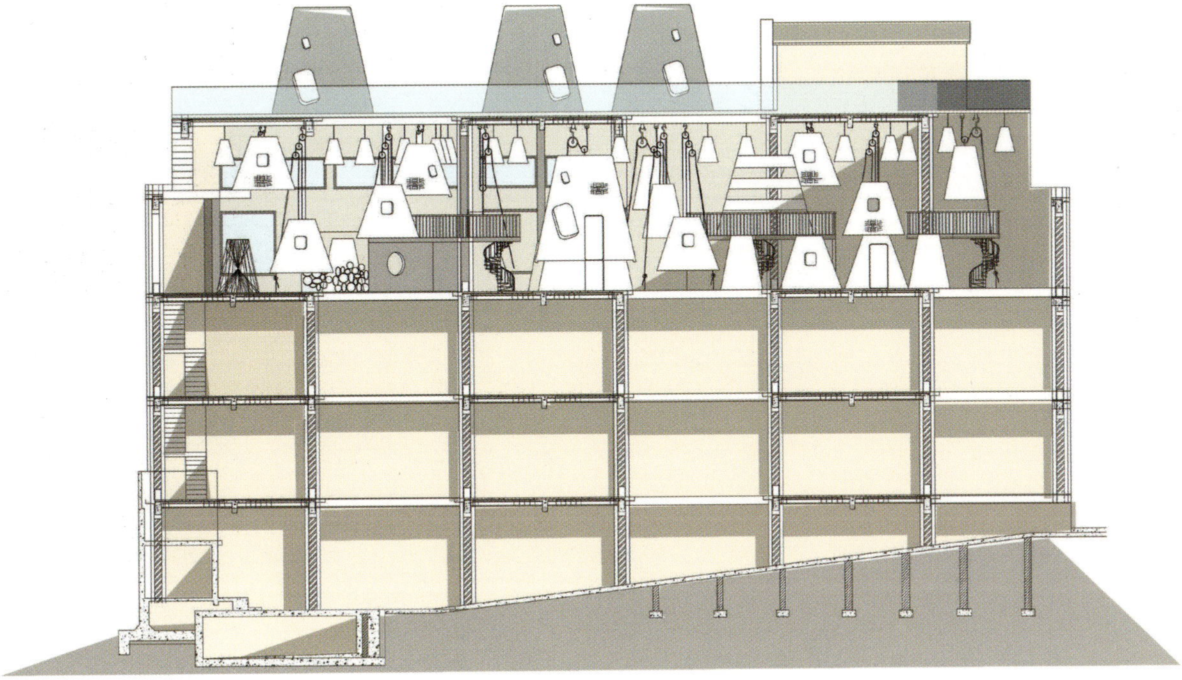

Section

Section

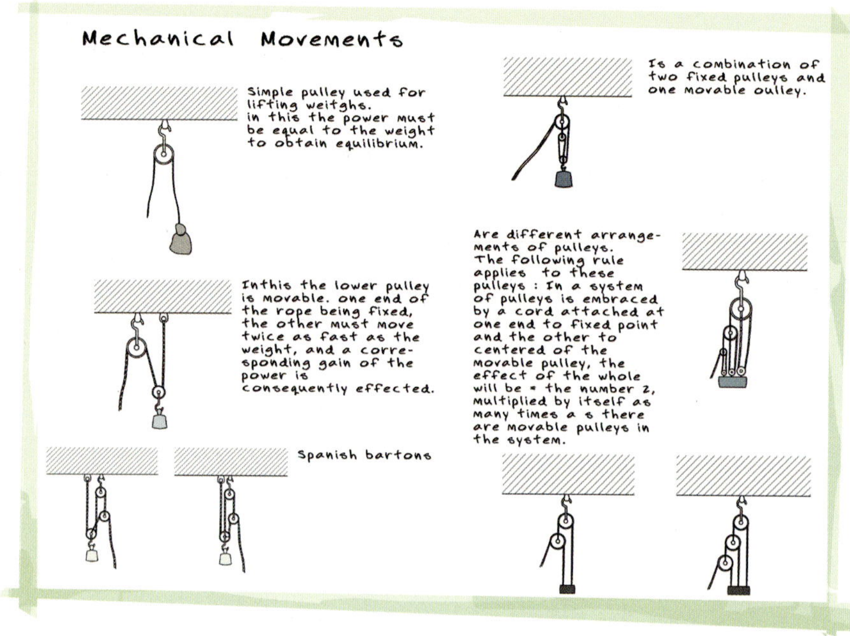

Mechanical pulley

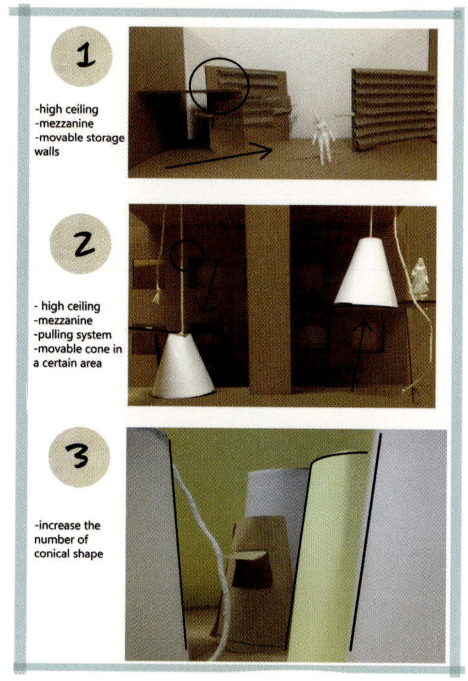

Sketch model

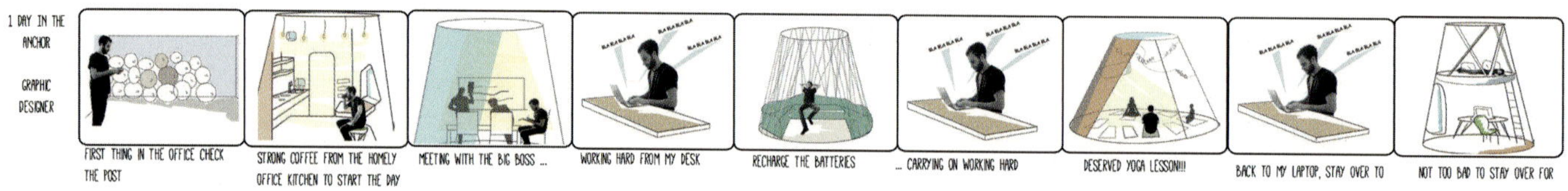

One working day inside the anchor

The cones are moving up to the mezzanine floors though a hand pulling system, the cones have different sizes, different functions and are made with different materials according to the location within the space to provide privacy and light.
The anchor office space offers a vary of rooms that can be rented per hour, day, week and month. The office space offers from hot desking option, open space desks, single office, meeting room, vip rooms, the communal area provided are video room, yoga room, re-charging room, printing room.

원추들은 수동식 당김 시스템을 이용해 중이층까지 들어 올릴 수 있으며, 원추는 크기와 기능이 서로 다르며, 프라이버시와 조명을 공급하는 공간 내 위치에 따라 서로 다른 소재로 만들어져 있다.
기본 사무실 공간은 시간별, 일별, 주별, 월별로 임대할 수 있는 다양한 공간들이 있다. 사무실 공간에는 핫 데스크옵션, 개방형 공간 데스크, 싱글 오피스, 회의실, VIP실 등을 제공하며, 영상실, 요가실, 재충전실, 프린트실 등의 공간도 제공된다.

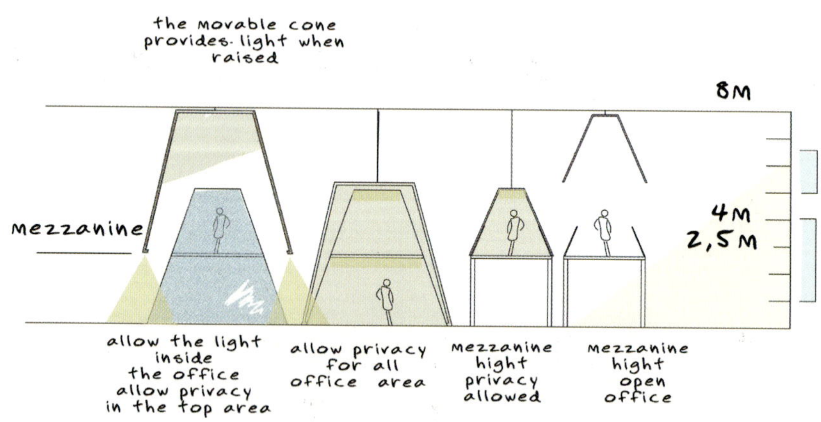

Vary of 'cone' office structures

The single office

The single office provides to the worker of the anchor ,a private working station with desk and storage. The single office offers a re-charging area at the top of the cone reachable with a ladder with the option to stay over the night.
The cone is equipped of a second movable structure that provides privacy and light when is needed.

싱글 오피스는 책상과 보관 장소가 있는 사적인 업무 스테이션을 제공해준다. 이 싱글 오피스는 사다리를 타고 올라갈 수 있는 원추 꼭대기에 위치한 재충전 공간을 제공한다. 이곳에서 밤새 머무를 수도 있다.
이 원추는 프라이버시와 필요한 경우 조명을 제공해줄 수 있는 2차 이동식 구조가 갖추어져 있다.

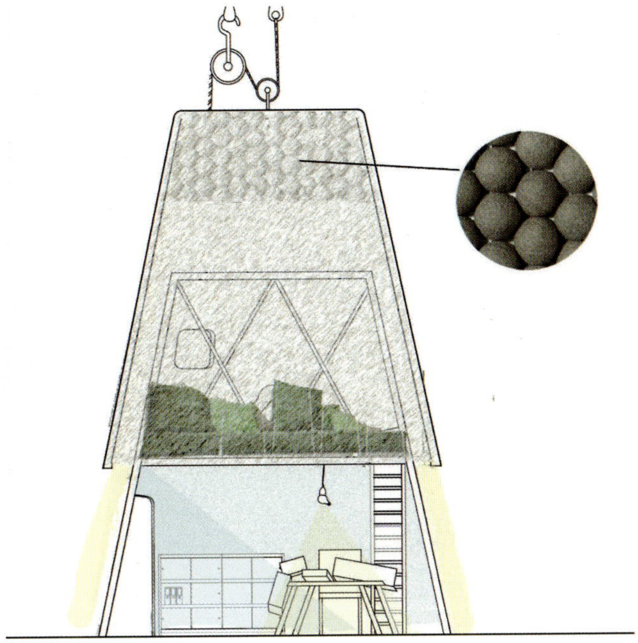

Single office type

Sketch of the space planning

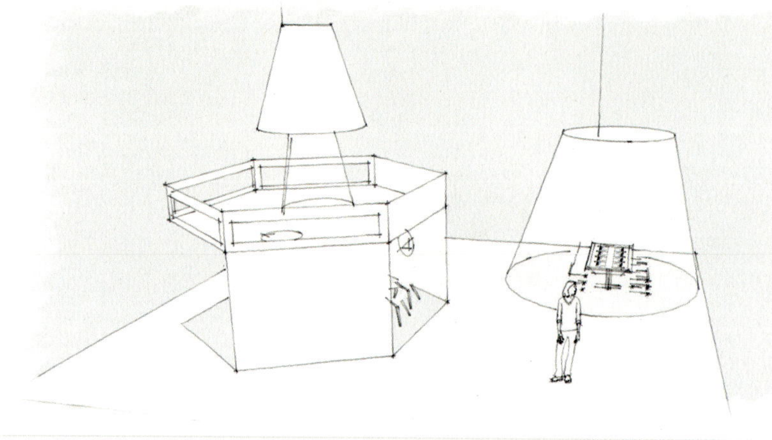

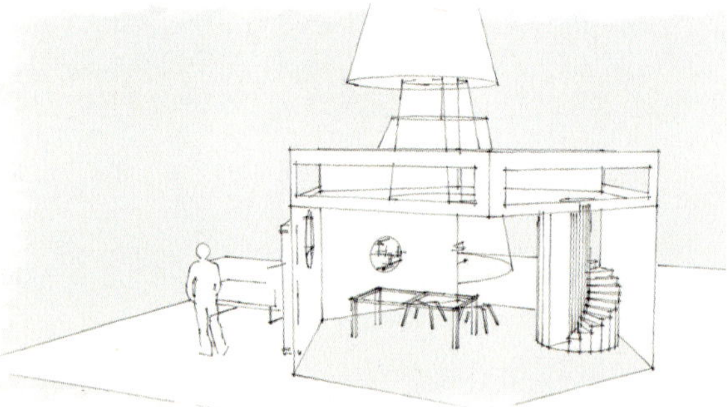

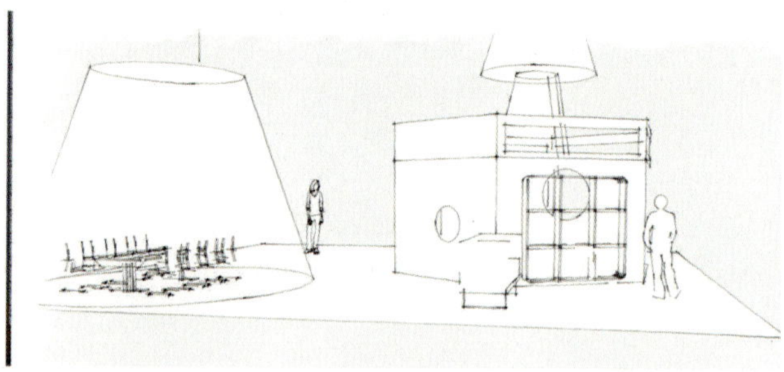

MEZZANINE AND MEETING ROOM

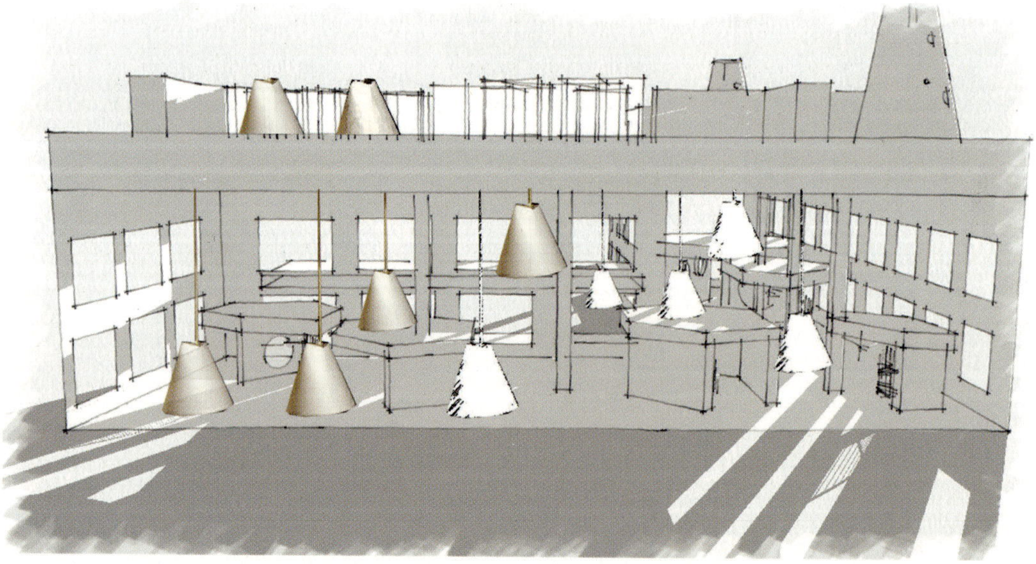

SECOND AND THIRD FLOOR

VISUAL OF THE VIEW FRONT

VISUAL OF THE OPEN SPACE OFFICE

Farida Hasan
faridahasan@yahoo.com

THE BUZZ at SHOREDITCH

THE BRIEF

Context & overview

Site: Shoreditch Exchange, Shoreditch High Street
Client: British Telecom Building: Erected by the General Post Office in 1956, utilitarian in structure, a construction of brick, concrete, steel and glass.
BT own the Shoreditch Exchange. It houses the telephone exhange which services the East End of London, and this occupies the left side on the ground and first floors of the building.
The Exchange is one 1000's of BT's redundant properties which came into being after the technology shift towards modernisation. This change began to take effect some 30 years ago.
The shift was from telecommunications services to more digital based services - Broadband/Wifi and TV sports, rendering their large number of Exhange and Office buildings redundant/vacant or unoccupied.
The project aim is to consider ways to bring the context of the Shoreditch community and BT's Shoreditch Telephone Exchange Building together, using the Exchange building space to work as a venue for small co-working start-up businesses.

Flexibility & adaptability

As a hub for start-up companies the nature of its remit is transient, & the interior landscape will reflect this changeability.
Components of the interior need to be adaptable - expanding/contracting according to the requirements of companies that will be coming and going.
Flexibility also extends to the idea of companies who wish to work from home or on a location basis depending on client meetings etc, for which the flexible spaces may only need to be booked for hire for the event only.
The brief requires an environment where collaboration & connection, creating journeys, chance meetings, new encounters; community space where events, gatherings, lectures, promotions conferences and team projects can all be accommodated.

Private & Public Space

The Shoreditch Excahnge Building is overlooked - the facade goes unnoticed, the building remaining incongruous to the vibrant community it is placed in. These issues need to be addressed to bring the community into the space whether for leisure or for business opportunities, thus considering allocation of private and public space to work together.

The exchange building

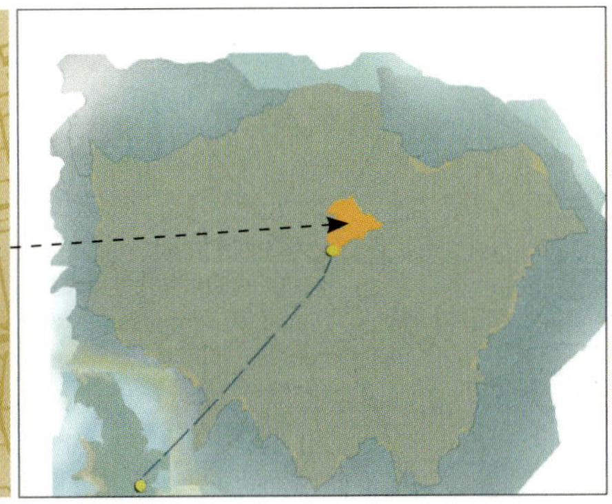

Shoreditch - locatded within London Borough of Hackney, England.
A built-up part of the inner city immediately north of the City of London

Location

Arnold Circus - located at the centre of Boundary Estate. A contrastingly quiet leafy green residential area where independent local, organic and artisan stores servicing the Estate can be found just behind the busy high street bustling with traffic and people.

Tech-City -technology cluster - occupying London's East End between Old Street and the Queen Elizabeth Olympic Park in Stratford, its main density around the Shoreditch area. Such businesses are continuing-to thrive and expand, thus making Shoreditch a prime location for continued expansion.

The Tea Building - Houses 7 Floors of indepenent start-up businesses, home to busy community of marketing agencies, media companies, fashion brands, retailers, artists & desginers, &the Pizza East restaurant. The closest located competitor to The Exhange Building's proposed business start-up plan.

Well serviced for transport links: National rail international links from Kings cross St Pancras - within 19 minutes & Just 2 hours to Paris.
Underground - Metropolitan, Hammersmith & City, District & Circle, Northern lines all within 10-15 minutes walking distance
City Airport in 33 minutes by public transport.

Client brief

As the client, BT wish to use the Shoreditch TE Building as a flexible space to accommodate small start-up companies reflecting the needs of the community. Coming from a technological background, BT has seen the steady development of tech based companies in the area known as Tech-City or the Silicone Roundabout, and envisage that a proportion of the Lets would do well as tech based companies.
BT are happy to remain involved as the Landlord and as such, do not seek to have a building that will require their branding.
Finally, they are keen to see the external spaces of the courtyard and rooftop being made use of as a functional part of the community as well as the co-working hub.

클라이언트로서 BT는 Shoreditch TE 건물을 지역사회의 니즈를 반영한 새로 시작하는 소기업들을 입주시킬 수 있는 융통성 있는 공간으로 사용하기를 원하고 있다. 기술적 배경을 지니고 있는 BT는 Tech City 또는 실리콘 로터리로 알려져 있는 이 지역 내 기술 기반 기업들이 꾸준하게 발전하고 있는 것을 지켜보았으며, 일부 Lets는 기술 기반 회사로서 잘 해내갈 수 있을 것이라 예상하고 있다. BT는 지주로서 관여하는 것에 만족해하고 있으며, 따라서 브랜딩이 필요한 건물을 가지려 하고 있지는 않다.
마지막으로 협업 허브 뿐 아니라 지역사회의 기능적인 부분으로 활용되는 야외 안뜰과 옥상을 보고 싶어 한다.

Potential Hub users

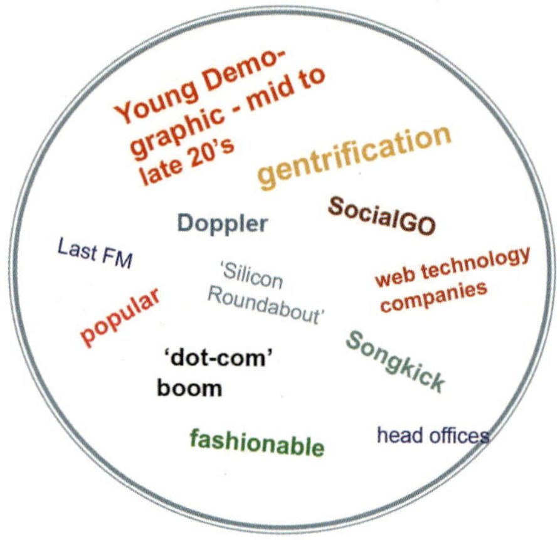

Flexible adaptable

THE CONCEPT

The concept is about 'Sociality' of the Co-working hub environment.
When thinking of a space for shared disciplines, it leads us to think about how best to optimise its efficiency for the co-workers, making the space work hard for them, honing in on making communication & collaboration effortless.
Kandinsy's Composition 8, which has a geometrical vocabulary in its elements of line and form is the inspiration to my Conceptual design premise for the TE building.
It shows a sense of movement and suspended balance and transfers to the idea of an efficient and functional space.
I intend to use the colour palette to be representative of the young, colourful and vibrant energy of Shoreditch and its community, and the geometrics as representative of the flexible-adaptable space within the hub.

컨셉은 협업 허브 환경의 "사회성"이다.
공유 공간을 고려해 우리는 동료들을 위해 효율성을 극대화 할 수 있는 방법과 동료들이 열심히 일할 수 있는 공간을 만들며, 의사소통과 협력을 용이하게 할 수 있는 공간을 생각해보았다. 라인과 형태 요소에 있어 기하학적 어휘를 담고 있는 칸딘시의 구성 8이 TE 건물의 개념적 디자인에 영감을 주었다. 칸딘시의 구성 8은 움직임과 균형 감각을 보여주며, 이를 효율적이고 기능적 공간에 대한 아이디어로 전환시켜 준다. 난 칼라 팔레트를 이용해 Shoreditch와 이곳 커뮤니티의 젊고, 다채로우며 활력 넘치는 에너지를 표현하고자 했으며, 기하학적 형태를 이용해 허브 내의 탄력적이고 적용 가능한 공간을 표현하고자 하였다.

Stair Theatre
- Under stair space
- Enclosure Box

Play space
- Informal meeting. Social/waiting. area Stage storage
- Lecture. Theatre space. Conference. Presentation

→ Tree house box pod structures in Court Yard area. Accessed from 2nd Floor or up from Courtyard.
Play spaces on courtyard level also
Entance threshold to Rooftop cafe/Restaurant.

→ Recharge relax
Play - ping pong, quiet space, play lists
Informal meeting
Event space - (Private & Public)
Team building

2ND & 3RD Floor Work Pods

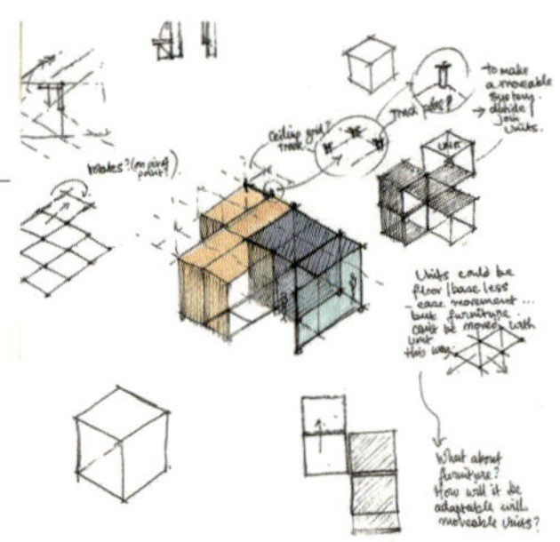

Team work
Group work
Project work
Conference/meetings

Gallery, Restaurant & Theatre 'Entrance boxes' -threshold experiences

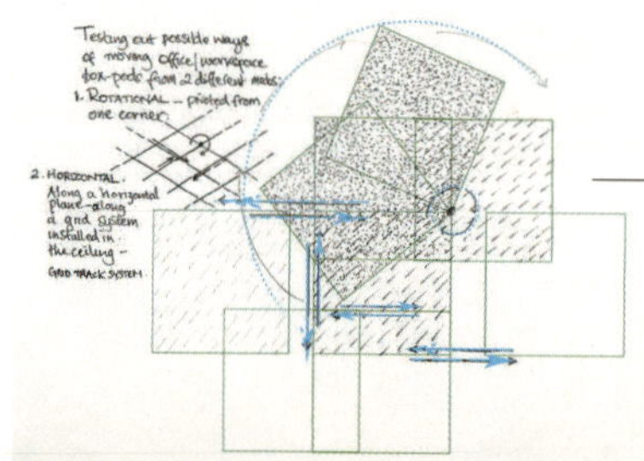

Box pods adaptable to varied requirements - adding/taking away units, rotating on axis point or sliding on horizontal plane - functioning from grid system on ceiling.

Social co-working - Testing ideas through sketch modelling

Cut out on 3rd floor - testing the idea of social co-working by way of vertical connectivity
visually & spatially. I have looked at ways to illustrate the idea & feel of a glimpse of 'suspended objects' connected through the different floor levels.

3층 커트 아웃–시각적 및 공간적으로 수직적인 연결을 통해 사회적 협업이라는 아이디어의 테스트. 여러 층을 통해 연결되어 있는 '부유 물체'를 본다는 아이디어와 느낌을 보여줄 수 있는 방법을 모색해보았다.

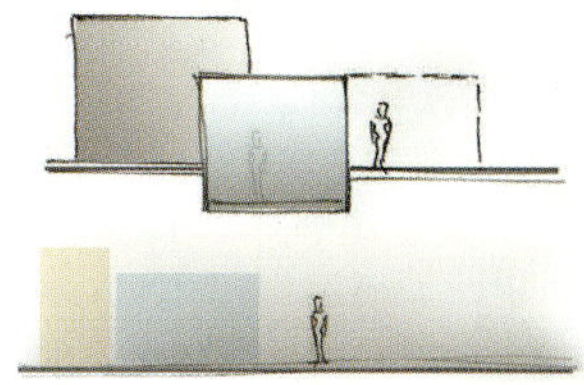

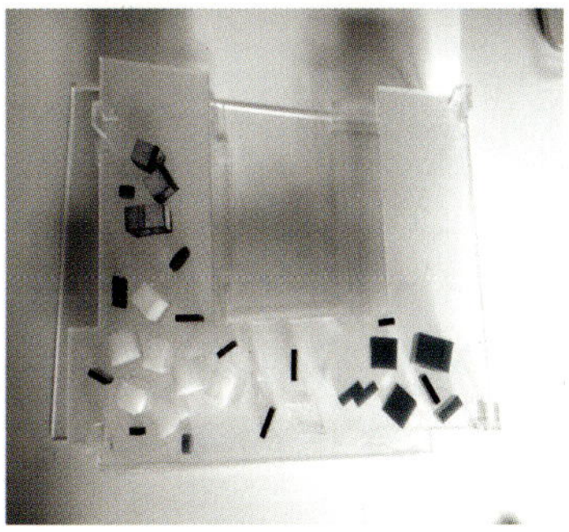

Left: testing the landsacpe of teh box pods in areas, dividing spaces with screens - keeping teh concept of suspended balance with functional adaptable spaces.
Right: Testing teh idea of 'dropped pods' giving teh feel of floating spaces between floors.

좌측: 박스 포드의 테스트, 공간을 스크린으로 구분, 기능적으로 조정 가능한 공간과 균형이라는 컨셉의 유지
우측: "떨어진 포드"라는 아이디어의 테스트, 층간 부유 공간이라는 느낌 제공

Above: Sectional sketch showing idea of adaptble box-pod office sapces and how some are sunken or 'suspended' between 3rd and 2nd floors

좌측: 박스 포드의 테스트, 공간을 스크린으로 구분, 기능적으로 조정 가능한 공간과 균형이라는 컨셉의 유지
우측: "떨어진 포드"라는 아이디어의 테스트, 층간 부유 공간이라는 느낌 제공

Space planning - ground floor

Bike Repair shop and Bike park area

Gallery & Cafe

New Rear entrance - Public and private access into building directly in to gallery and Cafe.

Courtyard Garden with area for recharge tree house box pods.

Driveway to basement car park servicing BT Openreach Vehicles

Cafe and Garden - servicing staff and public

Boxed stairs and disabled wheelchair lift - front entrance.

BT Exchange rooms -area not for redisign for Co-worker hub

Courtyard Recharge pods - bookable spaces for art/craft activities/short courses for community - flower arranging, painting, textiles etc, for adults and children.

'Stair theatre' and stage - moveable to join & function as conference or lecture space. Seating & waiting space in under stair area making it a multifunctional unit.

First Floor

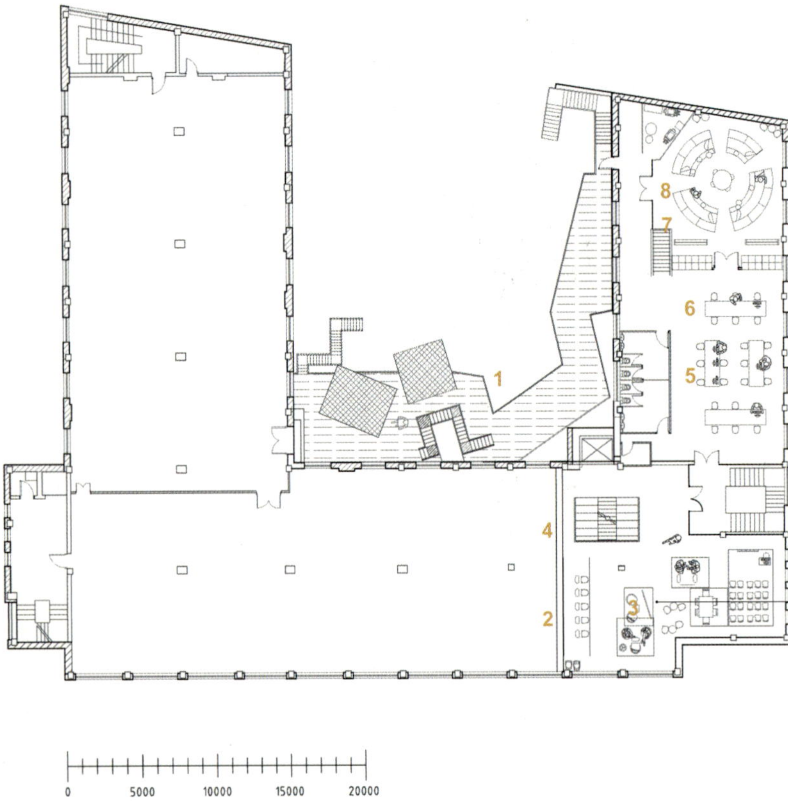

1. Recharge/play tree house box pods
2. Whiteboard corridor area - group work - brain storming - move able so screen can be pushed away to wall to make space.
3. Flexible adaptable box pods
4. Stair theater leading up from ground floor
5. Hot desk work area
6. Lockers
7. Conecting stairway from gallery to first floor
8. Resource/material library
9. Print room

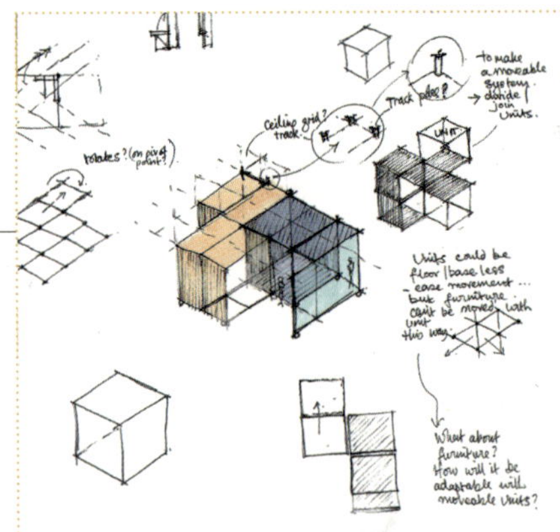

Second Floor

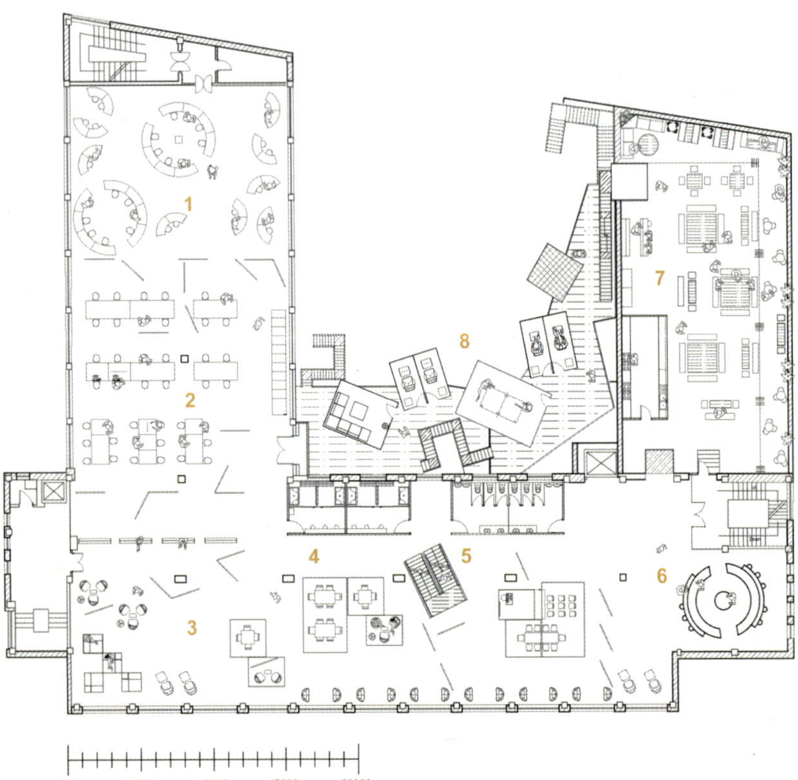

1. Quiet working space - situated in North East side of building - cool quiet area. Semi screened work stations.
2. Work studio space - benefiting steady North light for creative industries work.
3. 'Break & Connect zone' - Space to regroup - facilitaed by phone booths & lounge chairs for individuals/ coffee table areas or benches for groups - allowing chance meeting,collaboration or even informal brainstorming.
4. 'Floating Work Pods' - elevated from floor with central riser giving illusion of floating, gravity defying work pods - enabling an experientially alternate space. Work, meet, conference, lecture, presentation - join or divide - vadaptable to size and function.
5. Stairs connecting 2nd to 3rd floor through central void.
6. Kitchenette bar - self serving beverage and snack area with bar seating.
7. Rooftop cafe/restaurant for staff, community and as destination dining
8. Tree House Recharge Pods - 1st & 2nd Floors connected by stairways to mesh platform.
Reconnecting idea of play back to the outdoors - Wooden boxes opened to the elements with seamless glass panels, replenishing mind with greenery of the courtyard.
Re-energise activities: ping pong room, basketball longe room, quiet rooms.

Third Floor

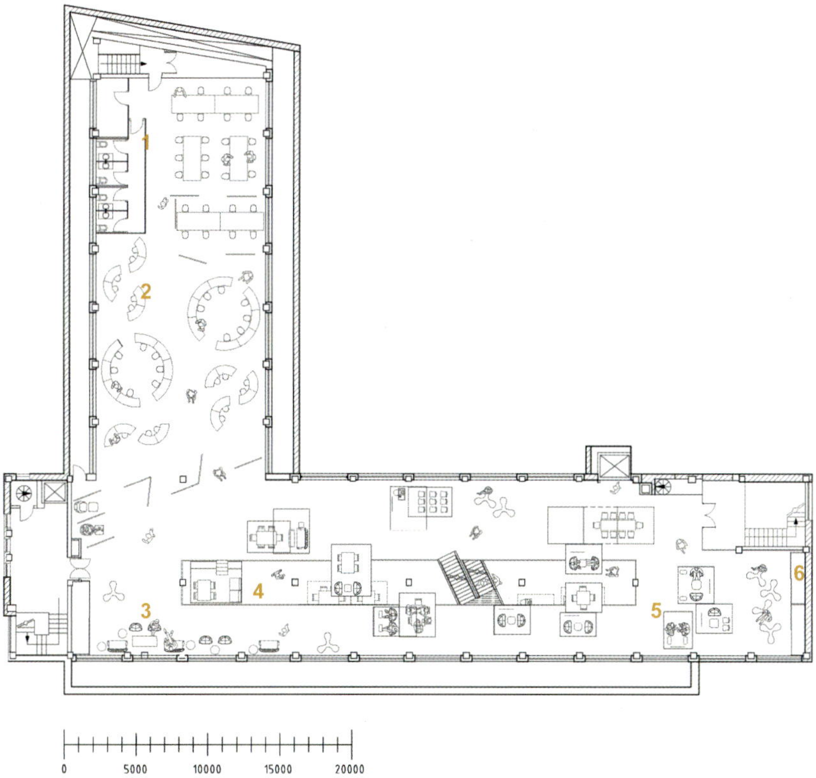

1. Work Studio space
2. Quiet working space
3. Break & Connect zone
4. Void overlooking work floating pods zone on 2nd floor. Populated with cantilevered and suspended work pods over and in the void continuing theme of floating work spaces.
5. Floating work pods
6. Storage with sliding doors - for chairs and tables - stowed away when work pods are arranged taking up more floor space, and vice versa when grouped together opening up more floor space.

The stair theatre

Model showing use of Stairs as Theatre seating - larger outside steps have riser and tread both at 500mm.
Stage enclosed in moveable screen with transparent glass sides and solid back wall.

극장 좌석으로 계단을 활용하는 것을 보여주는 모델-큰 외부 계단들은 500mm 간격으로 riser와 tread가 있다. 투명 유리면이 있는 이동 가능한 스크린에 내장되어 있는 계단과 단단한 후면 벽

When not in use as Theatre space, enclosure box can be moved closer to create more private space where lectures conferences or presentations can take place. Access to and from upper level when in use for private space is restricted.

극장 공간으로 사용하지 않을 때에는 밀폐 박스를 가까이 이동시켜 강의나 컨퍼런스 또는 프레젠테이션을 할 수 있는 좀 더 사적인 공간으로 연출한다. 사적 공간으로 사용할 때 상층부로의 접근은 제한된다.

Under stairs becomes social seating area - for waithing, meeting , talking, with accessable space of around 8m2.

계단 아래에는 8m2의 사회적 공간이 자리하고 있어, 사람을 기다리고, 만나고, 대화할 수 있는 공간으로 활용된다.

Hunan University
Department of Industrial Design, in School of Design

School of Design, Hunan University
Yuelu Montain, Changsha, Hunan, P.R.China, 410082
http://www-en.hnu.edu.cn/

Renke He

Professor & Dean
School of Design, Hunan University

Hunan University (HNU) is one of the oldest universities both in China and in the world. It was called 'Yuelu Academy' when it was built in the year 967, more than a thousand years ago, as an educational institution of Confucianism. Recent years have seen the University's dynamic move to extend its global reach, based on one thousand years' established scholarship and capacity to embrace diversity of creative thinking and talents from all around the world.
School of Design at HNU (HNU Design) has consistently taken the initiative in the University's innovation. The School attaches great importance to international exchanges. More than 30% graduates go to the prestigious universities worldwide to continue their study. Since HNU joined Cumulus (the International Association of Universities and Colleges of Art, Design and Media) in 2006, School of Design has established close cooperation relations with international design schools as well as multinational corporations. In the same year, the School was nominated as one of the world's top 60 design schools by Business Week.

In 2014, the School hosts the Fourth LotusPrize, the International Design and Innovation Competition, the biggest design competition in China. With active participation from enterprises, higher education institutions and designers, the competition has been transformed into a platform to share creative design ideas and design industry and society leadership, as well as to discover and nurture future design stars.

In the perspective of our long-standing missions above, we see a good opportunity to extend HNU's network in the participation of our students in London Festival of Architecture 2014 in collaboration with international Universities in U.K and Korea.

I wish HNU Design, together with the 3 partner universities, to create a great synergy for achieving the mission of cultural exchanges and co-evolution.

허렌커
산업디자인과 교수
후난대학교 디자인대학 학장

중국에서, 그리고 세계에서 가장 오래된 대학들 중 하나인 후난대학은 지금으로부터 1천년이 넘는 967년에 공자의 유학을 가르치기 위한 교육기관으로 '열류서원' 이라는 이름으로 설립되었다. 최근 몇 년간 후난대학은 적극적으로 글로벌 네트워크를 확장해 왔는데 이는 전세계로부터 다양한 창의적 사고와 더불어 그런 생각을 가진 창조적인 인재들을 포용할 수 있는, 1천년간 축적된 학문의 힘과 잠재력에 힘입은 것이다.

후난대학교의 디자인대학(HNU Design)은 동 대학의 혁신에 있어서 항상 앞장서 왔다. 국제교류에 큰 기여를 해 왔으며 이에 따라 졸업생의 30% 이상이 학업을 계속하기 위해 전세계의 유수한 교류대학으로 진학하고 있다. 2006년에 예술, 디자인, 미디어 분야 대학들의 국제적 연합체인 Cumulus (the International Association of Universities and Colleges of Art, Design and Media)에 가입한 이래 디자인대학은 상위급 대학들 및 다국적기업들과 긴밀한 협조관계를 구축해 왔다. 그리고 2006년에는 비즈니스위크에 의해 전세계 최상위 60개 디자인스쿨 중 하나로 선정되기도 하였다.

올해는 후난대학에서 중국과 아시아에서 가장 큰 규모의 학생과 영디자이너를 대상으로 하는 국제디자인공모전인 제4회 LotusPrize, the International Design and Innovation Competition 을 개최하고 있다. 수많은 기업들, 대학들, 디자이너들이 적극적으로 참여하면서 로투스프라이즈는 창의적인 디자인 아이디어와 디자인산업 및 디자인 커뮤니티의 리더쉽을 공유하는 한편 미래의 디자인 스타들을 발굴하고 양성하는 플랫폼으로 변신해 왔다.

상기 언급된 장기적 목표의 관점에서 볼 때 후난대학 디자인대학은 우리 후난대학의 학생들이 영국과 한국의 국제적 대학들과의 협조 하에 2014년 London Festival of Architecture 에 참여하게 된 것이 좋은 기회라고 생각하며 후난대학이 글로벌 네트워크를 확대할 수 있기를 기대한다.

후난대학 디자인스쿨 학장으로서 이번 전시에 참여하는 파트너 대학들과 함께 우리 디자인대학이 문화적 교류 및 공진화를 달성할 수 있는 커다란 시너지를 창출할 수 있기를 기원한다.

Jun Zhang

Dr. , Assistant Professor
School of Design Hunan University

Currently China is suffering from the severe negative consequences of over-developed manufacturing industry and unheeded public facilitates & service from large and densely populated cities to smaller one. "Sustainable City" has become a shared view within all government levels in China. Designing for sustainable urban space has become the hot issue where ever institutional or practical circle these years.

System Design is a course in Hunan University which combines Product Service System (PSS) Design and Sustainability in Sustainable urban & community context. The target of this year is exploring comprehensive system design concept in Changsha, the capital city, in order to find concrete solution towards environment-friendly, economically cost-effective, a way to keep fit and healthy and, on occasions, an enjoyable social activity.

Product Service System Design share us a holistic design thinking & method from a system design perspective which offers opportunities to dematerialize the economy and reduce the environmental impacts of industrial activity and has the potential to accelerate the transition to a sustainable society by providing intangible services as a functional economic model.

MEPSS (Methodology for Product Service System) was adopted as a framework to guide the design process which helps student designers to think 'How is today's value chain organised' and actively use visualization tools and a modular method to analysis stakeholder management and service processes. In which, you can see the main analyzing and visualizing tool, System Map, a useful diagram facilitates the organization of the data and an understanding of how key elements in the each public system interact, support or conflict with each other.

A PSS and innovative design perspective could be seen from all these design works. The teaching and learning process offers a "Use Orientated" approach. Aging group, blind people, migrant workers or junior students are becoming target user and research object. Students need to study an intensive research and design process which involves working collaboratively with different stakeholders, such as institutional investors, system operators, and users in "Solution-Oriented Partnerships".

With a view to deriving influences on the sustainability, these eight public design proposals could not only be seen as a valuable approach to foster clean & energy-efficient modes of lifestyle. Here they conceptualize such systems as being sustainable in the sense that they are well-established and making a contribution to the meeting of environmental and social sustainability goals, they are also seeking to meet new conceptions of value from both public sector and private sector perspectives.
In urban areas, specifically in China urban context, we like to see in the future diverse well- designed and well-organized public system may provide a complete solution to sustainable consumption and production.

장준
박사, 후난대학교 디자인대학
산업디자인과 조교수

현재 인구밀도가 높은 중국의 거대도시들은 물론 여타 도시에 이르기까지 과도하게 개발된 제조업과 관리가 아직 부족한 공공시설과 서비스들의 결과로 나타난 문제들로 어려움을 겪고 있다. 이런 맥락에서 지속가능한 도시(Sustainable City)라는 개념은 중국 정부의 각 계층이 공유하는 대안으로 받아들여지고 있으며 이러한 도시 공간 디자인은 공적 기관 회합 및 실무적인 모임에 있어서 열띤 토론의 주제가 되고 있다.

시스템디자인 교과목은 도시와 서스테이너블디자인 그리고 공동체 맥락에서 제품과 서비스시스템 (Product Service System; PSS) 을 디자인에 결합한 것이다. 이 수업의 금년도 목표는 후난성의 수도인 창사에서의 도시 스템디자인을 탐색해 보는 것인데, 이는 도시공간의 제반 사회적 활동들이 적절하고 건강하고 즐거운 그 무엇이 될 수 있도록, 환경친화적이고 비용 대비 효율성이 높은 방법을 찾기 위한 과정이다.

PSS디자인은 시스템디자인 관점에서 전일적인(holistic) 디자인 사고와 방법론을 공유할 수 있게끔 도와준다. 시스템디자인 관점은 과도한 산업활동에 의해 유발되는 환경 피해를 감소시키고 물질적인 사고에 의해 점령당한 경제를 비물질적인 가치로 전환시킬 수 있는 기회를 제공한다. 또한 PSS디자인은 실제로 작동 가능한 무형적인(intangible) 서비스를 제공함으로써 유지가능한 사회로의 전환을 가속화시킬 수 있는 잠재력을 갖고 있다.

MEPSS(Methodology for Product Service System)은 디자인프로세스를 안내하는 틀로서, 학생 디자이너들이 “현재 가치사슬(value chain)이 어떻게 구성되어 있는가?” 생각하는 것을 도와주는 한편, 가치공유자(stakeholder) 관리 및 서비스프로세스 분석을 위해 시각디자인 도구들과 모듈 방식(modular method)을 적극적으로 활용하는 것을 도와준다. MEPSS의 중심인 시스템지도(System Map)는 분석과 시각화에 활용되는 도구로서 시스템지도의 다이어그램은 데이터를 편리하게 구성할 수 있게 해주며 각 공공시스템의 핵심 요소들이 어떻게 상호작용, 협조, 충돌하는지 이해하는 것을 도와준다.

혁신적인 디자인과 결합된 PSS는 제반 디자인 작업에서 활용되는 것을 볼 수 있으며 그 교육 및 학습과정은 사용자지향(user oriented) 접근법을 제시한다. 노령화 인구집단, 시각 장애자들, 이주 노동자들 또는 미성년 노동자들은 PSS가 지향하는 사용자이며 연구대상이다. 학생들은 연구 및 디자인과정을 강도높게 공부해야 하며, 이에는 해결지향적 파트너쉽(Solution-Oriented Partnership)에 참여하는 기관투자가, 시스템관리자 및 사용자 등 가치공유자들과 협력하여 일하는 것이 포함된다.

유지가능성을 높이는 요소들의 추출이라는 주제하에서 이번 전시에 출품되는 8개의 공공디자인 제안들에서 학생들은 효율성이 높은 서스테이너블 디자인 접근법을 제시하는 한편, 이 개념을 구체화시키고 있는데, 유지가능성에 대한 평가는 이 시스템들이 얼마나 잘 구성되어 있고 환경적, 사회적인 유지가능성 목표를 얼마나 잘 달성할 수 있는지의 관점에서 이루어진다. 여기서는 학생들이 공공부문 및 민간부분 관점에서의 새로운 가치개념에 어떻게 호응하려고 노력했는지도 볼 수 있다. 도시지역에 있어서, 특히 중국의 도시문제에 있어서, 우리는 멋지게 디자인되고 훌륭하게 구성된 다양한 공공시스템들이 서스테이너블한 소비 및 생산을 위한 미래의 해결책을 제공해 주는 것을 보기 원한다.

Ahn, Sung-Hee

Dr. , Assistant Professor
School of Design Hunan University

The works of Hunan University students is the outcome from the collaborative program between system design class, sustainable design workshop and public design workshop, which were linked to each other under the sole theme – 'City Changsha and the Sustainability'.

In China, the basic approach dealing with urban agenda is still governed by development mentality. The starting point of the collaborative program was a proposition that reorienting urban design and regeneration toward sustainable growth would be a better option for the human-centered bright future of city development.

5 weeks' system design class included the workshop which was masterminded by Karen Blincoe, a Danish expert on sustainable design. The workshop started from the students' seeing Changsha, their everyday living space, from a totally different viewpoint. They concentrated on how to achieve social and cultural sustainability as well as functionality of physical space in the existing community-centered perspective. There were many times for the students to enjoy workshop process which invited them to the brainstorming sessions for the purpose of finding the right track to the sustainable city, imagining everything possible and sometimes referring to and learning from the lessons from history and tradition. They went out of the campus to comb the city, making interviews with many residents and watching the behaviors of people. They showed proactive attitudes in the process so respectable as for me to regret not being able to display their beautiful field study results.

5 weeks were rather short for the students to do so many things from identifying and analyzing problems of existing spaces to designing new spaces equipped with organic system and service to overcome the constraints and limits. As a result, they couldn't allocate enough time to the final stage of the workshop - design proposal of the sustainable city.

Nevertheless, I believe the workshop presented invaluable experiences and lessons to the students who actively participated in it. I am also thanks for the other students who worked so hard together but couldn't join the exhibition and I appreciate their concerted efforts to achieve the excellent performance. I send my special thanks to Professor Karen Blincoe for leading the exciting workshop, and to Professor Eunji Jo for taking charge of service design sessions.

The city is an organization tangled with people, products and services in which any part cannot survive without links and relationships with other parts. But, we can find concentration of networks, and, on the contrary, deserted spaces and alienated people in the city. Through the workshop, we had the opportunity to be trained as a designer responsible for identifying those abandoned elements and members of the city and giving helping hands to them.

안성희
박사, 후난대학교 디자인대학
산업디자인과 조교수

이번 후난대학의 작품들은 시스템디자인 수업과 서스테이너블디자인워크샵 그리고 공공디자인 워크샵의 협력수업의 결과물로 나온 것이다. 이 수업들은 도시 창사, 그리고 서스테이너빌리티 라는 하나의 주제를 가지고 연결되어 있었다. 중국에서는 아직까지 도시에 대한 기본개념이 발전 위주의 개념이 지배적이다. 이러한 시점에서 개발이 다 끝나고 반성하는 유럽의 과정을 반복하는 대신 개발의 중간에 '지속가능한 발전으로 도시개발의 방향을 틀어 줄 수 있다면' 사람들을 위해 보다 나은 미래를 제공할것이라는 전제에서 시작했다.

5주동안의 시스템 디자인 수업 안에는 서스테이너블 디자인의 대가인 덴마크의 Karen Blincoe 교수와의 '도시 그리고 서스테이너빌리티' 워크샵이 포함되어 있었다.
학생들은 창사라는 익숙한 도시를 낯선 관점에서 바라보는 연습에서부터 출발하였고, 도시공간을 그곳에 살고 있는 커뮤니티 중심의 시각에서 물리적인 공간의 기능보다는 사회,문화적인 관점의 서스테이너빌리티를 달성하는데 집중하였다. 나은 방향을 위해 개선 되어야 하는 부분들에 대해서 폭넓게 때로는 상상력을 동원하며 역사의 시간을 넘어서는 즐거운 워크샵의 과정들이 있었다. 캠퍼스 밖으로 나간 학생들은 도시의 곳곳을 다니며 주민들과 인터뷰를 하고 행인들의 행태를 관찰하였고 아름다운 필드스터디의 결과물들을 자세히 나열하지 못하는 것이 아쉬울 정도로 칭찬받을만한 학습과정을 보여주었다. 다만 기존의 공간이 가진 문제점들을 찾고 분석하고 객관화 하고 그 한계에 맞서 새로운 유기적 시스템과 서비스를 가진 공간을 만들어내는 과정이 5주간의 학부수업으로는 시간적으로 벅차서 최종 디자인에서 시간을 많이 할애 하지 못해 아쉬움이 남는다.
같이 전시하지 못했으나 열심히 작업해준 나머지 학생들 그룹, 워크샵을 진행해준 Karen Blincoe 교수 그리고 서비스 디자인부분을 담당해주었던 Eunji Jo 교수의 노고에도 감사를 전한다.

도시는 사람과 제품 그리고 서비스가 서로 복잡하게 얽혀있는 하나의 유기체라고 할 수 있다. 서로의 연결과 관계를 벗어나서 어느 부분 하나가 단독으로 존재하지 못하는 모두의 것으로 향해 연결되어 있는 유기체 같다. 그러나 이 유기체 안에는 그 연결이 한곳에 집중되거나 소외되어 간혹 버려진 공간이 되거나 지나쳐버리는 요소들이 분명 존재한다. 우리는 디자인의 힘을 빌려 이러한 곳을 찾아내고 손을 내밀어 줄 수 있는 책임 있는 디자이너가 되는 과정을 이번 수업을 통해 연습해 보았다.

project members

Co-reading Corner
Chen Xiaoqing, Qi Yi, Gong Kai, Zheng Yaoming, Zheng Zhong

FORESH
Zou Yue, Cao Yuan, Yu Weiguang

Fangboo
Yi Huan, Xiao Anqi, Zhao Shikuan, Zhang Haolin

Public Kitchen
Ma Kang, Ding Kebi, Zhou Linlang, Zhen Hongxia, Xu Jiayang

Cool summer
Tan Butian, Ni Qiji, Gu Sujun, Yang Xiangjie

Street Music Sharing System
Zheng Yuling, Zhang Shiqi, Tan Butian, Gu Sujun

Fresh Line
Liu Yue, Hu Xiaoyun, Cai Zhiwen, Wang Yidan

Autarkic Roof Vegetable Garden
Zhu Zhaoqiang, Xu Shihui, Ke Junxue, Zhang Xiaoyu, Xie Zhenzhen

Project Process image

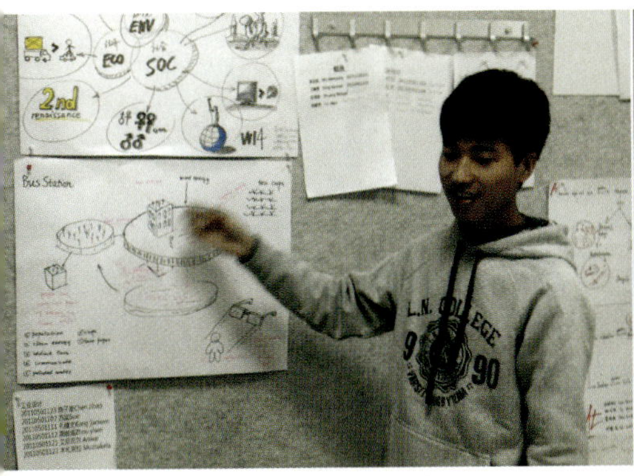

CO-READING CORNER
-Book Sharing Space

Team members | Chen Xiaoqing, Zheng Yaoming, QiYi, GongKai, ZhengZhong

Background

In China community residents' reading rate is low. Although chinese urban residents get richer,but a personal or family is not rich enough to pay for many books in daily life expenditure. And the public library is very few.But the amount and quality of books of the community as a whole is considerable. At the same time, community residents' communication, however, is quite lack,which causes few of them would have access to sharing the books. So we want to establish a system to integrate the community book resources, making the books flowing, making community residents obtain more knowledge, and promoting the mutual communication and understanding, and the harmonious development of the community.

Set up a book-sharing space with automatic book-selling machine in the community, combined with a network platform providing information on borrow or lend, and communication, to make it possible to read without going too far.
The biggest characteristic of Co-reading Corner is to improve the overall reading rate through the way putting the books on the door to reduce the distance between residents and books. At the same time, by sharing books it can achieve higher utilization rate of books, promote the exchange of neighborhood and break dilemma between community people now, back to childhood habit of everyone acquaintance community.

중국에서 주민들의 독서량은 낮은 편이라고 할 수 있다. 비록 중국의 도시들이 경제적으로는 부유해지고는 있으나 개개인으로 볼 때는 일상의 소비문화 속에서 많은 책을 소비하기에 여유로운 것만은 아니다. 그렇다고 커뮤니티 도서관들이 활성화되어 있지 않다. 이런 이유로 커뮤니티를 위한 책의 공급이 양적인 면에서나 질적인 면에서 충족될 필요가 있고 주민간의 커뮤니케이션 역시 원활하지 않다는 배경이 이 프로젝트의 출발점이다. 이러한 이유로 우리는 커뮤니티의 책이라는 자원과 그 책의 커뮤니티 내부에서의 순환을 이용하여 소규모의 지역주민들이 지식과 정보에 노출될 수 있으며 서로 이해하고 소통할 수 있는 공간환경을 조성하여 향후 커뮤니티의 조화로운 발전에 도움이 될 수 있는 시스템을 구축하고자 하였다.

Field Work: Interviews —research in cummunities

Character

We selected five typical characters for a interview, from their reading requirements, the actual reading situation(reading amount, kinds of books, etc.), access to books, as well as how to deal with the used books, view of sharing books and views on the community social status of lack of communication between people, to inner emotional appeals.

We do research on the geographical environment structure and layout of communities.

Design Develpement

System Map

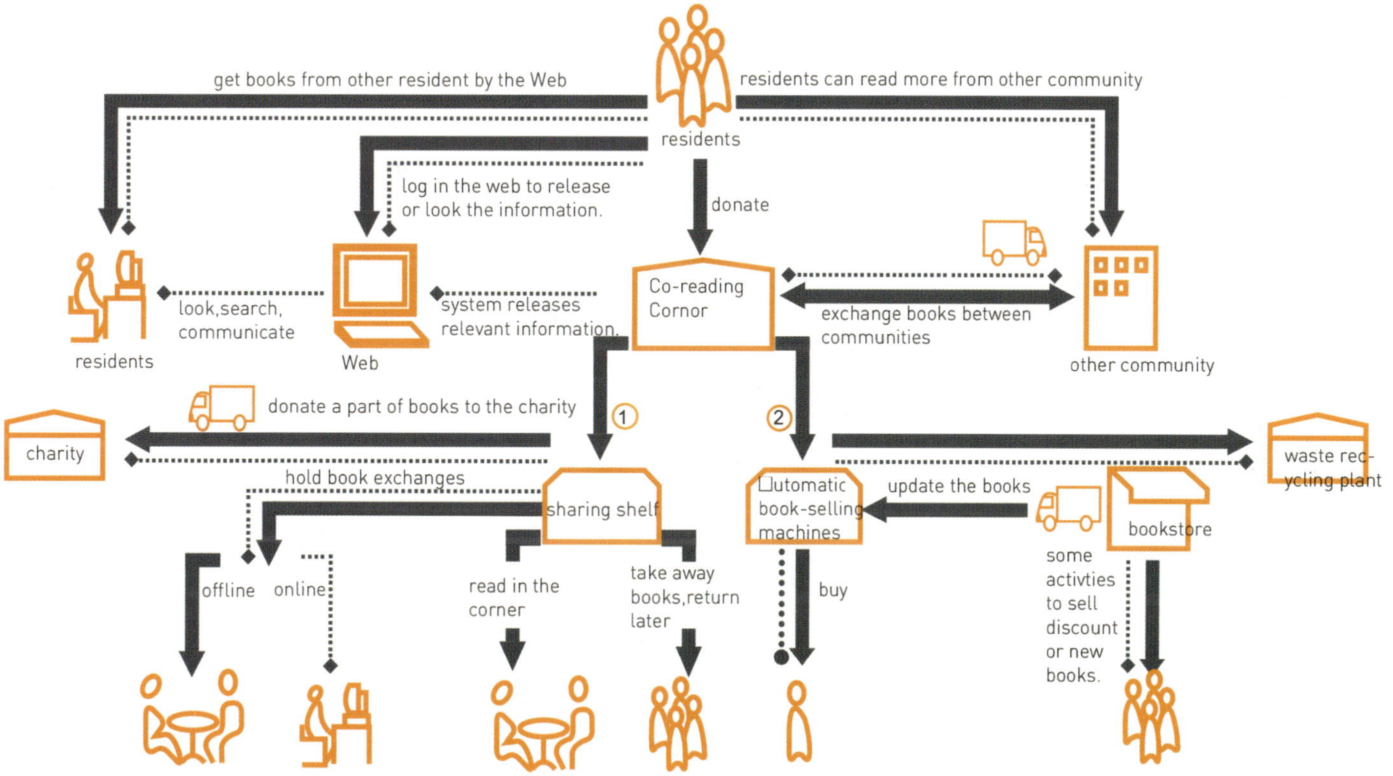

Story Board

When relaxing in the square of community, she finds the Co-reading Corner.

There are many books .They can be read and taken away too. And she finds an interesting book.

She takes the book home and read for several days.

The other day, she comes to return and donated some books too.

Tom released a book-sharing information on Co-reading Web .And Marry saw it,interested.

They chat online for a while, and get agreement to share the book.

They meet at Tom's home. and have a happy talking.

Later, Mary returns home with the book.

Radar Diagram

System life optimisation

To repair some used books, prolong the life of a book. Through system optimization, to increase the value of the system.

Toxic reduction

Waste recycling plant tansforms the damaged books as waste paper into recycled writing paper.

Transportation/distribution reduction

Transform Individual transportation into unity transportation, maximize reducing individual mobility.Improve the transport efficiency and save energy.

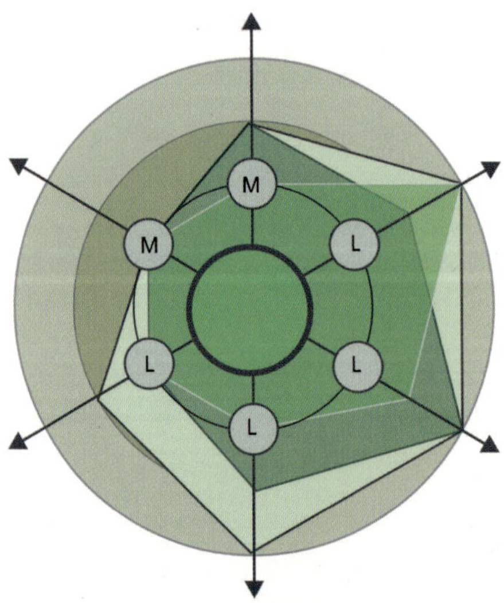

Conservation /biocompatibility

Whether buying books or borrowing, system reduces the energy consumption by reducing the transport directly.

Through book sharing, fundamentally improve the resource utilization of books.

Transportation/distribution reduction

Integration of resources, recycling, between individuals, exchange between community , etc., and, in collaboration with waste paper recycling plant to overall process waste books.The generation of recycled paper can be used in the production of new books.

The Main Views of Product

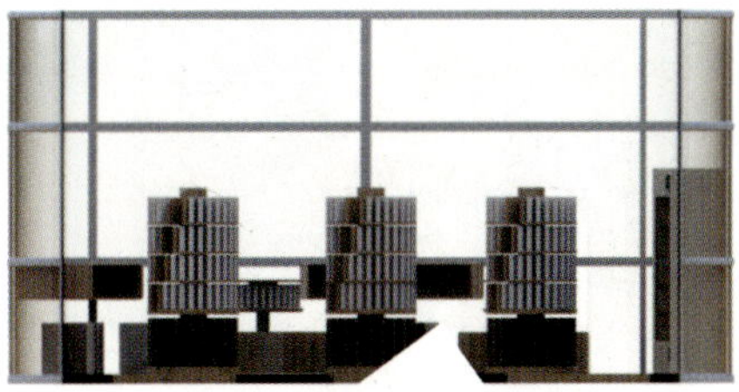

Front view

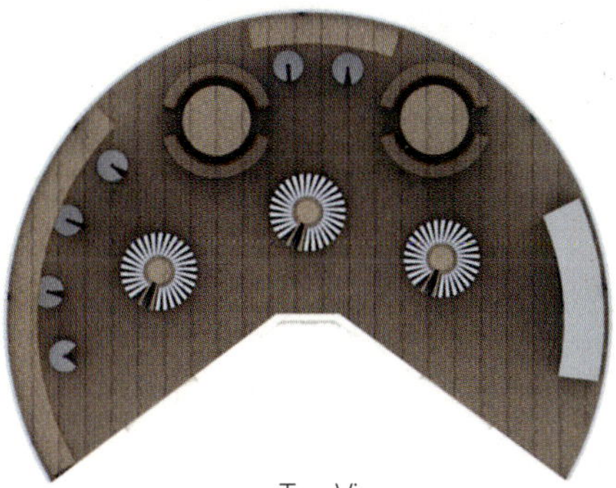

Top View

The size of Product (Unit:mm)

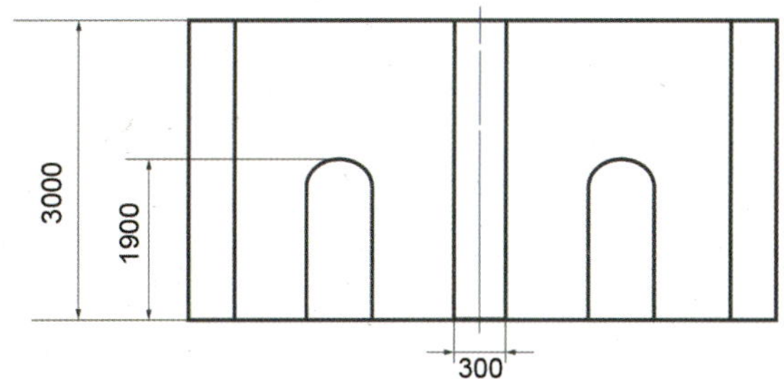

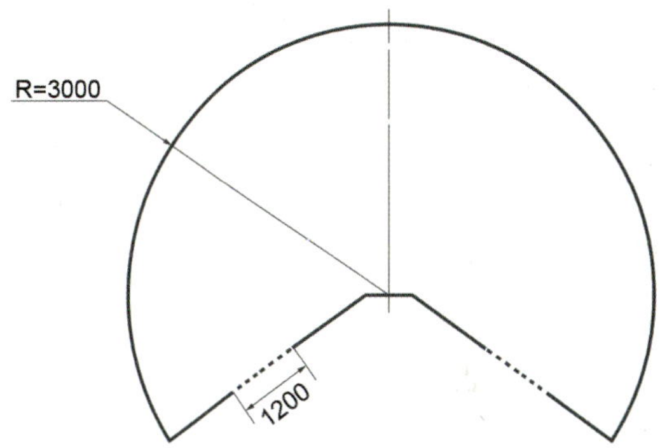

Interaction Board

			34		56		78		9	1
Community residents	The community residents donate books to the Co-reading Corner at any time.	When resting in the center of the community ,residents can read in the Corner.	The residents can buy books in the automotic book-selling machines.	When leaving,you can borrow books on the shelf and read it at home.	After the reading,return the books next time going there.	Or after reading ,return back books to the bookshelf downstairs.	Chat online ,ex-change informaton on sharing and activities. Also they can buy books on the website.	Reach agreement on lending or borrowing books online ,they meet each other offline.	Residents go to get the pickup.	Residents can be volunteers helping arrange-ment
Bookstore	Bookstore puts new books in machines.		Bookstore regularly checks and updates the books in the machines					Bookstore prepares the books residents buy online and post to them.		
Charity and poor villiages								At stated periods,donate part of books to charity.	The charity transports the books to poor villiages.	
Terminal devices	Supply space to store the donated books and the automotic book-selling machines.	Supply space for residents to rest.			Supply space to store the returned books .					
Service personnels					Regular tidy up the books in the Co-reading Corner.	Regular tidy up the books on the bookshelf, return to the Co-reading Corner.	Every 2 months, exchange the books between different communities	Do equipment maintenance at regular.		

The process to use the fan-chair

Pull a fan-chair out from the botton of the bookshelf.

Then unfold the fan-chair to the degree you like.

Sit down on the fan-chair to read books.

Reading details

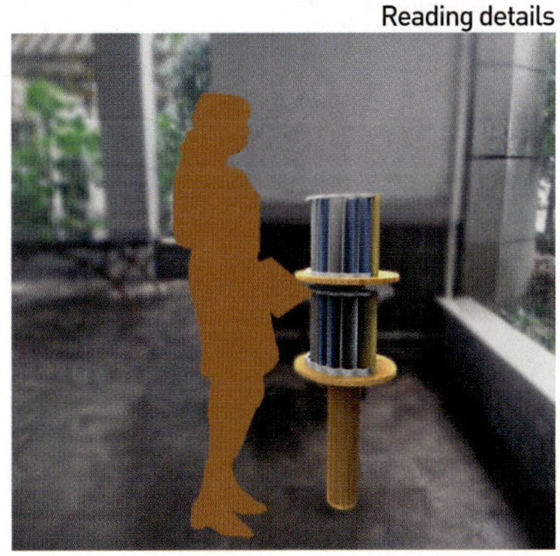

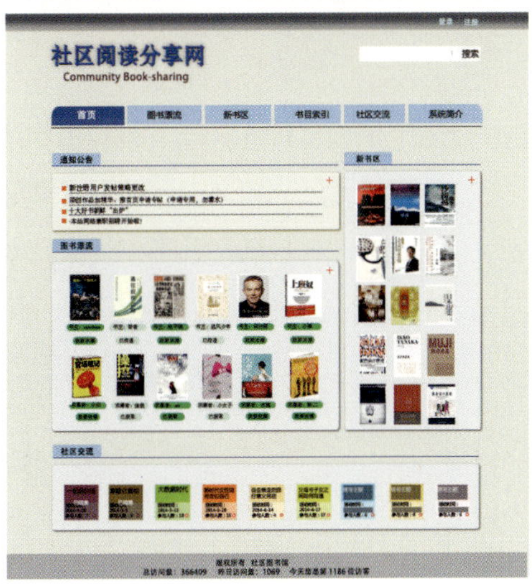

The web of book sharing for community

The Inspiration of the fan-chair

We named the chair for fan-chair because it's inspiration comes from the 360° fan, and it can be extended to angle you like.

Team members | Zou Yue, Cao Yuan, Yu Weiguang

foresh

FORESH is a cultural space design through innovative exhibition system design which modulelized by a range of sizes and forms.This exhibition system modules can achieve sustainable stage by locating itself in various kind of occasions in the public space.It is designed for a showcase of young designers, a promotion platform of sustainable product companies and a public cultural space of museums. Changsha citizens can enjoy cultural experience through the FORESH (Forest+Fresh)and have a feeling of refreshment by creative air.

Foresh는 다양한 크기와 형태로 모듈화 되어 공공 공간의 다양한 상황에 적용시킬 수 있는 혁신적인 전시 시스템 디자인을 통해 문화공간 디자인을 진행하는 프로젝트이다.
이 전시 시스템은 퍼블릭 스페이스에서 기업이나 미술관이 공공문화공간에 젊은 디자이너의 쇼케이스와 같은, 지속가능한 프러덕트를 전시하고 프로모션하기 위해 디자인된 것이다. 창사 시민들은 이 foresh (forest + fresh) 의 의미처럼 '창의적인 공기' 에 의해 환기되는 문화적인 경험을 즐기게 된다.

CHANGSHA BRIEF

Surrounded by the beautiful Yuelu Mountain where has 1,000 years of historical academy and the Xiangjiang river , Changsha, is one of important city in central China, is among the first group of Chinese cities which were selected as famous historical and cultural cities in the country and becoming to open-up to the outside world. With the rapid social and economic development, Changsha has laid the foundation for the local tourism boom. Changsha is now time to create some new public space which can express the unique culture code and show the creativity of city.

Problem

1.People are too busy to see some exhibitions.
2.Young Designers have no chance to promote their talent designs.
3.People have no time to go to museum.
4.Sustainable product company have little money to do a advertisement.
5.The city is lack of culture code.

Show a new world

Focus on

1.Remind the public of splendid culture.
2.Promote talent young designers and give them a right design direction in future design.
3.Recommend good green products to the public and help promote sustainable product companies popularize.
4.Make people easy to see a show anywhere.

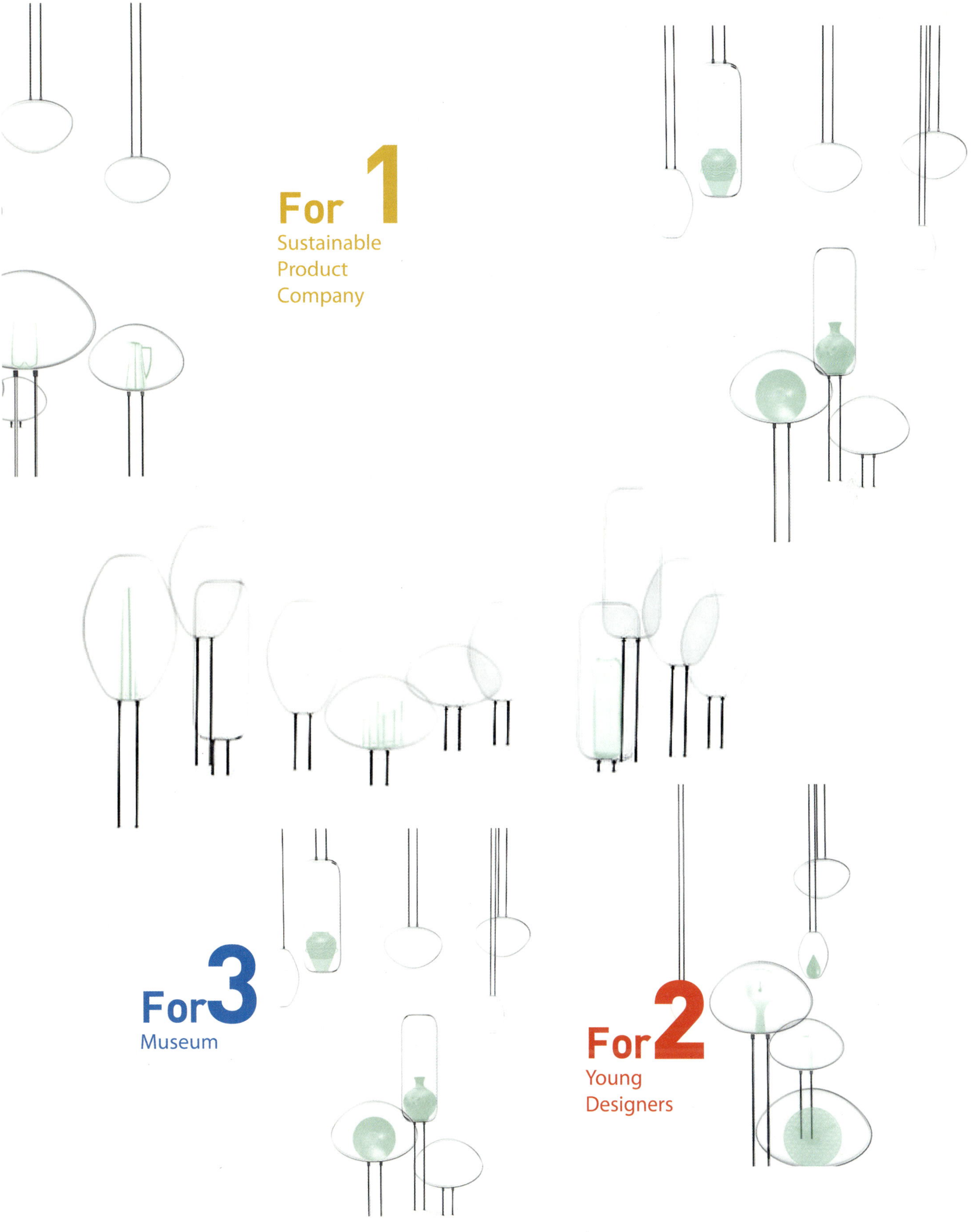

For 1
Sustainable Product Company

For 2
Young Designers

For 3
Museum

System map

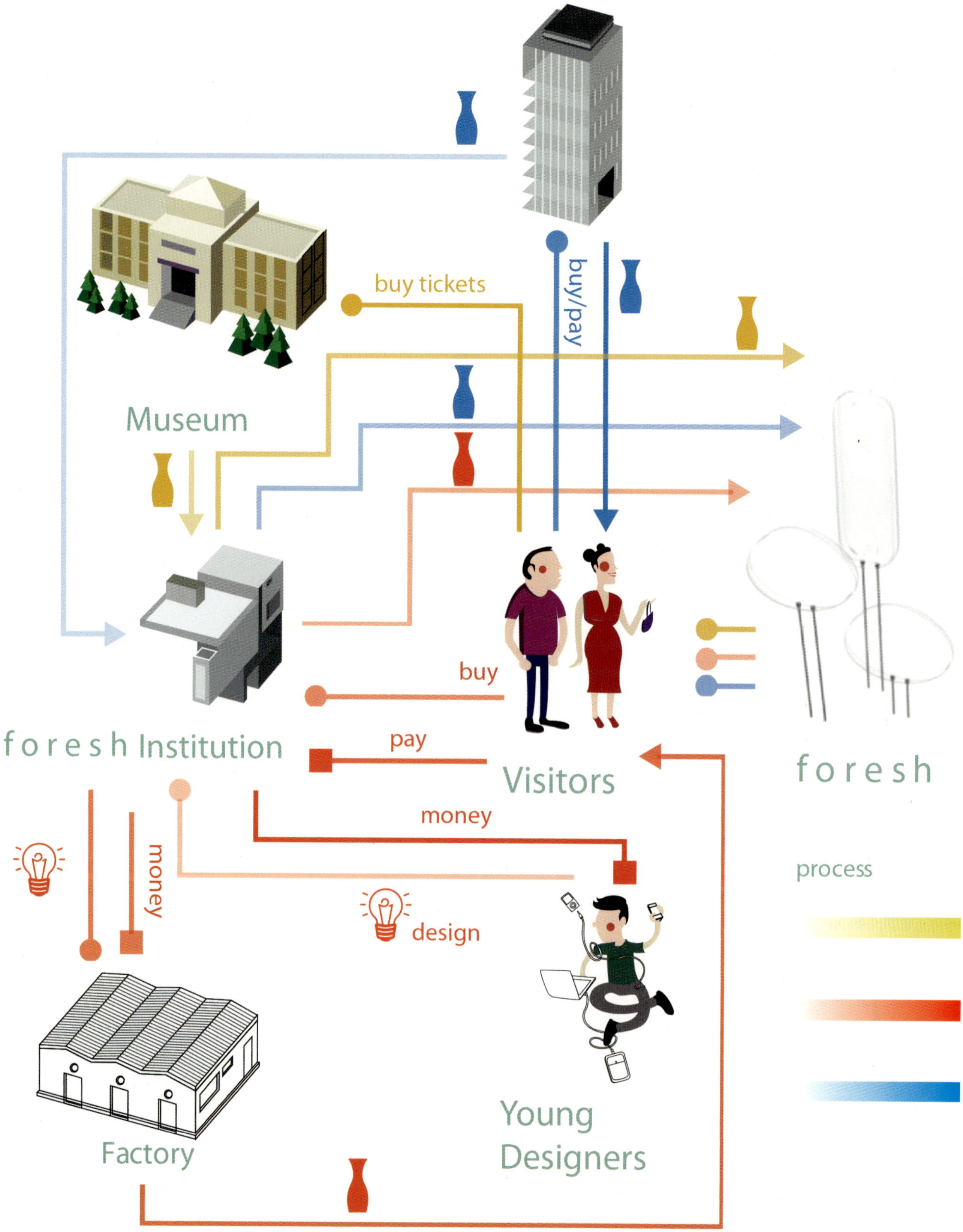
buy tickets
buy/pay
Museum
buy
pay
f o r e s h Institution
Visitors
f o r e s h
money
money
design
process
Young Designers
Factory

Sustainable structure

form | **founction**

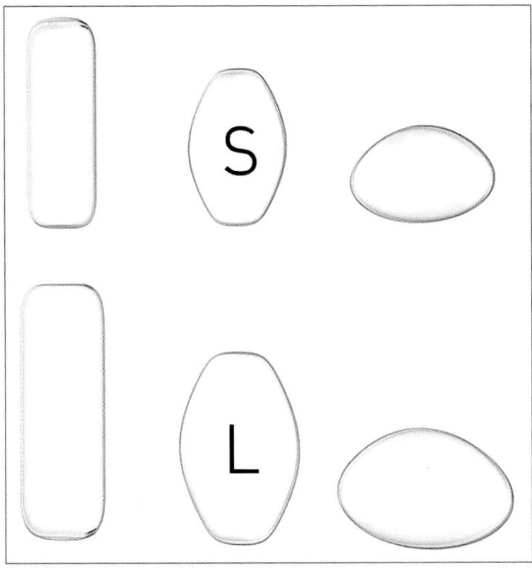

Different sizes can be used for different objects.

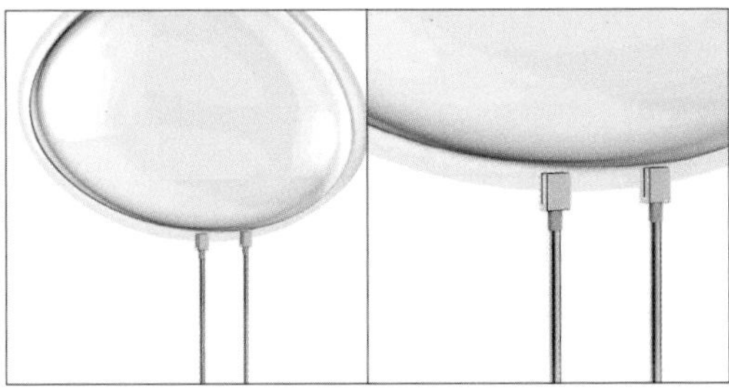

The smart clamp structure can link to any size container.

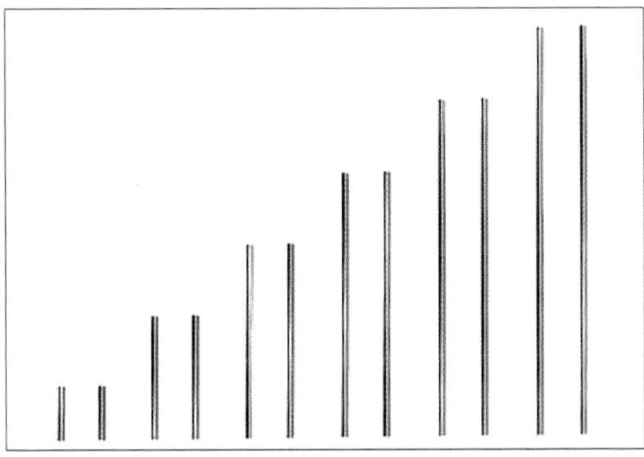

The supporting rod structure with different heights creates a fluctuation,which makes space designs more flexible.

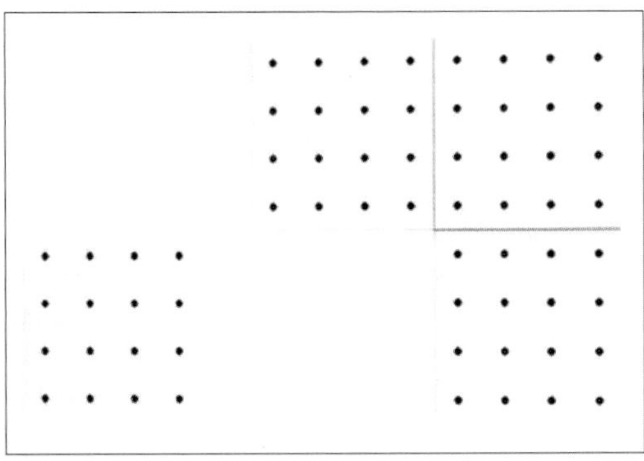

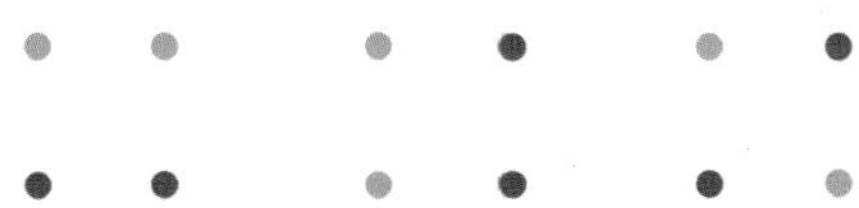

The base can be used in floor or ceiling, and the supporting rod can be positioned on the base in 3 ways.

Team members | Yi Huan, Xiao Anqi, Zhao Shikuan, Zhang Haolin

FANGB

FANGB : a sustanable waiting wi-fi zone

Introduction

Fangboo is a public space design for citizens.It attaches to the plants,especially the vines, they are both dependent on each other. On one side, it provides supports for the plants.On the other side, individuals can have a rest under the shady and enjoy the cool and the smell of natural. What's more, it provides free Wi-Fi and charging devices to convenient people's life. Also,sitting here will be a great enjoyable memory.

The whole design focuses on sustainability and engineering, leading to a very flexible structure and shape.

Background

Changsha is located in the central China and has a typical subtropical monsoon climate. Spring and autumn move fast, while the summer is not only long but also sizzling hot. People usually need shelters to protect themselves from sunshine.

Electronic products, like telephones, computers, ipad and so on, have gradually became a part of human being's daily. People need it to get contact with others and to have fun and so on. There is a need to get free charge and Wi-Fi, especially for tourists and visitors.

The camphor and vines is prevalent and popular in Changsha.

Fanboo 는 시민들을 위한 공공 스페이스 디자인 이다. 이것은 등나무 같은 덩굴식물을 위한 지지대 이면서 때로는 반대로 다른 나무에 의지하기도 하는 쉼터로서 이용자들에게 자연의 향기와 시원함을 제공한다. 더욱이 와이파이 존의 역할을 하기 때문에 즐거운 기억을 선사하게 될 것이다.

창사는 중국의 내륙남부에 위치해 있고 전형적인 아열대 몬순 지역이다. 봄가을은 짧고 여름은 길고 무덥기에 모두에게 그늘진 쉘터는 절실하다. 전화기나 컴퓨터 등의 기기들은 이제 사람들의 삶의 일부가 되었고 이를 통해 외부와 연결된다. 특히나 방문객과 여행객들에게는 충전장비와 Wi Fi 가 있는 곳이 도시 곳곳에 필요하다.

Use Scenes

Design Aesthetics

We try to create a emotional silence space against the bustling and busy city, using natural beauty to remind people the voice of soul and remember the essence of life. We got inspiration from the forest semantic and the elements from China.

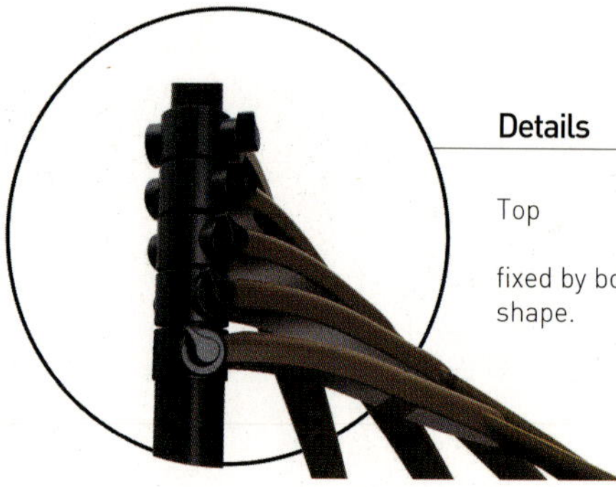

Details

Top

fixed by bolts. People can easily decide the shape.

Mental Joint

This part is movable joints, therefore the angle of the two pillars can change.

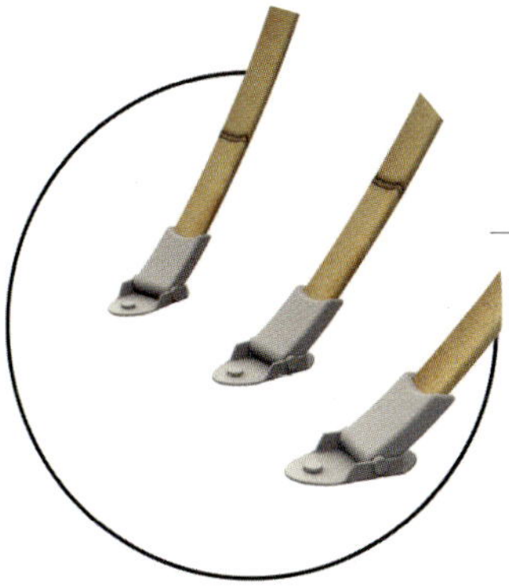

Legs

It adapts itself into two different materials -mud and concrete.

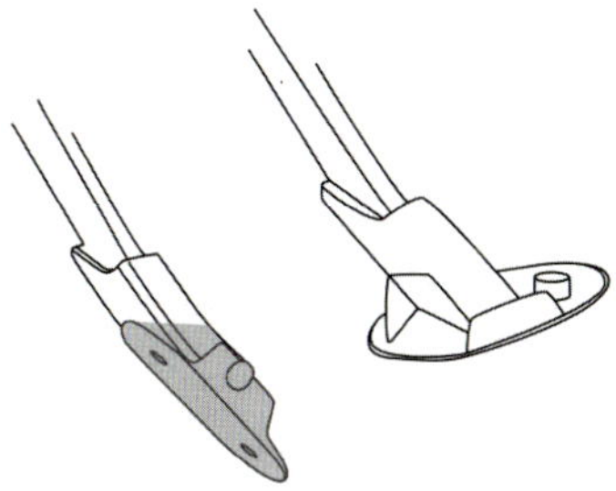

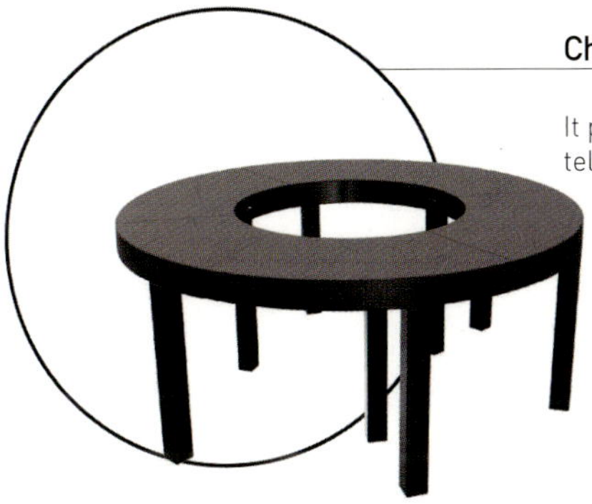

Chair

It provides individuals a place to have a rest. What's more important, your telephone can get free charge and access to wifi.

Users' Needs

After a long time sightseeing ,the visitors need a place to have a rest as well as do a charging.

Some fashion ladies just hate to bring the heavy things like charger when shopping.

Some people are addicted to the mobile phone, they play with it all the time.

Some workers have a tense schedule and go out in a hurry , so the battery is always insufficient for they usually forget the charger.

When waiting the superstar , the fans need a desirable Wifi support to share the exciting experience.

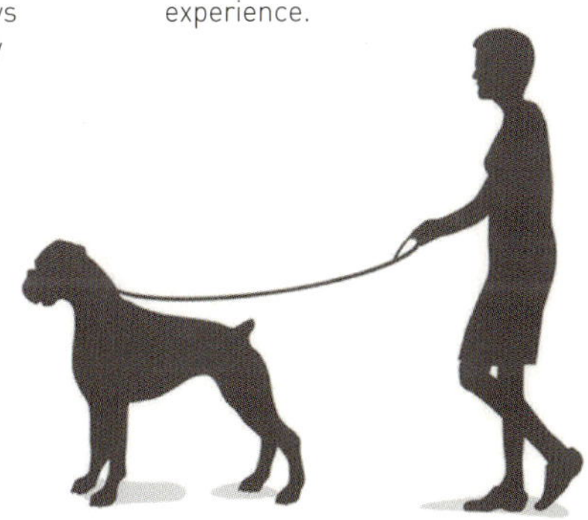

Variability & enginering & sustainalbility

The inspiration comes from the combination of hand fan. It has a basic and removable structure, which makes it possible to create variable sizes that fit with all different conditions and can be flexible-adjusted to serveral fractional parts. The convenient disassembly way makes it easy to transport and remove.

One of the 'green' uses of bamboo is as a building material for alternative energy installations. In our design, bamboo forms the most part of construction.

디자인의 영감은 접었다 펼칠 수 있는 손에 드는 부채에서 왔다. 이것은 상황에 다라 다양한 크기로 변화된 모습을 만들어 낼 수 있으며 몇 개의 작은 유닛으로 나뉘어져 운반과 조립, 조정이 용이하도록 구성되어 있다. 또한 친환경적인 지역에서 생산되는 재료인 대나무를 기본재료로 사용하여 조립이 용이하고 유연한 대체재로 사용하였다.

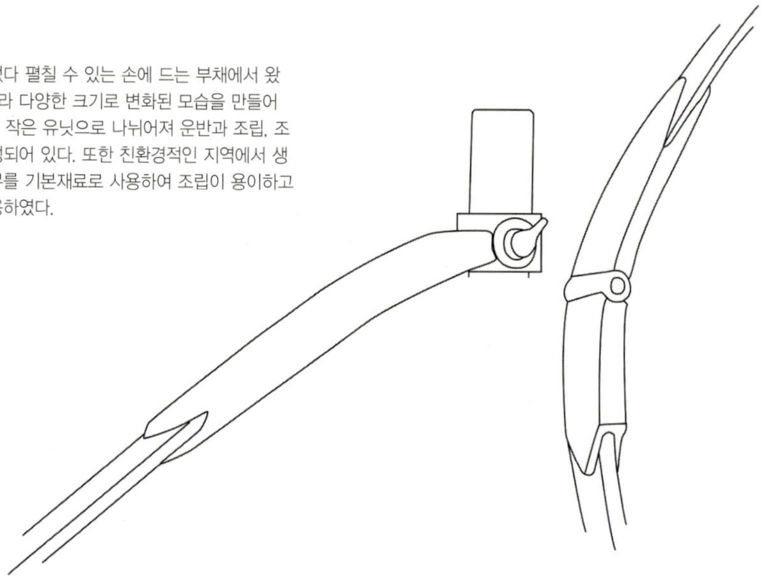

Variabe

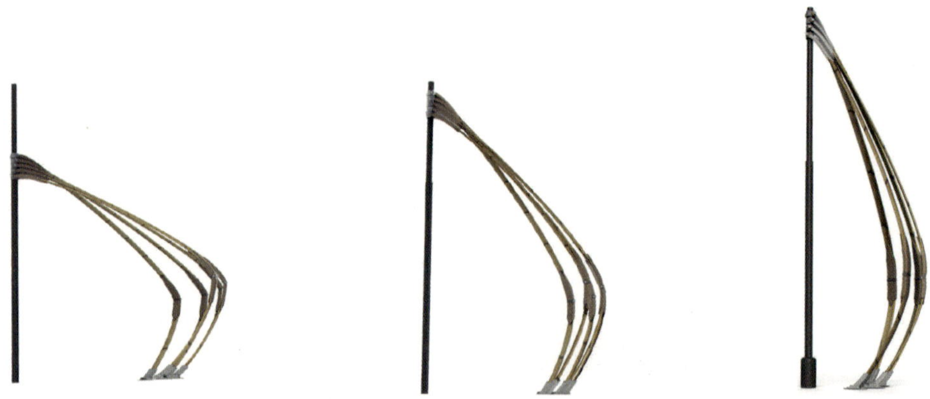

Flodable

Modularized

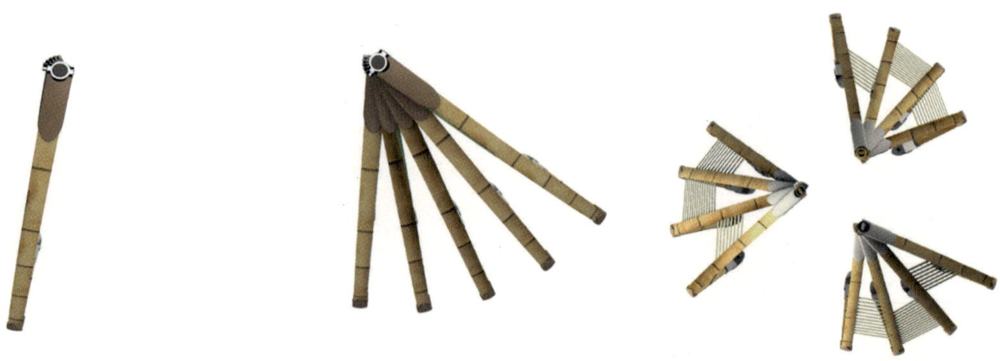

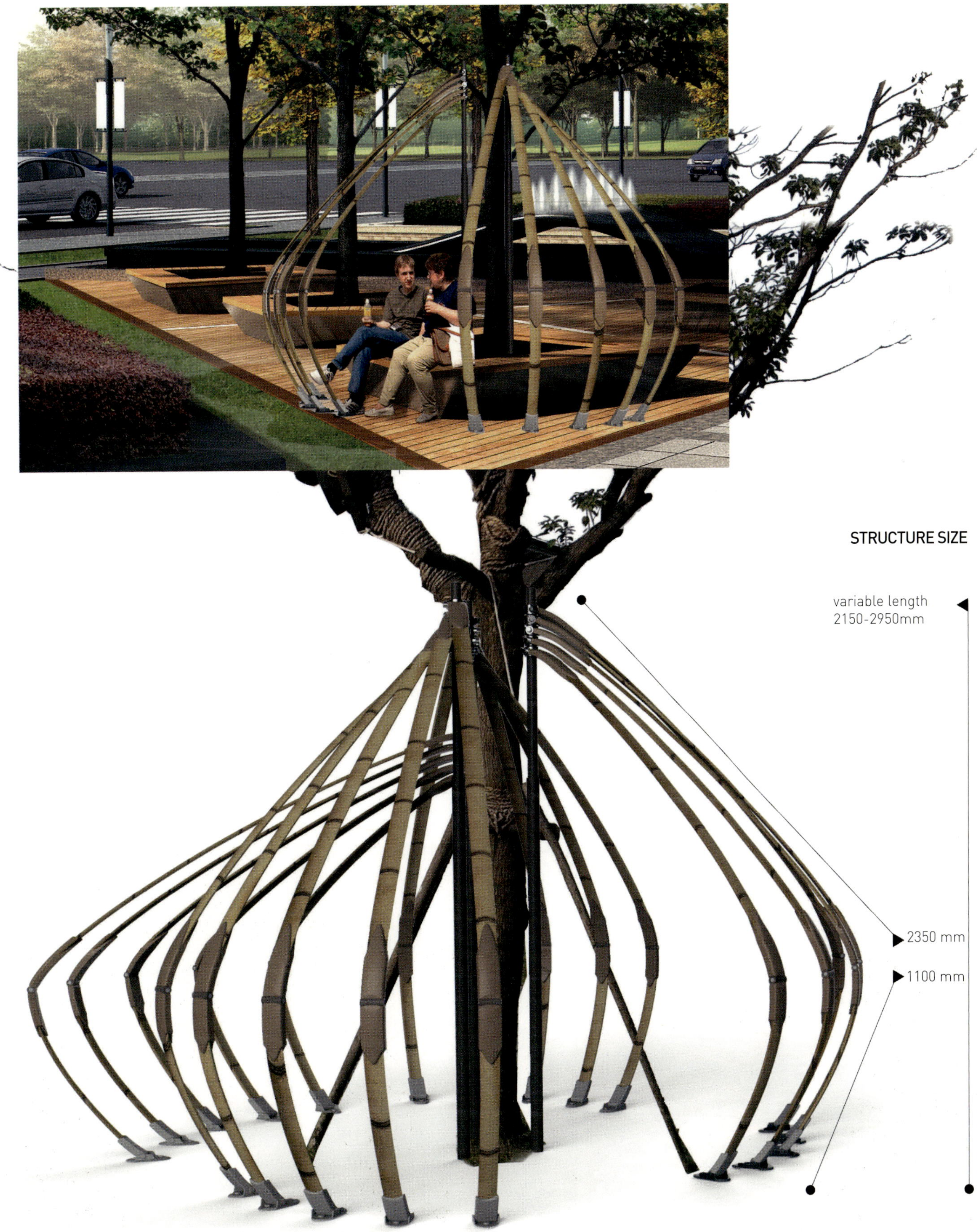

STRUCTURE SIZE

Enjoy Public Kitchen

Team members | Ding Kebi, Ma Kang, Zhou Linlang, Zheng Hongxia, Xu Jiayang

Background

China is a Country with splandid food culture. But nowadays, the young generation in China gragually lose the skill of cooking,because student in school have no space to cook.However, there are still some student who is wild about cooking, they choose to use the simple cooking device such as a electronic pot, but these things are not safe enough in the narrow dormitory room.

중국은 역사적으로 매우 다채로운 음식문화를 갖고 있는 나라이다. 그러나 오늘날 중국의 젊은 세대들은 조리공간이 부족하고 이로 인해 요리에 대한 관심과 그 전통적인 기술 등이 점점 희미해 지고 있는 실정이다. 요리에 관심이 좀 있는 학생들도 간편하지만 사고의 위험이 내포된 조리기구를 사용하거나 학생들끼리 모여서 식사를 하는 경우도 요리법을 모르기 때문에 외식을 즐기고 있다. 우리는 이러한 문제점에서 출발하여 가장 즐겨먹는 중국의 음식을 조사하고 그 음식의 조리방식을 패턴화하여 간편하게 요리를 즐길 수 있는 공동 부엌 겸 식당을 젊은이들이 거주하는 공동건물에 설계하고 이를 나누어 공동으로 사용하는 시스템을 디자인하고자 한다.

The simple cooking devices which are used by student in domitory room.

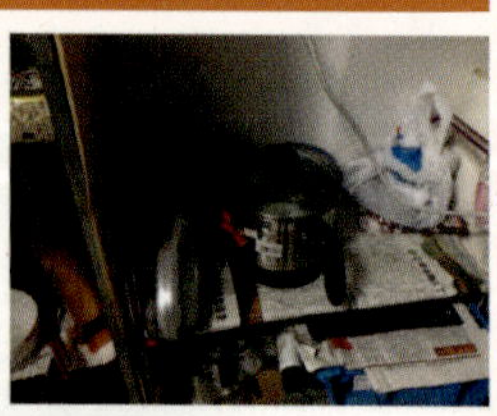

When dine together, the young generation in China seldom cook themselves but always eat in a restaurant, some young people even don't know how to cook.

Field Work & Research Process

Research 1 | User Interview about the Dinner Time

1

Lin

Q:What do you have dinner normally?
A:17:00
Q:What time is your earliest and latest dinner time in last month?
A:16:00 and 22:00
Q:When do you think is the most appropriate dinner time?
A:19:00

2

Hao

Q:What do you have dinner normally?
A:16:30
Q:What time is your earliest and latest dinner time in last month?
A:16:30 and 24:00
Q:When do you think is the most appropriate dinner time?
A:16:30

3

F.B

Q:What do you have dinner normally?
A:18:00
Q:What time is your earliest and latest dinner time in last month?
A:17:00 and 20:00
Q:When do you think is the most appropriate dinner time?
A:18:00 and 20:00

4

Wang.KQ

Q:What do you have dinner normally?
A:17:00
Q:What time is your earliest and latest dinner time in last month?
A:17:00 and 22:00
Q:When do you think is the most appropriate dinner time?
A:17:00 and 20:00

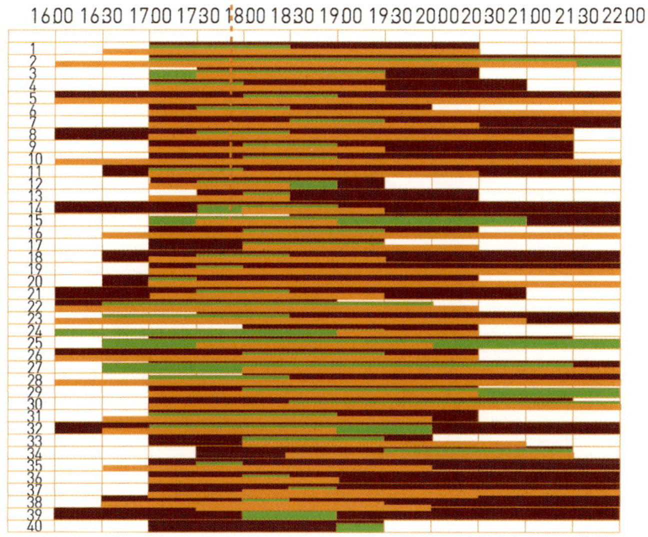

Research 3 | Kitchens around Hunan University Campus

Research Place: the Family Restaurant in HNU Community Interviewee: the chef in the Canteen

Q:What do you use the kitchen every day?
A:11:00-14:30 and 16:30-1900
Q:Could student come and use your kitchen when you are not using the kitchen and pay you some money?
A:No,it's troblesome to clean up the kitchen.
Q:What if the students clean up the kitchen after they use,and there is someone supervise them?
A:That is ok
Q:So they can come between 14:30-1600 and 19:00-22:00?
A:Yes.

Research Place: the 6th Canteen of Hunan University Interviewee: the chef in the Canteen

Q:What do you use the kitchen every day?
A:Around each mealtime,6:00-8:00,11:30-12:30,17:00-19:00
Q:Could student come and use your kitchen when you are not using the kitchen and pay you some money?
A:No,anyone without a health certificate for catering industry can not get into our kitchen,but you an come and make dumoplings here,and we can help you to boil them
Q:When can we come and make dumplings?
A:Any time between 6:00-21:00

Research Place: "FEB"Apartment Kitchen

Add: Building2,Flat2,301,Tianma Community Business
Hours: 12:00-16:00, 16:00-20:00,20:00-24:00
Free Scale:20 yuan/Person/Time

Q:Do you always have many customers?
A:Not always,only a few people know this place,most of them are students.

Research 3 | about Disposal of Leftovers

Interview Place:Food Area at 1F of Jiarunduo Shopping Mall Interviewee : An Employee of the Food Area

Q:Is there anyone who collecting these leftovers?
A:Yes.
Q:What do they collect these for?
A:Feeding pigs.
Q:Do they pay for these leftover?
A:Yes,800 yuan per month.

Interview Place:the Sixth Canteen of HNU Interviewee : An Employee of the Lianhe Leftover Recycling Company

Q:What do you collect these leftover for?
A:We use these thing to produce biologic diesel.
Q:How ofter do you come?
A:Twice a day, the noon and the evening.

Design Brief

Enjoy Public Kitchen System is a system which supplies kitchen service for student, the system content the cooking needs of student by build a kitchen on the empty space of dormitory roof and supplying the service of booking the idle(unused) kitchen around the campus, meanwhile, it increase the usage of the kitchen around the campus.Besides, New Public Kitchen System will reduce the transprotation by build a consolidated food material,leftovers,fertilizer transportation system.And the energy consuming can also be reduce

Benifits

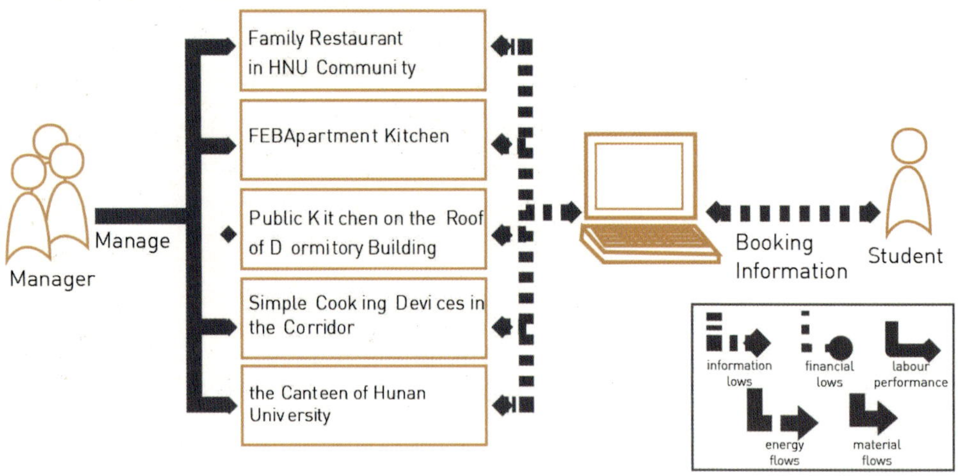

Student can choose to use the kitchens(include the kitchens around campus)online

Comparison with the old system

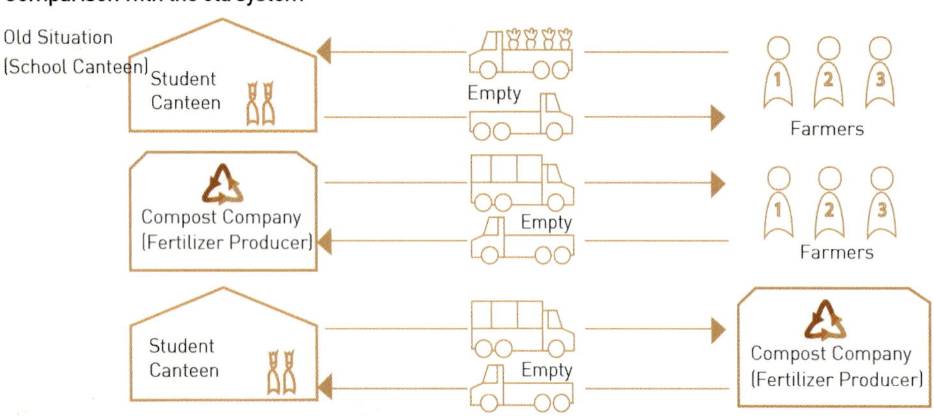

3 Trucks, Halftime Empty, Wasting of Transportation

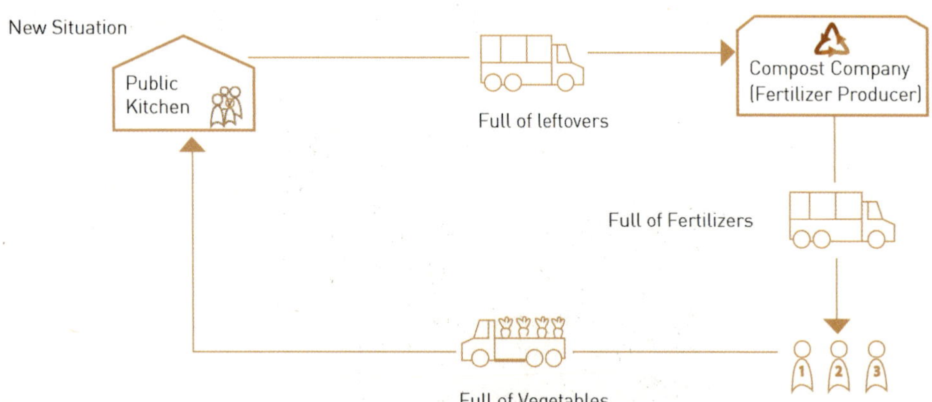

1 Trucks, Always Full, Saving of Transportation

Sketches

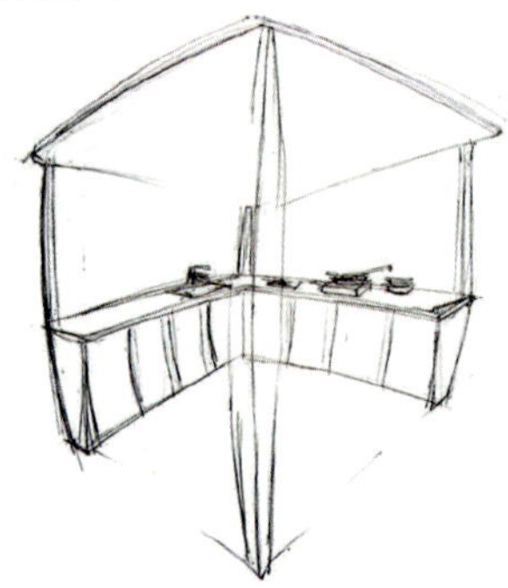

a Single Public Kitchen on the Roof

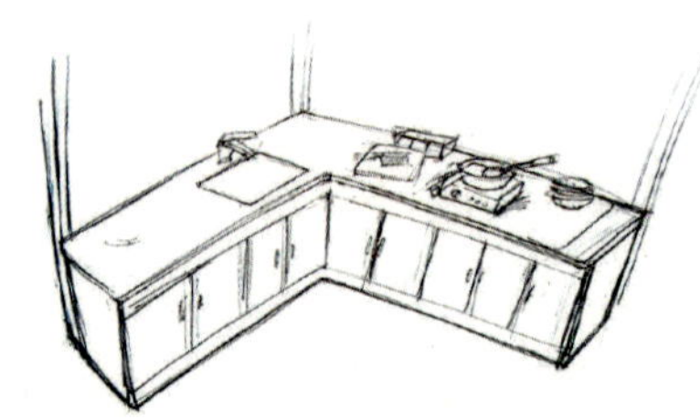

inside the Public Kitchen on the Roof

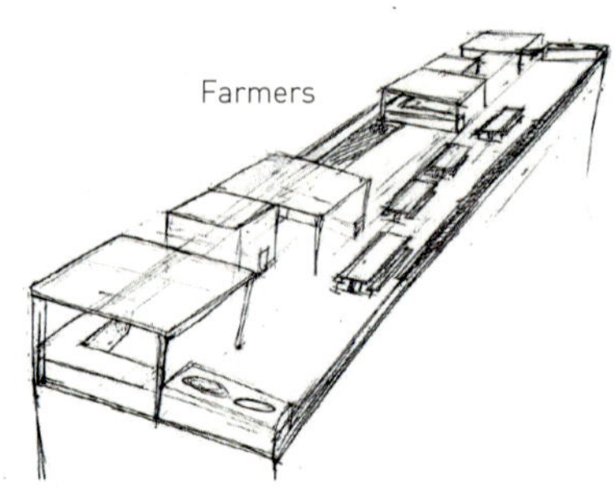

the Whole Roof Space

the Simple Cooking Devices in theCorridor of Student Dormitory

The Scenarios of the Transprotation of Food Materials, Leftovers and Fertilizers.

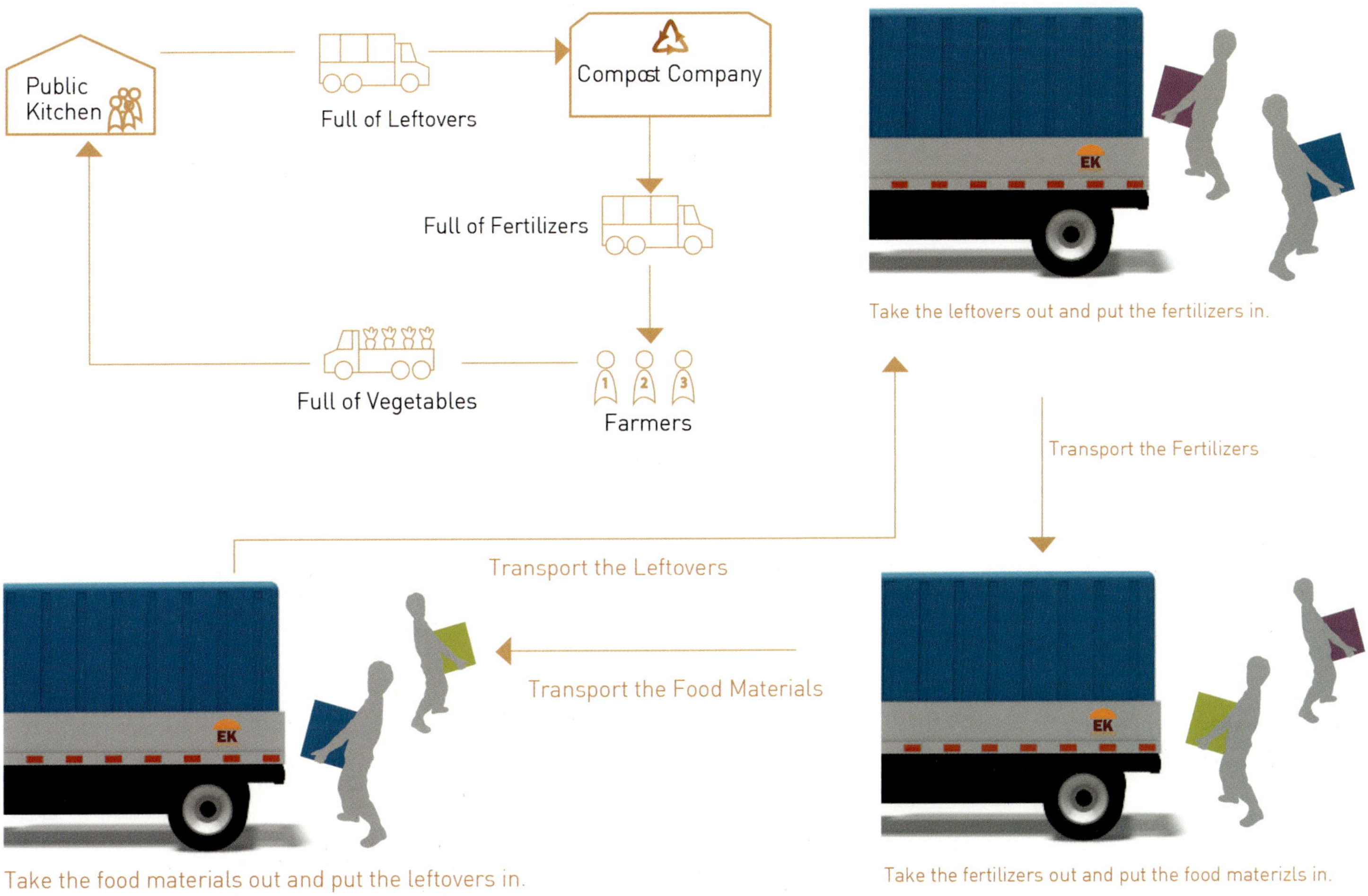

Plan & Size

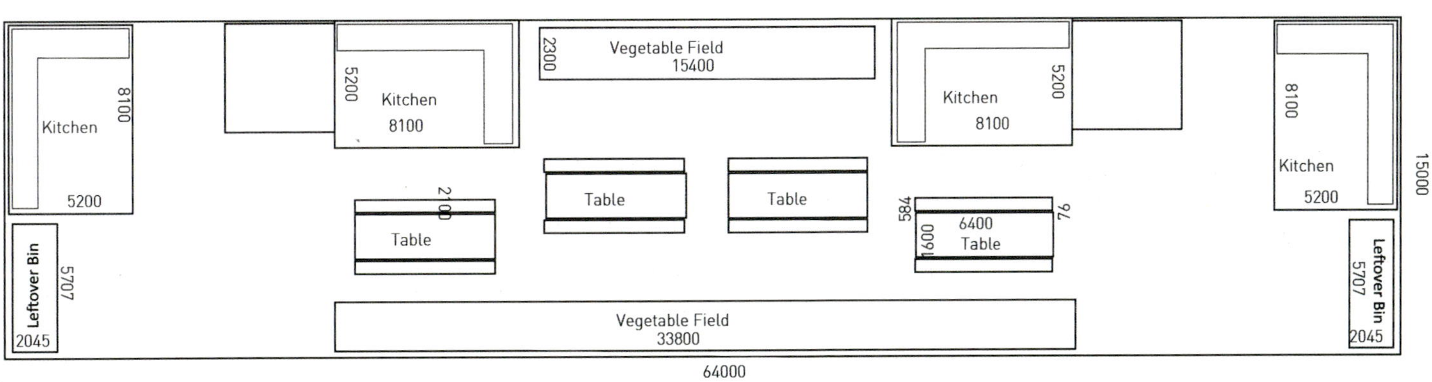

The Scenario of the Whole Roof

User can have different choice

Student can cook together with their friend in the kitchen on the roof and dine together there.

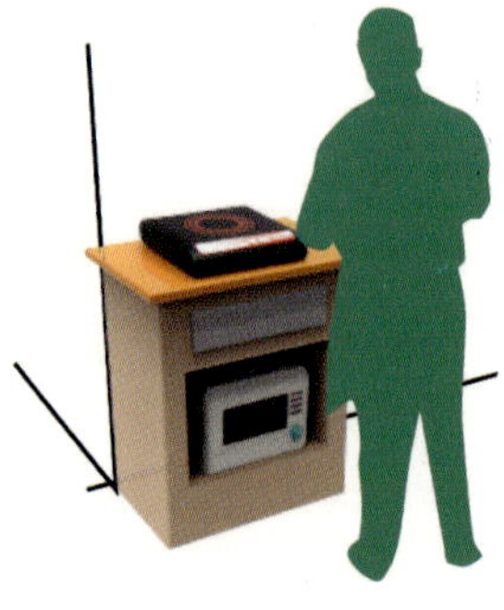

Or use the simple cooking devices in the corridor of dormitary building.

Or book a idle kitchen around the HNU campus

The Scenario of Enjoying the foods on the roof

The Scenario of Enjoying the foods on the roof

Cool summer

System Design for Sustainable Urban Space & Product Service System in Changsha

Team members | Tan Butian, Ni Qiji, Gu Sujun, Yang Xiangjie

Average temperature in July of Changsha

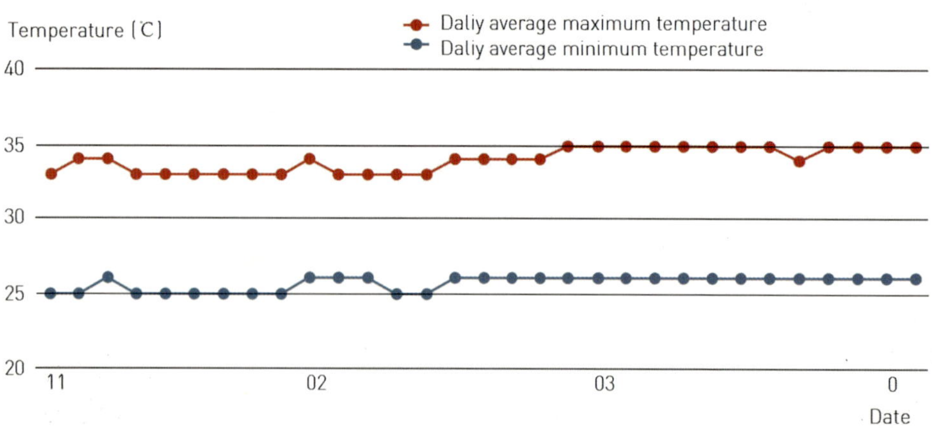

The summer of Changsha city is really very hot.Daliy average maximum temperature is 34 C in July. It will be even hotter in August.

There are few public relax places can make people feel cool in summer of Changsha.Nobody want to walk or play with friends outside. People stay at home and use air conditions to lower the temperature. But it can cause some air conditioning diseases.

Background

창사의 여름은 7월의 평균기온이 섭씨34도에 육박할 정도로 무더운 날씨이다. 8월의 기온은 7월보다 더 높은 기온을 기록한다. 그래서인지 창사에서는 더위를 피할 수 있는 외부의 휴식 공간이 많지 않고 사람들은 여름에 야외활동을 꺼려한다. 더위를 피하기 위한 에어컨의 사용빈도가 높고 냉방병을 유발할 수 있다. 본 프로젝트는 창사에서 가장 이용도가 높은 강의 동쪽 강변을 배경으로 하여 여름의 야외활동이 가능한 파빌리온 휴식공간 디자인을 진행하였다. 각 파빌리온은 약간의 냉방과 공기정화기능을 가진 공기순환시스템을 갖추어 주민들의 휴식을 돕는 공간이 된다.

Design research

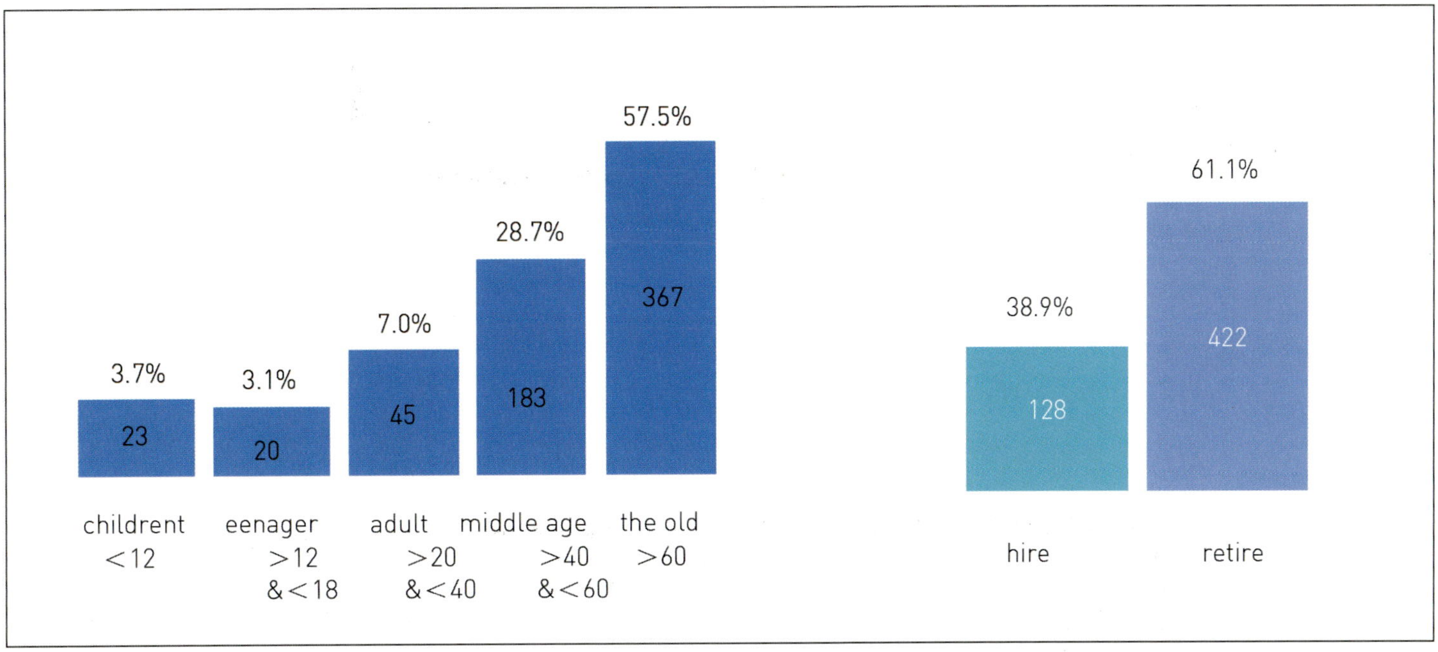

1.Enough free time.
3.Out of condition, need practice.
5.Enjoying exercise in the morning.

2.Most friends are the elderly, meeting place always in public place.
4.Having many interests, they all can find their hobbies in public relax places.
6.Lack of entertainments at home.(e.g. They are not good at playing computer)

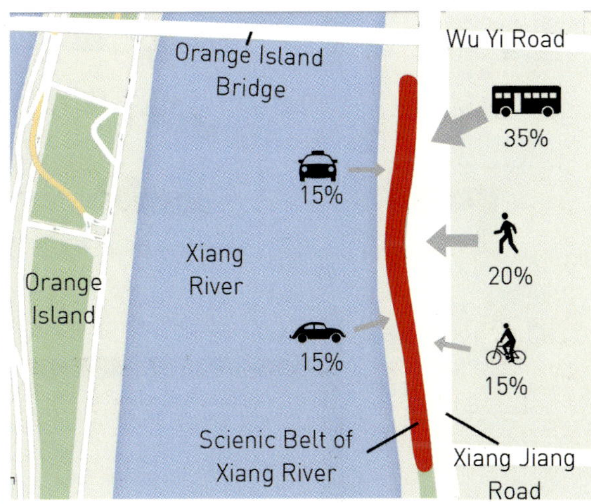

Scienic Belt of Xiang River is next to Wu Yi Road and Orange Island Bridge. Most people go there by bus. There is a very good scenery. People willing go there and relax themselves.

Design development

Water ciculation system

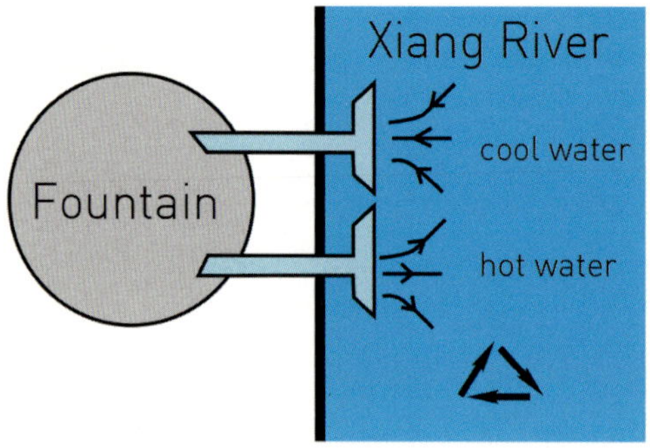

This is water circulation system.The system take use of the cool water in the Xiang Riverto lower the temperature. It can recycle and no pollution.

Map about stakerholders

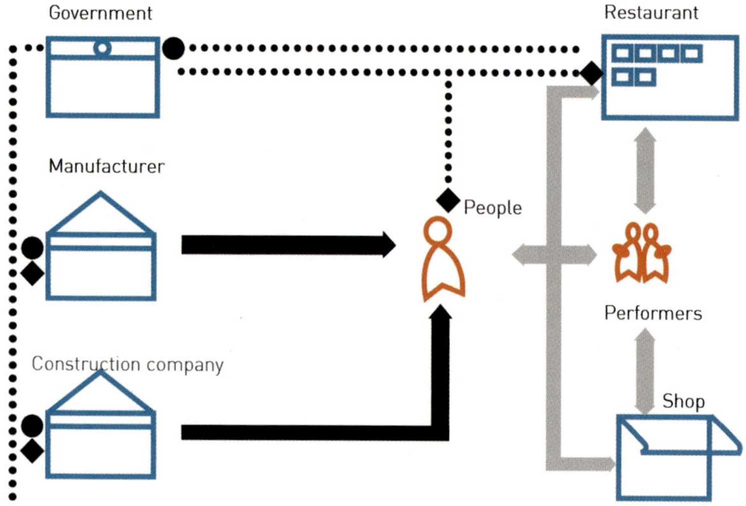

System Map

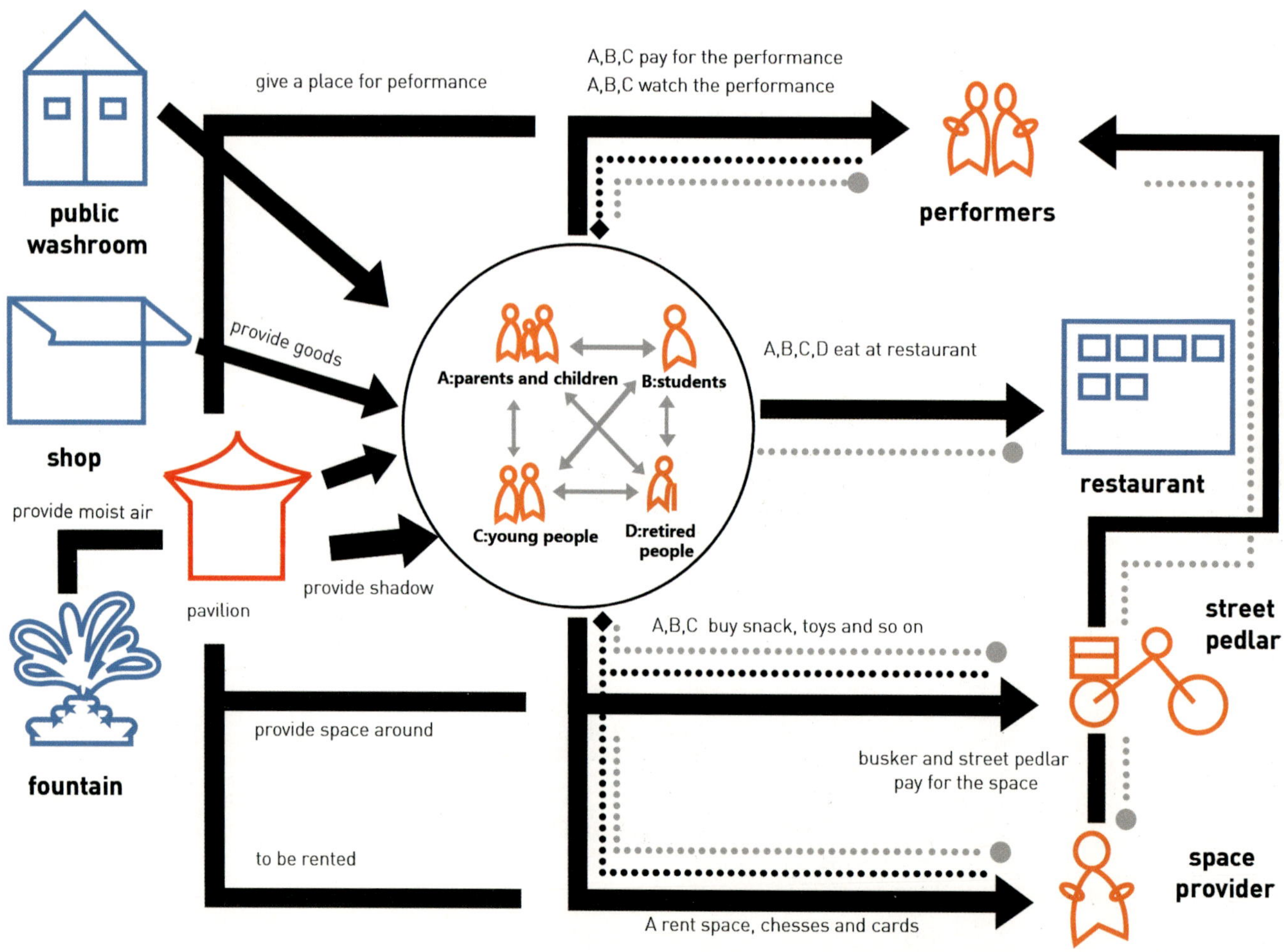

Design detail

Sketch

Final design

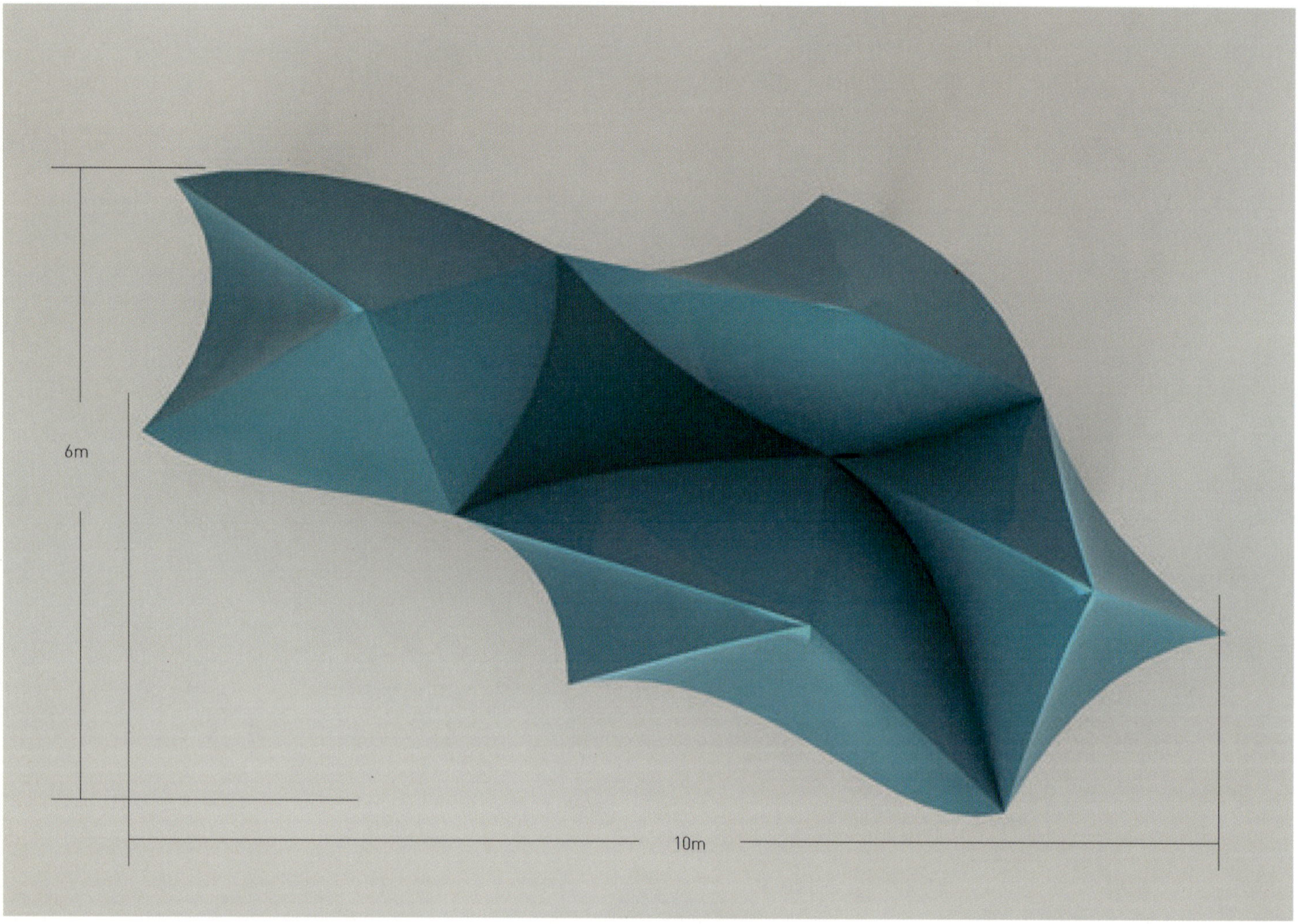

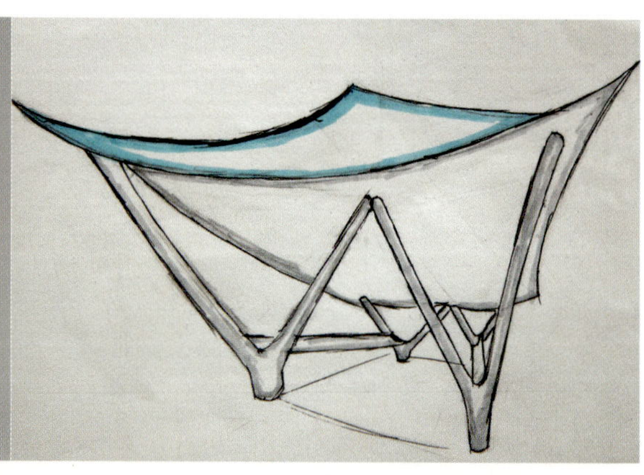

Images

Storyboard

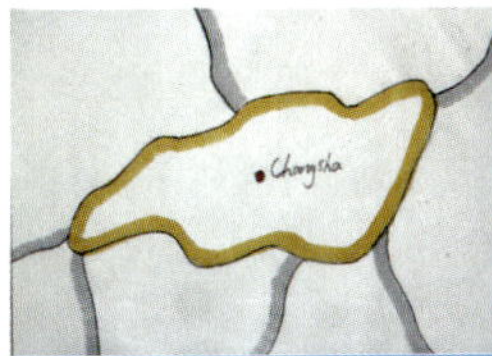

24th

August

2014

This is Changsha city where we live.

11:00am

34 ℃

62%

Xia decide to go to the pavilion to cool himself and watch others playing chess.Because there is a bladeless fan which can provide the cool wind.

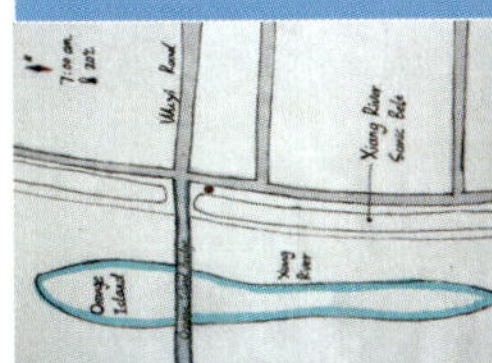

Xia is a retired people .He likes walking along the Scenic Belt of Xiang River.

13:00pm

35 ℃

62%

Xia is thirst ,so he buy some water from the street peddler.

8:00am

23 ℃

69%

Xia is a retired people .He likes walking along the Scenic Belt of Xiang River.

14:00pm

37 ℃

62%

Xia returns to the pavilion and sits on the bench. He is drinking water and watching people playing around the fountain while the bladeless fan blow the cool wind. Xia feels cool and enjoyable.

9:00am

26 ℃

65%

The sun rise, some people walk to the pavilions Thewater in Xiang River make the air moist.

15:00pm

35 ℃

62%

There are plenty of toilets. Peoplecan use the toilet conveniently if they need.

10:00am

30 ℃

64%

Xia finds a singer is singing. So he stops and listens.There are a lot of people listening to the song too.

16:00pm

30 ℃

63%

Xia takes the bus to go home.The moist air from the Xiang River and fountains.
The shadow which provided by trees and pavilions.The Scenic Belt of Xiang River become nice and cool.

Background

About Changsha :

1.Changsha is famous for its entertainment programme (including Hunan TV) especially the singing talent show which will be held every year in order to provide a stage fot the people who love singing but lack of the chance to be heard. It attracts singers from the whole country and keeps high audience rating.

2.Every year, there is Orange Music Festival in Orange Island of Changsha city. It lasts three days and gathers hundreds and, sometimes thousands of music lovers. It is the most exciting outdoor music party in south of China.

창사는 후난TV를 필두로 여러 문화와 엔터테이먼트로 유명한 도시이다. 또한 매년 도시중심의 강에 위치한 오렌지 섬에서 매년 3일간 개최되는 '오렌지 음악 축제' 는 수 천명이 넘는 음악 애호가들을 이 도시로 모이게 하는 중국 남부에서 가장 유명한 야외음악축제 가운데 하나이다. 이번 프로젝트는 이러한 창사의 문화적 환경에 부흥하여 거리음악가들을 프로모션 할 수 있는 공공 공간을 디자인하는 것이다. 각각의 거리음악으로 유명한 지점에서 거리의 음악가들의 영상을 다른 공공공간으로 송출하는 디지털 영상 시스템과 연결된 음악공연장 겸 휴게 공간을 디자인하려 하며 이는 거리 음악의 네트워킹 시스템으로 on-line 과 off-line 모두를 이어준다

Design Brief

Objectives :

It is to explore the new type of field of public space design in the social and cultural context.

We want to offer a stage to people who
- Love music
- Be willing to perform and show their talent in public
- Lack of professional equipment
- Lack of a good chance to be famous

We want to offer a place to the public where they can
- Stay and enjoy the time with music
- Share their feeling of the street music
- Feel the atmosphere of music festival everyday

Team members | Zheng Yuling, Zhang Shiqi, Tan Butian, Gu Sujun

Street Music Sharing System

It's not only a stage for street singer to show their talent but also a new kind of live show from which the whole city can feel the same music at one time

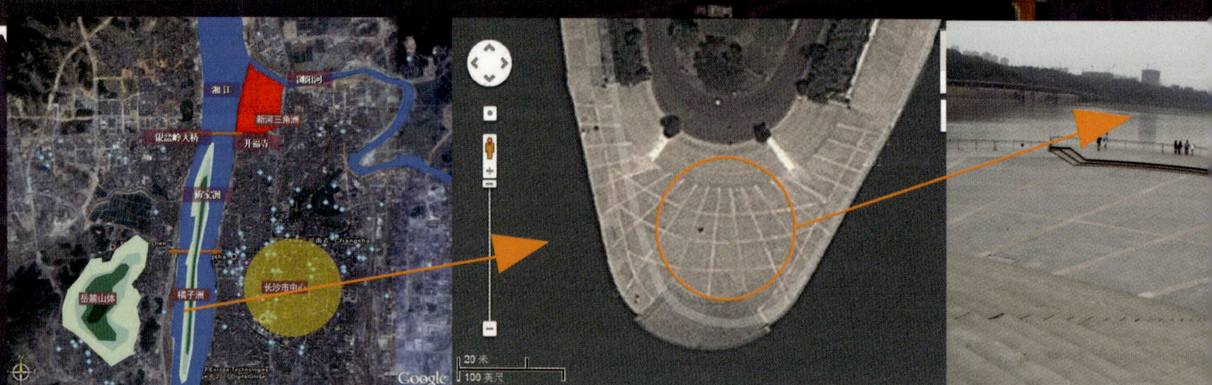

Research about the location ● **Location for central stage**

The Orange Island in Changsha city:

Research about the People

We randomly interviewed 100 people on the street of Changsha.

1. How do you feel about the street singers?

6.3% feel annoyed
22.7% enjoy
71% it depends

2. What kind of way do you enjoy in listening music?

11.7% listening alone
6.3% listening with other people
82% both

Concept

This is a music sharing system in Changsha city. With one centre stage and many holographic projection sites, it creates a new kind of live show from which the whole city can feel the same music at one time. It provides not only a stage for street singers to show their talent but also a music party everyday.

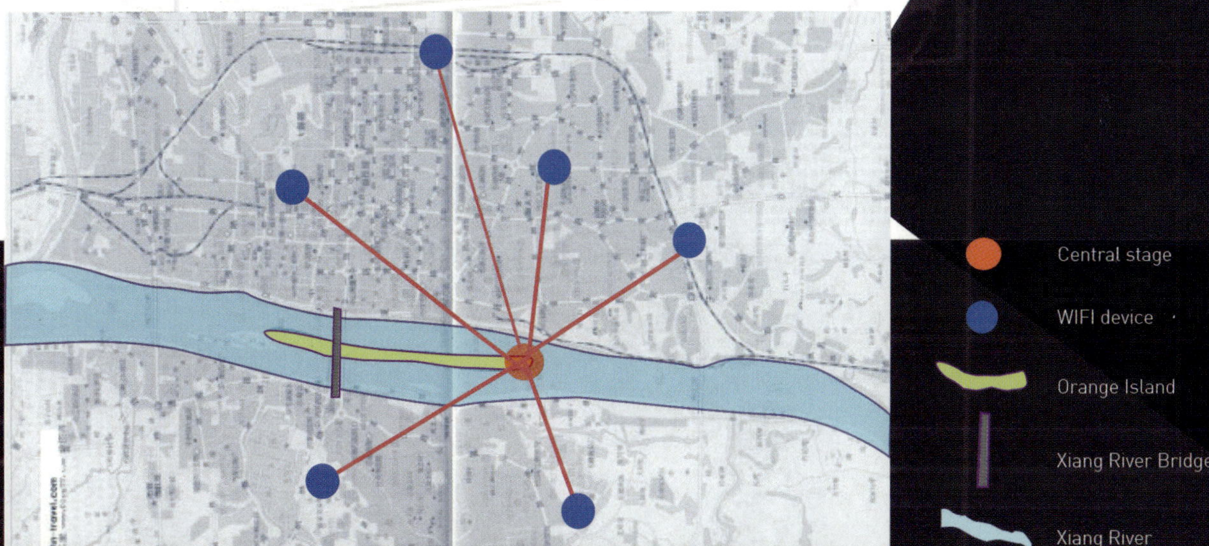

Site Candidate

1.Many people pass by these places everyday.
2.It's common that many street singers choose performing at these places now.

Walking street

Riverside road

Tourist attractions

Design

Central Stage | Layout

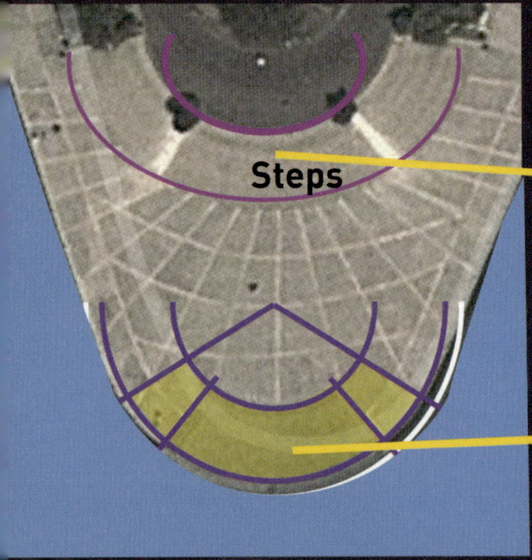

Central Stage | Size

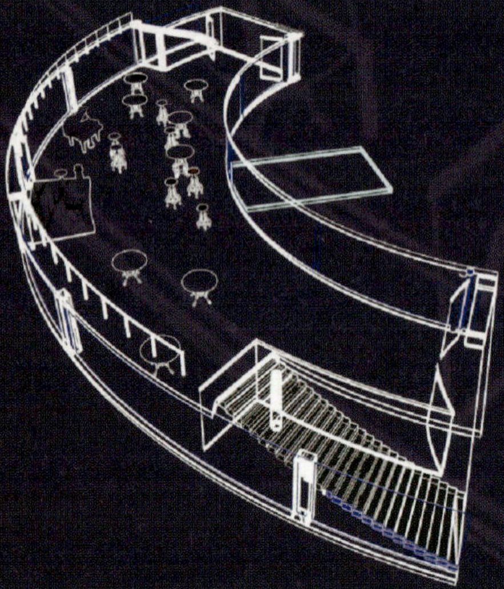

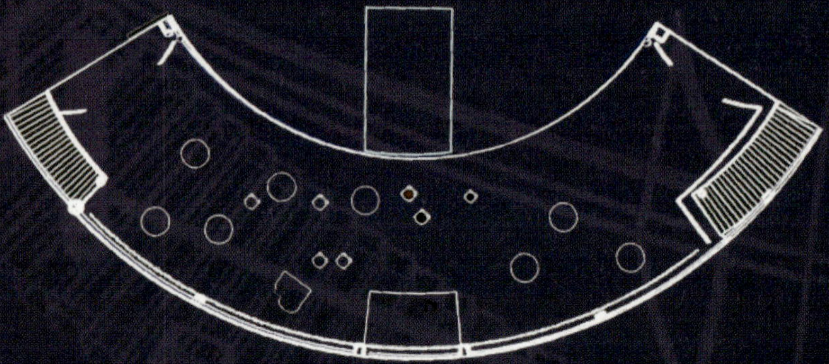

Central Stage | Area by Function

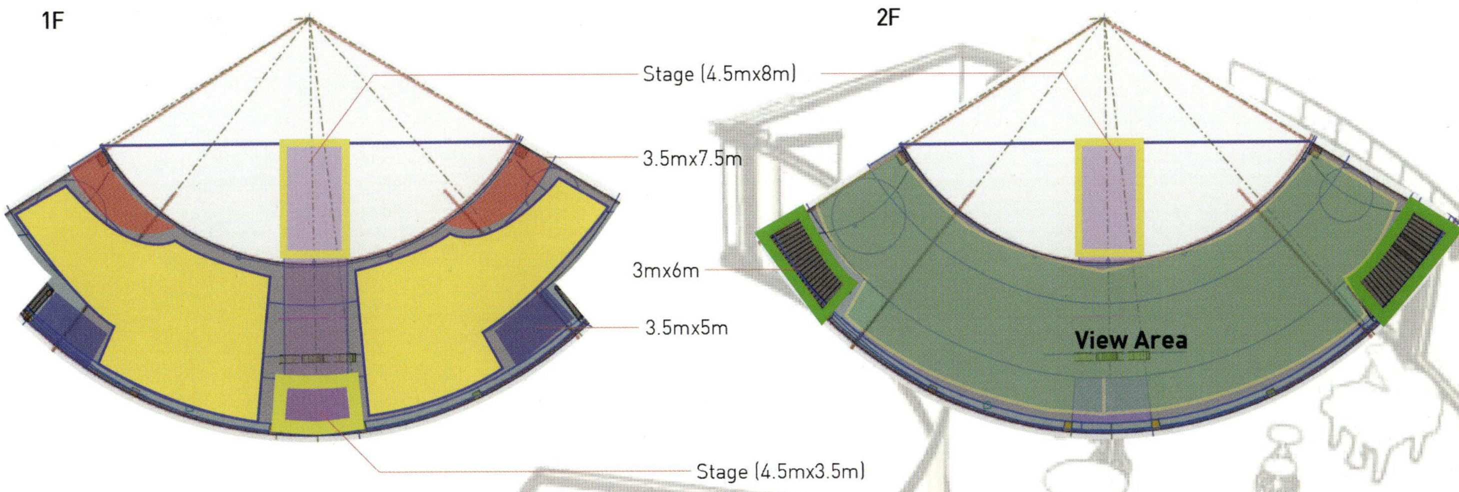

Stage: two stage, one is inside and the other is outside
Rest area: People can stay there drinking, chatting, relaxing, enjoying the music
Service station: Offering some drink like coffee, beer, juice (the income will support the fee of running the system)
Equipment room: One officer here will supervise the stage and system

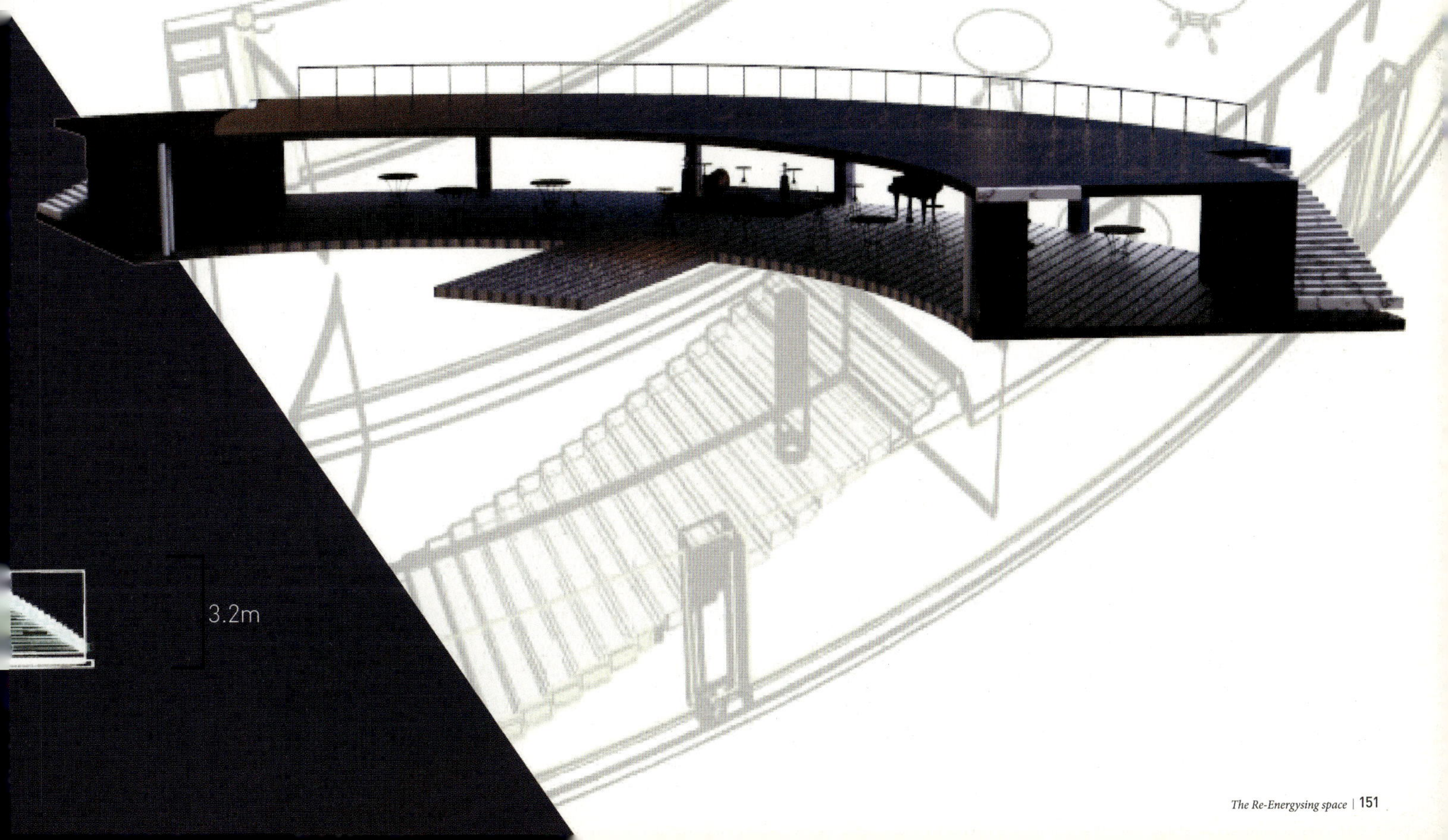

Central Stage | Show

Overview

Stairs

Glass wall

Live show everywhere

Final Design | Music System and Space for Street Music Culture

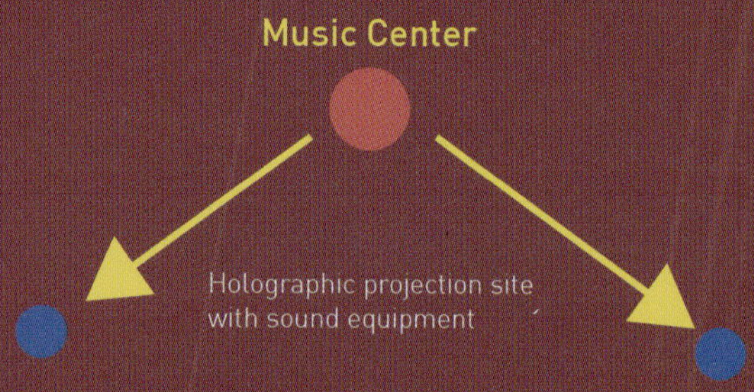

Final Design | Adaptability

Stage inside

When the weather is not good enough to stay outside, we use the inside stage.
The glass door will be shut down. People can stay in the rest area enjoying still without interference. It's also a place for the public to stay when storm comes suddenly or the weather is too hot.

Stage outside

When it's sunny, the outside stage is used and glass door will be open.
People can stay in the rest area enjoying themselves with beautiful scenery.
People also can stay on the stairs watching the performance.

Support | APP

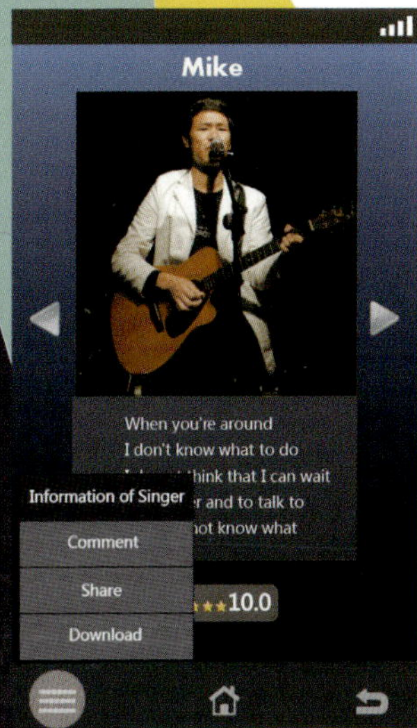

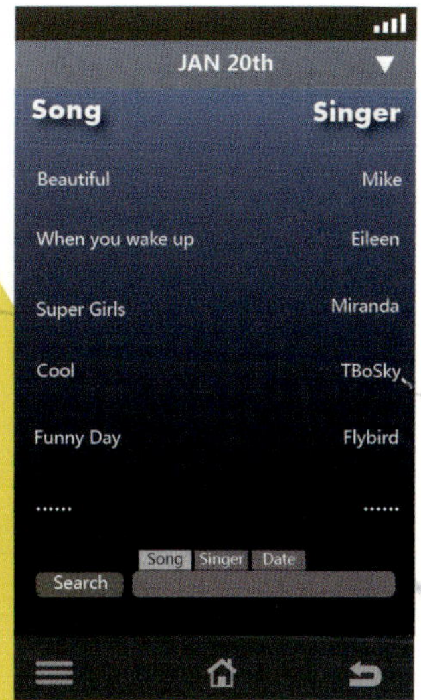

e can review and download the past songs
r it.
can comment and give favour to the
s they like.
n share the good songs to their friends.
tion of singers can be shown.
an earn a profit through this.

Design | Benefit

For music lover:
1.Offer a stage with professional equipment
2.A good chance to show their talent
3.Have nice audience atmosphere
4.Have a place to share and communicate with

For the public
1.Have a wonderful place to relax and enjoy music
2.A new kind of way to enjoy music when people are hanging out
3.Can feel the music festival every day

Fresh Line : Mobile Market in ChangSha

Team members | Liu Yue, Hu Xiaoyun, Cai Zhiwen, Wang Yidan

In China Changsha, most people prefer to buy vegetables in markets, there are many farmers who own some land will harvest their vegetables to the market to sell, some vendors will go to the wholesale market to wholesale vegetables themselves to the market for sale. However vendors cannot obtain the maximum benefit because of the simple and crude tools and transportation.

중국 창사에서는 대부분의 사람들이 채소를 시장에서 사는 것을 선호한다. 시장에는 농부들이 직접기른 채소를 거두어서 도매 혹은 소매로 내다 팔고 있지만 상인들은 근대화되지 않은 도구들과 교통수단 등의 여러문제로 많은 이익을 내지는 못한다.

Photogaraph From Wangjiawan Market

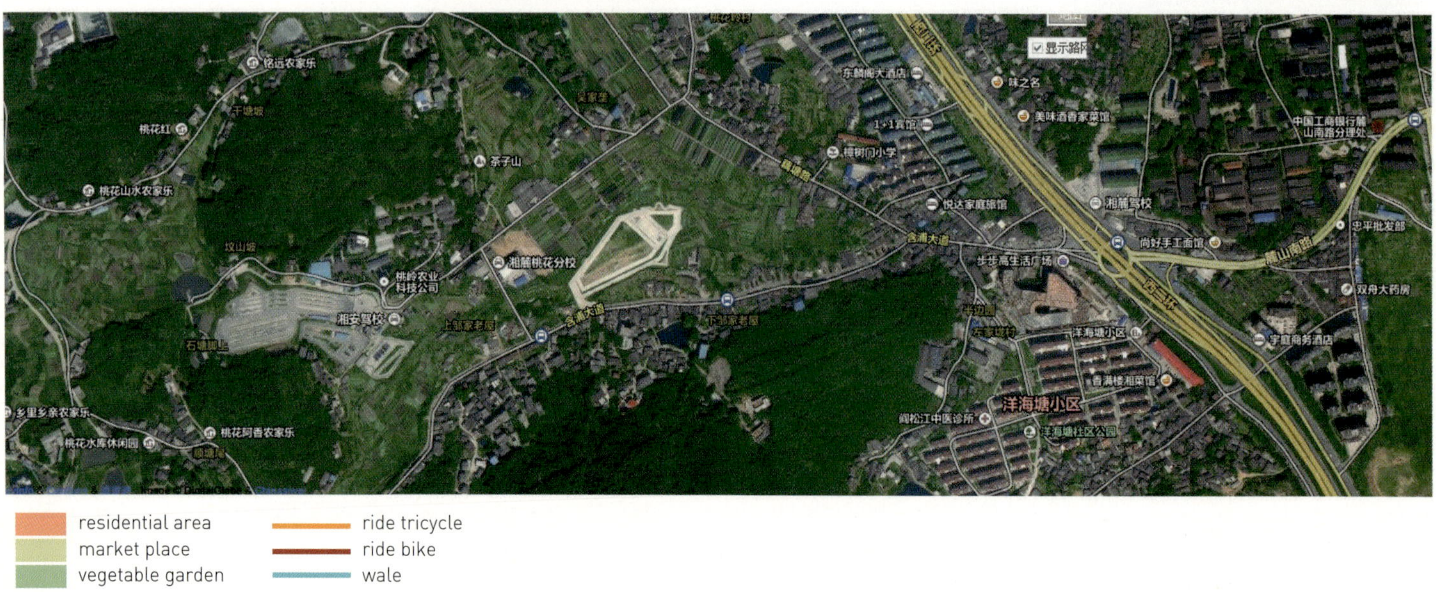

Field Work & Research Process

Respondents: Vendors(farmers, wholesalers) and Buyers(housewife, elders)

Background:

For vendors
*Limited saling place,like displaying, storage space and etc.
*Uncomfortable siting plece and ways.
*Long distance between the home and the markets.
*Inconvenient to carry the goods in transport.

For buyers
*Long distance between the home and the markets.
*No special services for customers.

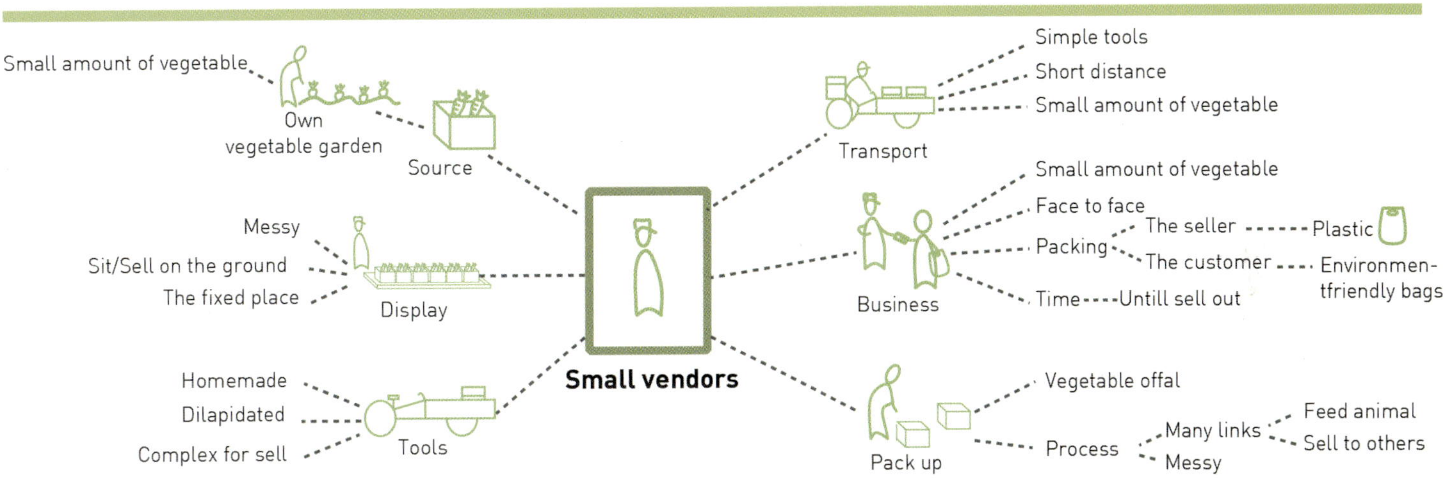

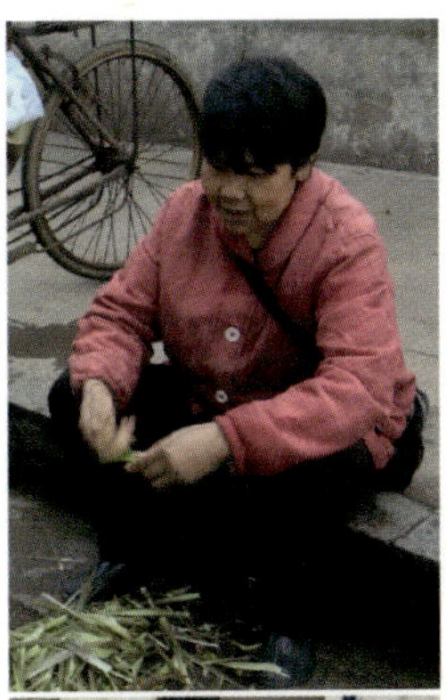

Grandma Wang
63 years old
A small vendor
Selling vegetable

Expectance
The delivery process can be more easy
Don't need to pay rent mobey
Sabe more time in disposal rubbish

Grandma Hu
59 years old
Local farmer
Selling vegetable

Expectance
Tools become more convinient
Don't need to pay rent mobey
Save more time in mange vegetables

Uncle Zhang
42 years old
A middle vendor
Selling vegetable

Expectance
Have some fun during the selling process
Don't need to pay rent mony
Save more time in disposal rubbish

Mrs Liu
49 years old
Housewife
Custormer

Expectance
Vegetable Markets can be more flexible
The market can be more organized and
cleaned some sustainable an
convenient packing product selecting
veges more comfortable

Design Development

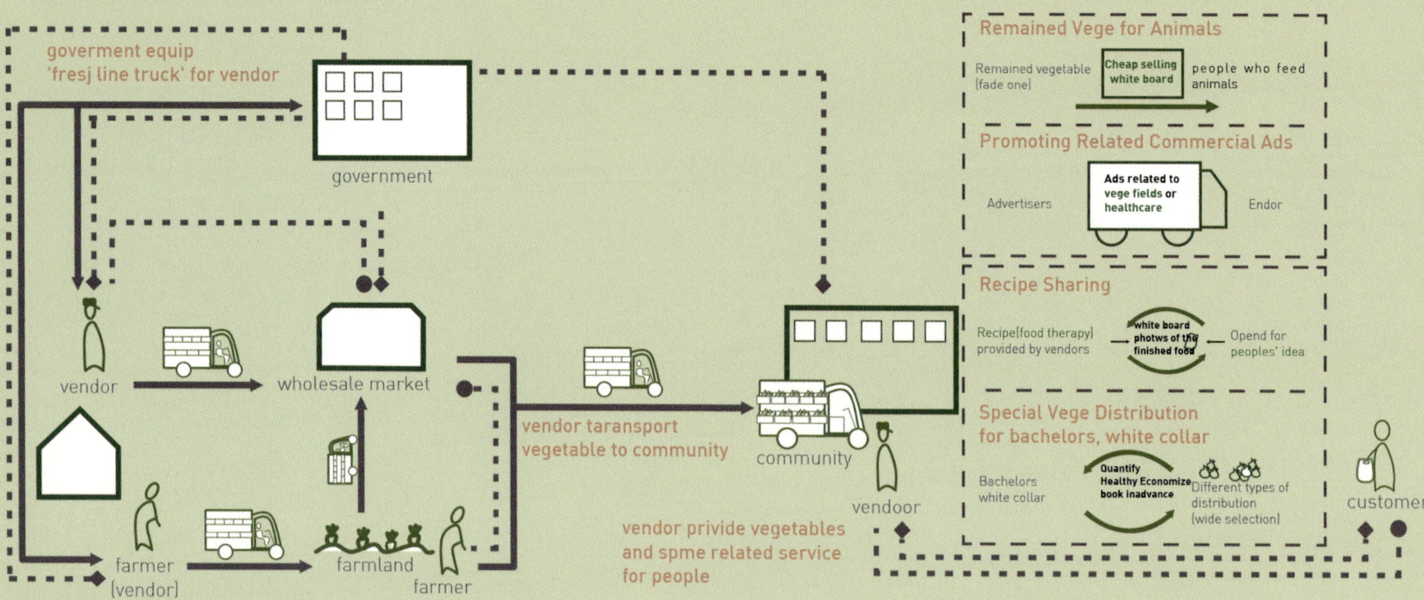

Design Interaction Blueprint

Design Comcept

The concept of Fresh Line focus on providing a set of scientific facilities which include distribution transportation and sales for local farmers and vegetable vendors. With the leading of government, this facilities keep the exchange trading form of traditional market sale mode then combine with modern vegetable cars which can meetvendors'needs in both bussiness and life. At the same time, the vendors provide the four service to to attract customers to buy vegetablesand make the sale process more convenient. Vendors also can flow dispersion sale, avoid the agglomeration in the same place caused by competition. Fresh Line Car contains : picking, transportation, exhibition, sales and provide services at the same time in each processwhich can make vendors to attract customers to buy vegetables in order to obtain the biggest profit.

Fresh Line이라고 이름 붙인 우리 팀의 디자인은 지역의 농부와 상인들을 위한 보다 과학적이고 능률적인 운반과 전시, 판매가 가능한 모발마켓 퍼실리티를 제공하는 것이다. 이 움직이는 공간은 기존의 재래시장이 가지고 있던 커뮤니티적 성향과 문화를 유지하면서 현대적인 전시 운반수단과 결합시킨 것이다. 즉 비즈니스와 생활을 모두 만족시키는 개념이다. 소비자는 재래시장에 간 것과 같은 문화를 느끼면서 집 근처에서 편리하게 채소를 살 수 있고 그곳은 하나의 커뮤니티 공간을 형성한다. 상인들은 지나친 경쟁 없이 효과적으로 사업을 운용할 수 있다.

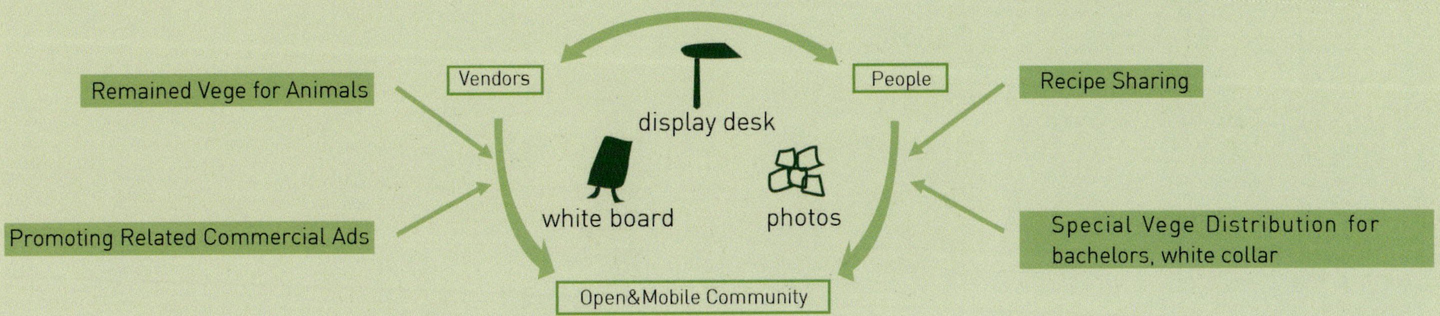

Benifit

For the people:

The vendor
- Save time during the deliver process
- Convenient to display the vegetables

The customer
- Acquire better services
- Get fresh vegetables in short distances
- Provide a place to communication

For the economic:
- Promote the development of new industries

For the culture:
- Preserve traditional sale mode

For the planet:
- Transform package material
- Improve the material utilization rate

For the government:
- Create a new emp;oyment model
- Increase employment rate
- Provide convenience for the residents
- Improve social choesion
- Standardize the vendors

Radars-Environemental (SDO)

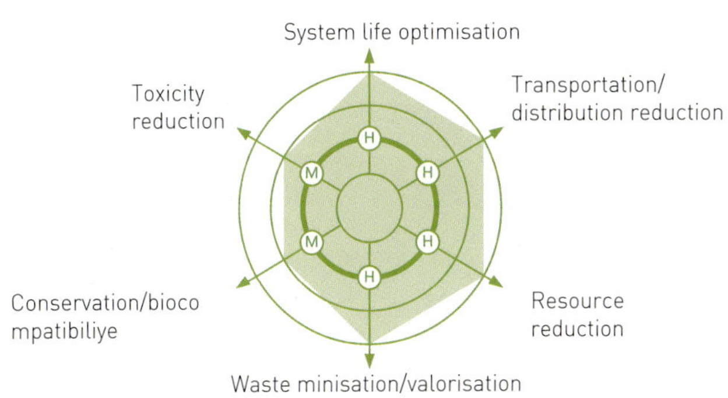

Ideas:

We think about design for creating community through a system include mobile vehicles and communication service with market lovers. For details,some will be like below:

Transport:Making it convenient&quikly,also maintain the vege steady.

Setting:Easy to classtify.

Communicate:Open&happy atmosphere presented in visualized ways.

Final Image

Design Details

Car size [3150mm* 1500mm*1650mm]

straw
use straw ropes as a packagemetierial

display board
menu
china dietotherapy
communication

traditional Chinese bamboo basket
bamboo material
breathe freely

Door become a roof

Autarkic roof vegetable garden

Team members | Zhu Zhaoqiang, Ke Junxue, Xu Shihui, Zhang Xiaoyu, Xie Zhenzhen

Background

In recent years, with the fast development of urbanization, more and more famers whose cultivating land was occupied by government move to the community provided by government, and this issue has brought to a lot of problems, government fulfil the financial demand of farmers but the new community in city is not very suitable for their inherited lifestyle in every way. the spiritual need of farmers is neglected largely, and not all of the facilities in city community facilitate their new life.
So it is necessary for us to gain the balance between modern city living environment and the inherited lifestyle of immigrant farmers.

최근 몇 년간 급격한 도시화로 인해 정부소유의 땅을 경작하던 농민들이 도시로 이주하게 되거나 살던 곳이 도시가 되면서 도시민이 된 농민들이 생겨나고 있다. 지역정부는 농민들의 경제적 요구를 충족시켜주고 있지만 도시 내 커뮤니티 환경은 농민들이 가진 본래의 문화와 전통을 지속시켜주기에는 적합하지 않은 실정이다. 도시 기반시설은 효율성이 주가 되어 이러한 이주민들의 삶의 질을 윤택하게 만들지 못하고 있다. 그러므로 우리는 현대 도시민의 삶과 환경 그리고 이주 농민들의 과거의 전통과 문화를 균형 있게 조화시키려는 노력이 필요하다.

292,000
immigrant farmers are living in
Chang Sha city

0.1%
population of Chang Sha city
are immigrant farmers

73
immigrant communities in
Chang Sha city

0.9%
communities in Chang Sha city
are immigrant communities.

"it is very bored for me to live without a private cultivating land"

"it semms that everything is changing.and we use some pots to plant"

"I followed my families menber migrate to city last year,even though I am so young"

"we have took many steps to prevent but it is useless"

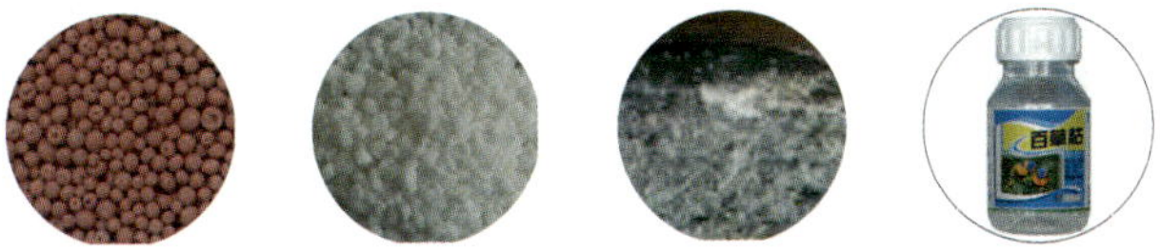

	compound fertilizer	urea	manure	medicine
source	fertilizer shop	fertilizer shop	kitchen	fertilizer shop
use	added a variety of elements	nitrogen supply	promote growth	kill insects
side effect	damage the soil	soil pollution	bad smell	air pollution

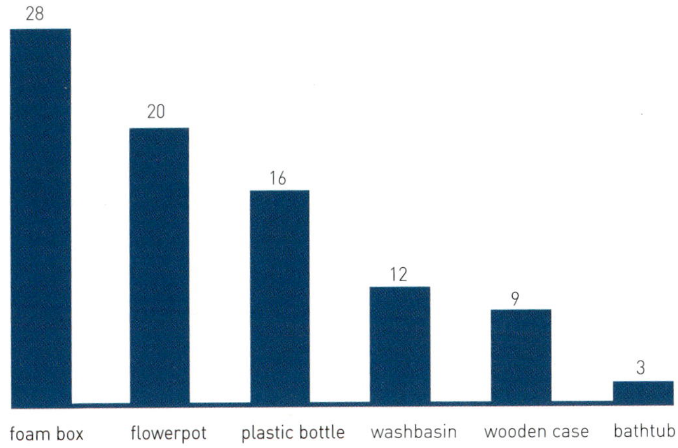

Design progress

Problems

classify the problems we found.

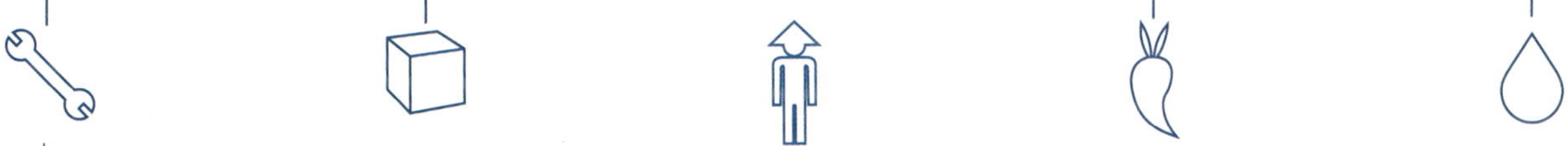

came up with 4 main solution trends base on the classified problems.

nutriment

nutrument is the core of planting the gress, and we should think about what is the sustainable and convenient way to solve the problem

space

too many cases can indicate that the limited space in migrant community is one of the giggest problem for resident, and the creative ideas should be doped out

information

there is no platform for residents to communicate message, and it also bring a lot of social problems,like the cold neighborhood relationship

tool

current phase is a transition phase, both advanced tools bring some problems to the planting progress

Final solution

- a rain collector sysytem without using running water.
- residents provide fertilizer material and regulator manage it.
- shelfs for placing planting containers.
- using the edge of roof of buildings.
- a information board to share imformation.

System map

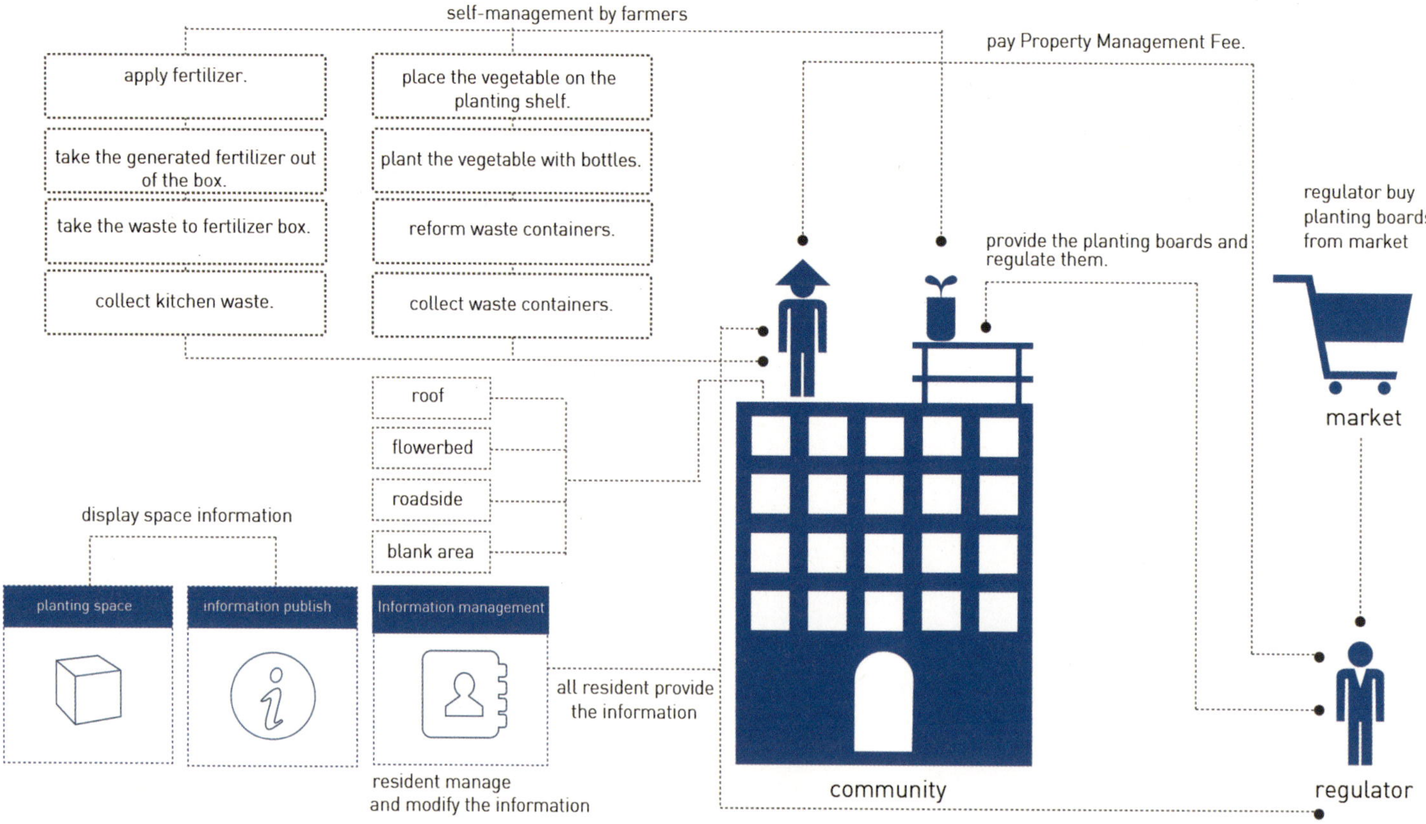

Research: vegetable

Interaction board

	1	2	3	4	5	6	7	8
migrant	residents want to plant vegetable and they need to know the space information.	apply the vacant seatof the shelf and write indentity imformation in the shelf map.	put their containers on the ac□ured seats.	cultivate their vegetables. •	provide kitchen waste to fertili□er collector	use fertili□e•		pump water from the rain collectors and water their vegetables.
regulator	provide the shelf map.	collect their indentity imformation and check in.	provide different services to shelfs like repairl inspection and ad□ustment.	give advices to residents like what kind of container is a better choise.	provide the survice like operation and repair of the machine and release the information about the fertili□er		install the rain collector on the downpipes to collect rain for residents to water their vegetables.	provide the survice like repair and installment of pumps.
e□uipment	a shelf map for residents to find vacant place •		there are some shelfs on the building roof for residents to place their planting containers.		there are some fertili□er makers □collector□and residents can turn their kitchen waste to fertili□er hrough it.		rain collectors are installed on the dowpipes.	the hand□operated pump without using energy resources and farmers are very kind and familiar with it.

Provide more space for planting vegetable and a better environment.
Collect rain for residents to water vegetable and it is a economical and eco-friendly way.
It is very economical for residents to use kitchen waste as fertilizer.
A imformation board for residents to communicate and it is a good way to continue the harmonious neighborhood relation.

Evaluation

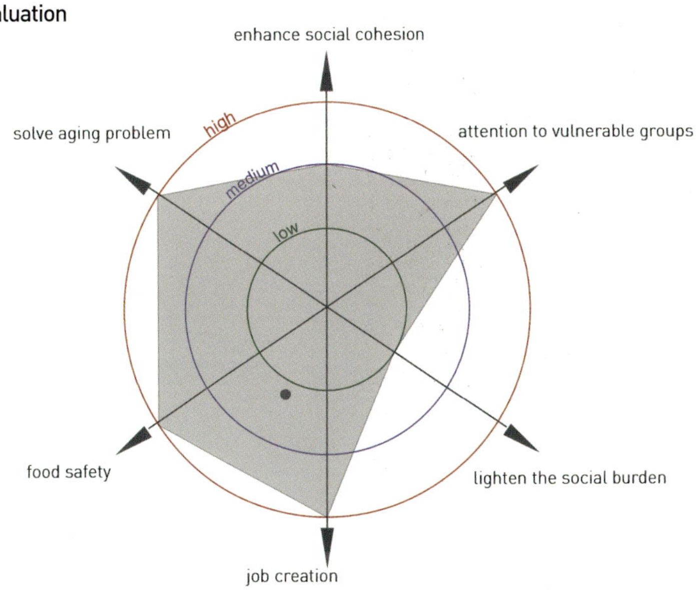

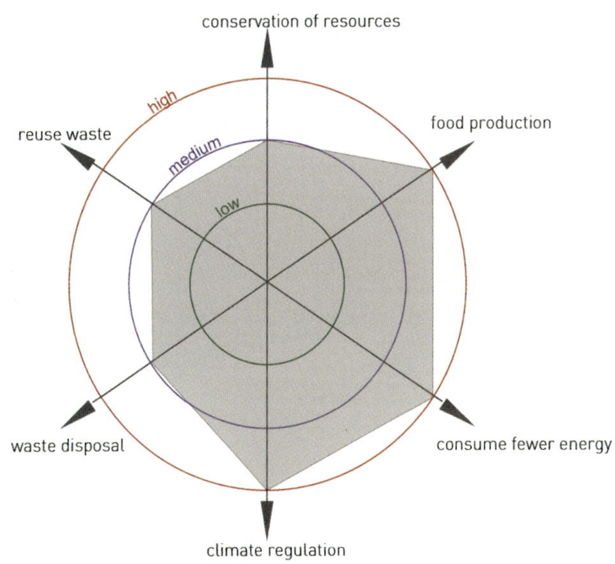

the procedure of food planting is under the supervision of residents, and the organic fertilizer is used in this system,so the food quality is very high as they plant in suburban area.

maybe the material life is not the biggest problem for them, but the mental demand of migrants also should be fulfilled, the system can help residents to find their sense of belonging.

That intense crowding can combine with the heat island effect in big cities, the roof garden provide more green space to city and it can regulate the climate of the community to some degree.

the kitchen waste can be one of the bigest problem of city like the sewers in London always be blocked by grease. the sysytem turn the kitchen waste into organic fertilizer, and take advantage of it.

Final work image

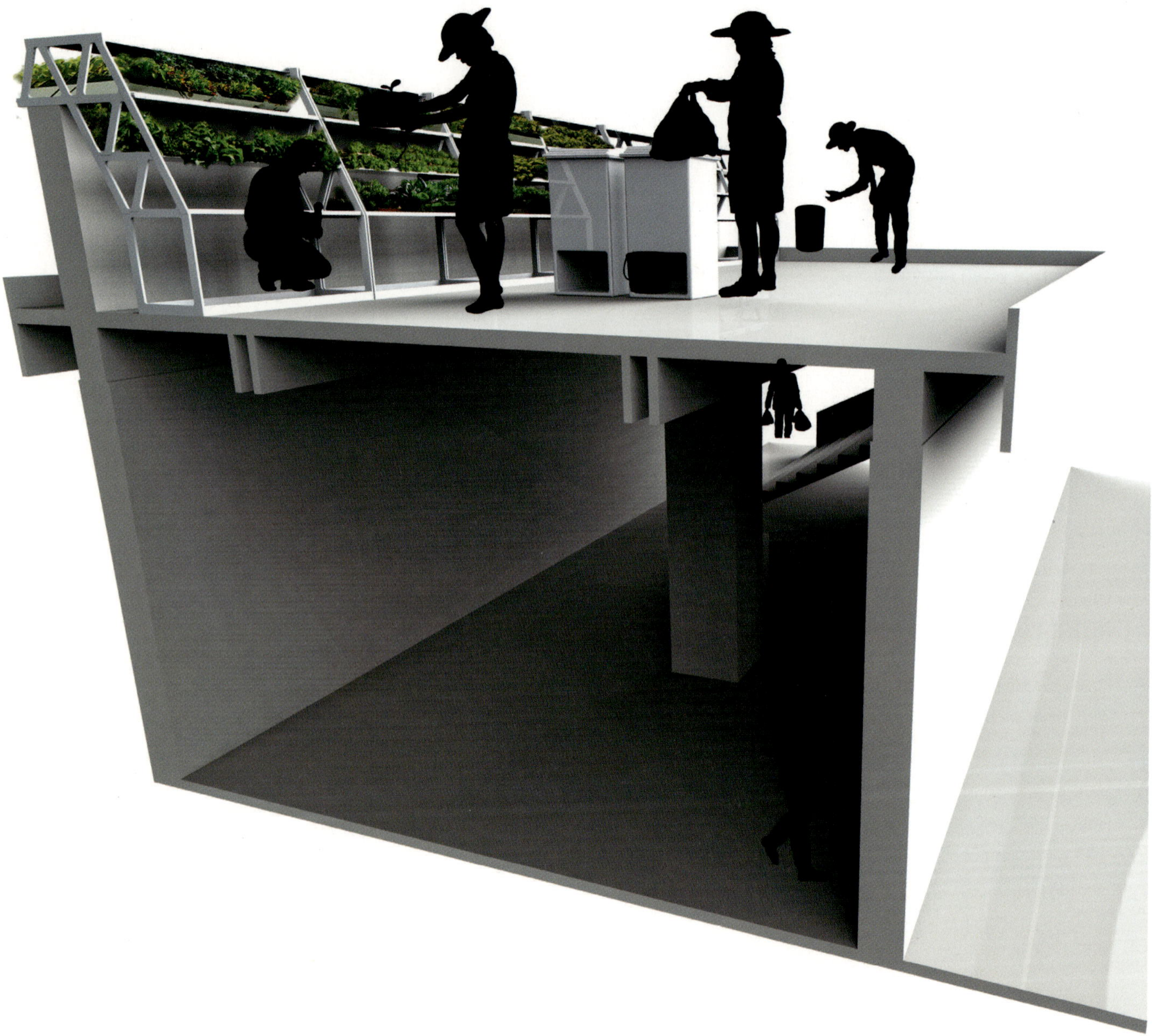

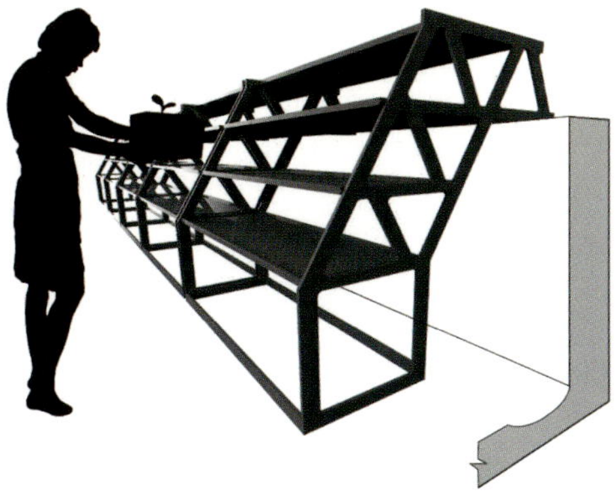

residents place their planting containers on the shelf, the shelf is made of waste steel products, and they can change container's positions casually in order to let sunlight light the vegetables evenly.

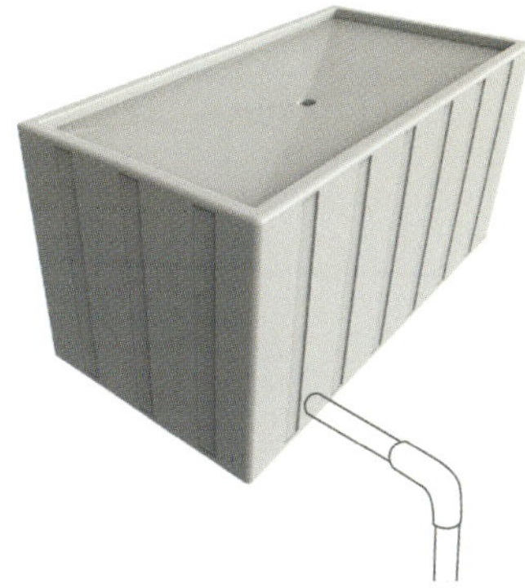

Funnel-shaped roof of the rain collector can collect rain efficiently through the inclined plane and its small hole can prevent the collected water from evaporation. the rain collector is installed on the top of the staircasewhere can not be used as planting area.there are some water pipes to connect the rain collect and the watering points where residents can get the water to water their vegetables.

It can be a bad experience for residents to take out the fermented fertilizer because of the bad smell, in order to solve the problem, the tap for outflowing fertilizer is installed in the bottom of the box,and there is a niche for barrels to accept outflowing fertilizer. residents collect their kitchen waste and take it to the roof of building, then lift lid of a waste box and throw waste in the box. it will cost a certain time to ferment the waste, the lid will prevent the odour from emitting out.

giving advices to your neighbour or exchanging information with them is a good way to develope a healthy and inherited lifestyle. resident occupy the planting seats by filling in the blanks of the roof plan with their house numbers. this area is also a place for regulator to issue notices.

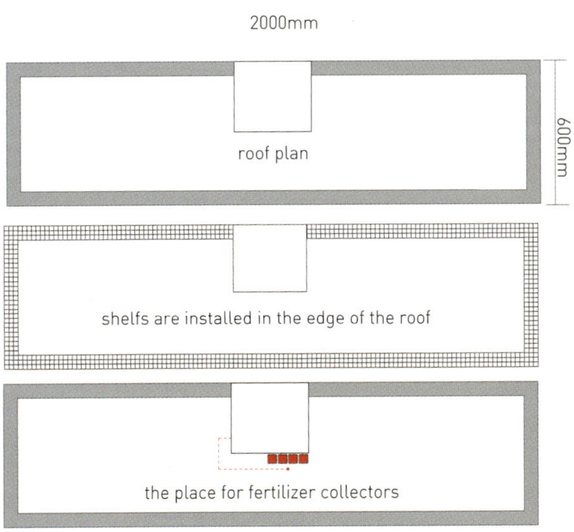

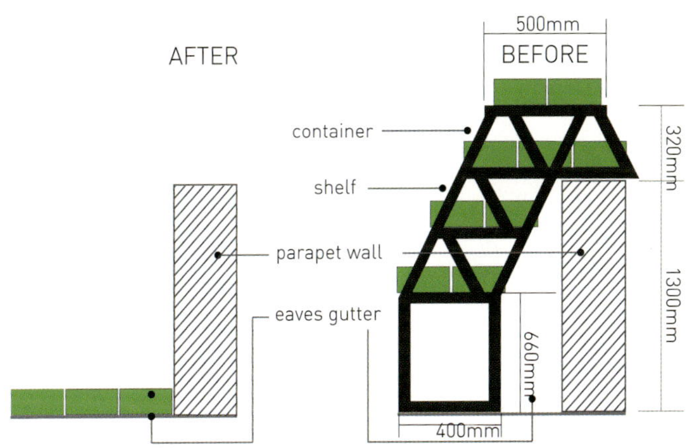

It is very obvious that the planting shelf really make sense, It provides more space to place containers, like plastic bottles and foam boxes, better light conditions can be provided because of the bank angle and higher height, at before the parapet wall is a disadvantage of planting vegetable because it blocks light and it is close to the eaves gutter witch can flood the vegetable and wash off the planting soil and now the side effects of the inherited facilities can be avoided.

3

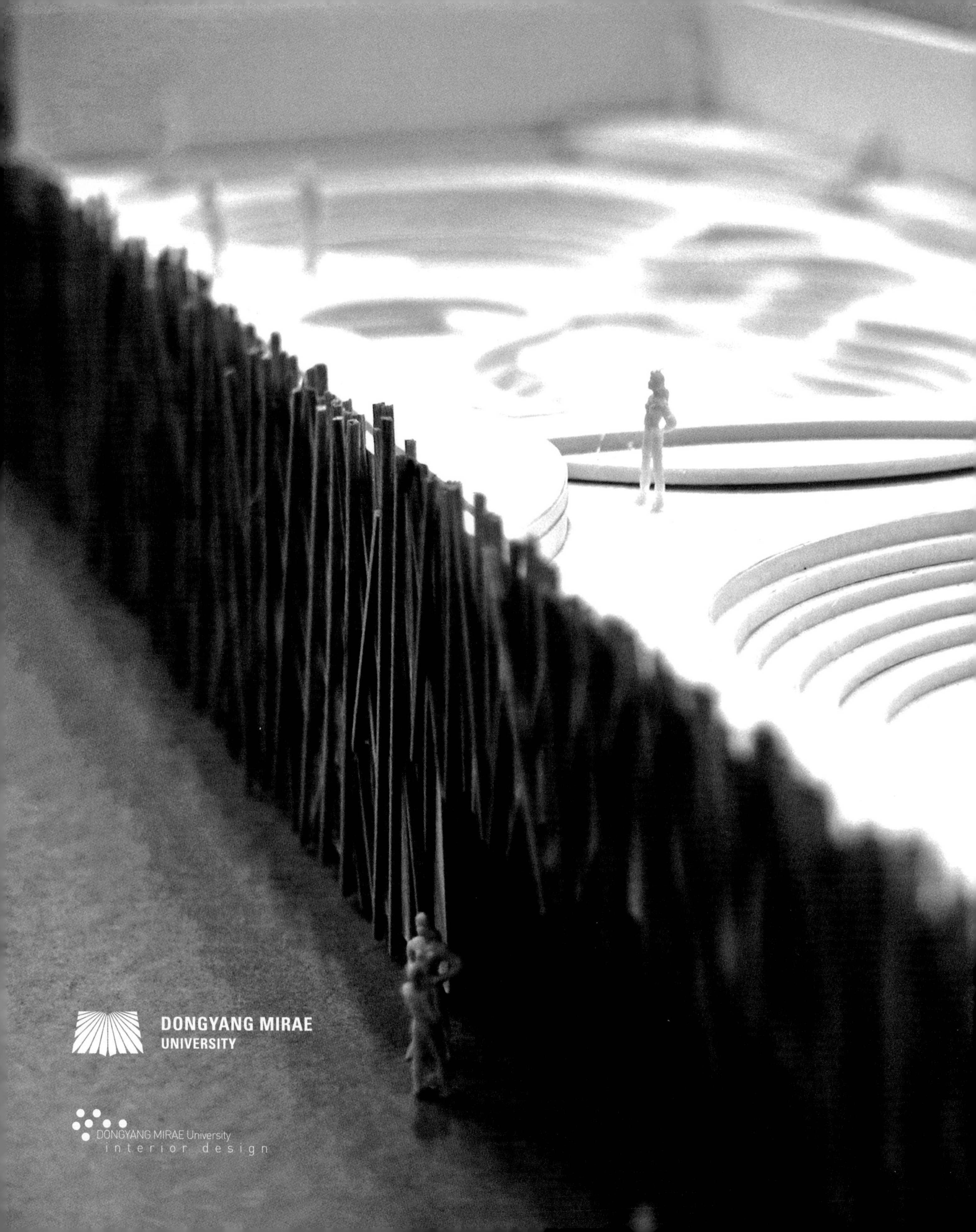
DONGYANG MIRAE
UNIVERSITY
DONGYANG MIRAE University
interior design

DONGYANG MIRAE UNIVERSITY
Department of Interior Design

2013 sophomore design studio

The 5th Hidden Space Project

THE RE-ENERGISING SPACE

동양 미래대학교 실내디자인과
2013 2학년 2학기 설계스튜디오 결과물

152-714
DONGYANG MIRAE University Dept.of Interior Design
62-160, gocheok-dong, guro-gu, seoul, korea
phone + 82 2 2610 1962
fax + 82 2 2610 1960

http://interior.dongyang.ac.kr

학부장 **김홍기 교수**
동양미래대학교 디자인학부 실내디자인과
hkkim@dongyang.ac.kr

서 문

런던. 후난. 서울
물리적 거리가 멀지만 생각의 사유는 멀지 않습니다.
지리적, 문화적, 역사적 배경이 다른 세 도시의 학생들이
같은 주제로 국제교류전을 펼칩니다.
산업화 시대의 도래 이후
도시는 끝없이 팽창해 왔습니다.
지역문화와 보편문화의 충돌 속에
개발과 보존의 충돌 속에
현대도시는 많은 문제점을 노정해 오고 있습니다.
자본주의에 있어 도시는 욕망의 배출구입니다.
기능주의적 플랜에 예속되어
버려진 공간이 탄생되기도 합니다.
친환경적 환경이 인공적인 공간으로
변질되기도 합니다.
집단화한 고층빌딩, 고층아파트가 즐비한 현대도시의 이면에는
인간성이 배제된 어두운 공간이 존재합니다.
이들 공간을 다시 인간에게 돌려주어야 합니다.
도시재생은 작은 공간부터 이루어져야 합니다.
리사이클, 업사이클,
재활성화할 수 있는 단서를 찾는 것이
국제교류전의 주제입니다.
요안 하위징아(Johan Huizinga)가 호모 루덴스를 외쳤듯이
인간은 놀이와 체험과 유희 속에 삶의 환희를 느낍니다.
런던에서 개최될 '2014 국제교류전' 은
동서 공간문화의 교류가 일어나는
아고라가 될 것입니다.
도시공간을 재활성화(Re-Energising)하는 프로젝트에 참여한 학생들과
지도교수에게 뜨거운 박수를 보냅니다.

Head of Interior Design Department, PhD | Prof. Kim, Hong-Ki
Department of Interior Design DONGYANG MIRAE University

Preface

London, Hunan and Seoul
These cities may be far apart from each other, but contemplation of thoughts cannot be too apart.
Students from these three cities with geological, cultural and historical differences come together to compete against each other at International Exchange Competition under the same topic.
Ever since the industrial age arrived, cities have never ceased to expand.
Throughout clashes between local culture and universal culture, throughout collisions between development and preservation, modern cities have been struggling with numerous problems.
Under the capitalism, the cities are outlets for desires.
And deserted spaces are born after being subjugated by functional plans, whereas eco-friendly environments turn into artificial spaces.
On the other side of modern cities filled with collectivized high-rise buildings and apartment complexes, dark spaces thrive where humanity is completed eliminated.
These spaces ought to be returned to the mankind.
Urban regeneration must be initiated from the small space.
The topic of the International Exchange Competition is to discover the clues which lead to recycling, upcycling and revitalization.
Just as Johan Huizinga professed Homo Ludens, man feels the joy of life through play, experience and amusement.
'2014 International Experience Competition' to be held in London is destined to become an agora where the exchange of space culture are made between the West and the East.
All students and professors participating in the projects to re-energize the urban spaces deserve a truly big round of applause.

박진호 교수
동양미래대학교 디자인학부 실내디자인과
jnpworks@naver.com

"공간에서 놀기"

놀이 또는 유희를 위한 공간으로 우리는 어떤 것을 떠올릴 수 있을까.?
아파트 단지의 놀이터? 각종 탈 것과 기구들이 즐비한 놀이동산, 테마파크?
자극적인 듯, 쉬운 듯 보이지만 막상 공간의 용도에 새로움을 갈구하며 그 무언가를 생각한다면 보일 듯 보이지 않고 잡힐 듯 잡히지 않는 애매함이 있는 듯 하다.
상념들이 많아서 일까? 논다, 즐긴다라는 것에 대해 주변환경, 시선을 의식해야하는 공공연한 불안감일까? 공간을 통해 즐기고 서로 다른 세대들이 향유하며 소통할 수 있는 그런, 상투적인 가치의식들이 무장해제된 일탈의 도시 속 공간을 상상한다는 것은 그자체로 즐거움일 수 있지만 현실도피의 출구 없는 피난처가 될까 하는 걱정이 들기도 했다.

사이트선정을 하며 구시가지의 모습을 가지고 경제사회적 이해관계로 인해 이웃간 소통이 단절된(?) 그런 현대도시의 '뜨거운 감자' 같은 곳을 찾아야 했다. 늘 그러했듯이 음과 양이 공존하고 공간의 주체가 좀처럼 사람이지 않은것 같은 복잡 다양한 곳, 그리고 그곳에서 놀고 즐기는 것을 상상하기...
상상의 날개란 우리 생활 속의 그런그런 보편성이 단서가 되어져야 한다고 믿었다. 그리고 그것에 대한 새로운 시각, 그리고 해석이 필요하다고. 짙은 일상의 암초가 삶을 불편하게 하고 있는 시공간에, 깊은 시름을 한순간에 잊게 만드는 이른바 '혁신' 이 필요하다고..

장소의 특수성을 감안하며 혁신을 주문한다는게 학생들을 이해시키기에 결코 녹록한 것은 아닐것이다. 어쩌면 이번 한학기의 프로젝트는 귀결과 해답을 얻기 위함이 아닌, 공간디자이너로서 시작을 알리는 출발의 신호음일지도 모른다는 생각을 했다.
중요한 것은 공간을 매개로 다듬고, 자르고, 붙이고, 잇고 하면서 즐거움을 발견해야한다는 것이다. 삶의 에너지를 재충천하게끔하는 그런 혁신의 공간은 결국 즐거움 속에서 출발해야 한다고..

Prof. ParK, Jin-Ho
Department of Interior Design DONGYANG MIRAE University

"Play in a Space"

What can we think of as a space for a game or a play?
Will it be a playground at an apartment complex or a theme park filled with all kinds of joyrides?
It may seem exciting and easy; however, it may also be elusive where it appears within a grip but cannot be reached when thinking of something while craving for a new use of the space.
Is it because we just think too much? Is it because of obvious agitation where we become self-conscious of the surroundings over having fun and enjoying ourselves?
It may be quite a pleasure just to dream of an urban space where we can escape from a daily life as conventional values are ignored while different generations come together to have fun through the space, but it also gave us a concern that it might turn into a shelter without an exit to avoid reality.
While searching for a site, we had to find a place which may be 'hot potato' of modern cities where communication with neighbor is completely severed because of socioeconomic interests while preserving the look of old cities. Just as it always has been, we were in search of a complicated yet diverse place where yin and yang coexists and the principal agent of the space is not a human. And we imagined playing and enjoying there.
I always believed that a complete imagination should be a clue for universality within our daily lives. And a new perspective and interpretation is required for the imagination. Within the space and time where casual barriers are making life uncomfortable, it requires so-called 'innovation' capable of wiping away all the sorrows at once.
It won't be easy to help students understand when they are asked for innovation while considering the distinctiveness of the place. I though that the project for this semester may be a signal for the students to begin their life as a space designer instead of trying to find an answer and conclusion.
The important elements is to discover a pleasure while refining, cutting, attaching and connecting the space as a medium. Eventually, such innovative space that recharges the energy of life needs to be initiated under the pleasure.

류숙희 교수
동양미래대학교 디자인학부 실내디자인과
chjrsh@hanmail.net

놀다, 어울리기.

요즘 학생들은 주로 컴퓨터나 스마트폰을 가지고 논다.
고요하고 적막하다.
하지만 사실 논다는 것은 북적거림이다. 내 기억 속의 놀이란 사람들과의 어울림이다.
이번 프로젝트는 사람들 사이의 어울림이 그려질 수 있는 놀이공간 구축에 주목하며 작업을 진행하였다.
서부 이촌동은 두 가지 얼굴을 가지고 있다. 한강 조망권의 잘 개발된 아파트 단지와 곳곳에 빈 집과 재개발 관련 플래카드가 널려 있는, 그곳에서는 아무도 즐거워 할 것 같지 않고, 아무도 뛰어놀지 않을 것 같은 지역. 우리는 이 낙후된 지역에 새롭게 활기찬 에너지가 넘치기를 바라는 마음으로 이곳을 이번 프로젝트의 사이트로 정하였다.

학생들은 먼저 현장 조사 및 설문 조사를 통해 놀이공간이 세워질 사이트의 특성에 대한 자료를 수집하고, 수집된 자료를 통해 사이트에 대한 SWOT분석을 하였다. 이를 근거로 팀별로 컨셉을 정하고 그에 따라 놀이공간을 표현하였다. 단순히 컨셉만을 제시하는 것이 아니라 구체적인 재료와 기술적 맥락까지 도입하도록 하였다.
프로젝트 초기, 학생들은 사이트에서 느껴지는 어두운 무게 때문인지 'PLAY(놀이)' 에 대해 쉽게 접근하지 못하였다. 답이 보이지 않는 상황도 있었다. 새로운 아이디어가 필요하였다. 함께 모여 논의하는 여러 번의 만남을 통해 자신들이 정한 컨셉을 다듬어 갔다.
모든 학생들이 제시한 프로세스를 따라 마지막 마무리까지 열심히, 그리고 잘 따라와 준 것에 대해 고마움을 느낀다. 어려울 때 집중하고 몰입해준 여러 얼굴이 떠오른다.

Prof. Ryu Sook-Hee
Department of Exhibition Design Kyewon, Kaywon School of Art & Design

Play, and socialize.

Today's students mainly play with a computer or a smart phone. It is silent and reclusive. However, playing is essentially based on crowdedness. As far as my memory serves me, playing relied on association with others.
This project was proceeded while focusing on the implementation of the play space where association among people can be materialized.
Ichon-dong has two different faces: well developed apartment complexes featured with prospect right of Han River and a flood of banners on redevelopment projects along with empty houses. The place seems as if no one would ever find any joy, and no one would play to have fun there, either.
We have decided on this place to be the work site of this project in hopes of revitalizing the deteriorated regions with new and lively energy.

Students first gathered information and data upon characteristics of the site where new play space is to be built through on-site investigation and questionnaire survey; subsequently, they conducted SWOT analysis of the site with collected data. Based on the analysis, each team developed a concept for the play space to be built. Instead of merely offering a concept, they specified exact materials to use as well as technological context.
At the early stage of the project, students appeared to experience difficulties to make an approach to a 'PLAY' perhaps because of the weight of dark force emitted from the site. And there were occasions where they simply couldn't find an answer. They were in desperate need of new ideas. After a series of meeting where everyone gathered to discuss further, the students were able to refine the concepts they chose.
I truly appreciate all students for investing their utmost efforts as well as fully complying with the given tasks until the end in accordance with the procedures that they presented. I still have vivid memory of everyone who focused on his/her works through the difficult times.

2학년 설계담당 **박영태 교수**
동양미래대학교 디자인학부 실내디자인과

responsible for sophomores, PhD | Prof. Park, Young-Tae
Department of Interior Design DONGYANG MIRAE University
robotomy@naver.com

Yongsan Dreamhub(the Yongsan International Business District)

The project was first proposed in 2006, and was described as Korea's largest and most ambitious property development project.Financing difficulties delayed it following the Financial crisis of 2007–2008.The Korean real estate market also did not seem appealing to the investors.In subsequent years, a growing number of companies, such as Samsung, have pulled out of the project.The project has been finally cancelled in April 2013.It is possible the venture will result in a number of bankruptcies.There are also speculations about a possible class-action lawsuit by the residents against the developers and the government. - http://en.wikipedia.org/wiki/Yongsan_Dreamhub

Project site - yongsan gu, seobu ichon-dong street view images

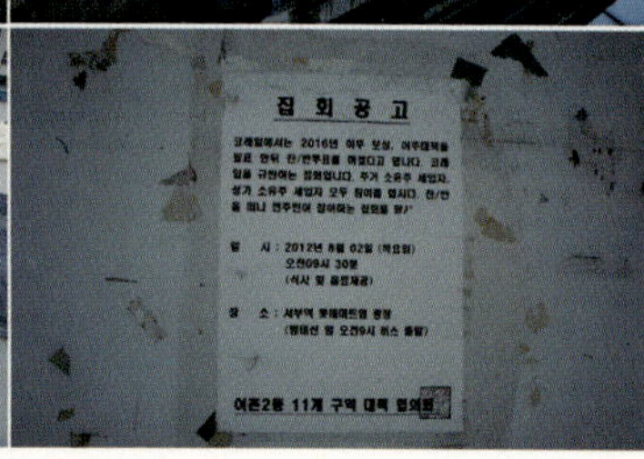

히든 스페이스 프로젝트 다섯 번째 에피소드 - "RE-ENERGISING SPACE"

2009년 이후 5번째로 진행되는 Hidden space project 이다. 이번 주제인 'RE-ENERGISING SPACE' 로부터 우리가 선택한 숨겨진 공간은 자본과 시장의 그릇된 욕망으로부터 씻을 수 없는 상처를 담고 있는 곳으로, 용산 재개발 예정지역 주거단지 내에 위치하고 있는 낙후된 커뮤니티 시설이다.

용산재개발 사업은 30조원에 이르는 대한민국 역사상 가장 큰 규모의 사업으로서 드림허브(DREAM HUB)를 꿈꾸며, 국제 비스니스 업무 지구를 개발하기 위해 2007년부터 시작되었다.

이 사업은 용산구 한강변 주거지역과 용산역 일대 철도정비창 기지를 통합 개발하는 것이었는데, 진행과정 중에 원주민들과의 갈등, 시민과 경찰의 대치로 많은 인명피해가 발생하여 잊지 못할 대형 참사로 기록되기도 하였다. 결국 불확실한 계획을 품은 채 접근한 불투명한 개발주체들이 문제가 되었다. 서울시, 코레일, 금융사, 부동산, 건설사, 개발사, 그리고 중앙정부간의 눈 앞에 보이는 작은 이익들에 대한 치열한 싸움과 갈등으로부터 무산되고 말았다.

현재는 진행 중인 모든 것이 정지된 상태이다.

특히 주거지역에 해당되는 서부이촌동 일대에는 이에 대한 경제적, 심리적 부담이 가속화되어 침체된 분위기로 새로운 활력을 필요로 하고 있는 곳이다.

Fifth Episode from Hidden Space Project - "RE-ENERGIZING SPACE"

This is the fifth Hidden Space Project since the year 2009. The hidden space that we selected based on the current theme, 'RE_ENERGIZING SPACE,' is the underdeveloped community facilities located in the residential district within Yongsan redevelopment district which is utterly filled with inexpugnable scars caused by unjustifiable desire from the capital market.

Yongsan Redevelopment Project, the biggest project ever in Korean history worth 30 trillion Won, was first initiated since 2007 in order to develop international business district in the hope of building a 'Dream Hub.'

This project was designed to comprehensively develop the residential area near Han River and the Rolling Stock Workshop around Yongsan Station at Yongsan-gu; however, it is now remembered as one of the worst tragedies where massive human casualties were suffered from the clashes between civilians and law enforcement agencies because of the conflicts with the residents during the course of the development. Eventually, this project was halted following fierce battles and conflicts amongst the principal agents, such as the City of Seoul, Korail, financial institutes, real estate companies, developers and the central government, which were blind with small gains without concrete development plans.

As of today, all ongoing projects have been ceased. More than anything else, the regions of Western Ichon-dong, the residential area, are in desperate need of new energy after economic and psychological burdens have been subsequently escalated to foster lingering depression.

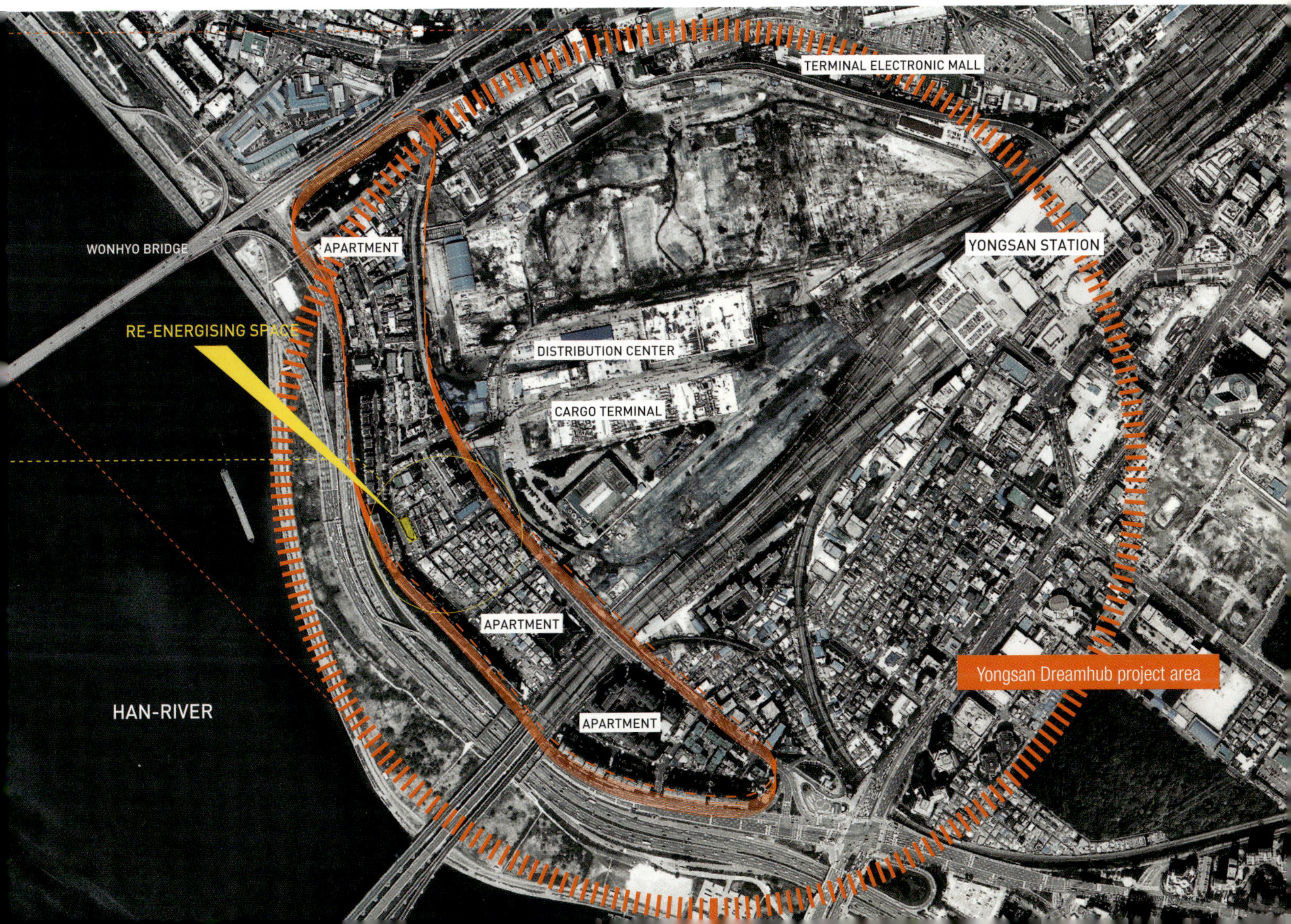

놀이, 유희로서의 재활성화

프로젝트 시작 즈음 RE-ENERGISING(재활성화)이라는 주제로부터 우리가 논의 끝에 선택한 주요테마와 방향은 놀이, 혹은 유희로서 지역민과 도시민들에게 새로운 (비)일상적 시공간의 제공이었다.

여기서 디자인 교육의 목적달성을 위해 전제한 것은 크게 두 가지인데 우선 재료와 그 재료로부터 획득 될 수 있는 구조에 대한 체계를 학습하는 것이었다. 재료의 입장에서 기존의 리사이클로부터 최근의 업사이클개념까지 모두 중요한 단서가 될 수 있음을 전제하였다. 그리고 이러한 체계를 지역민들이 경험할 수 있는 일상적, 혹은 비일상적 삶으로 프로그래밍 하는 것이었다. 즉 물질적 요인으로서의 질료(Matter)와 공간을 구성하는 프로그램과 기능으로서 비물질적 요인으로서의 질료를 놀이, 유희를 방법으로 매개하는 것이었다.

〈놀이〉

놀이는 문화보다 더 오래된 것이다. '놀이하는 인간(호모 루덴스)' 을 저술한 호이징거(johan huizinga)가 인용한 레오 프로베니우스가 말한 것처럼 원시인은 그의 정신 속에 새겨진 자연질서를 '놀이' 했다.

> *인간 사회의 중요한 원형적 행위들은 처음부터 놀이의 요소가 가미되어 있었다. 가령 인간이 만들어낸 최초의 중요한 도구인 언어를 살펴보자. 언어 덕분에 인간은 사물을 구분하고, 정립하고, 진술할 수 있었다. 사물들에게 이름을 붙여 말과 언어를 가지고 놀이를 하면서 물질과 마음 사이에서 끊임없이 ' 불꽃 '을 일으켰다. 모든 추상적 표현에는 대담한 은유가 깃들어 있는데, 이 은유라는 것이 실은 말을 가지고 하는 놀이이다. 이런 식으로 삶을 표현하는 과정에 인간은 자연의 세계 바로 옆에 제2의 세계, 즉 언어의 세계(시詩의 세계)를 창조했다.*
> *– 요한 호이징거, 호모 루덴스*

호이징거는 놀이가 가지는 주요특징은 '자발적 행위' 로서 이것은 자연과정에 덧붙여져서 개화(開化), 장식(葬式), 발현(發現)하고, '일상적인' , 혹은 '실제' 생활에서 벗어난 행위로서 시간과 공간의 제약을 받는데, 따라서 시공간의 특정한 한계 속에서 우리는 '놀아진다(played out)' 고 하였다.

놀이가 진행되는 동안에는 모든 것이 운동, 변화, 교대, 결합, 분리의 형태를 취하게 되는데, 여기서 개인에게는 생활의 기능으로서, 그리고 그 의미, 표현적 가치, 정신적 · 사회적 결속은 그 사회적 기능으로서 중요한 것이라고 역설하였다.

Revitalization as a Play and Amusement

After a series of discussion under the theme of 'RE-ENERGIZING,' we have decided upon the main theme and direction of the project to provide a new and casual space for local residents as well as people in the city with a play or amusements. Here, two items were predetermined in order to achieve the goal of design education, which were the materials and training upon the system of the structure that can be obtained from the said materials. Subsequently, it was to program the system into a conventional or unconventional lives that local residents can experience. In other words, it was to mediate the matters as physical factor, the program featuring the space and the matter as functional non-material factor through play and amusement.

'Play'

A play has been around longer than the culture. Just as Leo Frobenius, quoted by Johan Huizinga, the author of 'Homo Ludens,' once mentioned, primitive men played with the natural order that was imprinted in their spirits.

> *"Significant archetypal behavior in human society included play-elements from the very beginning. Let us look into language, a significant tool that humans first created. Because of language, humans were able to distinguish, define and describe the objects. While playing with words and language by giving name to objects, they constantly ignited flares between the matters and minds. All abstract expressions carry bold metaphors, and these metaphors are actually a play with words. In the course of expressing life as described above, humans created the second world right next to the natural world, namely the world of language (the world of poem)."*
> *- from 'Homo Ludens' by Johan Huizinga*

According to Huizinga, the main attribute of play is 'voluntary activity' which goes through enlightenment, decoration and manifestation in addition to natural process, and it is restricted by time and space as aa activity escaped from 'casual' or 'real' lives; therefore, we 'play out' under the certain restrictions within time and space.

Huizinga also asserted that everything turns into a shape of movement, change, alternation, union and separation while a play is in progress. And it performs an important function of life for individuals whereas it fulfills a social function with its significance, expressional value and mental/social unity.

어려움 – 재료, 구조, 프로그램을 통한 (비)일상적 놀이화

실제 작품을 진행하면서 곤란했던 부분은 비교적 쉽게, 혹은 즐겁게 접근할 수 있다고 판단된 '놀이' 라는 개념이 감각적인 질서와 규칙이 요구되는 지극히 진지한 행위라는 점을 뒤늦게 깨달았다는 점이다. 프로젝트가 심화될수록 어려움은 가중되었다.

호이징거의 말대로 그 흔한 놀이터의 내부에도 특정하면서도 절대적인 질서가 지배하고 있었으며, 놀이에는 사물을 지각하는 가장 고상한 리듬과 하모니가 부여되어 있음을 자각하게 되었다. 또 한편 이러한 놀이의 매혹에는 긴장의 요소가 중요한 역할을 하게 되는데 그러한 긴장은 바로 '불확실성과 우연성' 을 의미하는 것이었다.

자연스러운 재미와 즐거움이 내재된 놀이공간의 창조는 결코 쉽지 않았다. 따라서 요구되는 공간적 기능(프로그램)과 함께 이러한 차원의 새로운 놀이의 질서를 형성해야 함은 프로젝트의 구체화에 많은 고충이 함께했다. 애초 의도한 재료와 구조, 프로그램의 놀이화에 대한 집중보다는 새로운 '놀이' 자체에 필요한 스토리와 이를 구조화하는 질서와 체계의 형성에 많은 시간이 할애되었다. 따라서 여기에 거주자들의 일상 혹은 비일상과 공유될 수 있는 프로그램의 반영은 모두에게 많은 연구와 시간을 필요로 하였다.

대상공간은 지역 아파트 주민들의 커뮤니티 시설로, 향후 낙후된 곳들을 중심으로 주거지 개선사업 진행이 예정된 곳이다. 현재 사용이 뜸한 테니스 코트, 물탱크, 소규모의 주차시설로 활용되고 있는데, 주차공간과 물탱크는 이전될 것을 전제로 침체된 커뮤니티의 재활성화를 위해 놀이를 단서로 한 공간 구성을 요구하였다.

사실 '일상 혹은 비일상적 프로그램의 놀이화, 재료와 구조의 놀이화 '라는 이 두 가지 임무를 통해서 형식과 내용의 유기적인 조화와 자율적인 완성을 내심 기대하였었다. 하지만 기반지식의 부족, 놀이에 대한 다소 막연한 상상과 함께, 무엇보다도 무사무욕(disintere-stedness)을 전제로 '자발적 행위 '로서 진행된 놀이에 대한 경험과 및 인식의 부족이 나를 포함한 우리 프로젝트 구성원 전체에게 큰 장애가 되고 있음을 깨닫게 되었다. 섣부르게 단정할 수는 없겠지만 어찌보면 이는 지금까지 일상 속에서 놀이를 대하는 우리들의 자세에 대한 문제일 수도 있겠다라는 생각이 들었다.

Many Difficulties with 'Play' - (Non-)Casual Play through Material, Structure & Program

The toughest part while actually working on this project was that it took quite some time until I realized that the concept of 'play,' which I first thought to be relatively easy or fun to access, turned out to be rather an extremely sincere activity where sensitive order and regulation is required. As the project further progressed, it became more difficult. Just as

illustrator. Alessandro Gottardo, 2013

Project target place : abandoned community facilities(tennis court/parking lot/Water tank)

Huizinga pointed out, it was evident that specific and absolute order dominated even the inside of the play ground, and there was very noble rhythm and harmony recognizing the objects in the play. And while the element of tension assumed a pivotal role in the fascination of such play, such tension represented 'uncertainty and coincidence.' It was definitely not easy to create a play space that contains spontaneous fun and joy. Hence, the formation of the order for a new play at such level in addition to the required spatial function (program) demanded quite an agony while specifying the project. Substantial time was required to develop an order and system that structures a story necessary for new 'play' itself rather than focusing on turning the intended materials, structure and program into play. Thus, the implication of the program that can be shared with casual or non-casual life of residents demanded everyone for plenty of researches and time.

The target site is a community facility for the residents of the local apartments where residential improvement projects were scheduled mainly on the underdeveloped locations. Those locations are currently being used for rarely-used tennis courts, water tanks and small-scaled parking facility. Under the assumption that parking facility and water tank will be relocated, a demand was made for the space composition based on play for the purpose of revitalizing the depressed community.

As a matter of fact, organic harmony and autonomous completion of the formality and contents were discreetly expected through the two given tasks: making casual or non-casual programs into play and making materials and structure into play. However, it later became clear that the lack of experience and awareness upon play that was proceeded as 'voluntary activities' based on 'disintere-stedness' turned out to be a big obstacle to me as well as the members of the project, in addition to somewhat vague imagination of play and the lack of base knowledge. We should not rush to conclusion; however, I began to think that our existing approaches toward play in daily life may be the true cause of the problem.

놀이적 상상, 결실

지역적 맥락, 사이트의 물리적 조건의 차원에서 놀이적 상상이 최대한 자유롭게 표현되도록 허용하였다. 현재 지역의 정서상 침체된 주민들과의 인터뷰는 최소화하면서도 기존의 테니스 코트, 주차장, 물탱크라는 잠재된 기능적 단서를 새로운 사용가치로 재접근할 수도 있도록 하였으며, 또 전혀 새로운 시각의 아이디어와 개념이 표현되기도 하였다.

이번 히든 스페이스 주제 '재활성화-놀이'는 다른 해에 비해 조건상 가장 창조적인 제스춰를 많이 보여 줄 수 있는 것이었음에도 불구하고, 오히려 이러한 자유로움에 대해 왠지 모를 어색함을 느낄 수밖에 없었다. 하지만 이를 통해 '놀이'라는 행위가 우리들의 공간이라는 날실과 이를 풍요롭게 하는 일상과 비일상의 시간적 씨실의 조합에 대해서 고민해 볼 수 있었으며 무엇보다도 자연스럽고, 감각이 살아있는 자유로운 '놀이의 연출'에 대한 방법적 고민을 해 볼 수 있었다. 애초 목적한 재료와 구조에 대한 심화는 상당한 아쉬움이 남는 15주간의 프로젝트이기도 하다.

총 29개의 작품을 완성하기 위해 함께 지도한 박진호 교수와 류숙희 교수, 그리고 묵묵히 뒤에서 후원해 준 학과장 및 동료교수들에게 감사드린다.

또 이번에도 온라인으로만 교감할 수밖에 없었던 런던 메트로폴리탄 대학의 Kaye Newman 교수와 Janette Harris 교수와 학생들, 그리고 중국 후난대학교의 학생들과 안성희 교수에게도 감사드린다. 마지막으로 종합적인 감각을 소화하느라 수고한 2학년 구성원 모두에게 고마움을 전하며 5번째 히든 스페이스 프로젝트 과정속의 기억과 감정들을 이 작품집과 함께 공유하고자 한다.

Imagination of Play & Results

Based on the physical condition of the site and local context, the imagination of play was allowed to be expressed freely to the fullest. Considering the local sentiments, the interviews with the residents were minimized; yet, potentially functional clues such as existing tennis courts, parking facility and water tank were allowed for re-approaches toward new utility values. At the same time, ideas and concepts were also expressed under completely new perspectives.

Although 'Revitalization-Play,' the theme for hidden space, was capable of demonstrating a number of creative gestures considering the given conditions comparing to previous years, unknown awkwardness toward such freedom was rather compelled to be felt. Through this experience, however, I was able to fully examine the combination between the raw thread of our space and the time-based fillings of casual and non-casual lives that enrich the raw thread with the activity of 'play.' But most of all, I was able to thoroughly investigate the methods of 'reenactment of play' where the sense is free and alive. It was 15-week-long project that leaves quite a few regrets regarding the intensification of originally intended materials and structures.

I would like to extend my sincere gratitude to Professor Park Jin-Ho and Professor Ryu Sook-Hee who gave me guidances in order to complete all 29 works as well as those who silently supported me including the dean of the department and fellow professors.

Moreover, I would also like to convey my appreciation to Professor Kaye Newman, Professor Jannette Harris and the students as well as the students at Hunan University and Professor Ahn Sung-Hee whom I was merely able to communicate through online. Lastly, I thank all sophomore members who worked hard to gain comprehensive understanding; and I would also like to share all the memories and feelings throughout the fifth hidden space projects along with this collection of works.

Module Summary (2013.09.02)

01 REQUIRED COMPULSORY SUBJECTS

Sophomore "Inteiror Design Studio III"

4 Of 18 Credit Modules in a Semester
- 6 Hours (tuesday) / 1 Week
- Team Project - 2~3 Students / 1 Team
- Total 29 Projects

02 THEME

"THE RISE OF THE CREATIVE DISTRICT"

03 PROCESS & OUTPUT
- Research & Analysis (include FGI & survey, SWOT)
- Space Scenario making (like a cartoon presentation)
- Sketch(analog & digital method) process
- Complex Space Program(zoning, planning....)
- Drawing
- 3D Simulation
- Scale Modeling
- Panel Presentation

& Every Process(Photo & Text) Recording - for the Final Documentation

04 SCHEDULE
- 2013.9. 1st Week - Kick off & Orientation
- 2013.10.Last Week - Mid Term Critic (Small Scale Exhibition & Panel presentation)
- 2013 12.3rd Week - Final Exhibition

05 KEY WORD
- SOCIAL DESIGN
- COMMUNITY BASE
- SUSTAINABLE
- PROSTHESIS
- PERFORMANCE
- PROJECTIVE & REFLEXIVE
- RESONANCE
- SITE-SPECIFICITY

Project site - yongsan gu, seobu ichon-dong street view

Project site - yongsan gu, seobu ichon-dong road view

1st day_02, september ORIENTATION | Hyosung Bldg. 5th Hall AM 09:00~

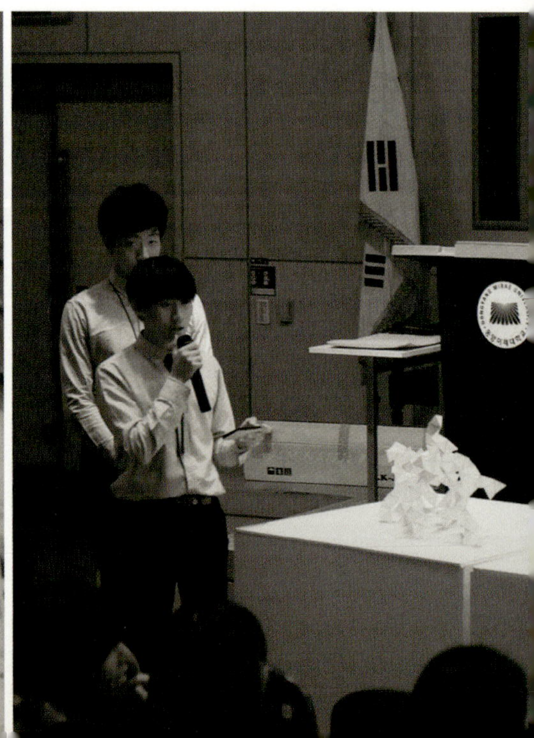

50th day_22, october MID-TERM CRITIQUE | Hyosung Bldg. 5th Hall **AM 09:00-18:00 29 WORKS**

102th day_19, december EXHIBITION

HYOSUNG BLDG.5th HALL | AM 11:00 | 29 WORKS with OPENING PARTY & AWARDS CEREMONY

102th day 19, december EXHIBITION

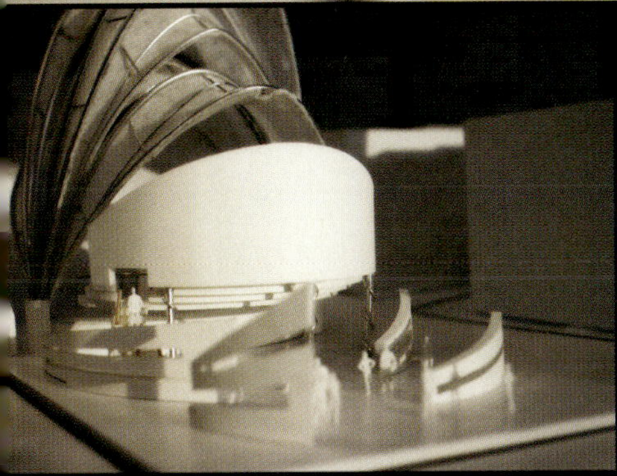

members

Studio_A
류숙희 교수 Prof. Ryu Sook-Hee

Studio_B
박영태 교수 Prof. Park, Young-Tae

Studio_C
박진호 교수 Prof. Park, Jin-Ho

DongYang MIRAE University
interior design

RE-FACTORY
이소라 Lee, so-ra | 정윤경 Jung, yoon-kyung

Urban Island
차민구 Cha min-gu | 정세양 Jeong se-yang

PIPE LINE
이윤정 Lee, yoon-jeong | 김진화 Kim, jin-hwa | 박수진 Park, su-jin

Colored with communication
장한별 Jang, han-byeol | 김아름 Kim, ah-rum

Solar eclipse
김혜미 Kim, hye-mi | 한수민 Han,su-min | 김태호 Kim,tae-ho

Movement
이인근 Lee, in-keun | 홍종무 Hong, jong-moo | 정한울 Jeoung, han-wul

Selling Ship
최은정 Choi, eun-jung | 황영지 Hwang, young-ji

RE;CREATE RECREATE
박용주 Park, yong-ju | 이은미 Lee, eum-mi

PUSH, PULL
심지윤 Sim, ji-yun | 이지애 Lee, ji-ae | 강민지 Kang, min-ji

Agritainment
박종배 Park, jong-bea | 안새봄 An, sae-bom

Butterfly Effect
송지은 Song, ji-eun | 윤설영 Yoon, seol-young

COMPOSE
곽민주 Kwak, min-joo | 이승은 Lee, seung-eun

Thinktank
유영주 Yu, young-ju | 김경희 Kim, gyoung-hee

Flow
김시은 Kim, si-eun | 이명섭 Lee, myeong-sub

Time control
이은비 Lee, eun-bi | 김윤애 Kim, Youn-ae | 채지연 Chae, ji-yeon

EO-UL-LIM
조경찬 Jo, kyoung-chan | 석진주 Seok, jin-ju | 이혜원 Lee, hye-won

FOLLOW ME
윤정인 Yun, jeong-in | 송연주 Song, yeon-ju

Assimilate With Nature
김미소 Kim, mi-so | 이혜경 Lee, hye-kyeong

Flag space inside
박종창 Park, jong-chang | 박소망 Ppark, so-mang | 최경천 Choi, kyung-chun

Freedom in the frame
전채희 Jun, chae-hee | 최수진 Choi, su-jin | 한어진 Han, eo-jin

Folding Space
장혜선 Jang, hye-sun | 이유리 Lee, yu-ri

interesting links
박소영 Park, so-young | 신기랑 Shin, ki-rang

Agritainment
김홍미 Kim, hong-mee | 한아름 Han, ah-reum

voyage of sound
노승희 No,seung-hee | 임정연 Lim, jeong-yeon

Break away
임대홍 Lim, dae-hong | 주민희 Ju, min-hee | 김아란 Kim, a-ran

Sliding PUZZLE
정호진 Jeong, ho-jin | 김해조 Kim, hoe-jo

EQ_III
최승민 Choi, seung-min | 여운비 Yeo, woon-bi

Silver Lining
이호림 Lee, ho-rim | 이성경 Lee, seong-kyung

Hiding in the Book
이기정 Lee, ki-jeong

RE-FACTORY

기존의 주차장을 이용한 Upcycling Community Carcenter제안

이소라 정윤경 | Lee, so-ra Jung, yoon-kyung

> Background

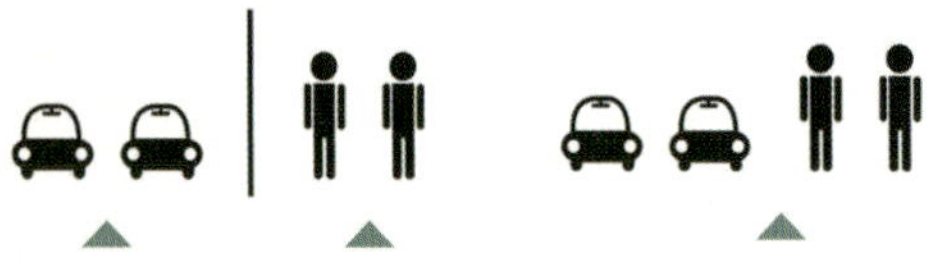

기존 카센터와는 다르게 자동차에 대한 학습과 놀이, 예방정비가 이루어지며 지역 주민들의 편의를 최우선으로 다른 지역과 교류를 활발하게 하면서 갑작스러운 편의시설에 지역 주민들이 거부감이 생겨 기피하지 않도록 세련되고 화려한 공간보다는 업사이클된 재료를 통해 새로운 가치가 부여되는 공간을 계획한다.

Unlike existing repair shops, learning, playing, and preventative maintenance are performed here. For the sake of convenience of local residents in top priority, it actively interacts with other regions. It plans a space of new values with up-cycled materials rather than stylish and colorful space, to avoid the local people feel aversion to the sudden facility.

> Target

자동차 매니아
Car enthusiast

자동차에 대한 이해를 돕고자 하는 학생
students who need understanding of cars

그와 함께하는 가족
his/her family

통합소통의 공간/
지역 주민들이 외부와 자연스럽게 연결될 수 있는 공간

자동차 라는 공통적인 관심사를 통해 새로운 소비자들과 기존 소비자들이 함께 어우러질 수 있는 공간을 계획하였다. 나아가 놀이를 통한 교육을 위해 찾아오는 어린이가 타겟이된다.

Space of integrated communication/
a space for community members to be naturally connected to external space

I planned a space to harmonize new customers with existing customers through a common interest in cars. Furthermore, children who come for education through play are the target.

> Site Analysis

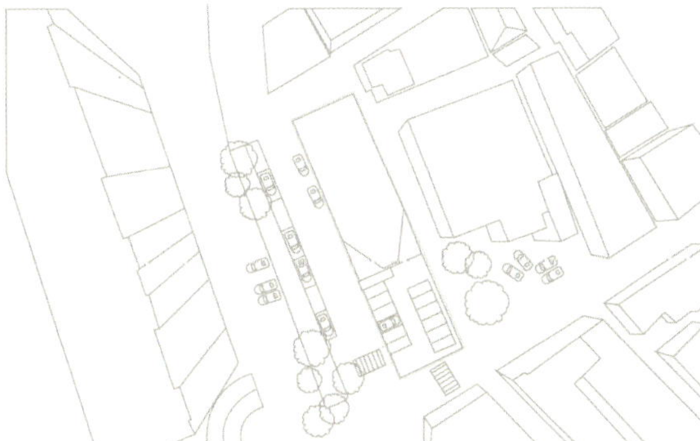

서울특별시 용산구 서부이촌동
이촌2동 대림아파트 주차장
737㎡ / 약 224평
지하철 1호선 용산역 / 4호선 신용산역
주차장과 테니스장으로 이용.

Location : West dong, Yongsan-gu, Seoul
2-dong Ichon Daelim apartment parking lot
Area : 737 m² / approx 224 ft
Traffic : Yongsan Subway Station Line 1/4
Line New Yongsan Station
Use : Used as parking and tennis courts.

신용산역 부근 이촌2동에 위치한 사이트는 한강과 용산역이 근접해 있으며, 용산전자상가와도 가까워 접근성이 높으며, 가시적인 성공의 기반이 마련되어있다. 그러나 현재에는 지역개발의 무산으로, 좁은 도로와 밀집된 주택들로 공간이 협소한 낙후된 환경을 볼 수 있다.

Located in Ichon 2-dong near New Yongsan Station, it is closed to Han river and Yongsan Station with high accessibility to Yongsan Electronics market providing foundation of tangible success. However, nowadays we can see undedeveloped environments with the narrow streets, dense housing, and limited space after the development of this region has collapsed.

자동차의 대중화로 인한
인간과의 밀접한 관계 /
자동차와 인간이 어우러질 수 있는 공간

Closely related with
human due to the popularity
of automobiles /
space where cars and human
can get joined together

Design concept

"움직임의 유희"

공간 안에 움직임의 경계를 모호하게 하여 움직임을 통한 유희 공간을 계획 하였다. 방문자가 움직임을 통한 능동적인 행위를 함으로써 다양한 즐거움을 선사하고, 놀이를 통한 다양한 움직임에 자동차에 요소가 더해져 누구나 찾아와 즐길 수 있는 공간을 마련하였다. 움직임을 반영한 공간은 이동성과 인간의 시점 변화에 따라 매순간 역동적인 투시효과를 자아낸다. 건물은 기호에 따라 시시각각 표정을 바꾸게 된다. 돌출과 함몰, 접힘과 펼쳐짐 그리고 빛의 받아들임과 차단의 대비적인 움직임으로 그 움직임을 밖에서 바라보는 사람이 느끼는 재미와 안에서 바라보는 사람이 느끼는 내부 외부의 모호성이 공간을 다채롭게 한다.

"The Game of the Movement"

In "The Game of the Movement" where the boundaries of the movement are obscure is a play space through the movement. Visitors' active behavior brings themselves a variety of pleasure, and these various movements through the play have elements of cars. Anyone can come here and enjoy themselves. The space that reflects the movement brings dynamic visual effects each moments depending on the movement and change in a viewpoint of human. The building changes its face minute by minute. Contrast movements of Projecting and recessed, folding and unfolding, and accepting and blocking of light make the space colorful with the people feeling fun from outside seeing that movement and the people feeling ambiguity inside.

Proposal

기존의 주차장을 이용한
RE_ENERGISING

+

UPCYCLING 재료를 통한
공간의 UPCYCLING

기존의 주차장을 이용한
RE_ENERGISING

UPCYCLING
COMMUNITY
CARCENTER

PROBLEM
주차시설의 낙후

Aging of parking facilities

SOLUTION
주차장을 이용한
RE-ENERGYSING

Reenergizing
using the parking lot

자동차의 대중화로 인한 인간과의 밀접한 관계 / 자동차와 인간이 어우러질 수 있는 공간
기존 카센터와는 다르게 자동차에 대한 학습과 놀이, 예방 정비가 이루어진다. 운전자들이 최소한이나마 자동차의 기본적인 구조 및 원리를 알고 안전한 운전이 가능하게 한다. 기존 카센터와 차별화를 두기위해 차를 활용한, 주차이외의 활동 공간을 마련하여 수리가 무대가 되며 정보를 공유, 습득 하게되고 공간의 중첩을 통해 수리와 놀이, 수리와 무대의 경계를 없애 "차" 를 활용한 RE-ENERGYSING SPACE를 만든다. 공장의 가장 밑자재인 파렛트를 주재로로 활용한다. 파렛트에게 건물의 중요 포지션인 벽체 역할을 준다. 이에 따라 공간 또한 새로운 의미를 띄며 UPCYCLING 하게 된다.

Closely related with human due to the popularity of automobiles / space where cars and human can get joined together
Unlike existing repair shops, learning, playing, and preventative maintenance are performed here. Drivers get to know at least the basic structure and principles of car, which allows them safe driving. Unlike existing repair shops, room for other activities besides parking is prepared with cars, so that the repair job is performed on the stage allowing information sharing and learning, which makes the place Re-ENERGYSING SPACE with 'cars' eliminating the boundary between repair and play, and repair and stage with space overlapping. Pallet, which is the bottom material of the factory, is the main material used. Pallet serves as a wall, the most important position of the building. Thus, this imparts a new meaning to the building allowing UPCYCLING.

DESIGN OF THE PALLET

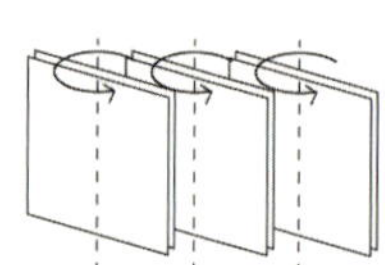

[Pallet, which rotates around the center]

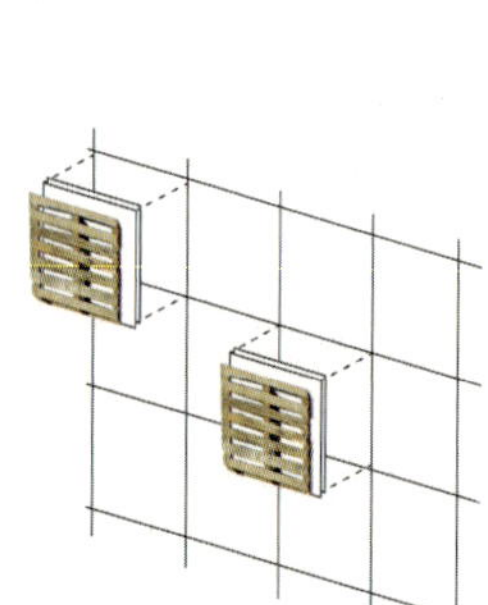

[Pre-made pannel system]

> Design process

01.
레벨차를 이용한 움직임의 유희
Game of movements using the difference in level

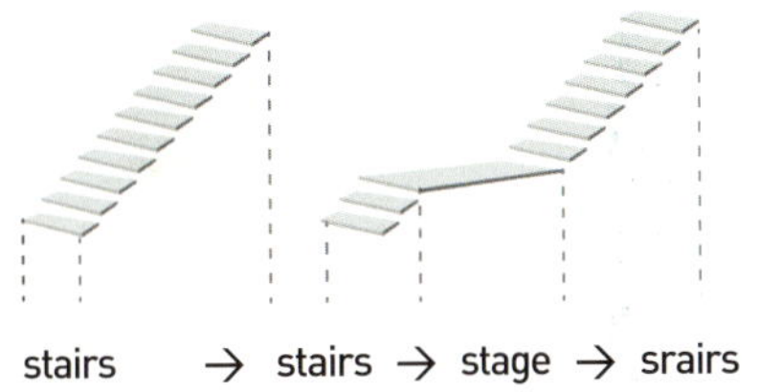

02.
리프트를 이용한 수리공간과 무대의 수직이동
Perpendicular movement of space for repair and the stage using a lift

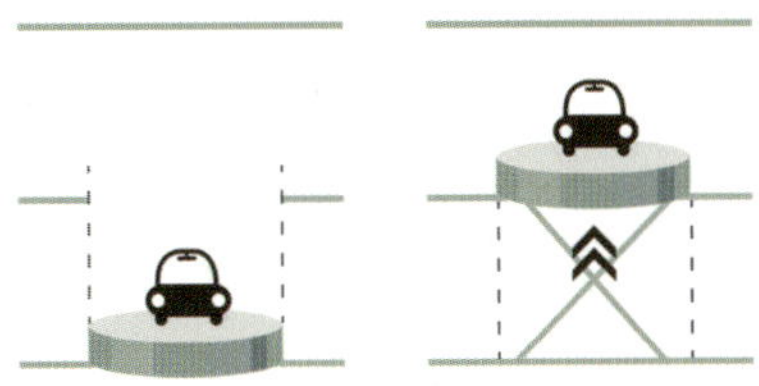

03.
컨베이어벨트를 이용한 sub stage 에서 main stage로의 수평이동
Horizontal movement from the sub stage to the main stage using a conveyer belt

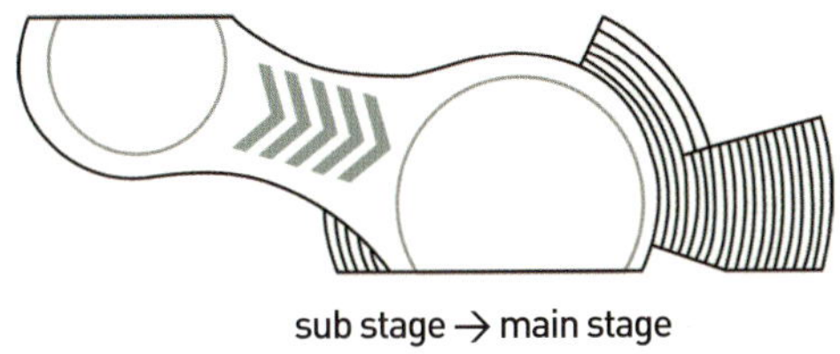

> Floor Plan

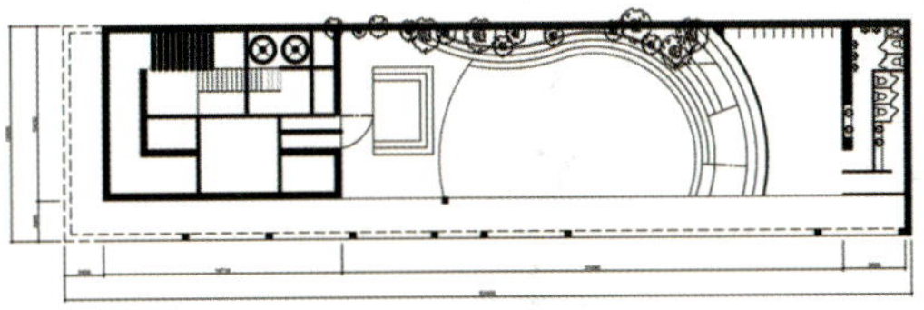
3rd FLOOR PLAN

/play_rollingball
/out door park
/terrace
/toilet

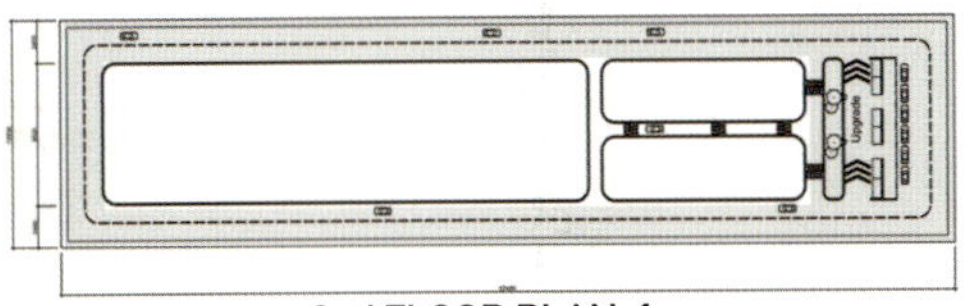
2nd FLOOR PLAN-1

/play_track
/rest space

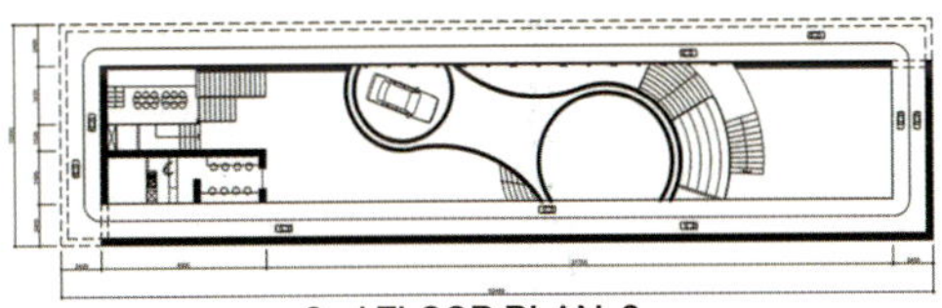
2nd FLOOR PLAN-2

/stage
/play_craft
/play_beer bar

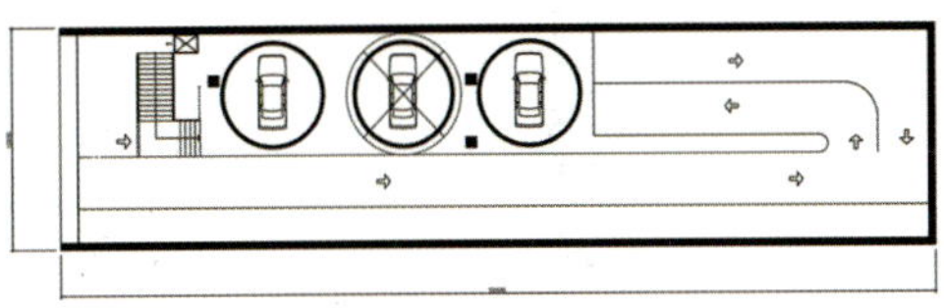
1st FLOOR PLAN

/entrance
/lobby
/repair

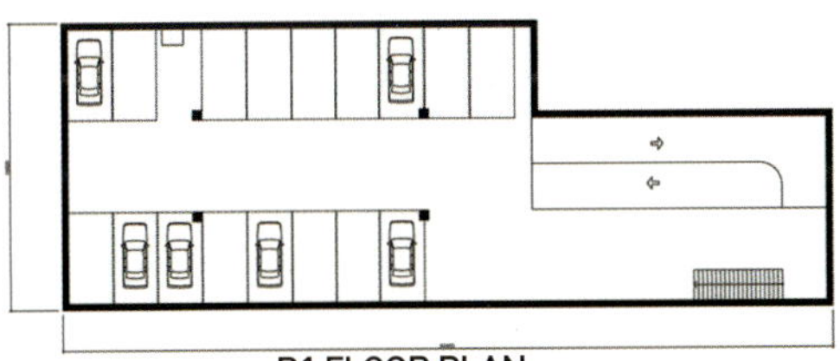
B1 FLOOR PLAN

/parking
/elevator

> Exploded View

2F

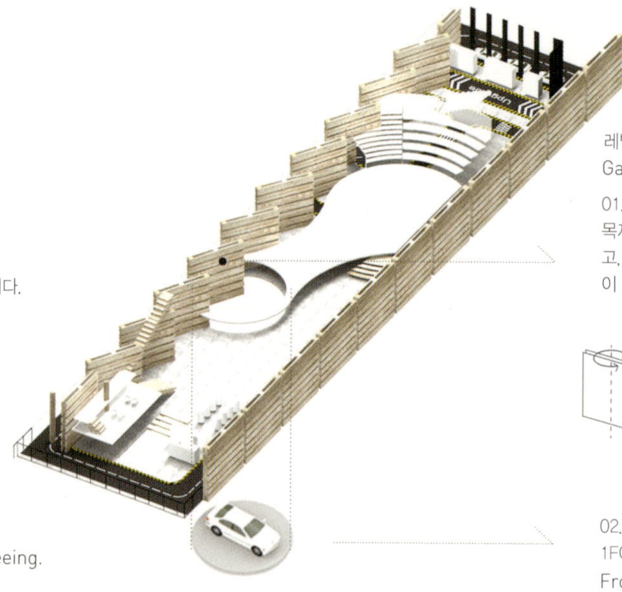

수리의 무대화

RE_FACTORY의 메인으로 자동차를 수리하는 공간이다.
무대 위에서 퍼포먼스 형식으로 수리가 이루어지며,
구경을 통해 학습이 이루어진다.
자동차 부품을 이용한 난타 공연 등
다양한 체험이 이루어진다.

Staged repair

It is the main space in the Re_FACTORY
where the cars are repaired.
Repair job is performed on the stage
as a performance, and the people learn by seeing.

레벨차를 이용한 움직임의 유희
Game of movement using the level difference.

01.
목제 파레트가 중심을 축으로 돌아가 파사드의 표정변화를 주고, 내부와 외부의 경계가 모호해지며 외부에서 내부로의 유입이 자유로워진다.

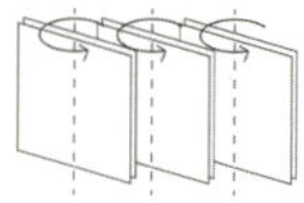

A variety of activities are provided including Nanta performance using auto parts. Wooden pallet rotates as the center axis making changes in the face of facade, obscuring the boundary of inside and outside, and making the inflow from outside to inside freed.

02.
1F에서 2F으로, 수리공간에서 무대로의 수직이동.
From the 1F to the 2F, Perpendicular movement from the repair area to the stage.

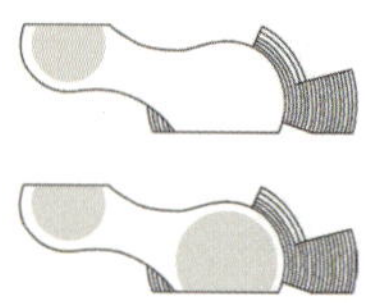

03.
컨베이어 벨트를 이용한 보조무대에서 주무대로의 수평이동.
Horizontal movement from the sub stage to the main stage using a conveyor belt.

탑기어 프로그램과 연계하여 차를 주제로 여러가지 실험을 방청하기도 하고 녹화도 이루어진다. 차를 잘 아는 이에겐 흥미를, 차에 관심이 있지만 어렵게 여기는 이에겐 정보를 주는 공간이다.

In conjunction with Topgear, people can watch various experiments related to cars, and there is also recording job. This is a space to provide interests for Car-savvy, at the same time, provide information for people who are interested in cars but feel difficulty with them.

04.
미니카 트랙을 2F무대 외곽에 둘러 무대와 놀이의 경계를 없앴다. 트랙은 공간사이를 비집고 들어가 터널의 느낌을 연출하였다.

Minicar track is placed around the stage on the 2F eliminating the boundaries of the stage and play. The track goes into the gap between spaces like passing through a tunnel.

05.
미니카 트랙을 2F무대 외곽에 둘러 무대와 놀이의 경계를 없앴다. 트랙 중간에 MINI REPAIR SPACE를 두어 재미를 더했다.

Minicar track is placed around the stage on the 2F eliminating the boundaries of the stage and play. MINI REPAIR SPACE on the middle of the track adds fun.

3F

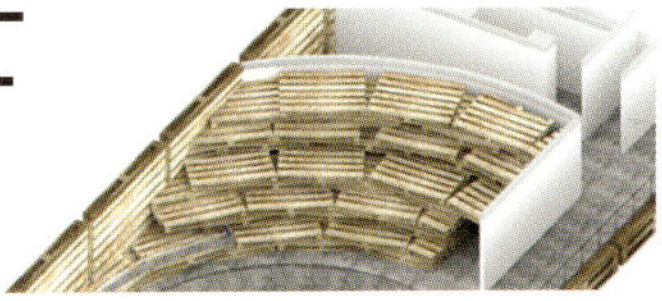

파렛트를 이용한 옥외공원

파렛트를 적층하여 계단형 의자를 형성했다. 유리로 마감된 바닥을 통해 2F의 무대를 볼 수 있다. 놀이를 하다가 지친 아이들을 위해 쉬는 공간이 되기도 한다. 외부와 연결되어 답답하지 않으면서 실내에 위치한 공원이다.

Outdoor Park using Pallet

The pallet is stacked to form a step-like chair. People can see the stage on the 2F through the floor with glass cover. Children who get tired after playing can rest here. The park is located inside, but not stuffy as it is connected to the outside.

차민구 정세양 | Cha, min-gu Jeong, se-yang

Background

There is strong communalism such as Hyang-Yak, Pum-At-I, Du-Re in Korean culture. Communalism which expresses ours: our family, our school, our neighbor and our corporation disappeared since egoism showed my family, my school, my neighbor and my corporation has emerged. For instance, people lived in western Ichon-dong in Seoul cooperated each other. However, they lost the faith because they followed just benefit for themselves, and Yongsan was redeveloped suddenly. It shows the gab between rich and poor that the number of apartments and education are rapidly advanced.

So we decided an alternative to remind the sense of community for people lived in western Ichon-dong. The project's name is 'Treasure Island'.

우리나라는 예로부터 향약, 품앗이, 두레처럼 공동체 정신이 담겨있는 문화가 있었다.
이후 서서히 오늘에 이르기까지 우리 집, 우리 학교, 우리 동네, 우리 회사로 부르던 공동체 정신이 사라지고 자기 가족, 자기 학교, 자기 동네, 자기 회사만을 생각하는 이기주의적인 모습이 많이 등장한다.
서부이촌동의 경우 과거 주민들은 공동체를 이루며 서로 협력하는 삶을 살았다. 하지만 갑작스런 용산 재개발로 인해 주민들은 서로에게 불신이 생기고 등을 돌리게 되었다. 또한 자기 이익 챙기기에 급급했고 그런 삶은 황폐하기 그지없었다.
급격하게 늘어난 도시 아파트 생활과 교육 문화는 빈부의 격차를 실감할 정도로 양면성을 보이고 이웃 간의 왕래 및 공동체문화 역시 찾아보기 어려운 게 현실이다.

우리는 서부이촌동 주민들이 잃어버릴 뻔 했던 공동체의식을 '보물섬'을 통해 되찾아주려 한다.

Issue

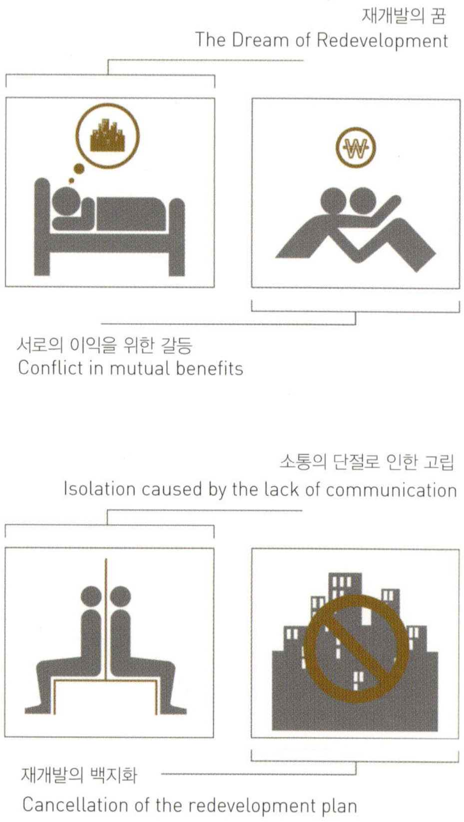

Urban Island

:THE HIDDEN TREASURE ISLAND

재개발 사업으로 인해 주민들이 잃어버린 가치를 찾다.
Value lost due to redevelopment projects to rehabilitate residents.

Site analysis

한강대로를 기준으로 서부이촌동은 '다리 서쪽', 동부이촌동은 '다리 동쪽' 이라고 불린다. 두 동네는 걸어서 20분, 차로는 5분밖에 걸리지 않을 정도로 가깝지만 동부이촌동은 외국인이 많이 사는 부촌, 서부이촌동은 서울의 마지막 미개발 지구로 인식된다. 60년대 말꿀꿀이죽 이웃이었던 동부이촌동에는 대단지 아파트단지가 세워진 반면 서부이촌동엔 판자촌이 밀집되었다. 40년간 동부이촌동만 부러워하던 서부이촌동에 한강르네상스 바람이 불기시작했다. 하지만 2013년 꿈은 무너지고 불어난건 세금과 빚뿐인 서부이촌동은 예전에는 남의 집 숟가락까지 알 정도로 재미난 동네였지만 이제는 서로 아무도 말하지 않는 동네로 변했다.

There are called "East side of bridge" and "West side of bridge" across the Hangang Bridge. There are closed to each town. It takes only 5 minutes by the car or 20 minutes on foot. However, there are known many wealth foreigners are living in eastern Ichon-dong, and western Ichon-dong is the last undeveloped district. In 1960's, they shared some meal like soup each other, but now eastern town has developed new apartments while western town still remain. It began Han River Renewal Project in 2006 to western town had not developed for 40 years. However, the project is canceled so only they have huge debts.

Concept

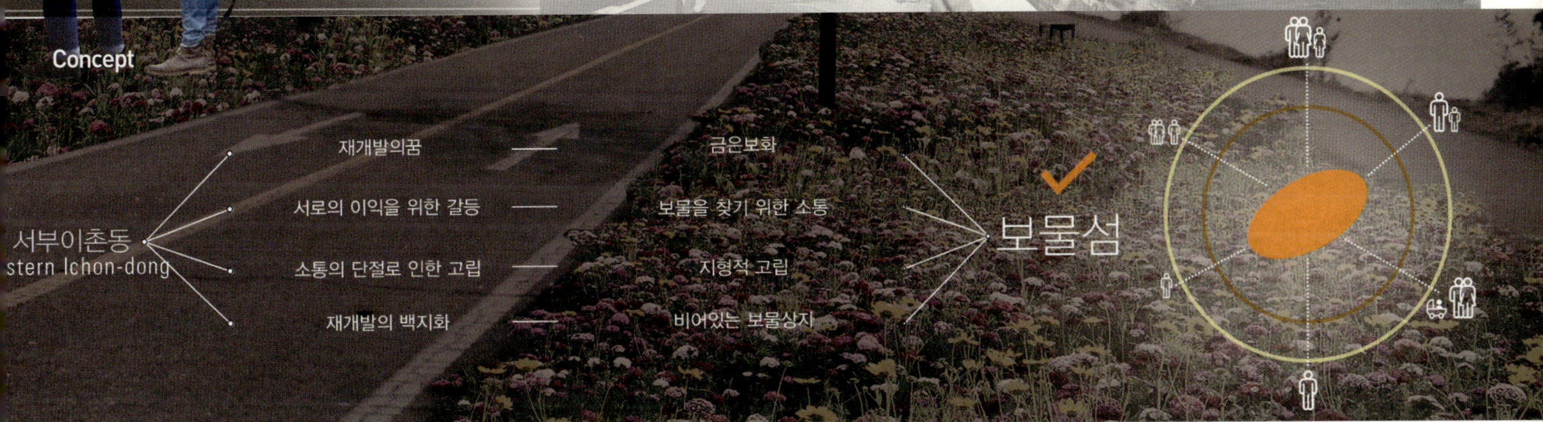

Communication between neighbors — 주민들간의 소통
Obstacle in Treasure Island — 보물섬 안의 장애물
Energy in Motion — 움직이면서 발생하는 에너지

Concept process

기존의 공간
The Existing

장애 요소
The Problem

기존 공간에
장애 요소의 유입
Flowing in the Problem in the Existing Place

잃어버린 가치
The Lost Value

기존의 공간에서
잊고있던 가치를 발견
Discovering the Forgotten Value in the Existing Place

외부 작용
The External Action

보물섬
The Treasure Island

Concept process

보물섬을 찾은 아이들이 보물을 찾기 위해 여러가지 놀이기구를 이용하며 공간을 즐긴다.

1. 보물을 찾아 나선 아이들은 보물섬에 들어선다.
2. 오래되어 보이는 밖과는 달리 안 쪽 멀리 반짝이는 무언가를 보며 더 가까이 다가간다.
3. 조금 더 가까이 다가가자 쉽게 닿을 수 없게 여러 장애물들을 만나게 된다.
4. 먼저 보물을 찾기 위해 온 사람들이 쳐 놓은 그물일까?싶은 그물로된 장애물을 만나게 된다.
5. 계단을 오르며 반짝이는 곳의 입구를 찾지만 쉽지 않다.
6. 주변을 둘러보니 다른 아이들도 쉽게 찾지 못한다. 서로 밀어주고 끌어주며 협력을 하니 한결 나아진다.
7. 마침내 서로 도와가며 목표점에 도달했지만, 각자가 원하던 보물은 어디에도 없고 함께 걸어온 친구라는 보물을 발견한다.

Children enjoy the Treasure Island with many kinds of rides for Treasure Hunting.

1 Children who leave for Treasure Hunting enter the Treasure Island.
2 Children approach something shiny. It is different from outside of Island where looks old.
3 Obstacle is appearing Children are more closed to something shiny.
4 Children meet the web trap first. It seems like people who already come here to find treasure installed before children come.
5 Children are looking for the gate on the each floor but it is difficult to find.
6 So, Children try to find the gate together.
7 Finally, They reach the goal, however there has no treasure. They realize that the real treasure is their friends by each others.

함께 한다는 것 자체가 어색해진 동네 서부이촌동. 다양한 놀이기구를 함께 즐기며 하하호호 웃는 아이들을 보며 어른들도 이웃들에게 마음을 열도록 도와준다.

Western Ichon-dong where people is weird to live together.
They open their heart while they are watching children's playing.

Plan process

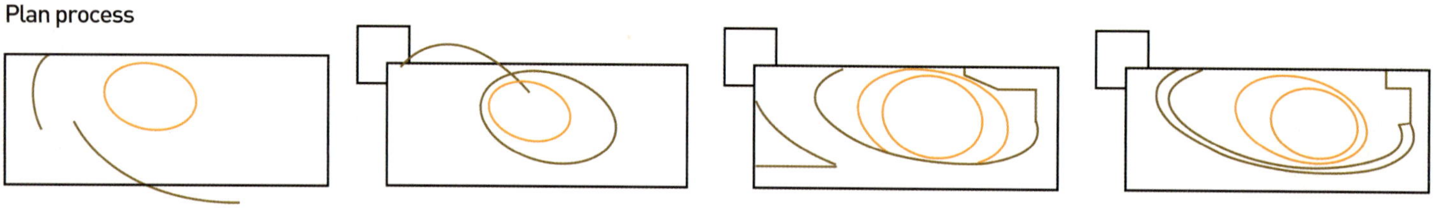

Elevation

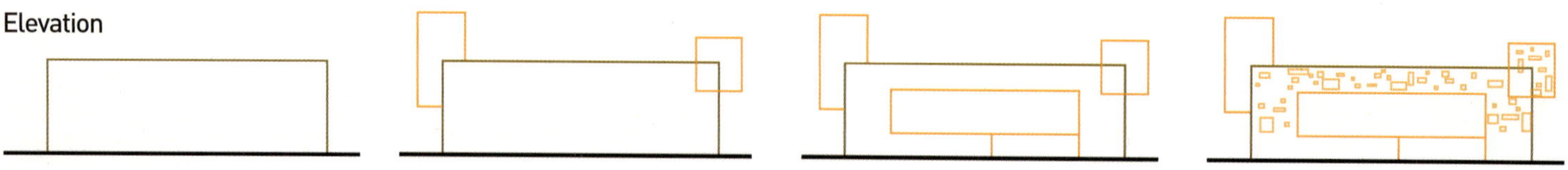

Plan

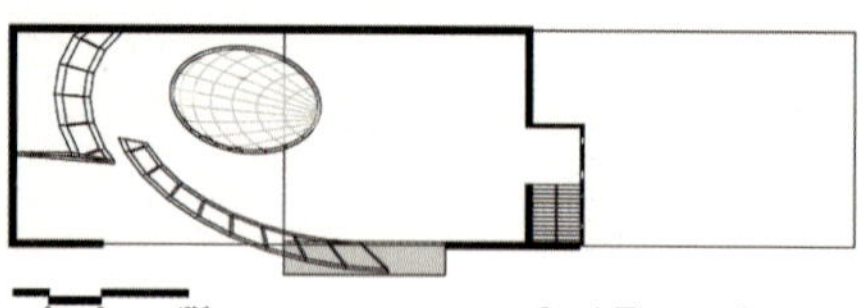

2nd Floor plan

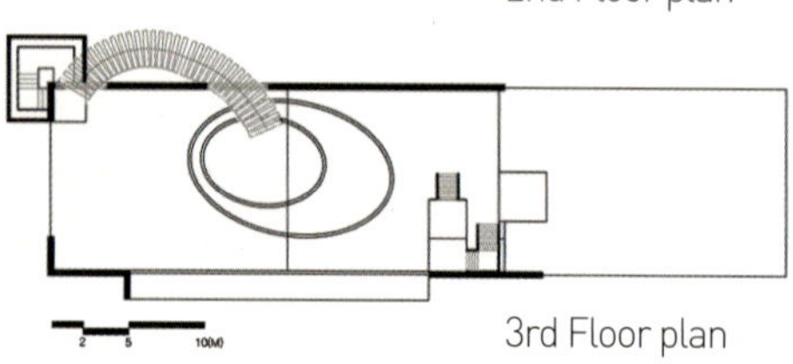

3rd Floor plan

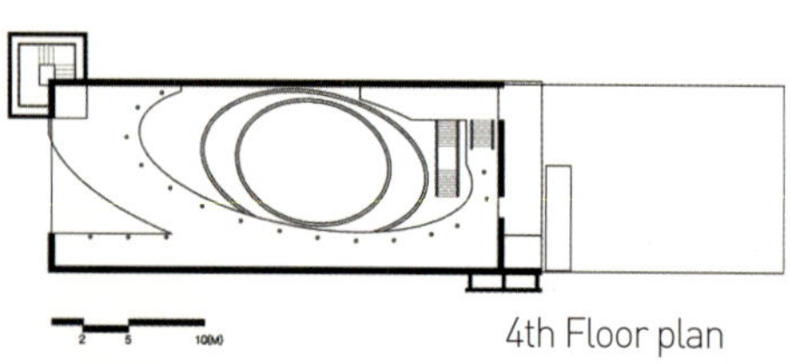

4th Floor plan

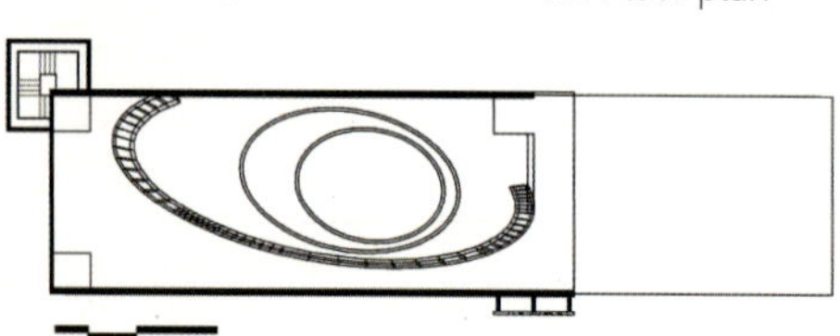

5th Floor plan

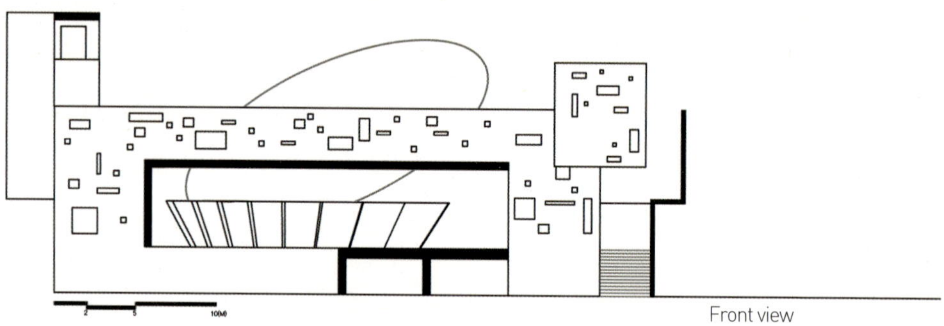

Front view

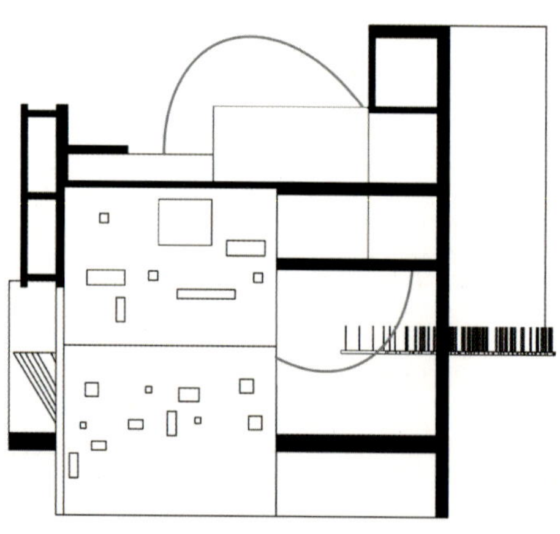

Right view

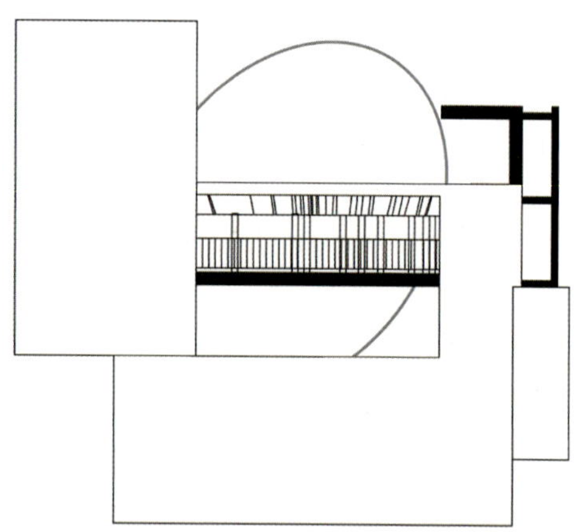

Left view

건물 입구에서 부터 타원체까지 이어지는 계단과 장애물을 통해 눈 앞에 보이는 보물이지만 쉽게 닿을 수 없다는 것을 느끼게 한다.

In the process from the gate to ellipsoid with obstacle, they notice that it is difficult to reach the treasure even so closed.

내후성 강판과 시멘트가 건물 외부 주요 재질이다. 코르텐 강판이라고도 불리우는 내후성 강판은 대기중에 방치하면 초기에 합금원소 작용에 의해서 표면부에 치밀한 산화 피막을 형성한다. 보물섬이라는 컨셉에 맞추어 생각하였다. 타원체를 감싸는 외부가 부식이 되면서 시간의 흐름에 따라 아름답게 변하는 것이다. 타원체안의 조경과 대조되면서 사이사이로 보이는 타원체 외부의 반사재질이 주위의 시선을 이끈다.

Weathering steel and cement is main material of outside. Weathering steel is also known, called Corten steel if left in the air by the action of initial alloying elements in the surface portion of a dense oxide film is formed. Treasure Island was thought to fit the concept. Surrounding the ellipsoid as external corrosion over time this will change beautifully. Contrast in the landscape as seen through the ellipsoid ellipsoid between the reflective material around the outside of the eye of the head.

섬에 있는 숲 속 이미지를 생각하여 놀이 시설을 디자인 하였다. 나무들이 울거진 거창한 숲 속 멀리 반짝이는 보물이 보이지만 숲 속을 풀어 헤치고 나아가야한다는 스토리다. 나무를 위에서 바라볼 때 나타나는 형태와 가지와 가지들이 얼기설기 엉켜있는 모습을 그물로 해석하였다.

We designed the rides which we imagined the forest of island. It is based on the find way to get the treasure in the forest. We make the web from the shape which sees the forest above the island.

4F 복도 전시실 각양각색의 창을 통해 서부 이촌동 밖을 내다 볼 수 있고 긴 의자에 앉아 잠시 쉬었다 갈 수 있다. 반짝이는 타원체와 스테인레스 재질의 기둥이 서로 반사되어 타원체에 대한 호기심을 더욱 더 자극한다.

On the 4th floor, it is served the long bench for the rest and various windows to see the western Ichon-dong's landscape. Shiny ellipsoid and pillar of stainless steel stir up people's curiosity though reflecting each other.

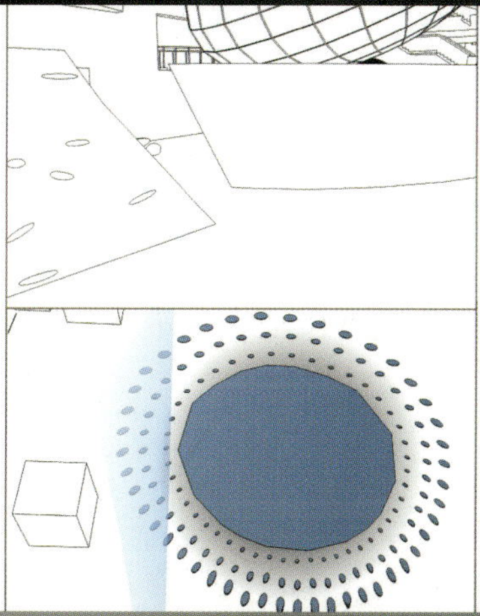

암벽타기 놀이를 통해 동선을 단축시킬 수 있다. 타원체의 모양대로 둘러쌓인 가벽이 타원체를 가리며 직접 볼수 없게 하지만 동선을 유도한다.

The line of flow can be reduced though rock climbing. The wall which surround an ellipsoid make it could not see the ellipsoid by blind, but the wall induce movement.

타원체 안의 구조이다. 위에 뚫린 구멍을 통해 자연 채광이 확보 되며 태양의 고도에 따라 달라지는 그림자가 지켜보는 이들에게 즐거움을 준다.

This is the structure in an ellipsoid. It is well-lighted though a hole of the top, and a shadow which is changed by sun light give children a pleasure.

PIPE LINE

그들의 눈인사의 얽힘과 소통으로 Re–energizing 되다.
Communicate with their eyes as the Re-energizing become entangled.

이윤정 김진화 박수진 | Lee, yoon-jeong Kim, jin-hwa Park, su-jin

Site

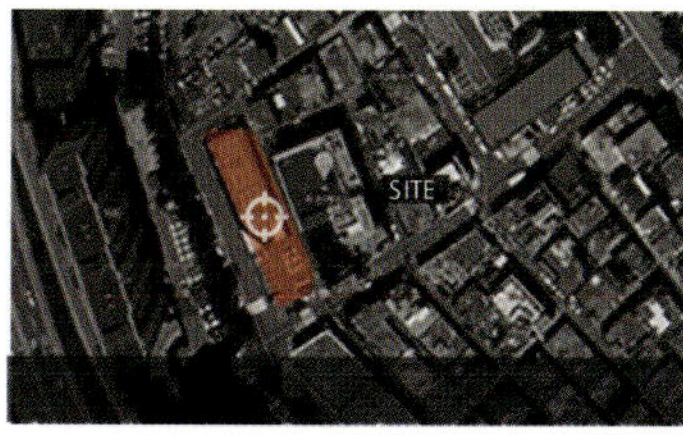

서울특별시 용산구 서부이촌동 이촌 2동 대림아파트 주차장

Daelim apt parking lot, Ichon 2-dong,
West Ichon-dong, Yongsan-gu, Seoul

History

여느 동네와 같았다. 재개발 이라는 정책이 이 지역에 다가오기 전까지 재개발 이라는 정책은 주민들이 서로 편을 가르게 되는 계기가 된다. '그것은 그들의 개인적인 이익과 관련된 일이었기에 그들은 상대의 의견을 잘 들으려 하지 않았고, 서로를 향한 눈빛들이 달라졌다.

웃음이 없고 대화가 없었다. 뚝 끊겨 버렸다. 마치 한겨울에 동파되버린 수도관 파이프처럼. 그들은 서로를 향한 마음을 닫아 버렸고, 그들사이의 오고갔던 인사들과 웃음소리는 사라졌다. 겨울이 지나 봄이 오듯, 이곳에는 봄을 맞이할 곳이 필요하다.

There was no difference as other village. The policy of redevelopment was the turning point to divide residents into outsiders before it was adopted. It was related to personal interest. That's why they didn't listen to other's opinions.

There was no smile nor talking. Everything was disconnected. Just like a water pipe broken in the middle of cold winter, people closed their mind and then sound of laughing disappeared. As spring comes after winter passed, people need a place to welcome spring.

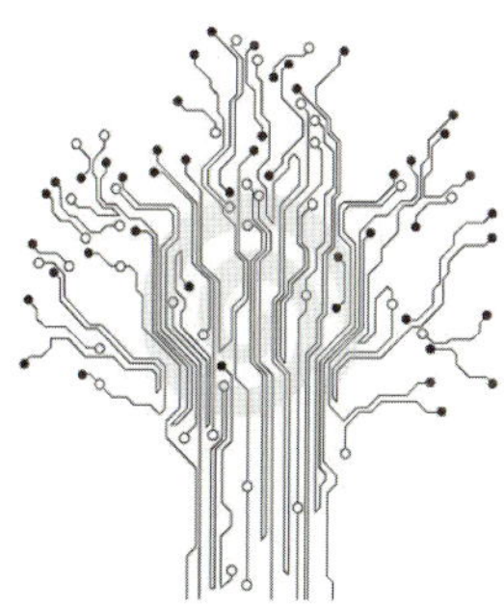

Background

용산 개발사업이 추진되고 집값과 주민들의 기대감 그리고 희망은 급격하게 오르기 시작했다. 하지만 갑작스러운 용산 개발 사업의 백지화로 주민들 간의 소통은 단절 되었으며 분위기는 점점 침체되어 갔다.

Since Yongsan redevelopment plan was pursued, the price of houses and expectation as well as hope rapidly began to rise up. However, with the suspension of Yongsan redevelopment plan, communication among them was stopped and it started getting worse.

Solution

Pipeline이 생기면서 주민들이 모여들기 시작한다. 주민 들간의 자연스러운 눈맞춤의 얽힘과 그로 인한 소통이 시작 되고 점점 순환 하면서 이촌2동 주민들의 분위기와 환경은 Re-energizing 된다.

With the coming pipeline, residents begin to get together. Thus residents started to communicate one another and get along with them, which gradually becomes an warm-hearted environment.

Design keyword

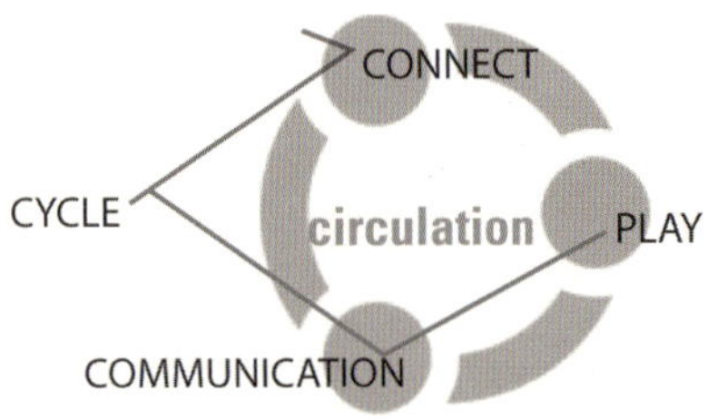

Design Concept

CPU Pipeline

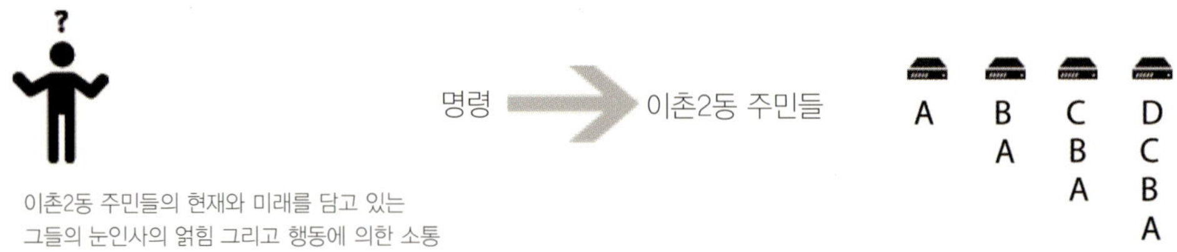

이촌2동 주민들의 현재와 미래를 담고 있는
그들의 눈인사의 얽힘 그리고 행동에 의한 소통

The communication that holds residents' hope, future, and present of Ichon 2-dong

생산 라인 등과 같이 여러 공정별로 생산 라인이 나열되어 있고 동시에 공정별 프로세서가 가능하게 하는 것으로, 시스템의 효율을 높이기 위해 명령문을 수행하면서 몇 가지의 특수한 작업들을 병렬 처리하도록 설계된 하드웨어 기법.

There is an assembly line of production such as the progress of work, which enables a working stage and include a hardware technology to deal with special jobs to elevate system's efficiencies.

Material

플라스틱과 비닐

재료 또한 순환하는 의미를 담고있다. 향후 이곳이 철거 되거나 리디자인 되더라도 이곳에서 나오는 잔해물 들은 무언가의 형태로 다시 재활용 될 수 있다.

그리드

이촌2동 주민 네트워크를 표현했다. 사람들의 어디로 튈지 모르는 그들 끼리의 연결 네트워크를 수평 수직의 일방향이 아닌 자유로운 곡선으로 표현하였다.

놀이의 순환

나이를 먹으면서 사람에게는 '놀이' 에 대한 인식이 변화한다. 놀이를 즐길 공간이 변화하고 방식도 변화한다. 그러나, 놀이를 즐기는 이유는 변함이 없다.

PLAY ZONE 에서 REST ZONE 으로 내려오는 놀이는 놀이공간의 순환을 보여주며 이는 세대 차이 없이 놀이의 공간이 중첩되는 것을 보여준다.

Plastic and vinyl

Materials include a circulation as well. Even if these are nullified later or redesigned, they can be recycled.

Grid

It represented those residents of Ichon 2-dong. It demonstrated their connecting network not just one vertical way but just a free curve.

Circulation of a playing

As people get older, the recognition of playing is changing. Thus the playing space as well as method is changing. However, the reason to enjoy playing is never changed.

Playing descending from PLAY Zone to REST ZONE not only demonstrates certain circulation but also proves an overlapping space.

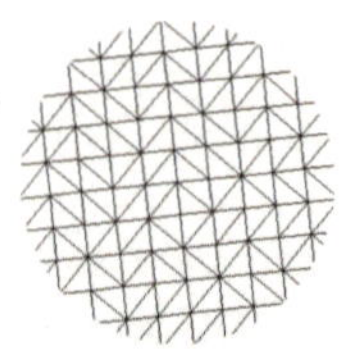

읽기(Fetch)
– 주민들을 공간으로 이끈다.
– 메모리에서 명령을 가져온다.

- Lead residents into a space.
- Bring a command from memory.

해석(Decode)
– 명령을 해석한다.
– 주민들의 연령 등 여러가지 상황을 고려

- Interpret a command.
-Consideration of several situations of residents' ages and others

실행(Excute)
– 명령을 수행한다.
– 자연스럽게 휴식과 놀이와 소통이 이루어진다.

- Execute command.
- Perform communication, rest, and playing.

기록(Write Back)
– 수행한 결과를 기록한다.
– 주민들의 행동과 소통이 공간에 흔적으로 남는다.

- Record performed results.
- Leave residents' action, communication and other remains.

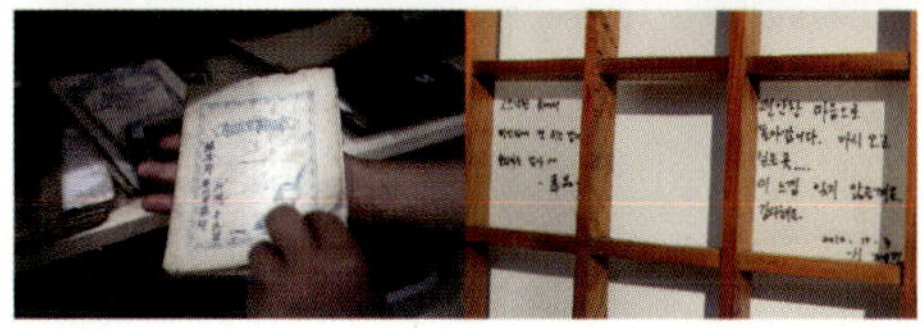

서로의 흔적이 쌓이면서 서로 공유하고 또는 다시 이 공간을 찾게 되는 매개체가 되기도 한다.

As traces are accumulated, they become a means of communication to find a space again.

Design Motive

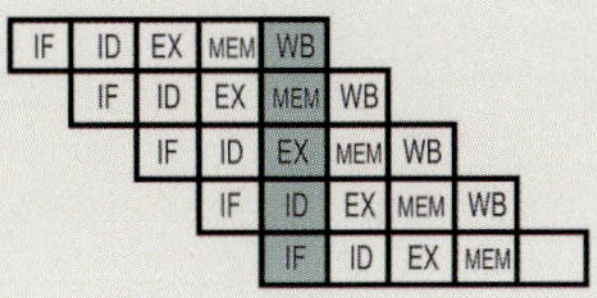

Design Process

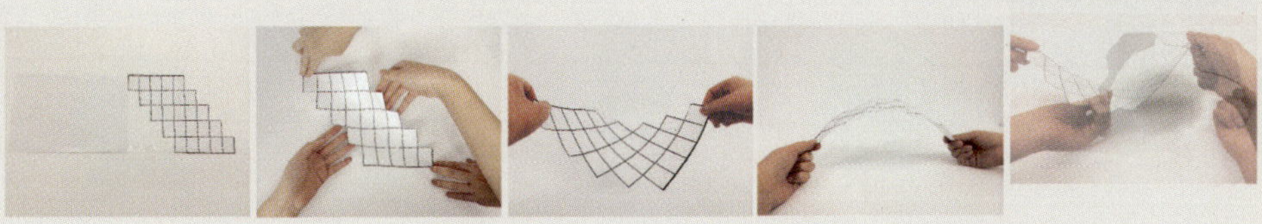

Zonning

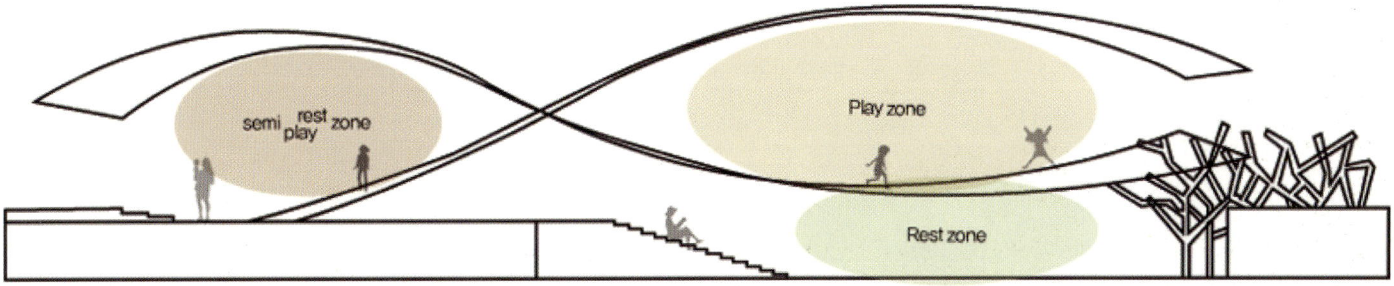

Floor plan

Elevation

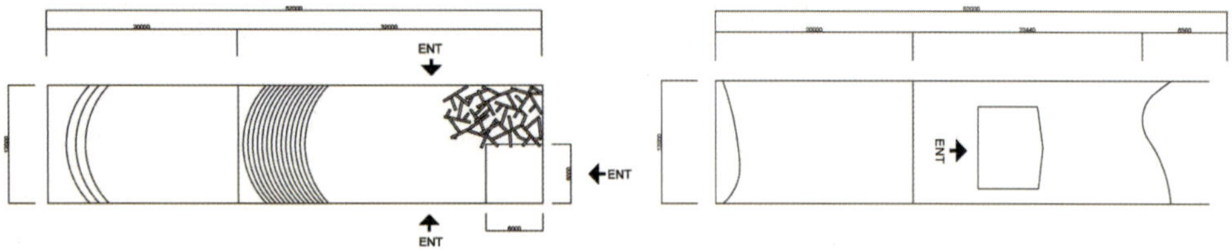

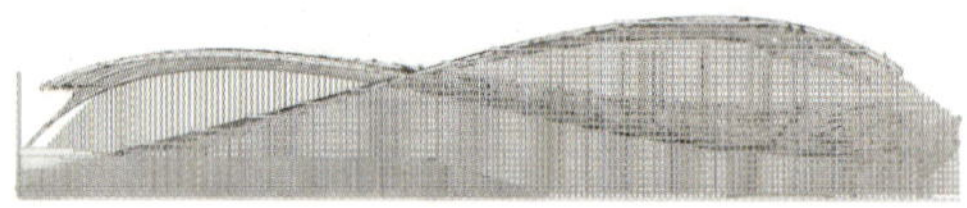

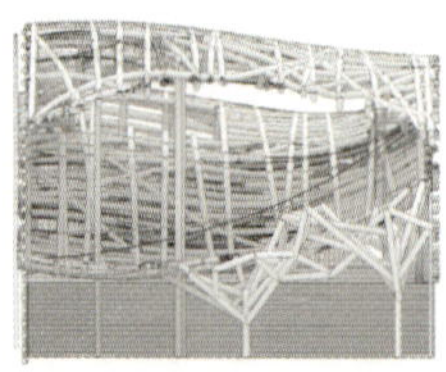

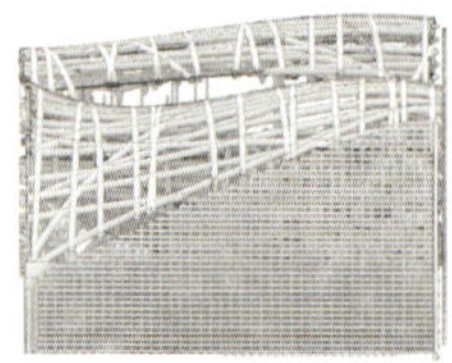

1F_Rest&play

타겟 : 영유아~초등학생과 보호자
파이프라인들이 벽을 타고 내려오다 의자를 만든다. 엄마들은 이곳에서 윗층과 연결된 나무네트에서 노는 아이들을 보면서 휴식을 취할 수 있다.

Target: Child-elementary school students and guardians. While coming down along wall, they make chairs. Mothers can take a rest, watching children playing on tree nets.

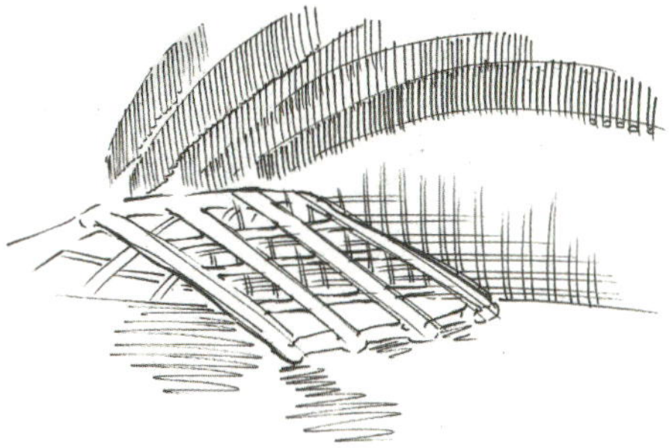

2F_semi rest&play

타겟 : 청소년과 어른
천장에 달려있는 패브릭의 소재는 시간의 흐름을 담는다. 이 곳에서는 어른과 아이 할 것 없이 놀이에 직접적으로 혹은 간접적으로 참여하게 된다. 끝 쪽에 있는 계단에서는 잠시 쉬어갈 수 있고 본격적인 놀이공간으로 가기 전 워밍업같은 놀이공간이다.

Target: Juvenile and adult
The source of fabric on ceiling holds the vestige of time. In this place, children as well as adult participate in playing. At the edge of a step, they can take a rest and it is like a warming space prior to playing.

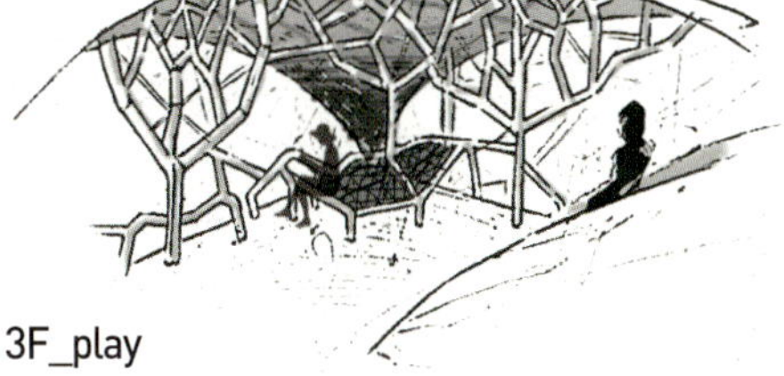

3F_play

타겟 : 전체
네트로 이루어진 놀이공간이 많다. 해먹의 형태를 빌려온 놀이와 네트 뭉치와 타이어를 타고 그네처럼 왔다갔다 거닐기도 하고 1층으로 내려가는 봉들도 있다. 이곳은 누구나 활발하게 즐길 수 있는 놀이공간이다.

Target: Entire
There are a lot of playing spaces. They are like a borrowed hammock, riding on a tire just like children swinging up and down. Some one comes down to the first floor. This is a playing space anyone can play actively.

장한별 김아름 | Jang, han-byeol kim, ah-rum

Background

용산 역세권 드림허브 프로젝트로 용산은 서울을 대표하는 랜드마크가 들어서고 서부 이촌동은 항구를 갖춘 수변도시로 재탄생 할 전망이었다. 그러나 사업의 금융난으로 인해 드림허브 프로젝트는 개발이 무산되었다. 오랜시간 개발 지체로 인해 지역 주민들은 지쳐갔고, 지역 주민들 간의 찬반 논란으로 가깝던 이웃들 마저 서로 갈등이 생기게 되었다. 마을에는 개발을 주최한 사람들을 맹비난하는 플래카드와 오랜 개발 미진으로 인해 낙후된 마을의 분위기는 침체되었고 지역 상권은 점점 죽어 가고 있었다. 그래서 우리는 침체된 지역의 발전과 주민들의 소통을 이어줄 수 있는 공간을 계획 하고자 한다.

Through Dream Herb project of Yongsan subway area, Yongsan was expected to rebirth as a waterfront city with ports in Seobuichon-dong and landmarks representing Seoul. However, the project foundered as a result of financial crisis of the business. The residents have been exhausted by long-time delay in development, and the neighbors, who had used to bond, had a feud, due to strong conflicts between them. The town became lagged because of the long-time delay and banners bashing people hosting the project, whilst the business districts have deteriorated more and more. Therefore, we are planning to make a space for residents to communicate and to revive the depressed area.

Colored with communication

fabric과 double skin을 이용한 전통 천연 염색 체험공간

Using a double skin fabric and traditional natural dyeing experience space

Site analysis

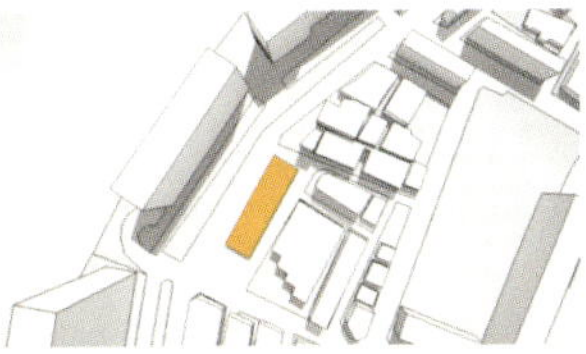

서울 특별시 용산구 서부이촌동 396에 위치한 대림아파트 공용 주차장 인근에는 용산역 1호선과 한강이 위치 해 있다. Site 주변에는 대중교통이 불편하여 유동인구가 적으며 오랜 개발 부진으로 상권이 침체되고 노후 된 건물이 많이 있으며 지역 주변에는 놀이시설 등 문화시설이 부족하다. 아파트 단지로 인해 한강에 대한 조망 확보가 어렵다.

Yongsan station line 1 and the Han River are placed in around public parking lot of Dae-lim apartment located 396, Seobuichon-dong, Yongsan-gu, Seoul. Around the area, the business districts deteriorated because there is not large floating population due to poor public transport system and the development has been delayed for a long time, moreover, there is a shortage of cultural and amusement facilities. It is also difficult to secure the view of the Han River, caused by apartment blocks.

Target & problem

서부 이촌동 아이들

우리 동네에는 놀 수 있는 곳이 많이 없어서 너무 심심해. 재미 있는 놀이 공간이 있었으면 좋겠어!

Children living in Seobuichon-dong

We're so bored that there aren't many places to play. We'd like to have some kinds of playing space!

서부 이촌동 주민들

예전에는 주민 모두 사이 좋게 지냈는데 개발 문제 때문에 서로 사이가 멀어졌어. 같이 어우러질 수 있는 공간이 생기면 사이를 회복할 수 있지 않을까?

Residents living in Seobuichon-dong

We used to get along well with all residents before we drifted apart because of some problems related to the development. Could we make the relationship better if there is a place we can socialize with all together?

어른들과 아이들

요즘 아이들과는 이야기를 잘 할 수 없는 것 같아. 아이들에게 좋은 것 을 알려주는 기회가 생겼으면 좋겠어. / 할머니, 할아버지랑 같이 재미있는 놀이를 하면 어떨까?

The olds and children

I think it's not easy to talk with kids these days. I hope I have an opportunity of providing something good for them. / Why not play fun games with grandma and granddad?

Space program

Natural dyeing activity

: 천연 염색 체험

전통 천연 염색공간을 통해 우리나라의 전통적인 염색 방법을 체험한다. 몸에 해롭지 않은 천연염색으로 아이들과 어른들은 염색의 과정을 통해 직접보고 만지는 체험을 하면서 서로 소통을 할 수 있고, 갈등을 해결할 수 있으며, 이 공간을 통해 외부 지역주민의 유입을 유도하여 지역 활성화를 기대할 수 있다.

We experience our traditional dyeing skills through the space for it. Through the space, we can expect to activate the area by inducing outside residents, experiencing the harmless natural dyeing process with children and adults all together, and communicating each other, which would resolve the conflicts.

Concept

천 [; Fabric]

천은 삭막한 도시와 대조되어 부드럽고 포근한 느낌을 가지고 있다. 천의 물질적 특성을 이용하여 공간을 계획하려고 한다.

Fabric has soft and cozy feeling, on the contrary of desolate cities. By using the material character, the space will be planned.

유연성 Flexibility

Solid한 벽체가 아닌 유연한 천으로 벽의 역할을 한다. 공간과 공간 사이에 자유로운 이동이 가능하다.

Flexible fabric plays a role as wall, rather than solid one. It is possible to move freely between spaces.

통기성 Ventilation

천을 통하여 빛,바람,소리와 같은 자연적인 소재를 느낄 수 있다.

Through fabric, natural materials such as light, wind and sound can be felt.

Design process

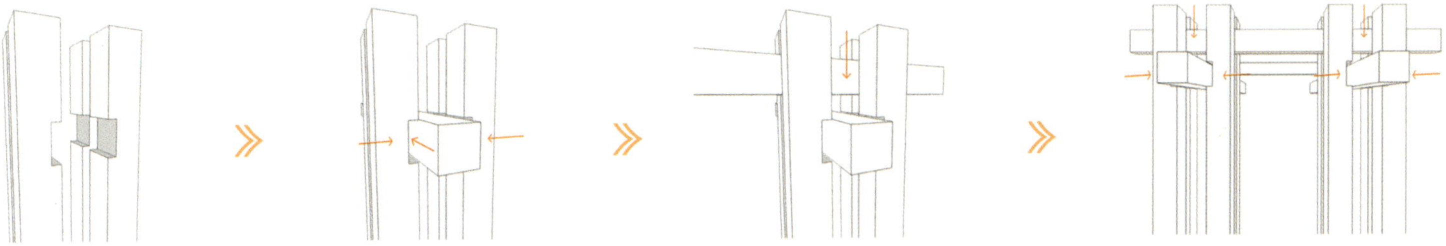

모듈러 방식을 가구와 파사드의 각재 접합방식으로 이용하였다.

Modulo type was used with a rectangular lumber type combining facade with furniture.

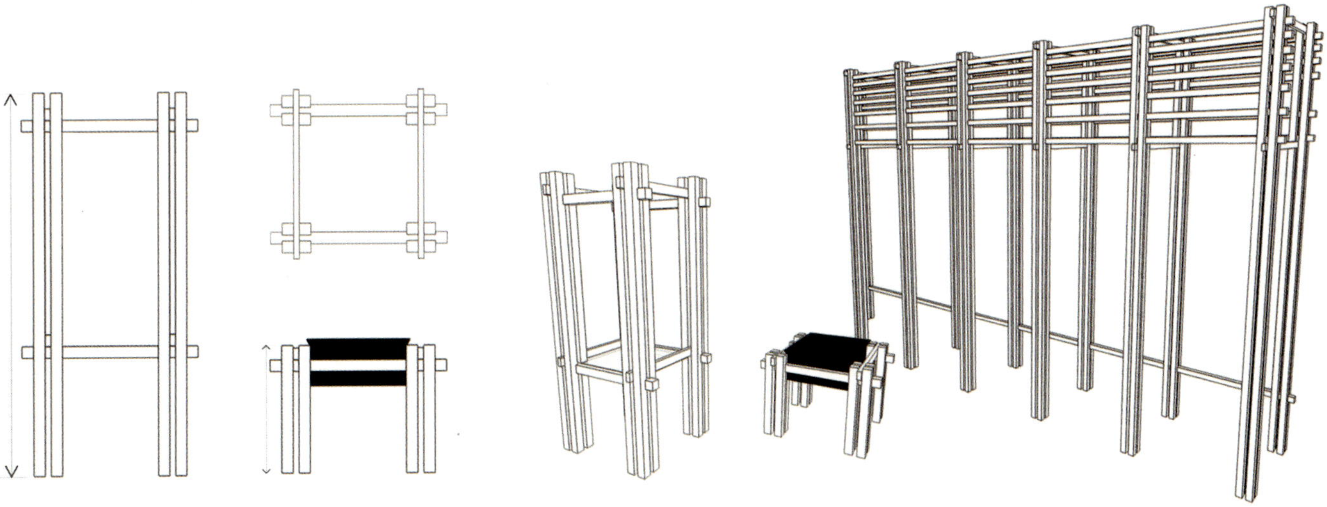

Furniture & Facade Frame

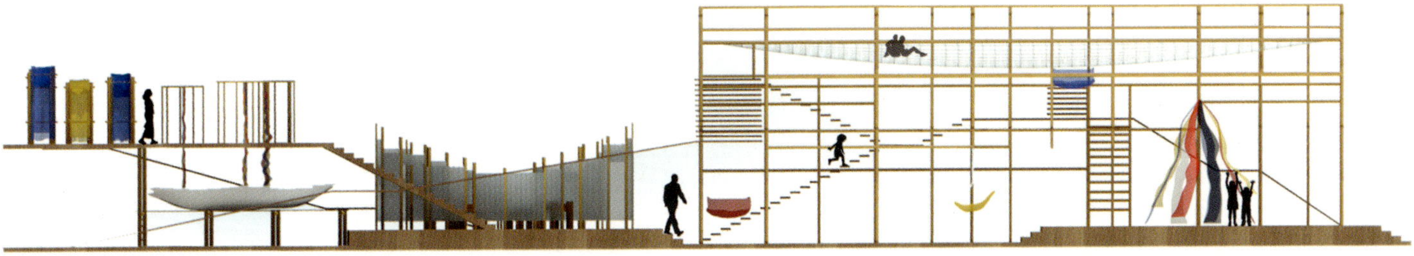

Exploded view

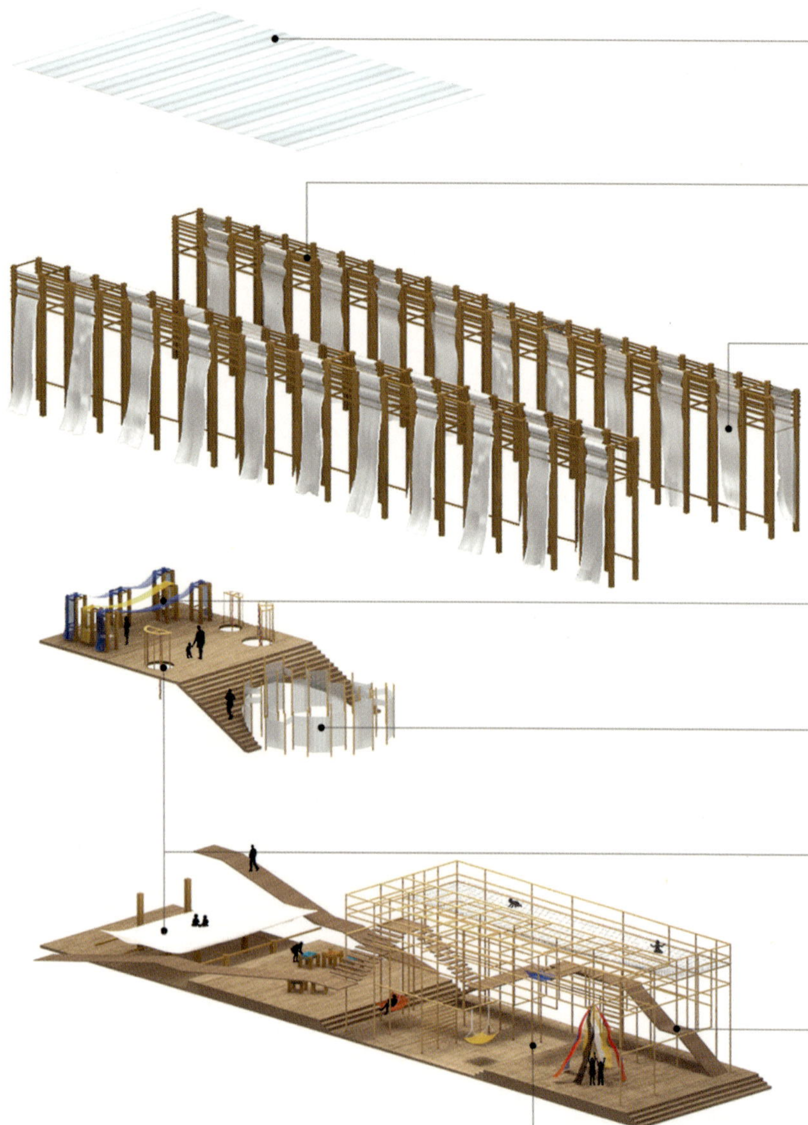

Polycarbonate [Thk 15]

지붕은 전통적인 천을 현대적으로 풀어, 값싼 물성인 폴리카보 네이트로 대신한다. 폴리카보네이트는 반투명한 소재로 자연광이 투과된다. 서로 엇갈리게 쌓아 공간에 빛이 유입되는 양을 다르게 하여 시간에 따른 자연광의 변화를 공간에서 느낄 수 있다.

The roof is made by low-priced polycarbonate which is translucent and permeating natural light, by interpreting traditional fabric in a modern way. According to time going by, changes of natural light can be seen in the place due to a variety of the amount of natural light by piling them alternately.

Wood [Thk 100, 50]

나무 각재를 사용하여 모듈러 방식으로 facade를 구축한다. facade frame을 double로 두어 두 개의 frame 사이에 통로가 생긴다. 이 통로로 내부 진입이 가능하다.

By using rectangular lumbers, facade is built with modulo type. With double facade frames, there are passages between them, where we can go inside through.

Facade에 천을 사용하여 공간에 유연성을 준다. 천이 휘날리며 빛과 바람 등의 자연적인 요소가 외부에서 내부로 스며든다.

The space has flexibility by using fabric on facade. Natural factors, such as light and wind etc, can pervade inside from outside though the fabric blown.

2층 체험공간. 천을 널고, 건조 시킬 수 있는 가구를 배치한다.
Experience the second floor space. Neolgo cloth, dry the furniture is arranged.

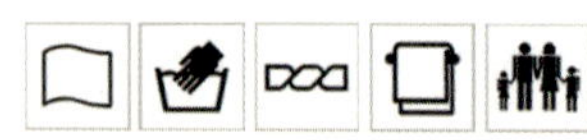

1층의 천연염색 체험 공간으로 open space. Double frame으로 통로가 된다. 벽체는 폴리카보네이트의 반투명한 성질로 외부에서 염색공간 안의 행위가 비쳐 보여지는 재미를 준다.
Open space as a natural dyeing experiential space on the first floor. Double frames make passages. The translucent polycarbonate walls show the inside dyeing space from outside, which is interesting.

천으로 만들어진 로프를 이용해 내려가, 1층의 천막구조에서 play가 가능하다.
Also, children can play in a tent-like structure, by going down with ropes made by fabric.

계단과 슬로프를 이용하여 체험공간과 놀이공간, 외부를 수직적, 수평적으로 연결 해 준다.
Experiential spaces, playing spaces and outside are all a by stairs and slopes horizontally and vertically.

나무 구조물 수직적, 수평적으로 연결해 구조물 안에서 계단과 슬로프를 이용해 이동하며 놀이를 할 수 있도록 하였다. 그네, 길쌈놀이, 사다리 놀이가 있다.

By connecting wooden structures horizontally and vertically, games can be played inside, moving with stairs and slopes. There are swings, Gilsam game(One kind of Korean traditional game), and ladder game.

Floor plan

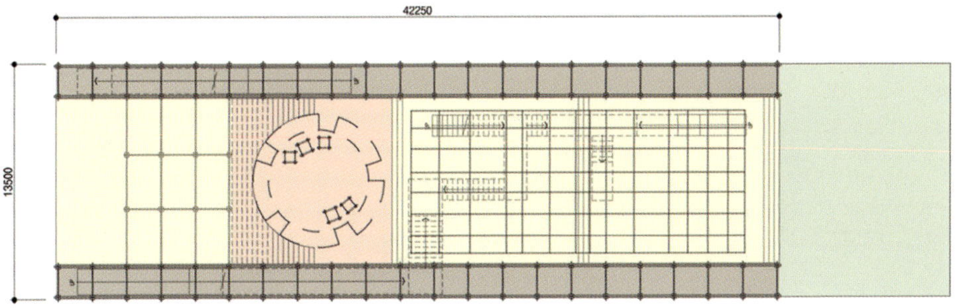

1st FLOOR PLAN

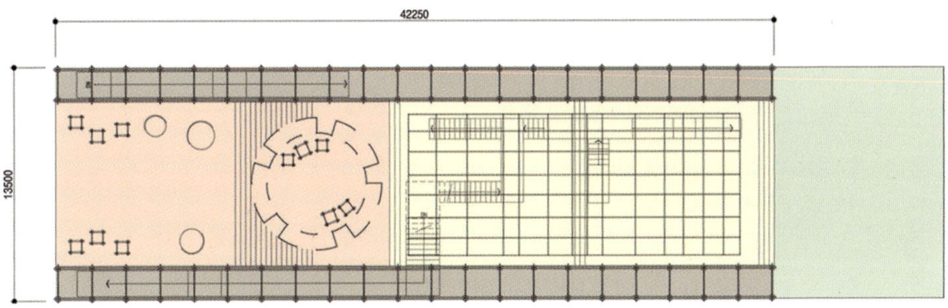

2nd FLOOR PLAN

Elevation

RIGHT VIEW

LEFT VIEW

- Play Space
- Experience Space
- Parking
- Facade & Path

김혜민 한수민 김태호 | Kim, hye-mi Han,su-min Kim,tae-ho

Solar eclipse

일식(日蝕) : 지구상에서 볼 때 태양이 달에 의해서 가려지는 현상

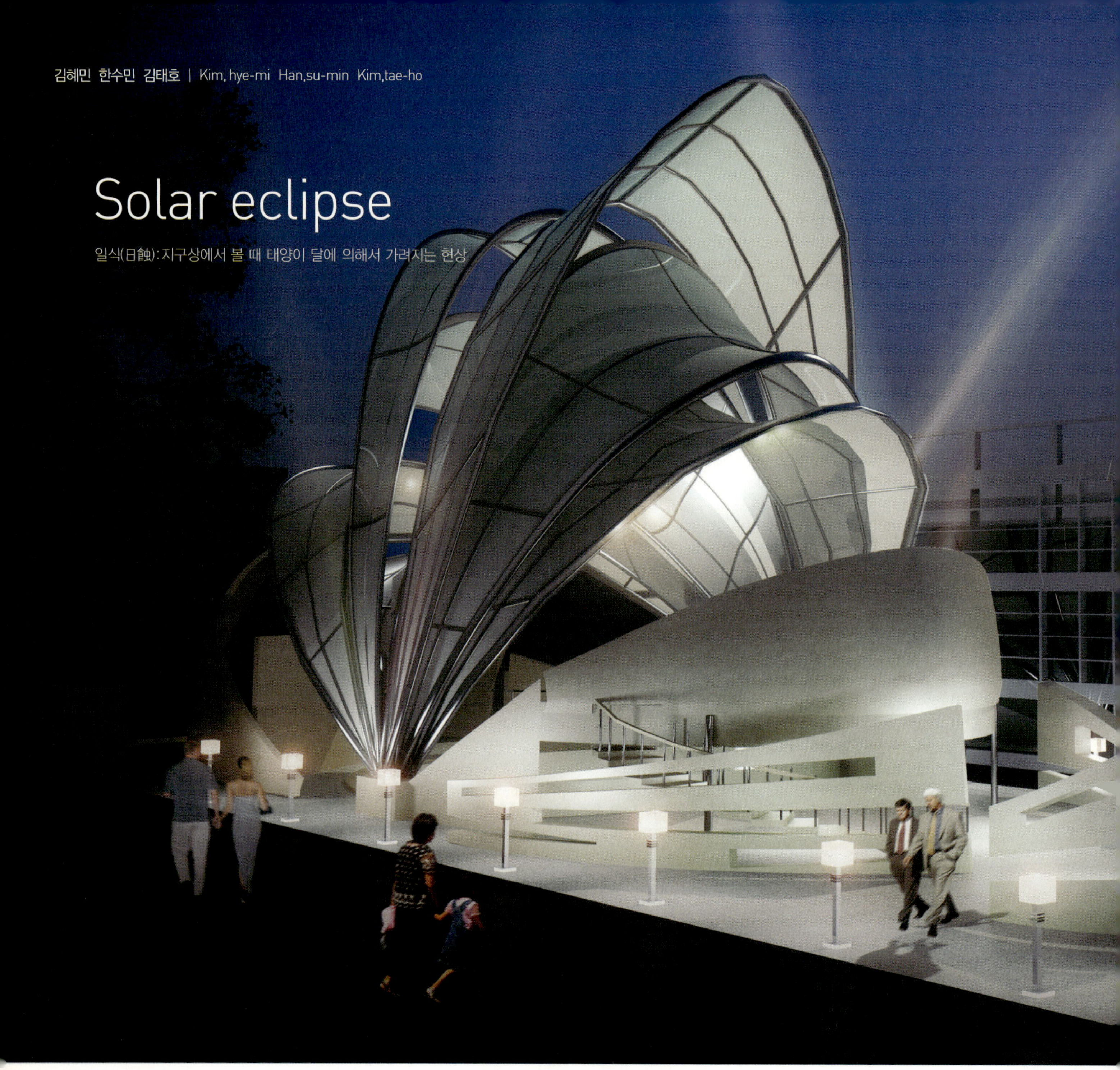

Site analysis

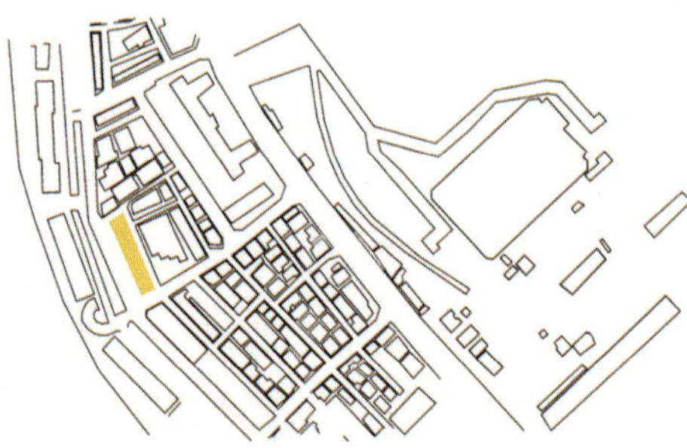

서울 용산구 이촌2동 주민센터 옆 테니스장

서부이촌동은 용산역에 근접해 있고, 한강대교와 원효대교 사이에 있다. 물리적으로 복지시설 미흡으로 인한 노인과 아동의 쉴 공간이 부족하며 환경적으론 근접한 도로로인해 미세먼지와 매연, 소음피해가 극심하다. 부족한 주차 공간은 좁은도로를 만들고 낙후된 주위환경과 건물로 인해 안전 위험에 노출되어 있다.

Tennis court next to the community center of Yongsan-gu, Ichon2-dong, Seoul. Seobuichon-dong is near Yongsan Station, and is located in between Hangang Bridge and Wonhyo Bridge. Senior citizens and children don't have enough place to rest due to the physical insufficiency of welfare facilities, and in environmental aspect, fine dust, exhaust, and noise damages are extreme because of adjacent streets. Insufficient parking space creates narrow streets and people are exposed to safety danger by underdevelopment of surrounding environment and buildings.

Background

사람들은 각자 자신만의 이상이 있고 그 이상향에 도달하기 위해 실패와 좌절을 겪으면서 끊임없이 도전한다. 그렇게 노력해서 얻어지는 이상이 있는 반면에 아무리 갖고싶어도 가질 수 없는, 욕심으로 찬 이상도 있다. 하지만 사람의 욕심은 끝이없고 가질 수 없는 이상을 잡기 위해 그 이상보다 소중한 많은 것들을 잃어가면서 더 큰 이상을 쫓는다. 이는 지금에 만족하지 못하고 이룰 수 없는 이상과 이익을 위해 소중했던 이웃간의 정과 소통의 상실, 커뮤니케이션의 단절로 갈등 중인 서부 이촌동에서도 볼 수 있다.

People have their own ideals and they constantly challenge themselves while experiencing failure and frustration to reach that utopia. While we have that ideal acquired through such effort, there is ideal filled with greed that we cannot have no matter how much we want to own. However, people chase for greater ideals by losing more precious things to reach that ideal that they can no longer have. Ideal of not being satisfied with current situation or unattainable desire, lose of attachment and communication between neighbors who were once valuable to each other because of their own benefits, and conflict from communication breakdown can be seen here in Seobuichon-dong.

Target

서부이촌동 주민들

서부이촌동 주민 구성들을 보면 아이들과 노인분들이 대부분이며 서로 공존하여 놀 수 있는 공간이 마땅치가 않고 부족하다. 그외에 다른 연령층의 주민들도 여유를 즐길만한 공간이 부족하다. 그래서 우리는 서로 공존을 하며 같이 놀이를 즐길 수 있는 공간과 휴식공간을 제공하고자 타겟으로 잡았다.

A greater part of the composition of Seobuichong-dong residents is children and seniors, and there are limited space for them to coexist and play together. Residents of other age groups do not have enough space to enjoy composure. So, we targeted them to provide space and resting area for them to co-exist and play games together.

Concept

일식이란 달이 해를 가리는 것으로 지구와 태양과 달이 일직선 일때만 나타나는 아주 드문 현상이다. 해는 사람들이 바라보는 이상으로, 욕심으로 가득찬, 가질 수 없는 이상을 말한다. 달이 해를 가리는 시간이 되는 일식은 가질 수 없는 이상이 잠시 보이지 않는단 의미로, 사람들로 하여금 자신의 이상이 허황된 욕망에 불과한 것이 아니었는지 되돌아 보게 되는 계기가 된다. 여기서 달은 유동적이며 움직이는 주체로 달이 지나가고 난 후의 해는 개인의 욕심으로 찬 이상이 아닌 모두의 바람으로 찬 이상향이 된다.

Solar eclipse is of moon covering the sun and it is a very rare phenomenon that only appears when earth and moon is lined in a straight line. Sun is an ideal that people admire which refers to an ideal filled with greed that they cannot possess. Solar eclipse of moon covering the sun signifies an ideal that cannot be possessed becoming invisible for a moment that it becomes an opportunity for people to look back on whether their ideal was just an absurd desire. Here, the moon becomes a fluid movable subject that the sun after moon passed by becomes utopia filled with everybody's wish, not an ideal filled with one's greed.

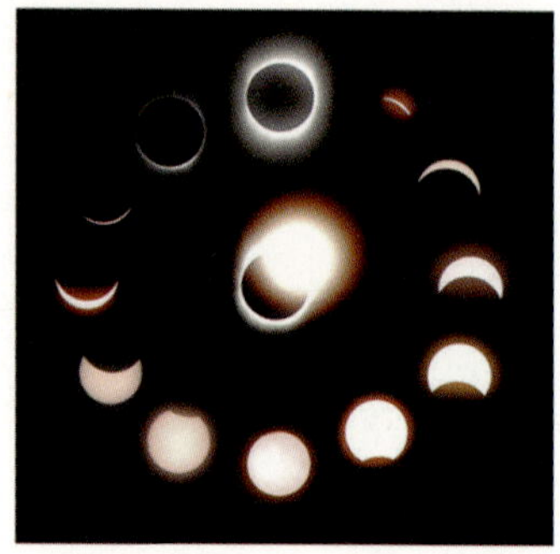

Motive

달의 움직임

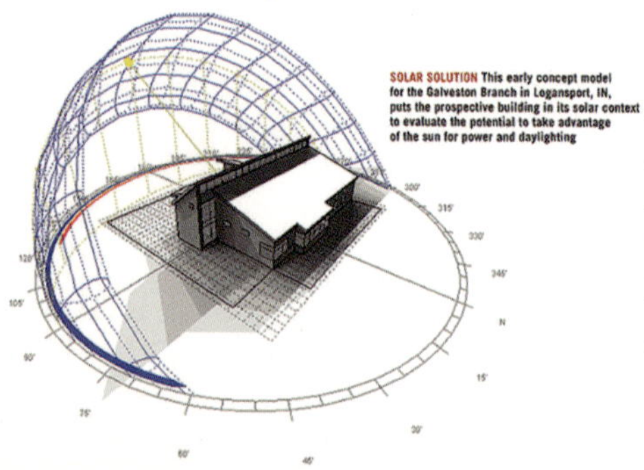

달이 초승달에서 반달, 보름달로 변해가는 과정에서의 움직임. 이 움직임에서 막구조의 형태와 천정이 이동하는 경로를 모티브로 가져왔다.

Movement of the process of moon changing from crescent moon into half moon and full moon.
The moving path of a shape of membrane structure and ceiling was set as a motive.

달은 한 방향으로 공전하지만 지구상 어느 위치에서 보느냐에 따라 움직이는 모양이 달리 보인다. 달이 이동하는 모습을 선으로 나타내면 직선, 대각선, 포물선 등 다양하며 그 각도 또한 여러가지로 나타난다.

Moon revolves in one direction, but the movement shape appears differently depending on which direction it is seen from. When movement appearance of the moon is represented into lines, the shape differs from a straight line, diagonal line, and an arc, and the angle is diverse as well.

Exterior process

식(蝕)이 일어나는 과정

Form of a total solar eclipse occurs

여러각도에서 본 달의 움직임

The movement of the moon as seen from different angles

User needs

넓은공간(extensive) 휴식공간(rest) 문화공간(enjoy)

Program

열린(Open) + 광장(Square) & 공연장(concert hall)

사이트 분석 결과 서부이촌동 사람들이 여가시간에 편하게 이용 할 수 있는 복지시설이 미흡했고 근접한 차도와 좁은 도로는 침체된 서부이촌동을 더 삭막하게 만들었다. 빽빽한 아파트와 주택단지, 주차문제로 인해 좁아진 도로, 인도에서 여유와 소통이라곤 찾아 볼 수 없다. 이처럼 보여지는 동네도, 보이지 않는 사람들의 마음도 꽉막혀 닫힌 서부이촌동에 탁 트인 만남의 광장 같은 공간을 만들어 여유로움을 주며, 공연장을 섭목해 열린 마음으로 서부이촌동 주민들이 서로 소통 할 수 있는 공간을 만들고자 한다.

As a result of the site analysis, welfare facilities where people in Seobuichon-dong can use in leisure time were insufficient and nearby steets and narrow roads made the depressed Seobuichon-dong more desolate. Composure and communication were nowhere to be found in streets that became narrower due to the tightly packed apartments, residential areas, and parking problems. The intention was to provide composure in Seobuichon-dong and to people where the town as well as invisible heart of people are confined and sealed tightly, and to create space where Seobuichon-dong residents can communicate with each other with open mind by grafting a performance arena.

Material

막 (curtain)

프레임 (frame)

막구조: 구조체가 휨, 강성을 갖지 않든가 또는 그것을 무시할 수 있는 부재로 구성되고, 외부 하중에 대하여 막응력 즉 막면 내의 인장 압축 및 전단력으로만 평형하고 있는 구조

Membrane structure: The structure is composed with member of framework that does not have the bending stiffness or can disregard the fact, and a structure which is balancing the external load solely by the membrane stress, meaning tension-compression and shearing force inside the membrane.

(휴식 + 즐거움)
rest + entertainment

공간안에서 사람들이 소통함
people communicate in the space

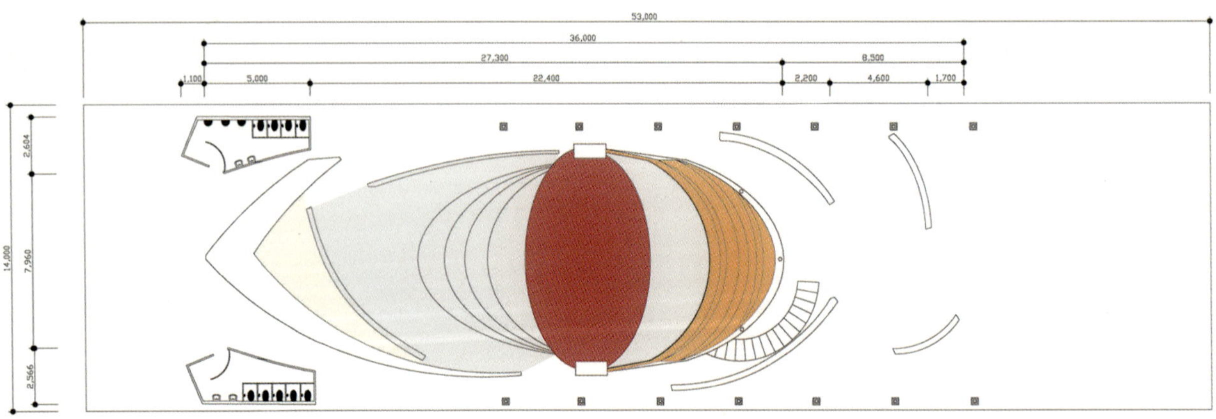

공연장(venue)

무 대(Stage)

대기실 (Staffroom)

관람석 (Stand)

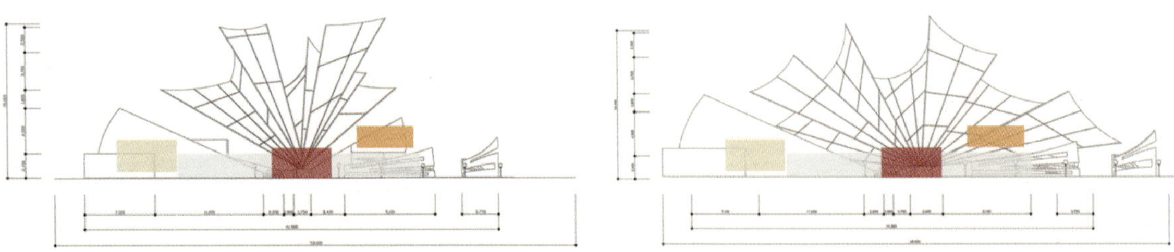

광장(square)

쉼터(reat area)

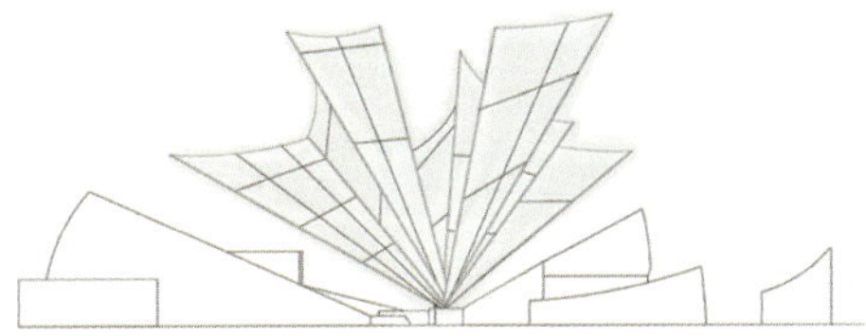

Open

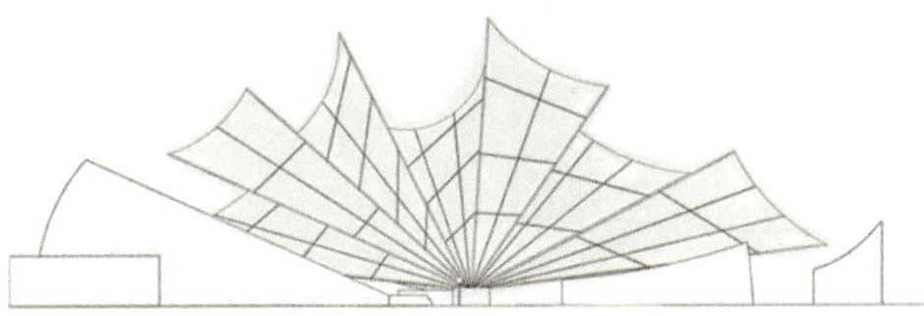

Colse

천정이 축을 중심으로 회전하여 열리면 오픈형 광장과 공연장으로 쓰인다. 공연은 각 타겟 연령층에 맞는 여러가지 공연이 열린다. 공연이 없는 평상시에는 누구든 들어와 여유를 즐길 수 있는 광장으로 쓰인다.
천정이 닫혀있을 때 는 비가 올 때 비를 피할 수 있는 피신처가 될 수도 있으며 날씨가 안좋은 날에도 천정이 닫혀있는 광장으로 쓰인다.

When ceiling revolves around the axis and opens up, it becomes an open plaza and a performance arena. Various performances are held according to each target age group. It is used as a plaza where anybody can come in to enjoy leisure time at ordinary times when there is no performance. When the ceiling is closed, it can be a shelter from rain and used as a plaza in very bad weather conditions.

이인근 홍종무 정한울 | Lee, in-keun Hong, jong-moo Jeoung, han-wul

Background

서부 이촌동 주민들은 용산 국제 업무지구의 재개발로 인하여 6년이 넘는 세월동안 재산권 행사가 불가능 하여 고통스러운 나날들을 보내 왔다. 그래서 현재 이촌동 주민들은 개개인의 각기 다른 상황으로 인하여 서로간의 마찰이 생기고 골이 깊어지는 상황을 맞이하고 있다. 이런 환경 속에 현재 이촌동은 생존권을 외치고 있다. 생존, 사람이 삶을 살아간다는 것은 무엇일까? 사람이 살아 있다는 것은 서로가 함께 살아갈 수 없다면 느낄 수 없다. 옛 선조들의 '잘 산다' 의 개념은 물질적인 풍요로움이 아니라 함께 나누고 공유 하는 삶을 통함으로써 '잘 살아감' 의 정을 나누었다. 우리는 과거의 삶의 모습을 통해 제시한다. 살아있음의 옛 이촌동을...

Residents of western Ichondong have been suffered over 6 years from redevelopment in Yongsan international practice area. Due to redevelopment, they couldn't exert their property rights and this made residents painful. Therefore residents of Ichondong are experiencing conflicts one another and situation is getting worse because of their conflicts.
In this atmosphere, Ichondong is now insisting their right to live. Survival, what is the meaning of 'live'? People cannot feel that they are alive if they don't live together. Concept of 'live well', ancient people thought, was not richness but sharing feeling of 'living well' through sharing life. We present 'alive old Ichondong' through life of past.

移
옮길 '이'

Concept

과거 이촌동은 원래 한강 변 모래벌판 이었다. 여름철 홍수가 들면 강 가운데의 섬마을에 살던 주민들은 강안으로 옮겨 살아야만 했다. 지금의 이촌동(二村洞)이 옛날에는 옮길 이 자를 쓴 이촌동(移村洞)으로 과거 삶의 방식이 그 이름의 유래이다. 홍수가 나면 항상 물에 잠기는 마을 이촌동. 하지만 이촌동 주민들은 자기 삶의 터를 옮겨가면서도 함께 살아갈 수 있었음은 물에 젖어 피난 온 주민과도 서로간의 삶을 나누고 공유하였기 때문이다.
옮겨 간다는 것, 올라간다는 것, 이촌동이기에 가능한 지역적 특성, 그 점을 활용해 우리의 놀이 공간 암벽등반을 계획한다. 이 공간은 옛 생존을 위한 오름이 아닌 지금, 자신을 극복하고 활력을 잃은 이촌동에 생기를 불어넣으며 주민들의 공유하는 놀이공간을 조성한다.

Ichondong was a sandy plain beside the Han River. when floods occurred in summer, residents who lived in the island at the center of the river had to move to the inside of the river. Current name of Ichondong originated from the meaning of 'move', which means lifestyle of past residents of Ichondong. Ichongdong was always flooded when floods occurred. However, while residents of Ichondong moved to avoid flood, the reason they could survive together was that they shared their life with refugees who escaped flood and helped one another.
Because Ichondong is it can has characteristics of moving and climbing. only Ichondong has characteristics of moving and climbing because of its regional factors. By making use of these characteristics, we plan to build climbing space. This space gives life to Ichondong which is losing vitality, makes Ichondong residents overcome selfishness caused by conflicts between them. As a place offering play, lots of situations and play programs allow residents to resolve conflicts and communicate one another. In these environments, they can feel sharing of feeling that they are connected and communicating.

Site

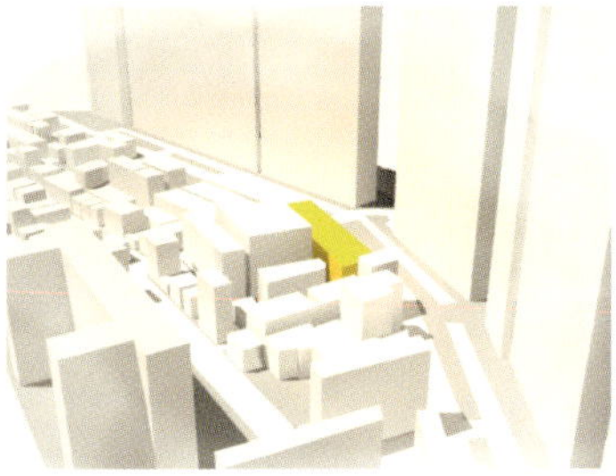

서울 특별시 용산구 이촌2동 주민 센터 앞 테니스장. 한강철교가 있는 우측에 위치한 이촌2동 서부이촌동. 서부 이촌동 주민 센터 앞 테니스장 은 현재 아파트 주민들의 주차장으로 활용 되고 있으며 운동시설로는 사용되지 않는 곳이다. 이러한 죽은 공간을 이용해 우리는 이촌동을 옮길 이[移]의 살아있는 이촌동으로 바꾸려 한다.

Tennis court located in front of Community Center, Ichon 2-dong, Yongsan-gu, Seoul, Korea.
Western Ichondong located at the right side of the Han River bridge.
Tennis court located in front of Community Center of western Ichondong is used as not exercise space but apartment resident's parking lot. by using this dead space, we intend to change from dead Ichondong to alive Ichondong which has meaning of 'move'.

Client

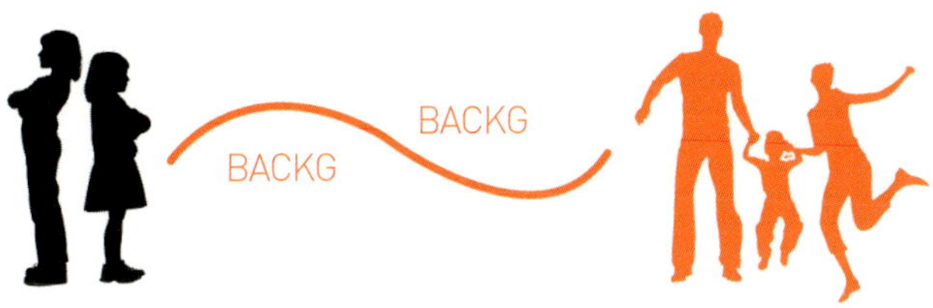

다양한 난이도를 가진 공간을 제공함으로써 모든 연령대의 시민을 유도하며 지루하거나 고정적이지 않은 다양한 경험을 제공하였다. 오르는 중에 있는 쉼터들로 난이도와는 또 다른 코스들을 이루게 되었다. 운동을 하며 이촌동 주민들 간의 몸과 마음이 동화 될 수 있도록 하였다. 또한 건축물의 특이성을 갖춤으로 외부 사람들의 방문을 늘려 단절 되어있던 이촌동 주민들에게 그 벽을 허물게 해주는 중요한 공간이 될 것이다. 그저 암벽 등반이 아닌 서로를 이어주는 옛 이촌동의 오름을 선사 할 것이다.

By providing diverse level of difficulty, we allow people in all ages to enjoy climbing and feel diversity of experience so that they don't feel bored. rest areas which can be discovered while climbing make people experience different course which is free from level of difficulty. while climbing up, residents of Ichondong can make their body and mind assimilated with one another doing exercise. Also by having peculiarity of structure, we aim to increase visit of external people. so that this will tear down a wall between closed residents of Ichondong and external people. therefore this place will play a key role in making residents be open.

Motive

모래 [Sand]
자연히 잘게 부스러진 돌 부스러기 [stone chips]
알갱이의 확대된 모습은 매우 다채로운 형태이다.
과거 서로살이가 살아있는 이촌동 이동의 원인이 되던 지형적 특성

The expanded shape of grain is very varied.
regional properties which was cause of custom of "two way street of helping each oher" of Ichon 2-dong.

Process

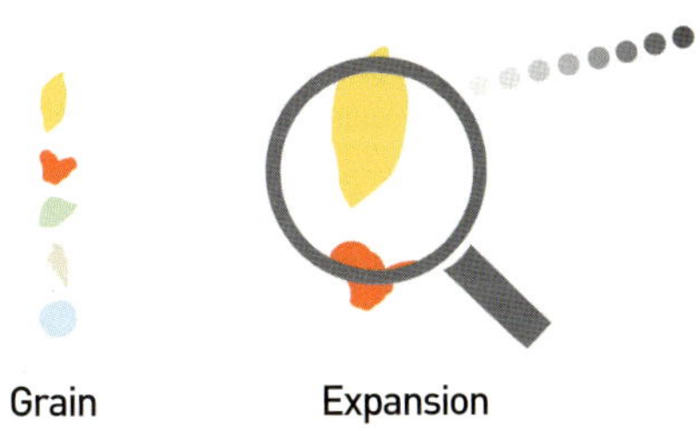

Expansion shape

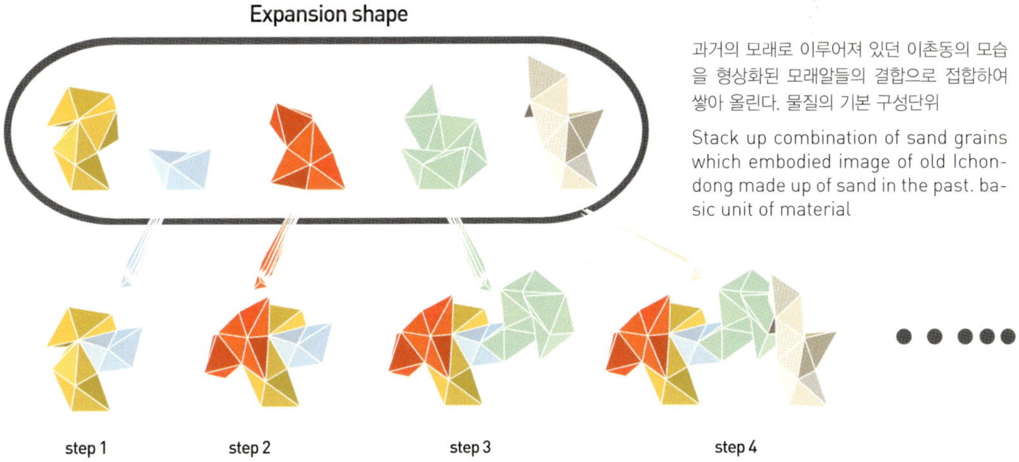

과거의 모래로 이루어져 있던 이촌동의 모습을 형상화된 모래알들의 결합으로 접합하여 쌓아 올린다. 물질의 기본 구성단위

Stack up combination of sand grains which embodied image of old Ichondong made up of sand in the past. basic unit of material

원자

물질의 기본구성 단위

The basic building blocks of material

정사면체

모든 면의 입체적인 결합이 가능한 네 개의 정삼각형 면으로 이루어지는 면이 가장 적은 다면체

least-sided polyhedrons consist of four regular triangle side which can be combined in three dimensions with all sides.

각각의 알갱이가 어느 면이든 같은 크기의 정사각형 면에서 만나 손쉬운 결합이 가능

each grains can be combined easily when met at any side of regular square.

Purpose

=

오르다

과거 홍수로 인해 오를 수밖에 없던 주민들
Residents who had to climb up due to flood

모래로 형상화된 구조 체를 오르는 주민들
residents who climb up structure embodied with sand

CLIMBING

몸의 전체 근육을 사용하는 클라이밍은 근력과 지구력 상승에 도움이 되고 근육의 발달에 도움이 될 뿐만 아니라 신체의 균형과 탄력 향상에도 큰 도움이 된다.
Climbing which requires whole muscles is not only helpful to strengthen muscular strength and endurance but also to make body supple and balanced.

실제 산보다 이용에 용이하고 더 안전한 방법으로, 관리되어지는 곳에서 취미활동을 할 수 있다.
because climbing center is administered regularly and appropriately, it is more safe and also easy to access than mountain. so people can enjoy their hobby in safer way.

초급자부터 상급자까지 자신의 숙련도에 따라서 다양한 난이도 코스의 선택이 가능하다.
From beginner to expert people can choose level of difficulty according to their ability.

Characteristick plan

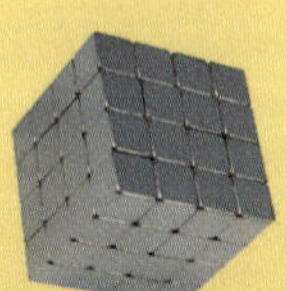

네오디뮴
ND (neodyminum)

네오디뮴은 17가지 희토류 원소 중 하나이다. 강한 자성을 띄고 있는 금속이다. 네오디뮴의 전성 성질을 이용하여, 암벽의 결합을 손쉽게 가능하도록 한다. 네오디뮴을 암벽 판 안쪽에 사용함으로써 안전하고 강한 결합의 형태를 만들었다.

Neodymium is the one of the 17 rare-earth elements. it is metal which has intense magnetic forces. we can make combination of rock face easy by using neodymium's character. we made strong and safe form of combination by using neodymium into the rock face's side.

자기장을 이용한 결합
Using magnetic coupling

자석이나 전류에 자기력이 작용하는 공간이다. 자기장에는 N극과 S극의 끌어당김으로 결합이 쉬워지며, 자기력선이 조밀하면 조밀할수록 더 강한 자력을 가지게 된다. 암벽 판 안쪽에 촘촘한 자기력선을 설치하여 자기장을 형성하고, 작은 암벽 형태끼리 쉽게 결합 할 수 있도록 하였다.

place where magnetic force works to magnet or current. in the magnetic field, because north-seeking pole and south-seeking pole attract each other combination become easier and the denser the magnetic line of force, the more intense the magnetic force become. we established magnetic field by installing dense magnetic line of force in the side of rock face so that small form of rock face can combine easily each other.

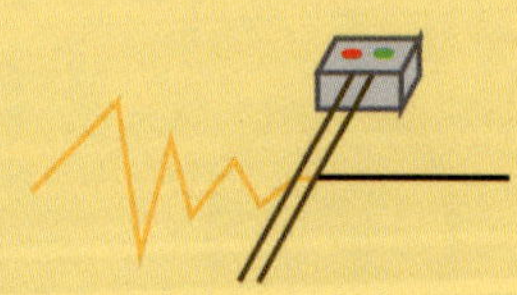

자화소거장치
Magnetization canceller

자성이 있는 부분을 소거하여 자력을 없애는 장치로, 직류인 상태의 자성 부분에 교류 장치를 넣어 자석의 N , S극을 교번 시키며 조금씩 자력을 줄어들게 하는 장치이다. 이 센서 장치를 사용함으로써 쉽게 다양한 변형이 가능한 암벽을 만들었다.

this device removes magnetic force by eliminating magnetic part. To be specific, it decreases magnetic force little by little by inserting device which makes alternating north polar and south polar in magnetic part when direct current is used. By using this sensor device, easily transformable rock face can be made.

Floor plan

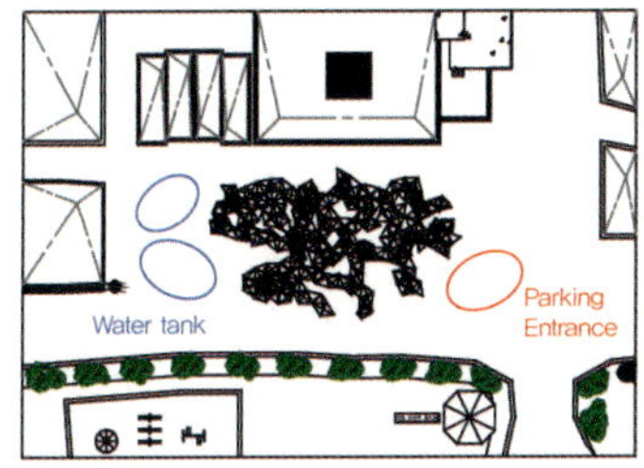

Elevation(before)

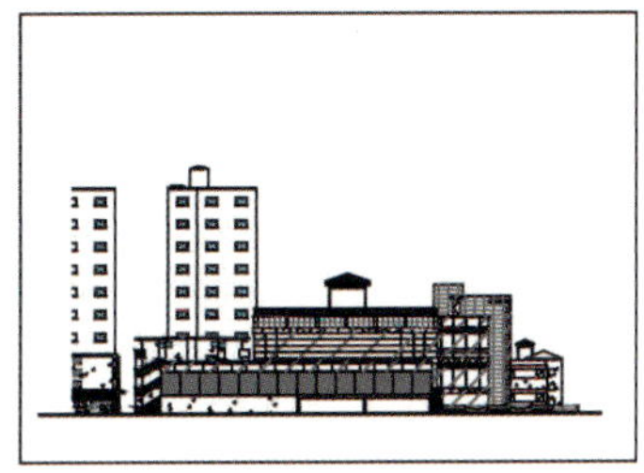

Material

리트라콘
Litracon

라텍스
Latex

우리가 이번 설계에 새로 추가한 새로운 재료는 콘트리트 중 요즘 각광 받고 있는 리트라콘과 안전을 생각한 라텍스이다. 리트라콘은 건물 외부 마감에 사용하여 자석 네오디뮴과 같이 빛과 잘 어우러지는 모습을 연상시키고, 라텍스는 암벽등반 밑바닥에 설치하여 안전을 위해 사용하였다.

new materials we added in this designation is Latex and Litracon ,as a kind of concrete, which is on the rise in the aspects of safety. Litracon is used to outside of structure. It shows image which makes harmony with light and magnet neo-dymium. Latex is used at the bottom of climbing area for safety.

Climbing hold

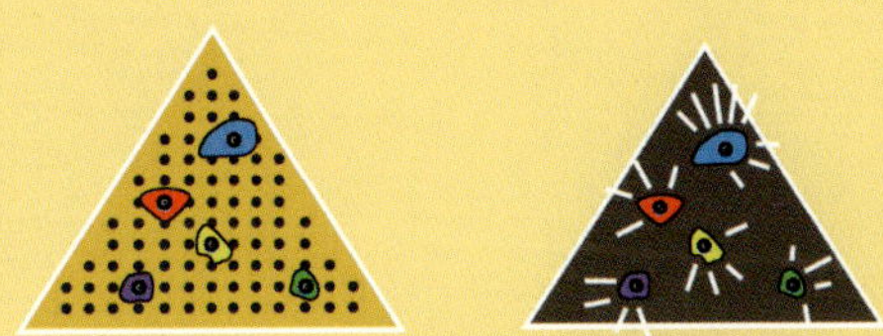

정사면체의 한 면에 규칙적인 구멍을 뚫어둠으로 어느 곳이든 홀드 부착이 가능하게 하였다. 그러므로 자유로운 변형이 가능해진 클라이밍 센터의 코스가 더욱더 다양해졌다. 또한 특별한 조명을 갖추지 않고 각각 면에 붙어있는 홀드가 직접 빛남으로 어두운 시각에도 동선 유도가 가능하며 어둠속에서 조형물의 시각적인 아름다움도 이룬다.

by drilling regular hole in one side of a regular tetrahedron, attachment of holds is possibile at any side. therefore because climbing center can be transformed freely, climbing course become varied. As attached holds in each side glow themselves, without any lighting, people can decide their route. In addition, people can see beauty of sculpture in the darkness with the help of light of holds.

Traffic line

코스 1(난이도 하)
Course 1 (difficulty doing)

쉬운 난이도 코스이다. 어른부터 어린이까지, 초보자부터 숙련자까지 아주 쉽게 클라이밍을 접 할 수 있게 도와주는 구간이기도 하다. 특히 우리 건축의 특징은 조형의 가변성으로 인한 여러 가지의 코스들이 존재 할 수 있다는 장점과 또한 그 하나의 구간에서도 여러 면으로 오를 수 있다는 장점을 가지고 있다.

Easy course - This course is such an easy section that all people, from children to adult and from beginner to master, can experience climbing. Especially, advantage of our structure is variability of sculpture so there are various courses. Also, in one course, people can climb up choosing various routes.

코스 2(난이도 상)

Course 2 (Level Award)

어려운 난이도 코스이다. 암벽등반을 어느 정도 즐길 줄 아는 사람들이 주로 이용하게 될 구간이다. 숙련자가 주로 이용할 코스인 만큼 오를 때의 근력 사용량이 가장 많으며 성취감 또한 최고로 느낄 수 있도록 설계 하였다.
하지만 아마추어나 초보자를 위해서 충분한 쉼터구간과 전망 구간을 설치함으로써 어려운 난이도 속에서도 클라이밍을 즐길 수 있는 구간이다.

Difficult course - people who enjoy climbing to some extent will use this course.
as mainly an expert will use this course, it requires high muscular strength when climbing up and is designated to feel high sense of accomplishment. However, for amateur and beginner, by building enough rest area and viewing platform, we allow them to enjoy climbing in hard course.

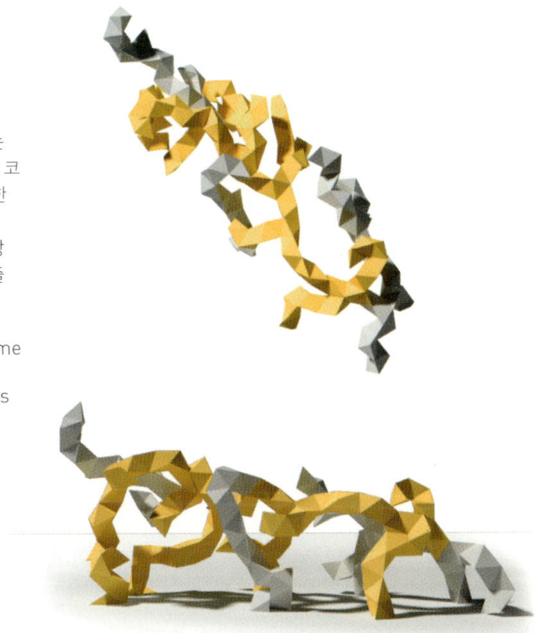

코스 3(난이도 중)

course 3 (Level Medium)

번외 코스이다. 이 구간은 상과 하를 구분지은 개념이 아닌 자신 외의 다른 사람과 같이 어우르고 소통하며 즐기는 구간이다. 가족과 같이 올라 와서 대화를 하고, 친구들과 올라오며 경기도 하고, 또 연인과 올라와 경치 구경도 하는 자유로운 코스를 유도하게 되었다. 이 코스는 올라가기에 가장 긴 코스이나 중간 휴식 공간도 많이 둠으로써 올라가는 중에 다양한 경험이 많이 일어나도록 설계 하였다.

Additional course - this course is not characterized by concept of up and down. this course is where people enjoy, join together and communicate one another. purpose of this course is to make people have conversation with family, play game with friends and see scenery with their lover. Although this course is the longest course, it has a lot of rest area. It is designated for people to experience a variety of experience.

최은정 황영지 | Choi, eun-jung Hwang, young-ji

주민의 소통과 관계회복을 위한 프로젝트 〈FLEA MARKET〉

벼룩시장은 세계각국의 서민들을 중심으로 한 노점상 문화가 정착되면서 역사적 의미와 함께 삶의 향수들을 느낄 수 있는 문화 소통 공간으로 발전하고 있다.

Flea markets have been developed as a space of cultural communication where people can feel nostalgia for life with historical significance as a culture of a street vendor has settled in around the people from all around the world.

Selling Ship

그물의 유연성과 다양성을 이용한 FLEA MARKET

Using the flexibility and versatility of the net FLEA MARKET

Background

서울특별시 용산구 서부이촌동 대림아파트

용산구 서부이촌동은 약 7년 동안 용산국제업무지구 개발사업에 거래가 묶여있어 주변의 편리한 교통환경에도 발전하지 못한채 낙후되어 있던 지역이다. 오랜기간의 투쟁으로 지역내는 침체된 분위기가 강하고 주민들의 외부인에 대한 경계심이 높다. 단절된 외부와의 커뮤니케이션의 상이 되면서 썰렁한 분위기를 바꾸는데 촉매제 역할을 해줄 RE-Energizing 장소가 필요하다.

Seobu Ichon-dong, Yongsan-gu, despite its convenient transportation, it has remained a backward and underdeveloped region for about 7 years due to its deal tied-down to Yongsan International Business District development project. With their struggle for a long time, the atmosphere within the region is depressing and the wariness against outsiders is high. It needs a place for RE-Energizing acting as a catalyst to switch the chilly atmosphere and bring communication with the outside world it has been cut off from.

Space program

FLEA MARKET

1.벼룩이 들끓을 정도로 오래된 것을 판다
2.사는 사람과 파는 사람의 행동이 액티브 하다.
벼룩시장은 세계각국의 서민들을 중심으로 한 노점상 문화가 정착되면서 역사적 의미와 함께 삶의 향수들을 느낄 수 있는 문화 소통 공간으로 발전하고 있다.

1. They sell the old stuff enough to be infested with fleas.
2. The behavior of those who buy and sell is so active as fleas.

Flea markets have been developed as a space of cultural communication where people can feel nostalgia for life with historical significance as a culture of a street vendor has settled in around the people from all around the world.

Program trend

단순히 즐기는 공간이 아닌 문화를 공유하는 곳으로 활용하고 '사고 파는 공간' 에서 '놀이 공간' 으로 바뀌고 있다. 또한 매상을 올리는 것보다 '나누는' 것에 더 큰 목적으로 두는 봉사활동 일환으로 활용되고 있으며 특정 주제를 정해 해당 마니아들을 불러 모으는 소규모 플리마켓도 늘고 있다.

It is changing not just a place to enjoy, but to take advantage of the shared culture becoming a 'play space' instead of just a 'place to buy and sell'. In addition, it is used as part of voluntary services for the purpose of 'sharing' rather than raising sales. A small flea market with a specific theme to bring together mania is also growing.

Motive

항구시장을 모티브로 한 유연한 플리마켓 계획안
Flea market that motivated port market

항구에 선박이 도착할 때 마다 선박에 실어왔던 물건들로 항구 주변에 열렸던 크고 작은 거래의 모습들인 항구시장의 모습과 매일 열리는 시장이 아닌 정해진 날 마다 혹은 특정한 주제나 물건만을 판매하기 위해 열리면서 판매자와 소비자나 끌어 모으는 플리마켓의 특성 사이에서 겹쳐지는 공통점을 찾았다. 제한되어 있는 시장이 열리는 날을 기대하며 기다리다가 사고파는 거래행위 속에서 소통한다. 이러한 소통을 통해 이촌동 주민들의 외부인에 대한 경계심을 풀어주는 RE-Energizing 효과가 나타난다.

I found a common characteristic between the two; the harbor market taken place whenever the ship arrives at the port with a large and small trading of the stuff from the ship, and the flea markets opened only on the specific day or for a particular theme or object. In terms of these two markets, people expect waiting until the open-day for these limited markets and communicate in the act of buying and selling. Through this communication, the people of Ichon-dong get to see Re-Energizing effect which lets their guard down against the outsiders.

Performance requirment

침체된 도시에 활력을 넣어줄 놀이공간

Play space to revitalize the depressed city

접촉의 계기를 만들며 시장의 정체성을 확보할 기능적 공간

Functional space to establish identity of markets creating an opportunity of contact

주민과 외부인과의 커뮤니티에 도움을 줄 수 있는 촉매 공간

Catalyst space to help the communication of the residents and outsiders

소재의 유연한 사용을 통한 시각적인 요소의 다양화

Diversification of visual elements through the flexible use of materials

Concept

그물의 유연성 다양성을 이용한 FLEA MARKET

NET

1. 그물 , 네트
2. 올가미 , 함정
3. 그물 모양으로 짠 것; 그물 세공;세공 레이스
4. 망상의 조직 ; 거미줄
5. 통신망; 네트워크(NETWORK), 방송망
6. 네트(볼)

NETWORK

1. 컴퓨터[통신망]를 연결하다
2. (방송망을 통해 동시에) 방송하다
3. 인적 네트워크를 형성하다

그물의 유연성, 탄력성, 가변성, 비침 등의 특징을 이용하여 소통과 관계회복을 위한 교류를 할 수 있는 활발하고 역동적인 공간을 구상한다.

Design the lively and dynamic space of exchange for communication and relationships recovery using the characteristics of the net; flexibility, resilience, variability, and see-through.

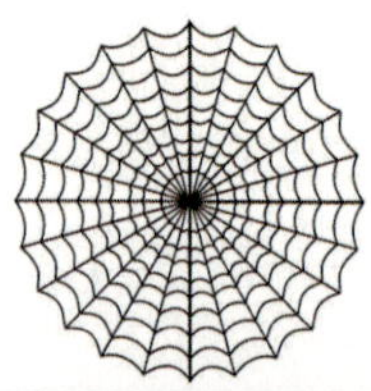

TRAMPOLINE

HAMMOCK

FLEA MARKET

ELEVATION

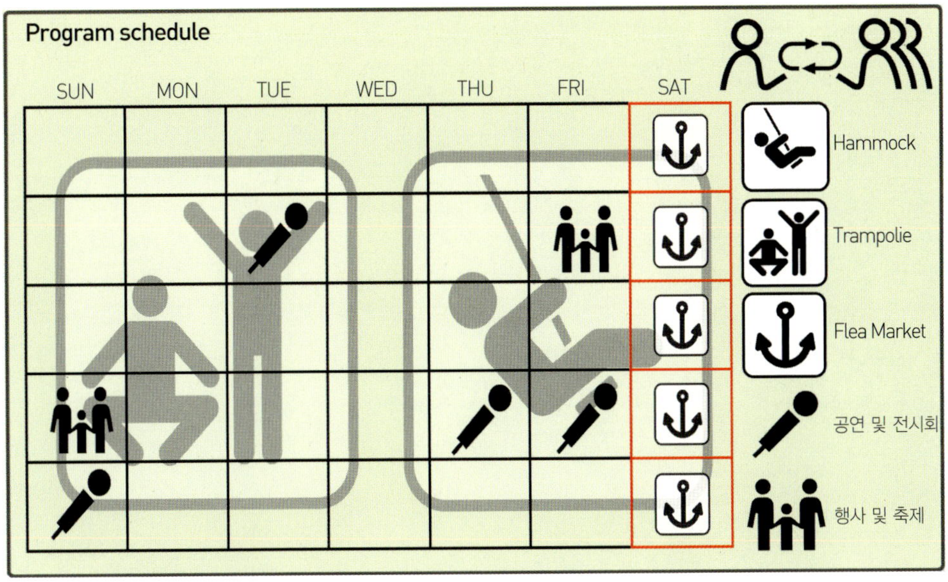

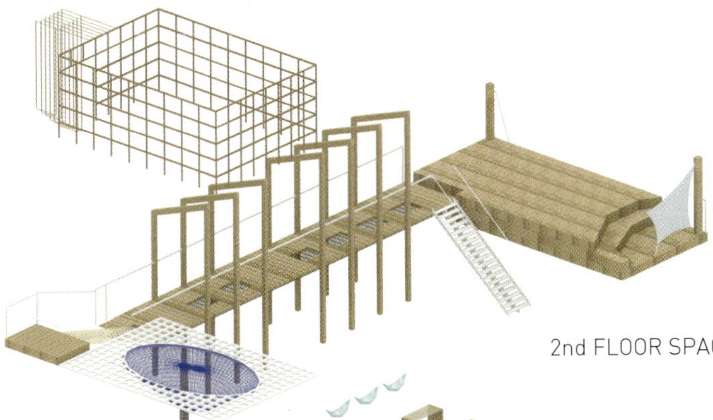

2nd FLOOR SPACE AXONO

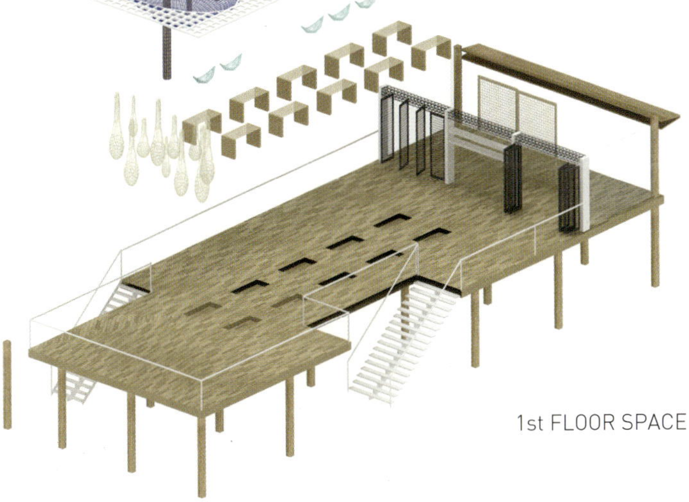

1st FLOOR SPACE AXONO

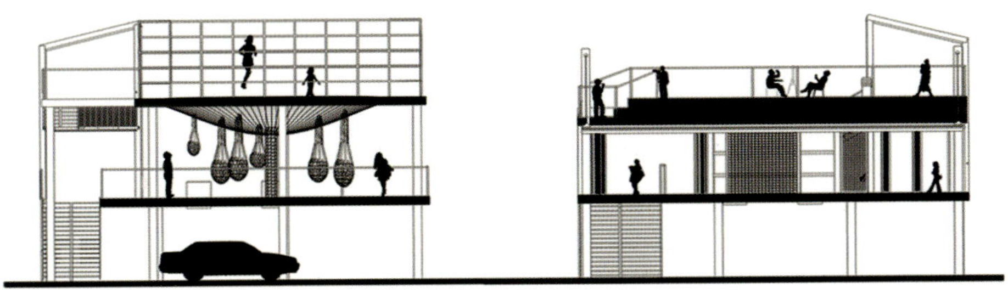

Floor plan

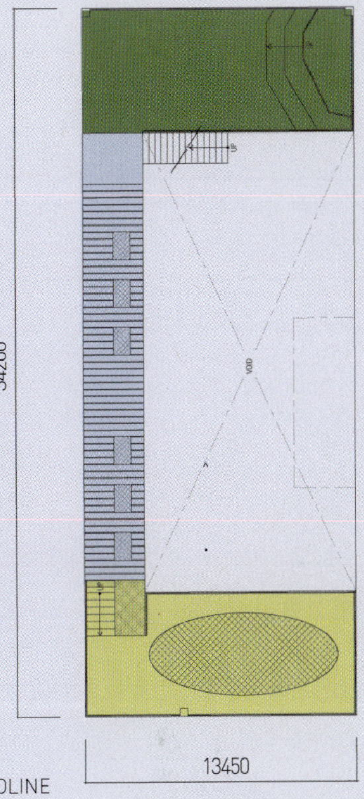

TRAMPOLINE

FLEA MARKET & BRIDGE

ARTIST ZONE

2nd Floor Plan

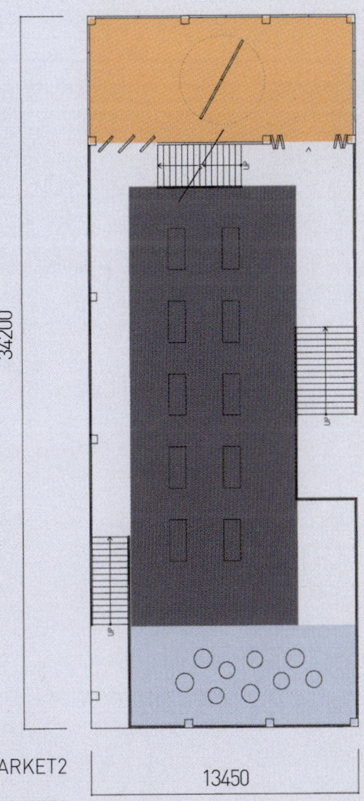

FLEA MARKET2

FLEA MARKET1

POP-UP STORE

1st Floor Plan

Form process

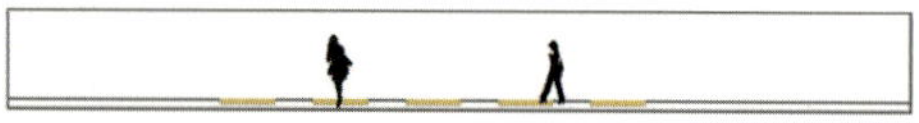

Empty space with a flat floor

Type of folding platform Installling the net

Designed a flexible sales

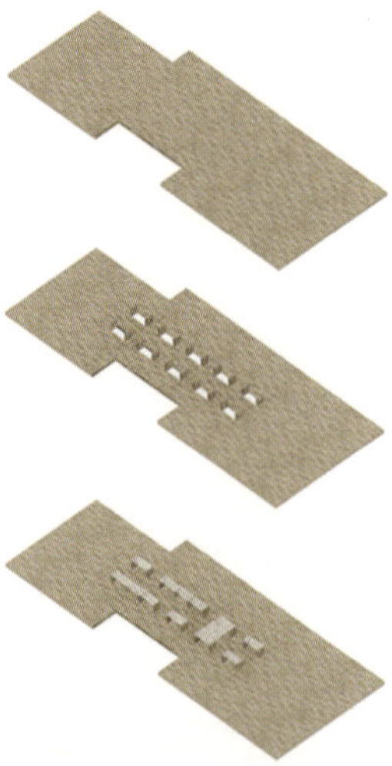

그물의 가변성, 유연성을 적용한 공간이다. 평소에는 그 바닥이 닫혀있어 평평한 형태이다. 플리마켓이 열리는 날에는 폴딩 형태로 바닥을 열고, 고정시킨다. 그 다음 그물을 설치하여 판매대로 이용한다.

플리마켓이 항시 열리는 것이 아니기에 플리마켓이 열리지 않을 때 공간을 유용하게 사용할 수 있다.

징검다리와 이어지는 프레임에는 해먹을 설치하여 플리마켓에 참여하는 모든 이들이 잠깐 쉬어 갈 수 있도록 하였다.

This is the place where the variability and flexibility of the net has been applied to. It is flat at other times with the closed floor. On the day of the flea market, it opens the floor in the form of a folding, and fixes it. And then, set up the net and use it as a sales counter.

Since the flea market is not opened at all times, when it is not opened, the space can be used for other purposes.

A Hammock is set up in the frame connecting to the stepping stones to provide the visitors in the flea market with a place to rest for a little while.

2층에는 그물을 이용한 판매대, 3층에는 트램펄린으로 구성된 공간 하나의 그물 구조물을 사용하여 두가지의 기능을 하고 있다. 3층은 외벽이 프레임으로 되어있어 현수막을 이용하여 광고를 하고있다.
두개의 기능을 같이 구성하면서 공간의 활발함을 표현하고자 하였다.

The sales counter using a net on the second floor and the trampoline on the third floor. A structure using a net has two functions.
The outer wall on the third floor is a frame, which is used for banner ads.
In order to express the active space, two functions was combined

FLEA MARKET -F3

징검다리 형태의 판매공간으로 보이드 공간을 이용하여 판매대를 설치하였다.

Sales stands are set up using a void space as an area of the merchandiser in the form of stepping stones.

박용주 이은미 | Park, yong-ju Lee, eum-mi

RE:CREATE RECREATE

Circuit Rear-car parking space

사람에 의해 만들어지는 다양한 교류의 구심점.즉 회복의 역할.아픔에 감싸져 있는 이촌동 주민들이 활력을 찾고 다시 모여 들 수 있는, 주체가 주민이 되고 사용자가 만들어가는 공간. 재활용센터와 리어카주차장을 이용한 re-energising.자동차 주차장은 있지만 주변 고물상에 비해 리어카를 마땅히 세워둘 곳이 협소한 상황. 물질의 순환이 이루어지는 재활용센터와 리어카를 주차할 수 있는 공간을 제안한다.

A center of various communication made by human beings. That is, a role of recovery. A space which attracts peoples who were once discouraged; the subject are residents in which they can make up a place to communicate by means of recycling center and handcart parking lot

Background

2007년 8월' 용산과 한강을 연결해 한강르네상스 시대를 열겠다.' 용산 철도정비장 부지와 서부이촌동 개발사업을 발표하였다. 용산 역세권에 150층 빌딩을 건립하고, 용산과 서부 이촌동을 하나로 묶어 최적의 주거, 상업 단지를 조성하겠다는 뜻을 밝혔지만 사업이 중단되면서 주민들은 냉소적 태도와 마음의 문을 닫아 서부 이촌동은 삭막하게 변해간다.

According to the development plan of Yongsan train repair site and West Ichon-dong, it was said that the renaissance of Han River was to be opened by connecting Yongsan with Han River. They said that 150th skyrocketing floor was to be built around Yongsan by connecting Yongsan with West Ichon-dong for the optimal housing environment as well as commercial complex. However, that plan was suspended, which discouraged residents living there and they became cynical.

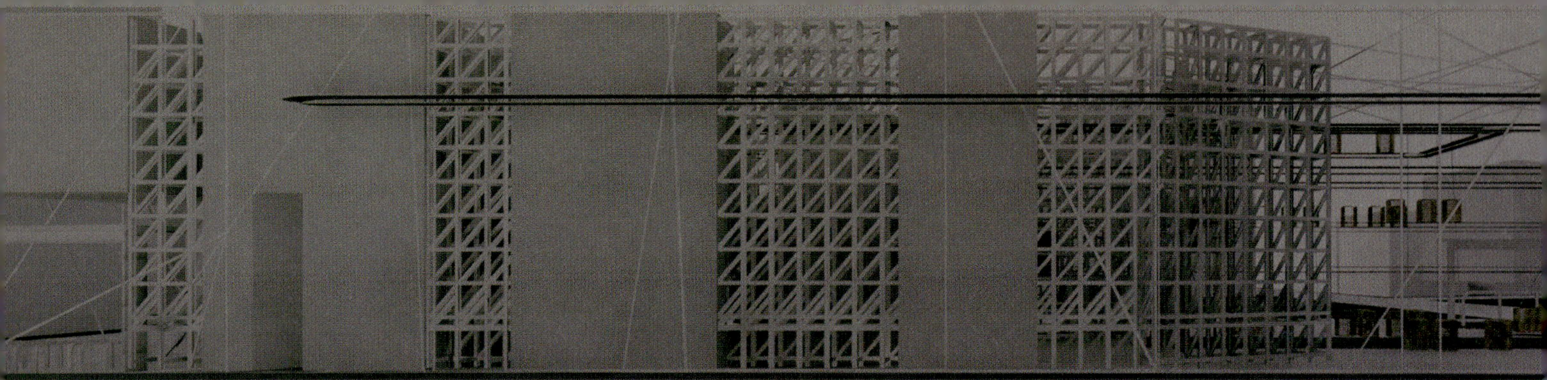

Site analysis (between public and privacy)

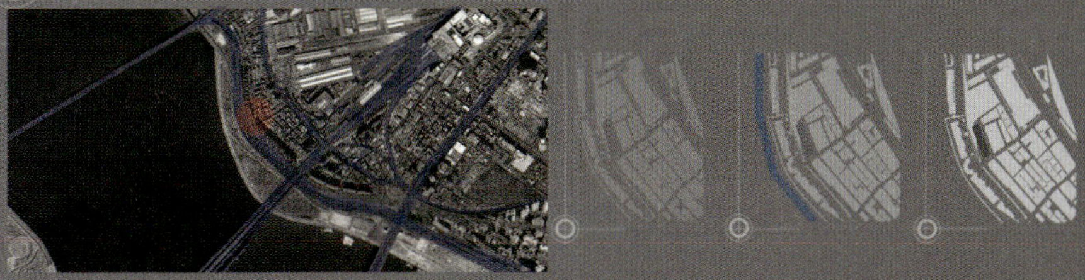

한강대교와 원효대교 사이 한강변을 따라 동서로 길게 위치하였으며 용산 역세권 및 한강로의 배후지–개발 기대지역.

It located along the East and West side of Han River between the grand bridge of Han River and that of Wonhyo. and As a hinterland of Han River Lane and Yongsan, it is expected to be developed.

Problem

주민들의 발걸음이 닿지 않는 골목사이. 그리고 주변거리와 아파트단지를 거쳐 수거된 상태로 손길이 닿지 않은 그대로의 고물상과 길 잃은 리어카들. 회복의 역할. 아픔에 감싸져 있는 이촌동 주민들이 활력을 찾고 다시 모여들 수 있는 공간이 필요하다.

There are many back streets where residents left; untouched handcarts are left around lanes. That's why a variety of centers of communication are need to communicate one another. That is, it needs a role of recovery. Residents of Ichon-dong need a space of center to activate its environment and to attract many people.

Solution + purpose

버려지고 관심 밖에서 손길이 닿지 않는 것들에 대해 다시 생각다. 대림아파트 부지에 둘러 싸여진 좁은 골목과 이리저리 자리를 못 찾고 널부러진 리어카들. 쓰레기들과 더럽혀져 있는 골목 골목 사이. 그리고 재개발이라는 갈등 속 해메이는 주민을 위한 공간을 구성하려 한다.

It is time for us to have more vitality than redevelopment. It is time to reconsider those who are isolated from society and the deserted. There are handcarts around backstreets surrounded by Daelim apartments. Dirty trashes are scattered around small lanes here and there. However, we are now trying to prepare a space for residents who are frustrated owing to the troubles of redevelopment.

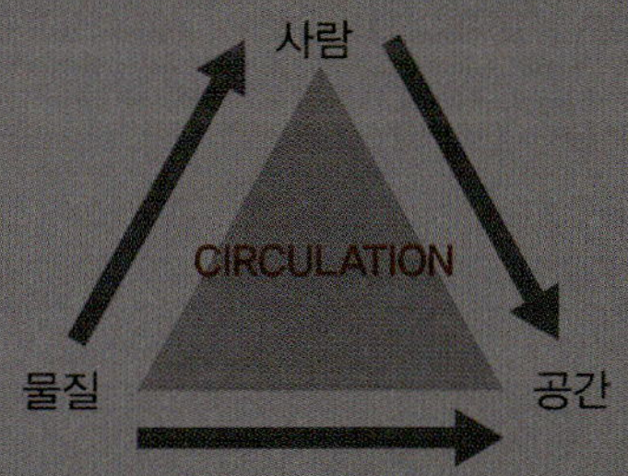

Needs analysis

아파트 주변 위치한 고물상은 님비(Nimby)로 풀이될 수 있지만 더 이상 폐부지로서의 분리수거용 재활용 센터가 아닌. 최대의 행복 보다는 최소한의 불편으로 접근하여 커뮤니티와 지속가능이 이루어 지는 공간이 필요하다.

Secondhand dealers around apartments are told as Numby; but they are not just a recycling center for a separate collection. In other words, they need a sustainable space connecting communities through minimum inconveniences rather than maximized happiness.

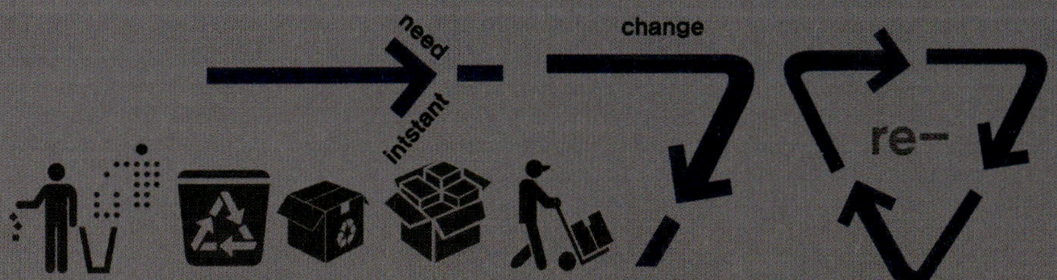

Process

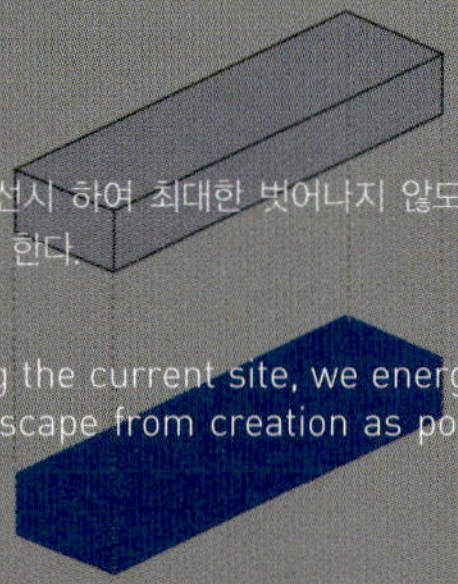

현 사이트를 우선시 하여 최대한 벗어나지 않도록 창조보다는 re–energising을 한다.

By prioritizing the current site, we energize the process not to escape from creation as possibly as we can.

물품별 분류를 하여 재활용이 가능하도록 하며 불가능 한 것들은 파기하여 재료/자재로서 재가공하여 사용한다.

By categorizing unusable items, we recycle and process them by tearing them to pieces.

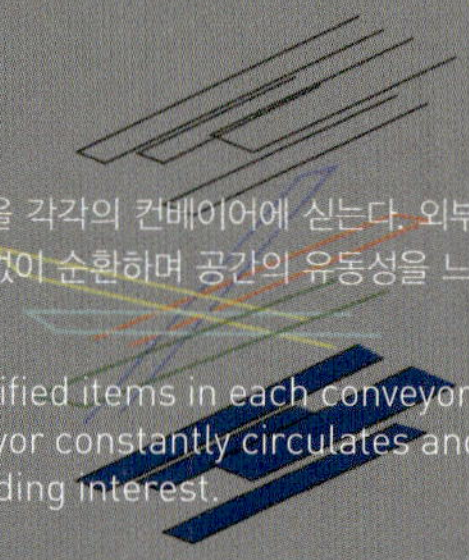

분류한 물품들을 각각의 컨베이어에 싣는다. 외부로 노출된 컨베이어는 끈임없이 순환하며 공간의 유동성을 느끼게되어 재미를 더한다.

We put classified items in each conveyor. Those external conveyor constantly circulates and add space fluidity by adding interest.

Space program & conveyor type

나무 물질의 순환 컨베이어 Circulating conveyor made of wooden materials.

지류 물질의 순환 컨베이어 Circulating conveyor made of paper materials.

철 물질의 순환 컨베이어 Circulating conveyor made of steel materials.

천 물질의 순환 컨베이어 Circulating conveyor made of fabric materials.

플라스틱 물질의 순환 컨베이어 Circulating conveyor made of plastic.

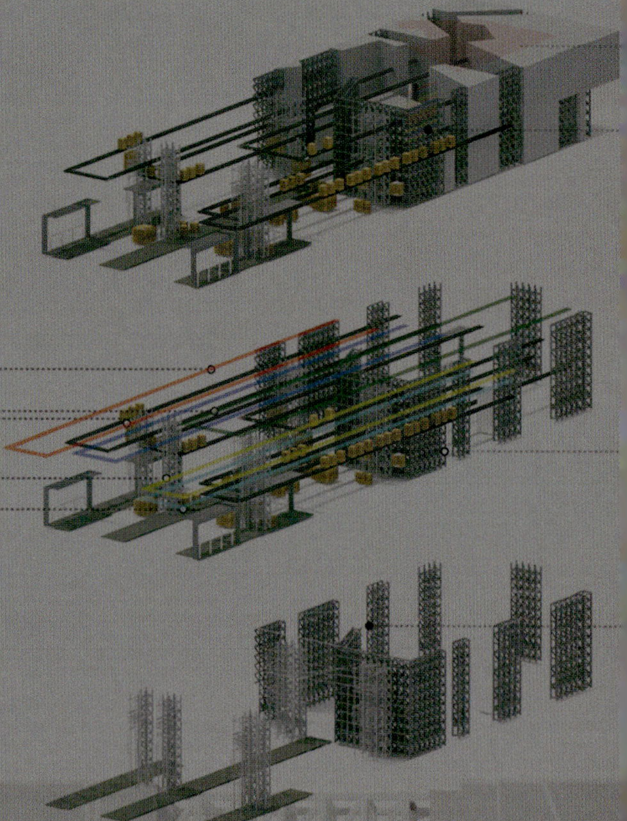

Circuit

회로, 서킷, 회선, 순회, 우회로, 빙 둘러서 감

의식의 순환과 물질간의 순환으로 구분되는 공간 안에서 멈춰 있던 에너지가 강한 시너지 효과를 더해 돌아오는 길. 회로를 통한 공간의 형성과 공간 속에서 반복적인 순환과 체험활동을 하게 되는데, 이는 의식과 물질이 소통하는 데에 목적을 둔다.

The purpose of this plan is to enable men to communicate and understand better one another through a repetitive cycling of experiences in a circuit lane by adding synergistic effects of energy which were stopped among materialistic cycling and process.

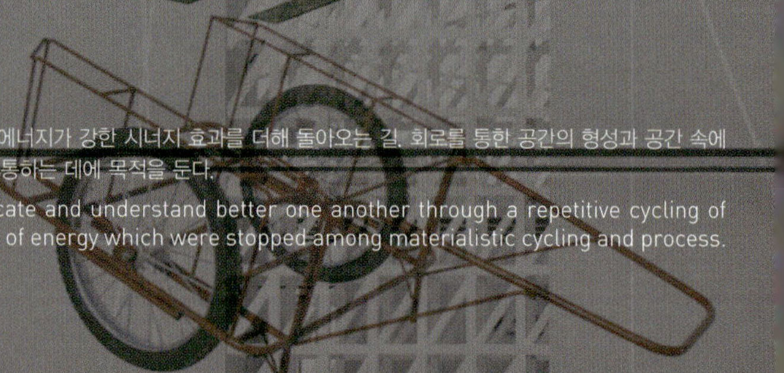

RE:CYCLE
RE:CREATE
RECREATE

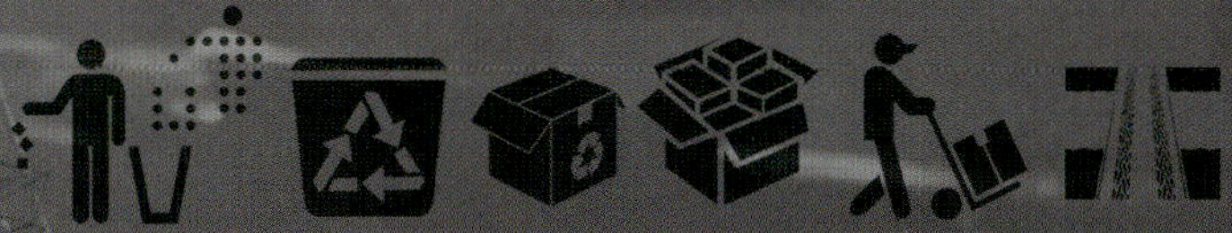

분류한 물품들을 공방으로 보내어 재생산하고 재생산한 것들을 토대로 순환이 이루어진다.

We put classified items in each conveyor. Those external conveyor constantly circulates and add space fluidity by adding interest.

물품들의 연속적인 순환을 통하여 재활용 센터의 기능과 동시에 반복적인 컨베이어의 얽힘과 지속으로 재:활력이 더해진다.

Vitality is added through the repetitive entanglement along with the recycling function through the continuous circulation.

Performance area

천상에서 볼 수 있는 소명 퍼포먼스
직접 조명을 달거나 재활용 조명에 대한 체험을 하는 공간.

Lighting performance which can be seen from a ceiling. direct lighting experience for lighting the space recycling.

색상별 의류를 통한 컬러 퍼포먼스
사용되어 질 수 있는 의류들은 따로 분류하여 1층과의 소통의 공간.

Color performance through clothes in color. Usable clothes are classified for easy communication with first floor.

목재 재사용 무대 공간 퍼포먼스
목재가 재사용되며 폐목재의 순환과 함께 공방으로 인한 사람들의 의식이 순환되는 공간.

Stage space performance by reusing wood.
Wood is reused and people can meet one another due to a vacant room.

플라스틱 재사용 퍼포먼스
천정을 폐 플라스틱을 재활용한 오브제로 장식하여 재활용의 퍼포먼스 속 새로운 공간.

Performance of plastic reuse.
By decorating ceiling with deserted objet of plastic, a new space is a space of recycling.

최대의 행복 보다는 최소한의 불편을 추구한다.

We seek the minimized inconveniences rather than the maximized happiness.

현 주민들은 당장의 갈라져 있는 벽과 새어들어오는 물줄기를 방치하고 있다. 재개발만을 기다리며..최대의 행복을 전해주기 보다는 최소한의 불편으로 다가서는 re-energising.

Current residents are disregarding soaking in water through cracked walls. They just wait for redevelopment rather than for greatest happiness by approaching them minimized inconveniences.

심지윤 이지애 강민지 | Sim, ji-yun Lee, ji-ae Kang, min-ji

PU SH LL

Push, pull and get your heart pumping

Site

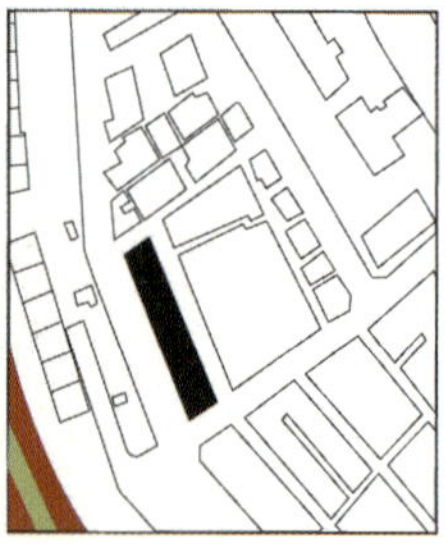

주소 : 서울시 용산구 서부 이촌동 396
용도지역 : 대림 아파트 주차장, 테니스장
인구 : 약 2300가구 1만 여명
주변시설 : 주민센터, 한강 시민 고원

Address: Ichon-dong 396, West Ichon-dong, Yongsan, Seoul
Use Zone: Daelim apt parking, , tennis court
Population: About 2300 households with 10,000 residents lot, tennis court
Surroundingn facility: resident hall, Han River Civil Park

대건로, 강변북로 등 도로시설이 있으며, 한강 시민 공원이 근거리에 위치

Daegun Rd, North Gangbyun Rd and other road facilities, which is closer to Han River Civil Park.

대단지 아파트 밀집 지역으로 유동인구가 많으며 근거리에 이촌2동 주민센터 보유

There are a lot of moving population as a congested area of apt, which holds Ichon 2-dong resident center nearby.

6년간의 재개발 중지로 다른 지역들에 비해 많이 낙후되어 있음

With the suspension of re-development for 6 years, it fell behind.

Background

서울시 용산구에 위치해 있는 서부 이촌동은 현재 재개발 중지로 인해 침체 되어있고 주민사이의 갈등 또한 고조되어있다. 혈액을 돌리고 보내면서 인체가 살아있도록 하는 데에 결정적인 역할을 하는 심장처럼 멈춰있는 용산을 Push, Pull 하여 활력을 찾고 순환하는 용산을 만들어 산소가 공급되는, 활력넘치는 서부이촌동을 만들고자 한다.

West Ichon-dong which is located around Yongsan, Seoul, became fall behind due to the suspension of redevelopment, which increased conflicts among residents. Indeed, Yongsan is just like blood which plays a crucial role in keeping a life by circulating blood throughout a body. However, it stopped now. That is why we try to resuscitate Yongsan so that it comes back to life.

User needs

REST

Senior citizen

– 소통 및 휴식 공간
– 상호 교류 및 친목의 기회 제공

- Communication and resing space
- Mutual communication and oppor tunity of friendships

PLAY

Youngster

– 놀이공간 제공
– 과학적 원리에 관한 배움의 기회 제공

- Offering a space of playing
- Offering a learning opportunity based on scientific principles

COMMUNICATION

Adult

– 자신이 원하는 형태 맞춤 휴식 가능

- Possibility of offering tailored rest

Concept

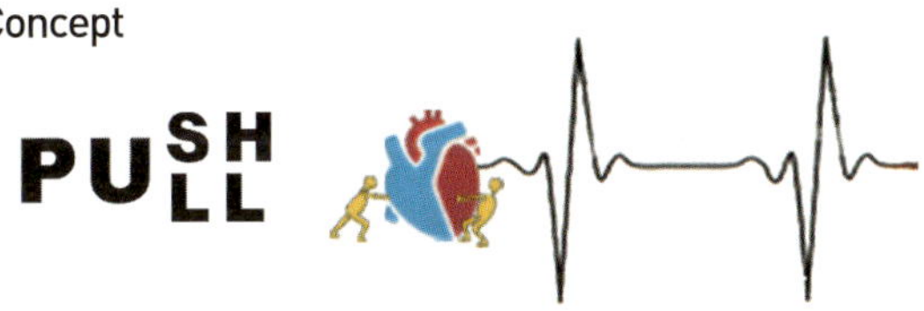

PUSH & PULL : "밀다" + "당기다"

밀어주고 당겨주는 Push & Pull로 서부 이촌동 사람들을 화합하게 하고 밀고 당겨 만들어진 공간에서 밀리고 당겨져서 작동하는 놀이를 통해 사람들의 심장을 Push, Pull하게 한다.

By pushing and pulling residents of West Ichon-dong, we can have them come alive.

Design process

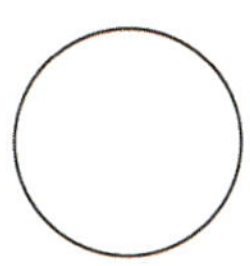

공간 분할
space division

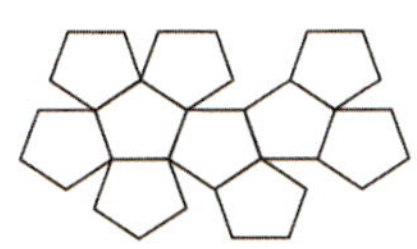

공간 펼침
space spread

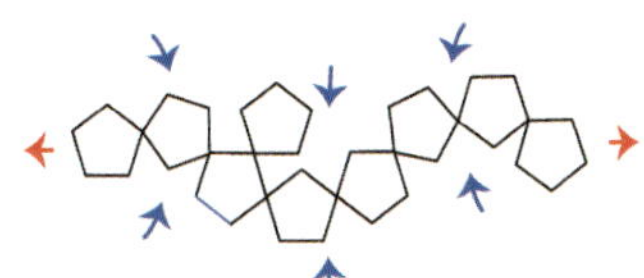

공간 PUSH & PULL로 재배열
space rearrangement through PUSH and PULL

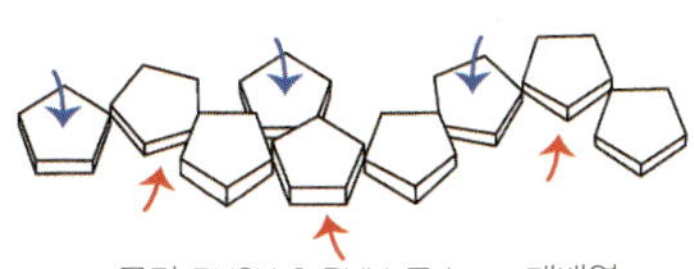

공간 PUSH & PULL로 layer 재배열
pace rearrangement through the layer of PUSH and PULL

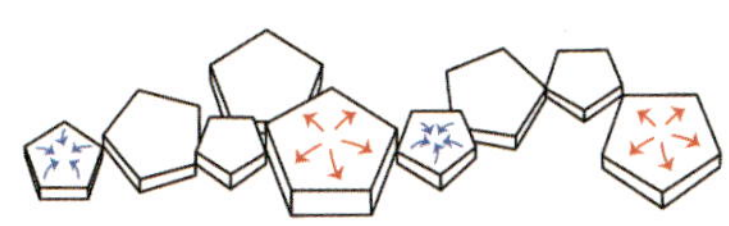

공간 PUSH & PULL로 공간 확장과 축소
Extension and Abridgement of space through PUSH and PULL

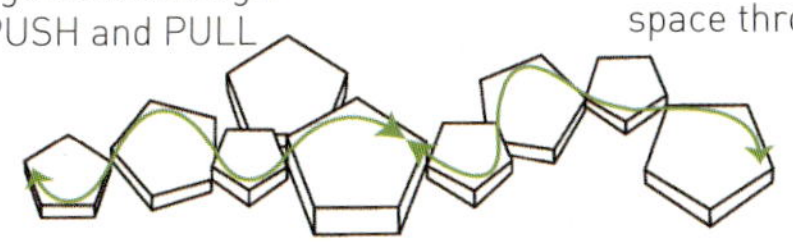

공간의 흐름
The stream of space

Background

Material : magnet

Basic Principles

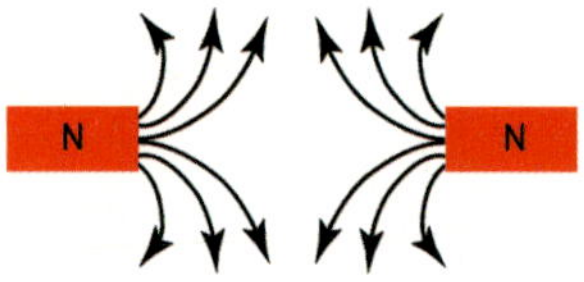

repulsive force

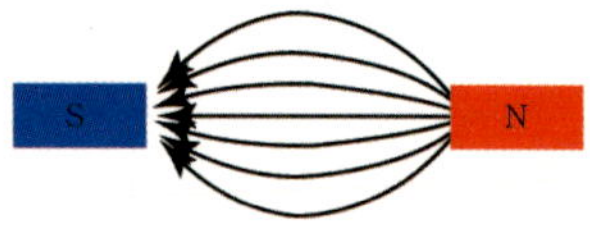

attractive force

같은 극끼리는 미는 척력과 다른 극끼리는 당기는 인력을 이용한 놀이
A play pulling and pushing.

Application 1

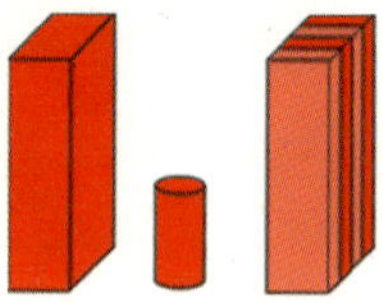

plat

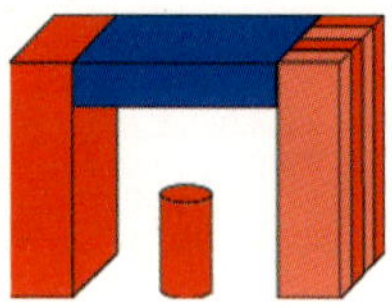

slope

한쪽 벽이 N극이나 S극의 힘을 가진 자석이고 반대편이 같은 극을 가지고 2가지의 다른 세기를 반복 할 때, 가운데의 물체가 움직인다.

When one repeats N negative magnet pole in one side and S positive one in another side, the matter in the middle of them moves.

Application 2

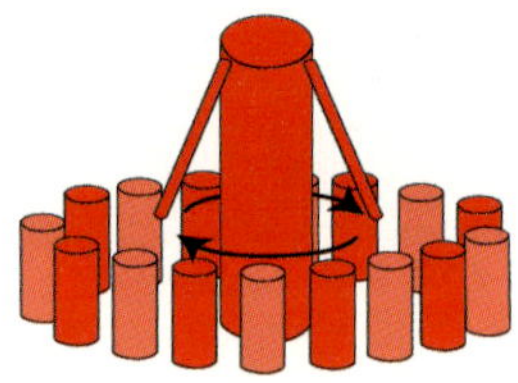

가운데 원이 N극이나 S극이고 그 원을 둘러싼 자석의 세기가 커졌다 작아졌다를 반복하면 가운데 원은 돌아간다.

If a circle in a middle N magnet or S magnet, repetition of those magnet enables a circle to return.

Application 3

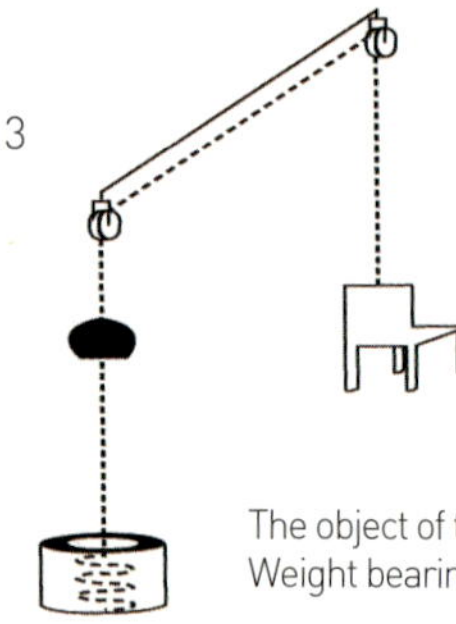

The object of the magnet
Weight bearing haejum.

도르래의 당김의 성질을 이용하여 개인공간이 되었다가 공공의 공간으로 쉽게 전환할 수 있다.

If using the nature of a pulley, you can easily convert a space freely.

Floor plan

PLAN 1

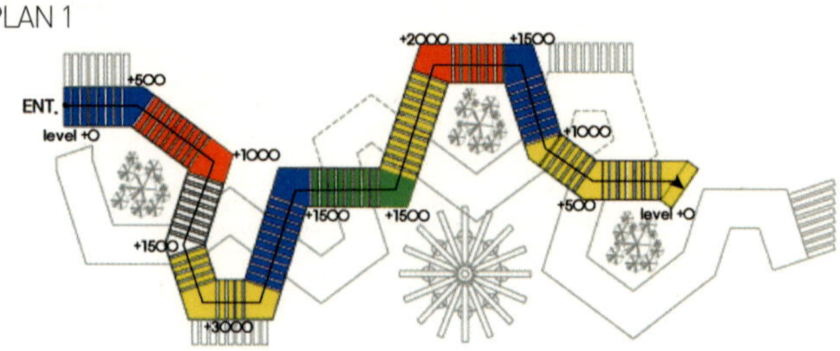

Elevation - 1

자석의 힘으로 움직이는 그네와 떼었다 붙였다 할 수 있는 자석 의자가 보이는 방향.
A direction easy to see magnet chair which can connect and detach just like a swing moving by the force of magnet.

PLAN 2

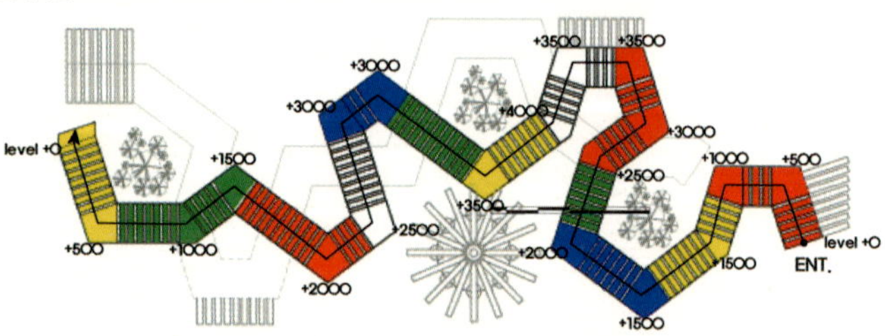

Elevation - 2

그네가 있는쪽의 반대방향으로 도르래 놀이가 있는 방향.
A counter-clockwise direction of a pulley play.

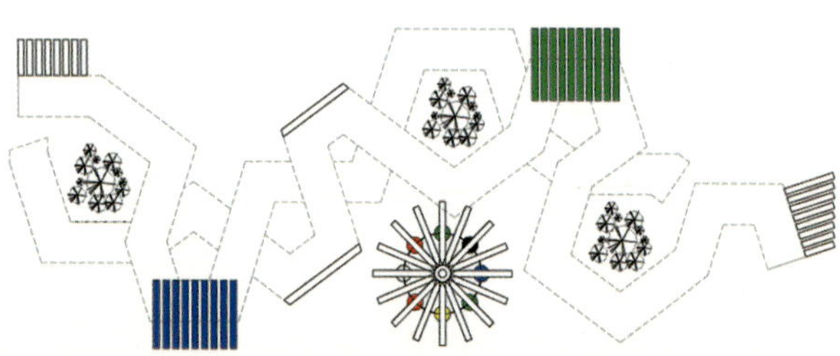

Process solution

Site 밖으로 미치는 자기장의 영향으로 생기는 문제점.
A problem caused by the influence of a magnetic field.

→ 자기장의 영향이 가장 큰 프레임 없는 공간 주변에 부도체인 나무를 조성함.
Cultivating trees around frameless spaces which never be influenced by magnetic field.

시간에 따라 달라지는 공간의 효율성의 문제점.
An issue of the efficiency of space based on time.

→ 자석의 원리를 응용한 도르래를 이용하여 이동식 놀이기구를 따로 보관하지 않아도 때에 따라 사용할 수 있도록 설치함.
By using the pulley through the principle of magnet, one can easily use regardless of time.

놀이 공간 안에서 휴식공간을 또 내는 것에 대한 협소한 공간의 한계점.
There is a limitation of cramped space in preparing for a resting room within a playing area.

→ 필요할 때마다 자석을 이용해 떼었다 붙였다 하는 의자를 만들어 휴식공간을 스스로 설치하게 유도함.
By making magnet chair which can be managed by connecting and detaching at one's conveniences, one can enjoy a resting space.

자석을 통해 정해져 있는 경로가 아닌 자신에 따라 변화하는 자신이 만들어낸 경로로 항상 새롭게 이용할 수 있다.
Not just a course designated by a magnet but just by a course of changing, one can always use them.

오전 시간 아이들과 중장년층보단 노인의 대화의 공간으로 주로 이용된다.

The old usually use it as a space of conversation for the old rather than middle ages.

오후 시간 아이들이 맘껏 뛰어 놀 수 있는 공간으로 이용된다.

It can be used as a space of running freely.

push & pull로 공적공간에서 사적 공간의 전환이 쉽게 이루어진다.

A formal space can easily be converted by a private space.

프레임으로 시선을 적절히 차단하여 긴장감을 유발한다.
It can bring about a sense of tension by blocking eyesight properly.

칙칙한 용산에 원색으로 이루어진 공간을 통해 생기를 부여한다.
It invigorates a vitality through the space performed by somber primary color in Yongsan.

박종배 안새봄 | Park, jong-bea An, sae-bom

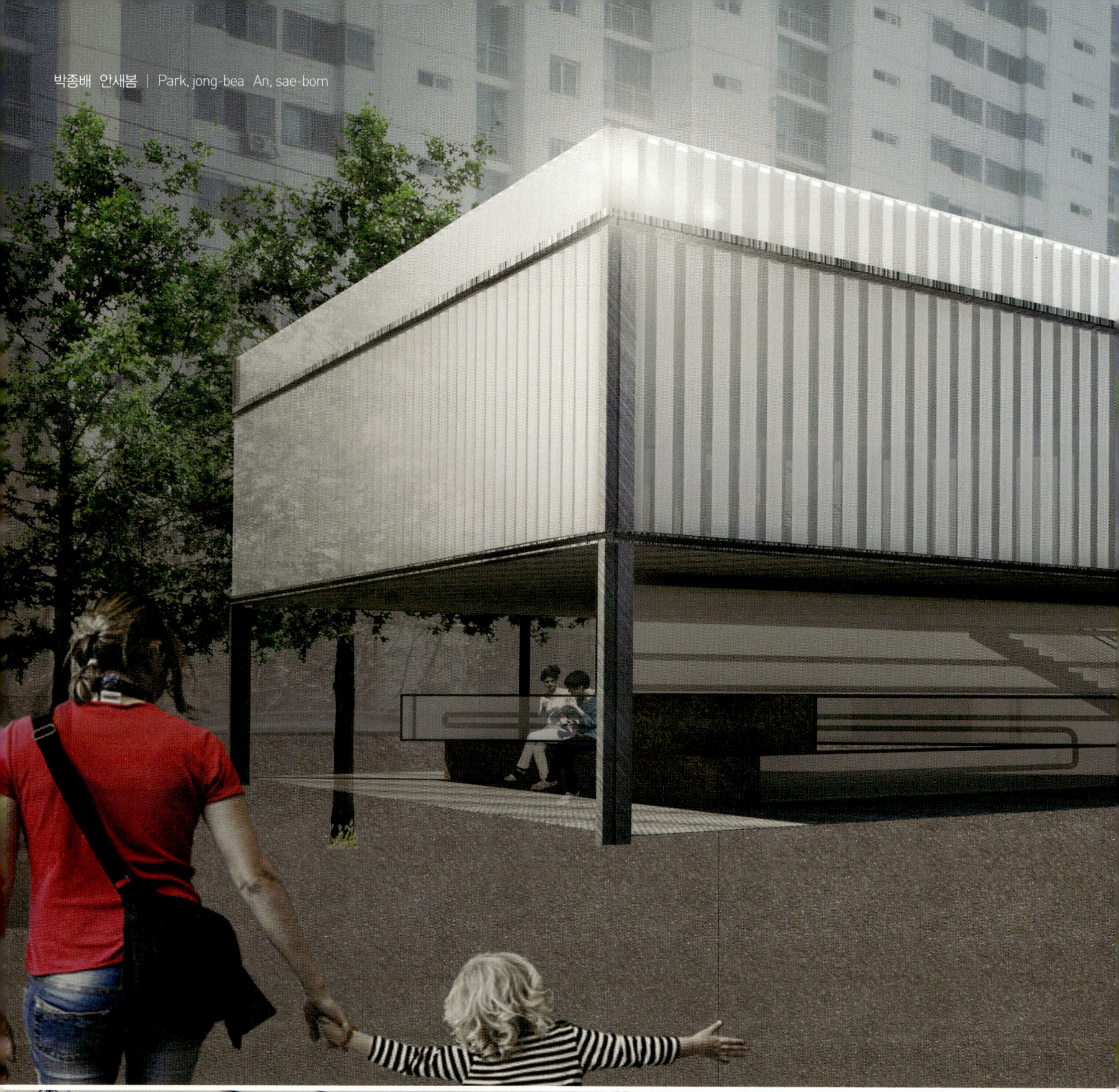

Agritainment

agriculture + entertainment, coined combined.

Background

도시 개발의 부푼 기대가 무너진 서부이촌동의 주민들 . 서부이촌동은 개발의 찬성파와 반대파와의 싸움. 정부와의 싸움으로 전쟁같은 시간들은 그렇게 흐른다. 사이트 조사를 위해 서부이촌동을 방문하였다. 조용한 동네 분위기와 사진을 찍는 낯선이에게 무서운 눈초리로 쳐다보는 아이들. 어두운 분위기에 그늘진 동네 전체를 자연의 빛으로 비추려 한다.

Expectations of Western Ichon Dong residents has changed to disappointment. Because of the fight between of supporters and objectors of development, between of resident and government, Every day in Western Ichon Dong is like a war.

Site analysis

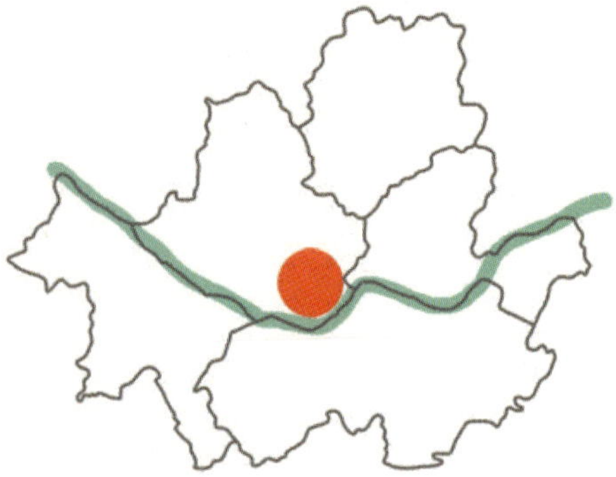

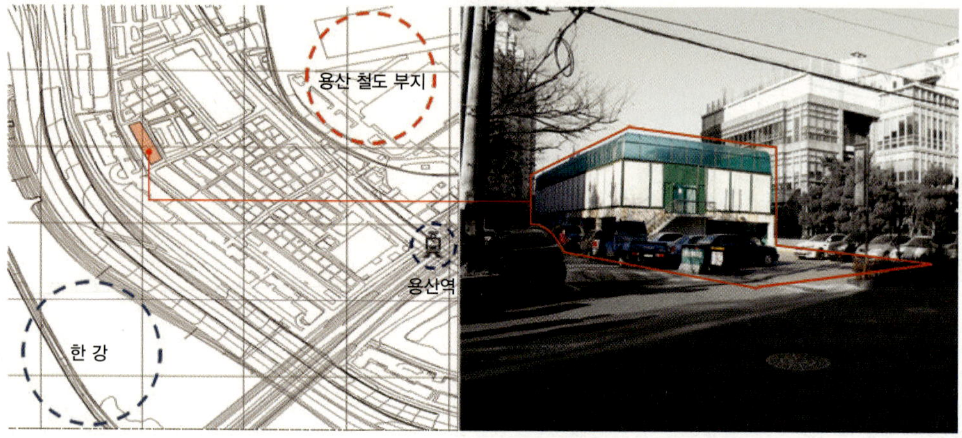

서울특별시 용산구 서부이촌동
규모 53255 X 12400 (mm) 약 200평
현 사용공간 1층 주차장과 물탱크, 2층 테니스장

서부이촌동은 도보 10분 경에 용산역이 위치해 있으며 디자인 될 건물 앞에는 바로 대림아파트가 있다. 본 건물 바로 앞에 바로 대림아파트가 있다. 본 건물은 대림아파트 소유의 주차장과 테니스장이다. 대림아파트 건너엔 한강이 위치해 있다.

Western Ichon-dong, Yongsan-gu, Seoul, Korea
Scale : 53255 X 12400 (mm) about 660m²
Present usage
- The first floor : Parking lot, Water tank
- The second floor : Tennis court

Western Ichon-dong's a 10-minute distance from Yongsan Station. There is an Daerim Apartment directly in front of the designed building. Daerim Apartment own this building of parking lot, tennis court. There is a Han-River on the other side of Daerim Apartment.

Problem & solution

서울시의 '용산 철도 정비창 부지와 서부이촌동 공개 개발' 추진 실패로 서부이촌동의 주민들은 많은 빚더미에 앉게 되었다. 그러면서 개발을 찬성하고 반대했던 사람들 사이엔 깊은 갈등이 생겨 났다.
갈등으로 인해 멀어진 서부이촌동 주민에게 서로를 만나게하는 계기를 마련하려고 한다. 작지만 같은 노동을 하면서 주민들은 서로 적이 아닌 같은 어려운 시기를 보내는 동료이자 친구를 느끼게 된다.

Due to failure of Yongsan railway workshop site and Western Ichon-dong opened development, residents fell into debt. So, conflicts arose between supporters and objectors. We'd like to give estranged people motivation to meet. By doing same, but not arduous, work, people could feel each other as friend and colleague who spends hard time together.

Target

성인이 스트레스를 받으면 많은 활동을 통해 풀려고 한다. 하지만 노인과 어린이들의 활동에는 제약이 많다. 그런 노인과 어린이들을 위한 공간이면서 스트레스를 풀 수 있는 공간을 디자인 하겠다.

Under stress many adults are trying to solve through the activity. The activities of the elderly and children, but there are many restrictions. Then, while the elderly and space for children to ease the design space will be.

FUNCTION

환경의 변화의 제약이 많아 사계절 내내 키울 수 있도록 온실의 기능을 하는 공간을 만들 것이다. 도시 한복판의 온실은 사람들에게 여름에는 서늘한 겨울에는 따뜻한 공간이 될 것이다.

Because of environmental constraints, We will make a space performed as greenhouse for cultivation throughout four seasons. Greenhouse of middle of city will be a cool space in summer and warm space in winter.

Programming

TARGET

수경재배
우리 공간의 주 타겟층은 노인과 어린이인 것은 감안하여 땅에서 재배하는 것보다 손쉬운 방법인 수경재배를 가져오게 되었다.

Water culture
Allowing for targets of this space are senior citizen and children, We take a water cultivation instead of a ground cultivation which is more difficult.

CONSERVATION

퇴비
많은 음식쓰레기가 배출되어 환경이 오염되고 있다. 음식쓰레기의 양을 줄이기 위해 음식물쓰레기를 퇴비화 시키겠다.

Compost
The environment is being polluted by immense food waste. To curtail food waste, We will compost food waste.

Concept

틈 :
우리는 가끔 돌 틈 사이에 자라나는 식물을 본 적이 있다. 저 틈사이에 씨가 들어간 것도 신기하지만 저 단단한 돌틈에 뿌리는 박고 계속 자라난다는 것은 우리에게 많이 의미를 준다. 돌 사이에 힘겹게 뿌리내려가는 저 식물처럼 서부이촌동이라는 돌틈 사이에 새싹같은 공간디자인을 제공한다.

Crack :
We often see plants that grow in cracks between stones. It is amazing to get into the crack. Also, putting down roots in crack and growing up continuously give us a lot of meaning. Like the plant growing up in a crack, We designed a space such as sprout in the crack; Western Ichon-dong.

Design motive

EXTERIOR
: cantilever,
H-bim

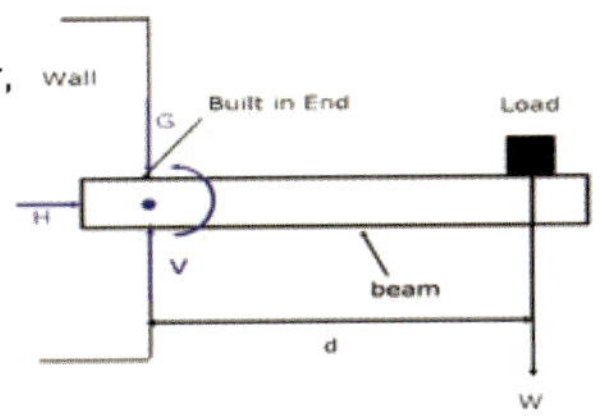

INTERIOR
:pipe,
plastic box

Design process

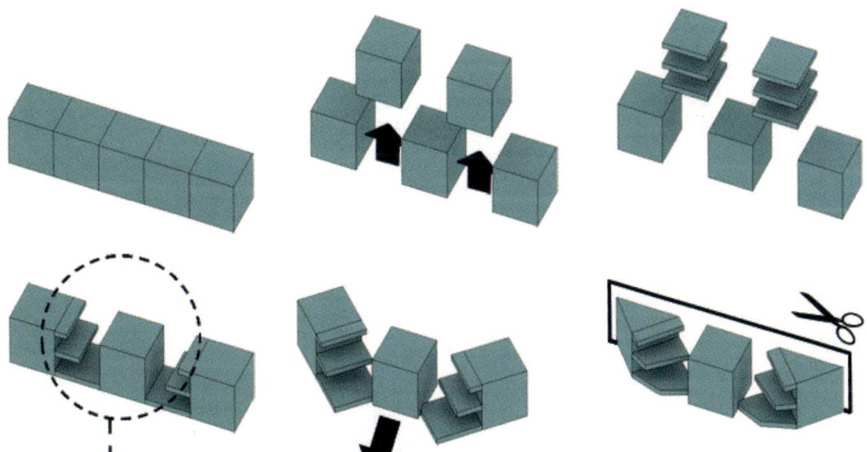

1. 사이트 전체의 공간이다.
2. 하나의 공간 사이 사이에 여백의 공간을 만들어 준다.
3. 나온 공간을 얇은 슬라브를 만든다.
4. 슬라브, 캔틸레버 구조로 건물에 연결 된다.
5. 서로를 마주보고 있는 건물을 대림아파트 주민들을 향한다.
6. 사이트에 면적에 넘어간 건물을 자른다.

1. Area of the entire site
2. Make blank space between each of spaces
3. Make ascended space with slab.
4. Slabs are connected to building in cantilever structure
5. Buildings on both sides face Daerim Apartment
6. Cut part in excess of the area.

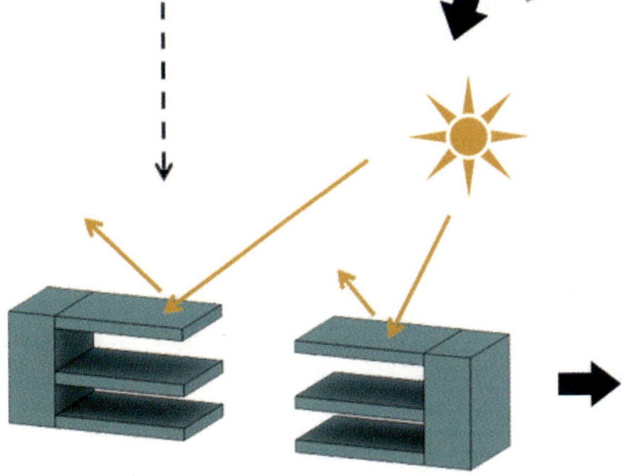

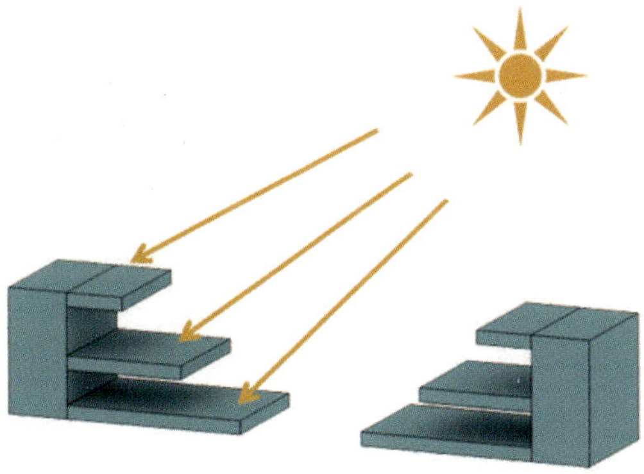

1St floor plan

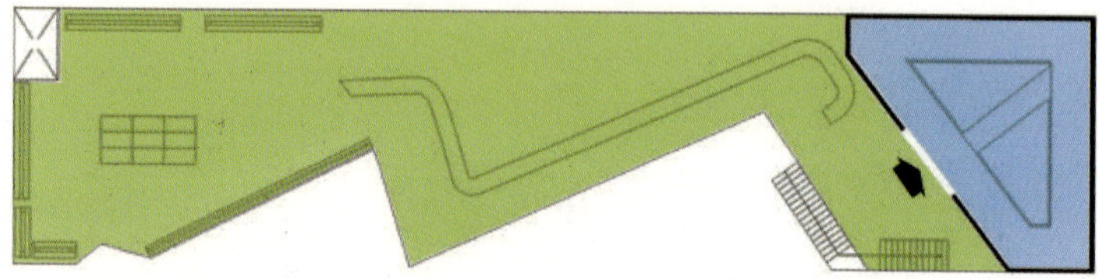

2nd Floor Plan

Plan

1

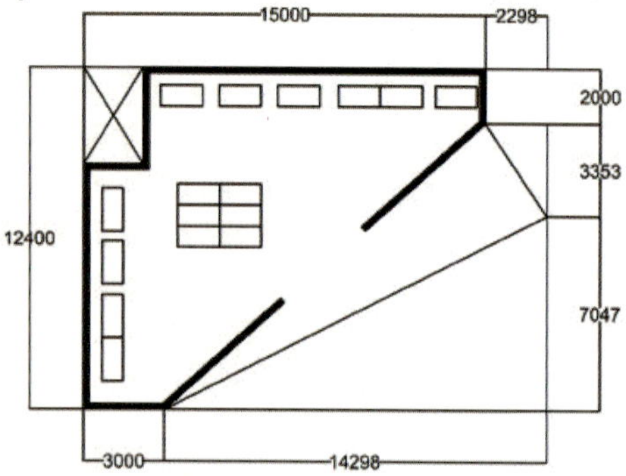

이 공간은 역시 플라스틱 박스를 통한 재배 공간이며 휴식 할 수 있는 공간이 제공될 것이다. 노출된 엘레베이터를 타고 올라가기 때문에 이 공간에서 사람들은 약간의 긴장감을 느낄 수 있다.

This place is also the cultivation space using plastic box and resting place is provided too. The elevator is exposed. So, people who use the elevator can feel a slight tension.

2

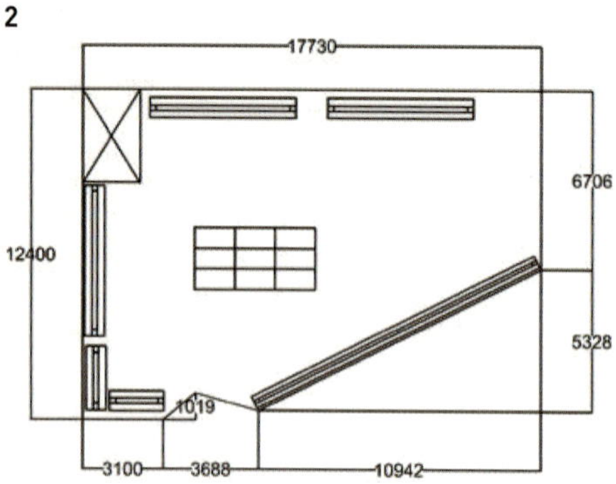

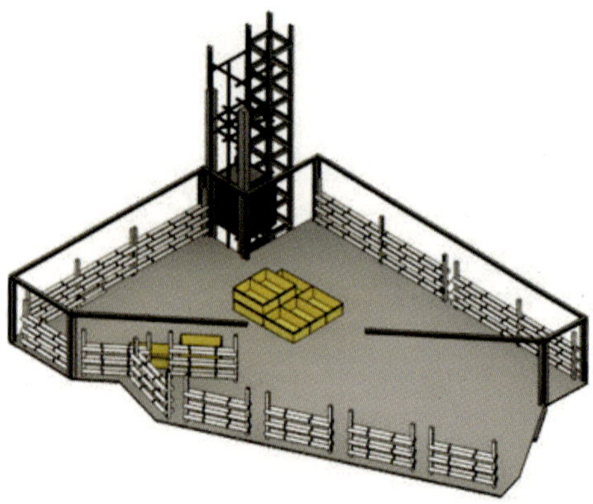

플라스틱 박스에 재배하는 공간이지만 오픈스페이스로 난간의 역할을 수경재배의 파이프가 대신 한다. 이 공간은 1층에서도 볼 수 있으며 1층에서 보게 될 경우 파이프를 통해 재배 되는 모습이 보여 사람들에게 흥미를 제공한다.

Plastic box cultivating section is main in this space. As open space, pipes of water cultivating serve as railings. This space can be seen in the first floor. Watching plants cultivated in pipes, people feel interested. This place is also the cultivation space using plastic box and resting place is provided too.

3

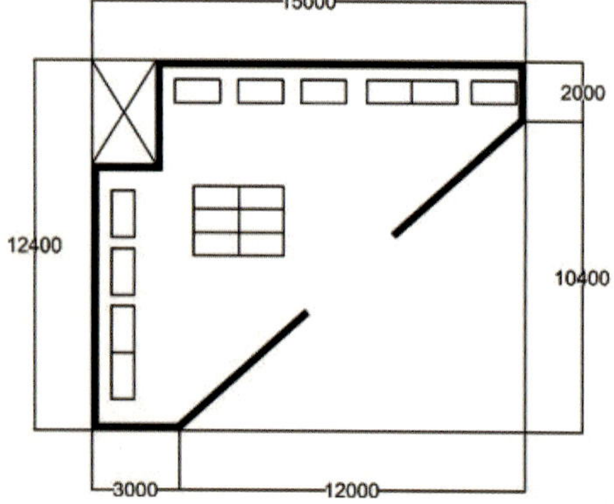

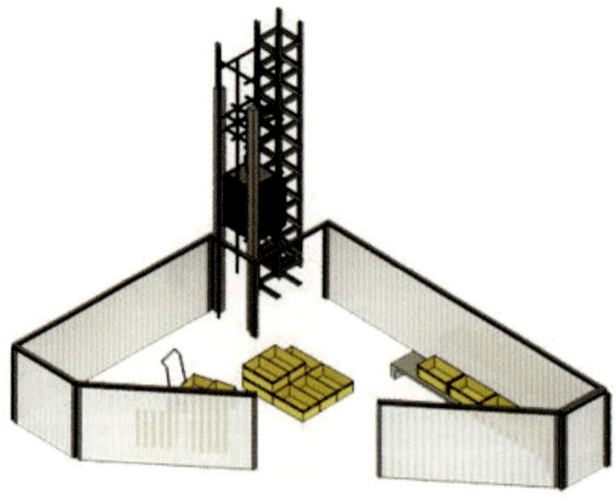

흔히 볼 수 있는 플라스틱 박스에 재배 할 것이다. 크지 않은 공간이기 때문에 한 가구 당 한 박스로 제공 될 것이며 박스는 수레를 통해 움직일 수 있다.

Plants will be cultivated in plastic box that can be seen easy anywhere. Because the space is not large, each family will be given one box. Box could be moved with cart.

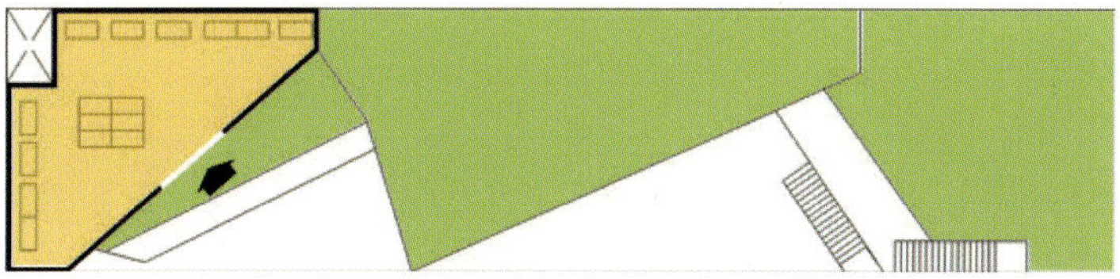

3rd Floor Plan

4th Floor Plan

Open space | Planting compost | Water culture

4

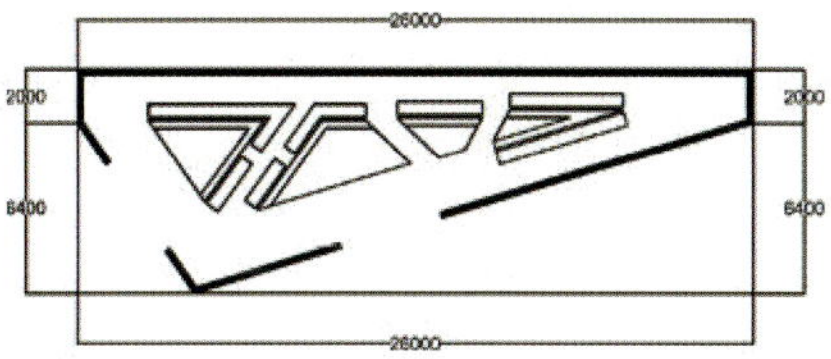

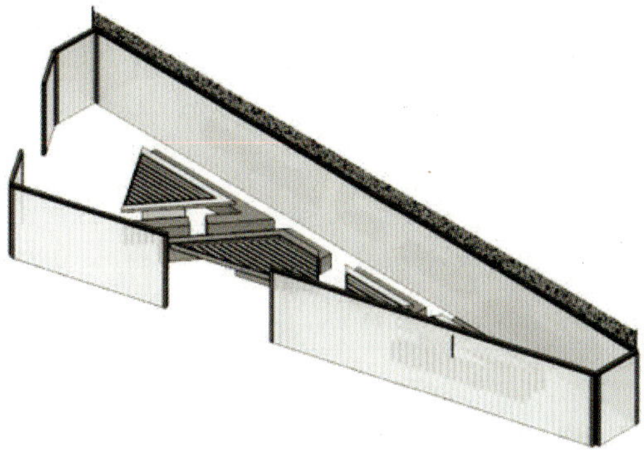

간단한 음식과 음료를 앉아서 즐길 수 있는 공간이다. 테이블 앞엔 수경 재배를 할 것이며 이 공간에 재배되는 야채는 식사를 하는 사람들에게 무료로 제공된다.

In this place, people can sit down and enjoy the food and drink. Water cultivating is performed in front of each table. The vegetables grown in here will be available free of charge

5

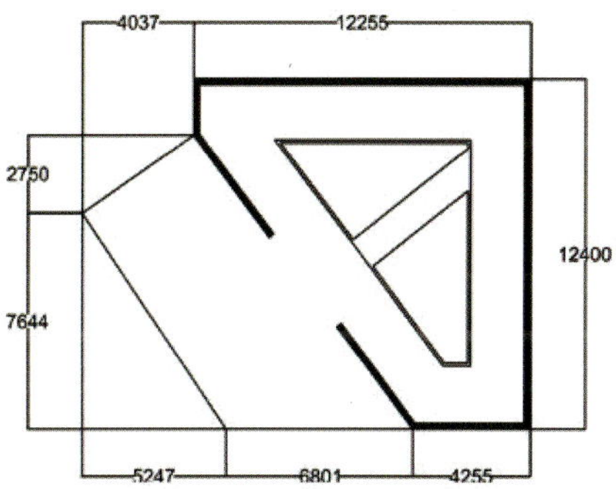

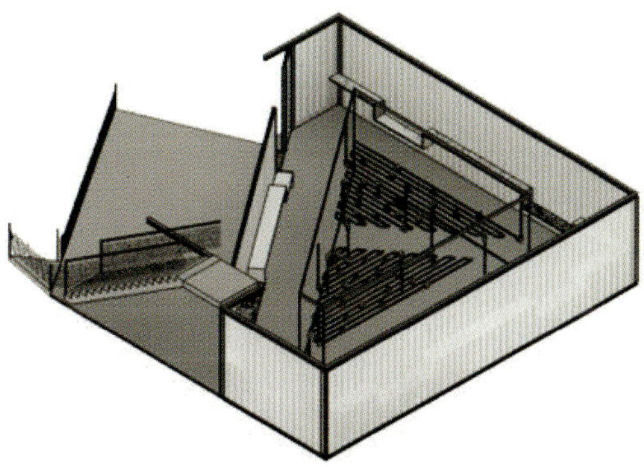

수경재배의 모습을 가장 뚜렷하게 보여주는 공간이다. 파이프를 통한 재배를 파이프의 수평, 수직 모든 모습으로 재배 될 것이며 위생을 위해 유리관 안에서 재배 된다.

Most clearly shows the appearance of hydroponic space. Cultivated through a pipe to a pipe in the horizontal, vertical farming will be all aspects of hygiene will be grown in a glass tube for.

6

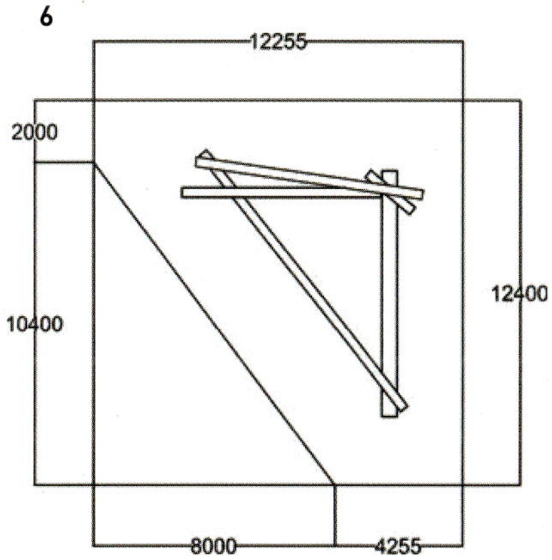

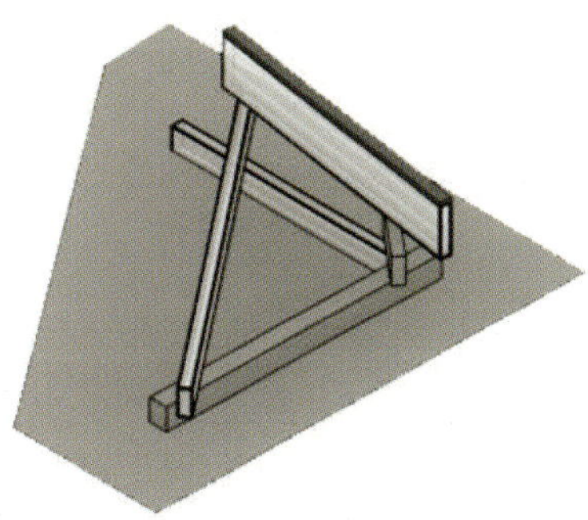

오픈스페이스의 공간으로 유리관 속에서 재배 되고 있는 모습을 방문한 사람뿐만 아니라 사이트를 지나가는 사람들도 볼 수 있도록한다. 파이프를 앉아서 볼 수 있는 벤치가 있다.

Because it is open space, all the visitors and passerby can see this cultivation performed in glass wall. There are benches people can sit and look at pipes.

1st Floor : Plastic box can be moved through the wagon

2nd Floor : hydroponic space

Skin & Structre: The buildings facing each other head Daelim apartment residents. Cantilever slabs which are connected to the building structure.

송지은 윤설영 | Song, ji-eun Yoon, seol-young

Butterfly Effect

숨겨진 공간으로서의 "Re energizing space"

서울시의 용산국제업무지구 개발사업의 지지부진으로 인해 정신적, 물리적으로 막대한 피해를 입은 서부이촌동 주민을 위로하는 충전과 유희의 공간을 창조한다.
주민들이 함께 대화하고 소통할 수 있으며 이 공간은 긍정적 생활의 원동력이 될 것이다.

Re-energizing space as a hidden space

The space of recharge and amusement will be created in order to soothe the residents in Seoboo Ichon-dong who have huge damage mentally or physically because of the delay of developing Yongsan International Business Center. This space will provide the residents with positive, dynamic and potential life, playing a role of connecting communicating each other.

주민의 소통, 화합의 공간이 되며 긍정적 생활의 원동력이 될 것

To overcome the reality that there is even no space for the residents who are devastated mentally or physically.

Purpose

물질적, 정신적으로 지치고 피폐해진 주민들을 위한 공간조차 없는 현실

this space will be not only a place for communication and harmony, but also a motive force for positive and active life.

Problem

Site analysis

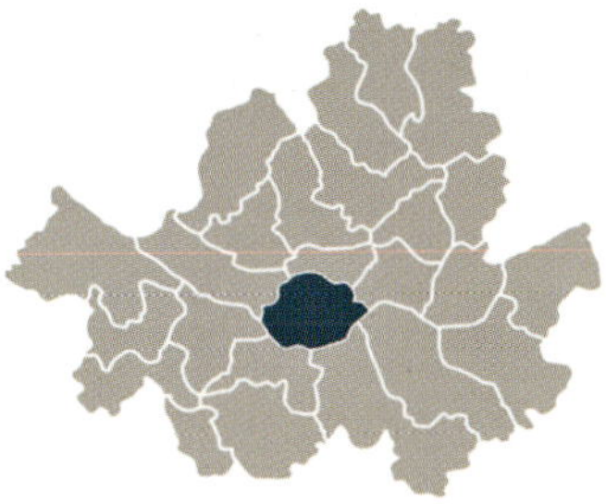

서울특별시 용산구 서부 이촌2동 대림아파트 주민센터 앞 주차장 & 배드민턴장 총면적 : 455m²
West dong Yongsan International Business District development Interna West dong

Concept | butterfly effect

어느 한 곳에서 일어난 작은 나비의 날갯짓이 태풍을 일으킬 수 있다는 이론. 초기조건의 사소한 변화가 전체에 막대한 영향을 미칠 수 있음을 이르는 말이다.
서부이촌동 주민이 모인 공간이 긍정적 생활의 시작점이 되어 서로 활동하며 소통하고 화합해 위기상황을 극복해 나간다.

The theory that the flying of small butterfly can cause the huge storm. This means that the trivial change at the beginning stage can have a huge influence on the whole result. As the space where the residents get together become a starting point for positive living, they can get over the difficult situation by living actively, communicating and harmonizing each other.

Keyword

Motive

나비효과의 연쇄작용
→ 톱니가 연쇄적으로 맞물려 돌아가는 톱니바퀴

a Chain Reaction of Butterfly Effect
a gear wheel where the teeth of a saw works properly and connectedly.

Space programming | domino game

작은 도미노 조각이 연쇄적인 반응을 일으켜 큰 그림을 만든다. 도미노게임을 하는 것 그 자체가 하나의 전시가 된다. 도미노는 다양한 연령대의 사람들이 즐겨할 수 있는 게임으로 어른들에게는 치매예방의 효과가 아이들에게는 창의성, 인내력, 협동심, 성취감 등의 효과가 나타난다.

왜 하는가 ?
주민이 협동해 블록을 쌓으며 대화를 통해 소통할 수 있고 성취감을 얻을 수 있다. 또, 도미노 체험공간을 통해 전시를 구경하며 색다른 재미를 느낄 수 있다.

A piece of domino triggers chain reaction and make big result. Playing a domino game can be an exhibition. Domino is a game that people in various age group can enjoy, that can prevent older adults from amnesia and contribute to enhancing creativity, perseverance, collaboration, and achievement for children.

Why we do a domino game?
While they are building blocks, they can communicate each other and achieve fulfillment. Meanwhile, they can feel unusual interests, while they are watching displaying of the space experiencing domino.

Floor plan

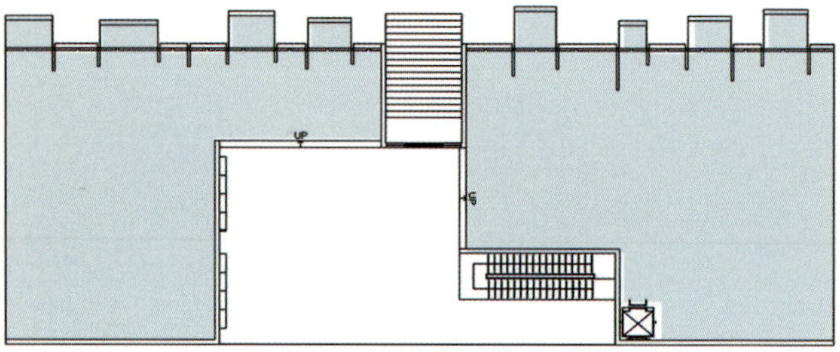

1St 2nd floor

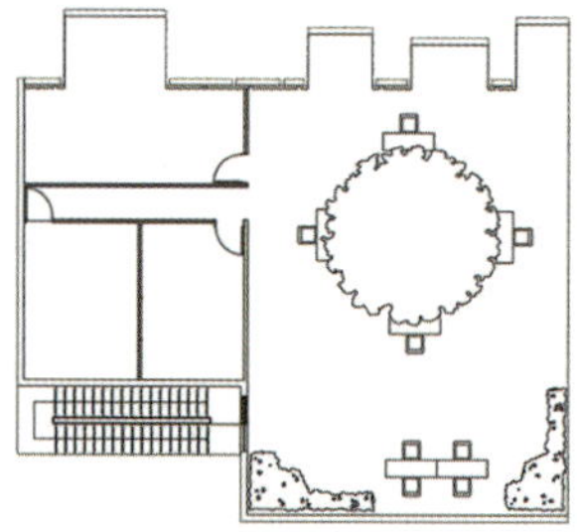

3Rd floor

Exhibition space

도미노의 연쇄작용을 연상시키는 십자형태의 전시벽을 돌려가며 이동한다. 공간마다 전시벽이 다르게 구성되어 있으며 사람들의 이동 방향에 따라 공간이 변화한다.

It can move through displaying wall of various cross forms associated with a chain reaction. From space to space, displaying walls are composed variously and the space can be changed according to direction of people's movement

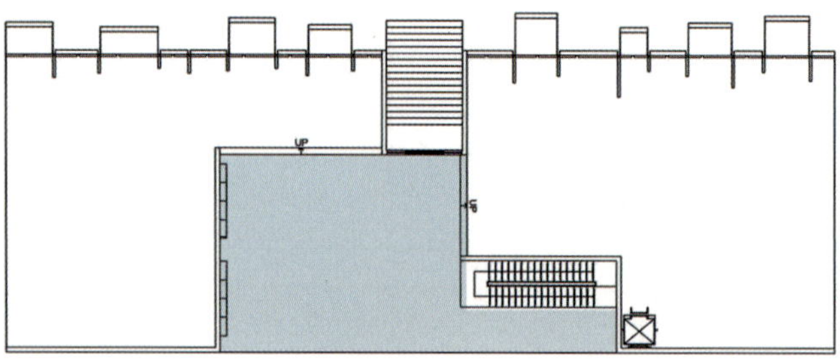

1St 2nd floor

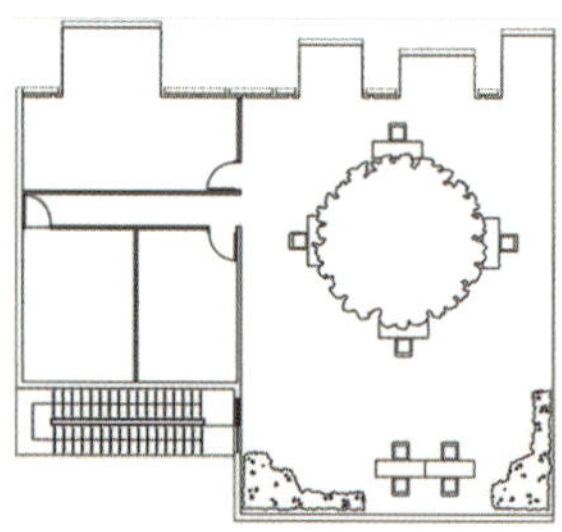

3Rd floor

Experience space

주민들이 함께 협동하여 블록을 쌓으며 게임을 직접적으로 체험할 수 있는 공간이다.

It's the space where residents can experience games directly, building blocks and collaborating each other.

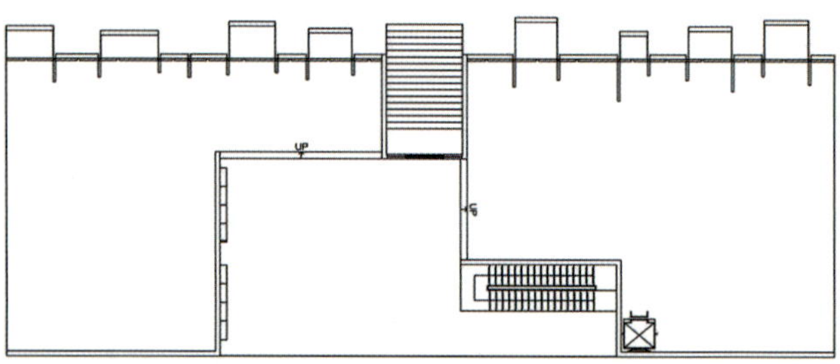

CIRCULATION

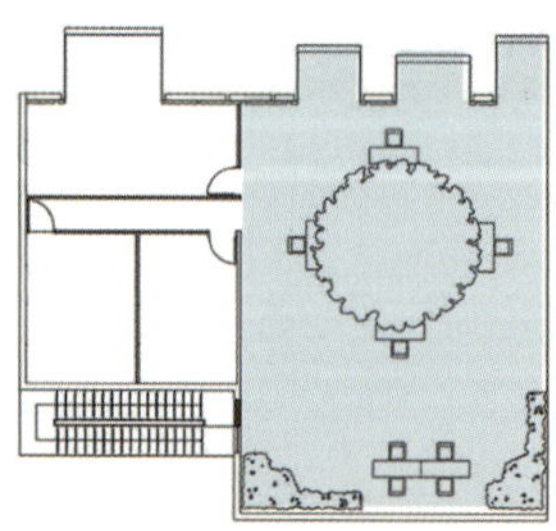

CIRCULATION

Rest Area

잠시 쉬어가기도 하고 앉아서 바깥 풍경을 감상할 수도 있는 공간이다.

It's the space where people can take a short break and appreciate outside landscape.

Variation of module

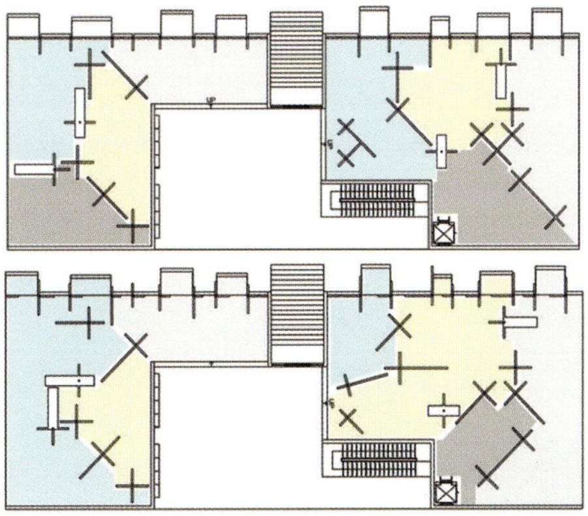

1ST FLOOR

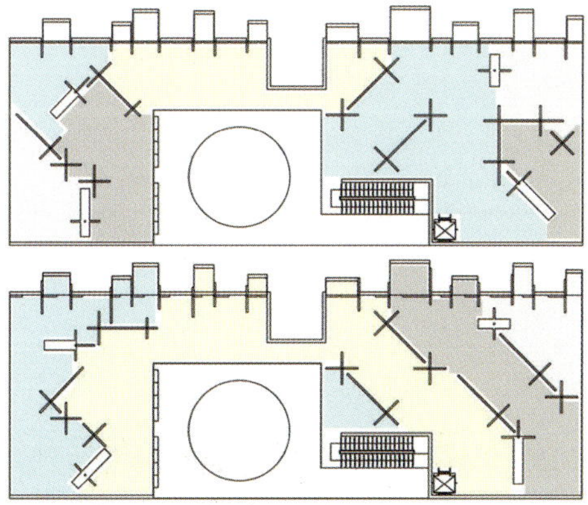

2ND FLOOR

다양한 형태의 전시벽을 이동시켜 공간을 이동한다. 전시벽이 움직이는 위치에 따라 공간이 변화한다. 전시벽과 동시에 의자로 사용할 수도 있다. 공간마다 다른 색상을 사용해 이동 할 때마다 다른 분위기의 공간을 접할 수 있고, 색상은 주기적으로 교체할 수 있다.

We can move various displaying walls and change the space. According to the location where displaying walls move, the space can be changed. It can be used as a displaying wall or a chair as well. We can be exposed to various spaces with different atmospheres whenever we move, and the color can be altered alternately.

Mass process

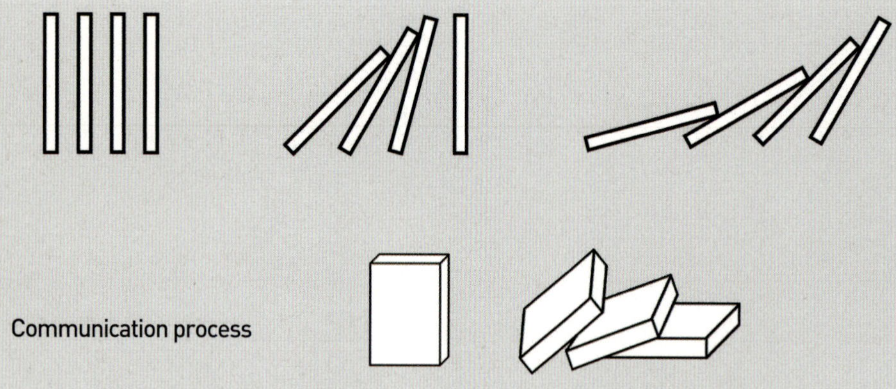

도미노 블록이 넘어지면서 다른 블록에 영향을 준다.
블록 하나를 주민 한명으로 설정하고 블록이 넘어져 다른 블록에 영향을 주는 과정을 주민이 서로 소통하는 과정으로 본다. (블록의 중첩)

Domino blocks influence other blocks, when they fall down. A piece of block symbolizes a member of residents and the course that the falling one affects another one is the same to the course that the residents increasingly communicate each other.

Design process

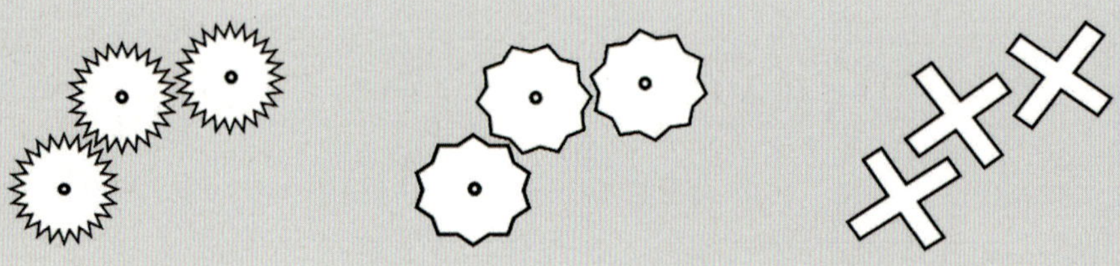

톱니바퀴 형태의 단순화를 통해 다양한 십자형태의 전시벽을 사용한다.

In order to make gear wheels simple, displaying wall with various cross forms is used.

Material

공간에서 발생할 수 있는 소음을 낮추고, 아이들이 부딪혀 다치지 않도록 하기 위해 전시벽에 패브릭 소재를 사용한다.
외벽에는 파사드의 디자인과 연결된 느낌을 주기위해서 대리석을 세로로 쌓아 사용한다.

In order to lower the noise that can occur in the space, and for children not to be hurt against the wall, displaying walls should be made of fabric materials.

곽민주 이승은 | Kwak, min-joo Lee, seung-eun

COMPOSE

공간을 작곡하다

Site analysis

서울특별시 용산구 이촌2동

한강대교와 원효대교 사이에 입지해 있으며 근처에 용산역이 위치. 다세대 연립과 아파트가 있고 주변에 어린이 집과 학교, 주민센터가 근접해 있다. 재개발 문제로 인해 상처받은 사람들의 마음을 위로받고 쉴 만한 힐링, 놀이 공간이 부족하다.

Seoul Metropolitan City, Yongsan-gu, Ichon2-dong

Located in between Han River Bridge and Wonhyo Bridge, and Yongsan Station is nearby. There are multiplex houses and apartments and daycare centers, school, and community service center are located nearby. There are insufficient healing and play space where people who got hurt from redevelopment issue can get comforted and rest in.

Concept

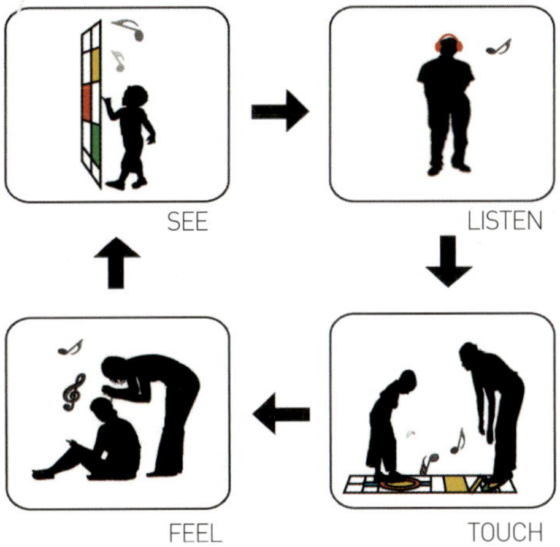

무엇을 지각할 때 우리들은 흔히 보고 듣고 만지고 느낀다. 보고 듣고 느끼는 오감(五感)에 사람들에게 마음의 안정을 가져다 주는 음악을 합해 '공간을 작곡하다' 라는 컨셉을 잡아 하나의 곡을 작곡하는 과정을 건물에 나타내었다.
도입부, 중반부, 클라이맥스, 후반부로 나누어 각 공간의 특징에 맞추어 공간의 컨셉을 잡았다. 높고 낮은 음을 이어서 하나로 길게 연결되어 멋진 음악을 만들어 내는 가락처럼 주민들의 화합과 힐링이 이루어지는 공간을 INTERACTION DESIGN을 접목시켜 만든다.

When we perceive something, we often see, hear, touch, and feel. The process of composing music was represented on a building with a concept of 'to compose space' by combining music which brings a state of balance to people with five senses of seeing, hearing, and feeling. The concept of space was set according to the characteristics of each space by dividing into introduction, middle part, climax, and the latter half. Space where unity and healing of residents take place is created by infusing the INTERACTION DESIGN just like a melody that creates wonderful tune by connecting high and low tones that are connected into one.

Background

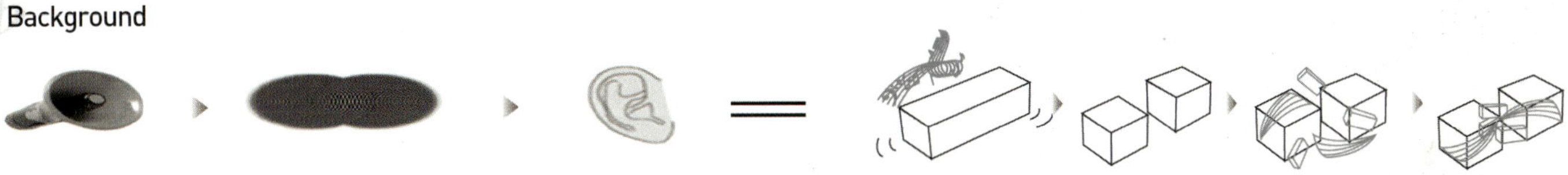

소리의 발생
the occurrence of sound

공기압의 진동
Pneumatic vibration

고막의 진동
Vibration of the eardrum

소리의 파동이 공기 중에 전달되어 진동해 분열된 것이 늘어놓아진 요소들을 '다시 작곡하다'
The sound wave gets delivered into the air and 're-compose' scattered elements of divided notes from vibration.

INTERACTION DESIGN

INTER + ACTION

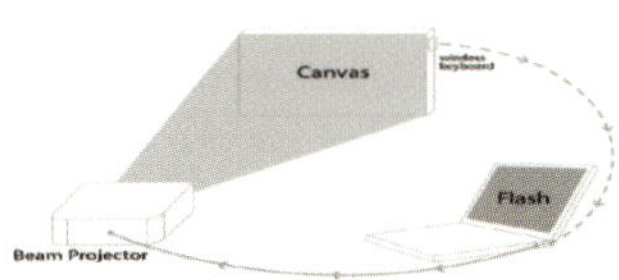

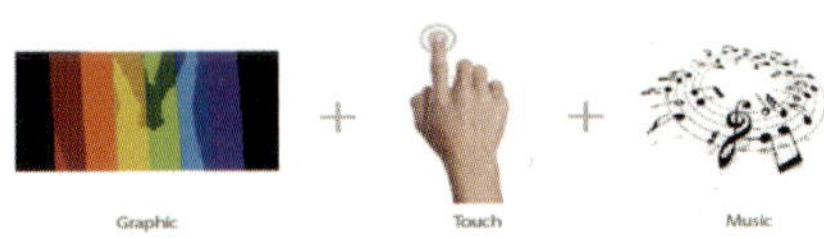

디지털 기술을 이용해 사람과 작품간의 상호작용을 조정하여 서로 소통할 수 있도록 하는 디자인 분야

The design field which helps people to communicate with each other by adjusting interaction between people, and works with digital technology.

Floor plan

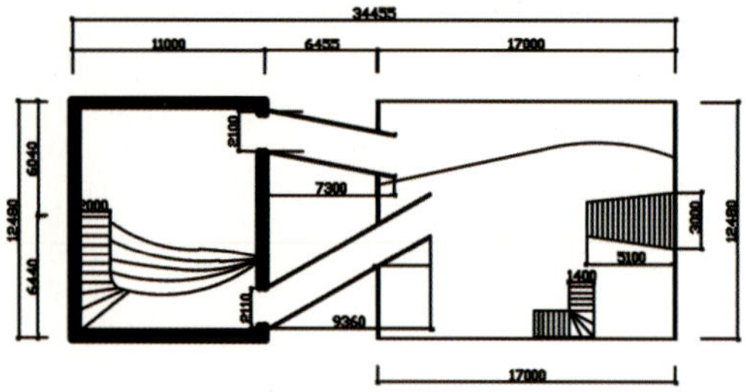

1F

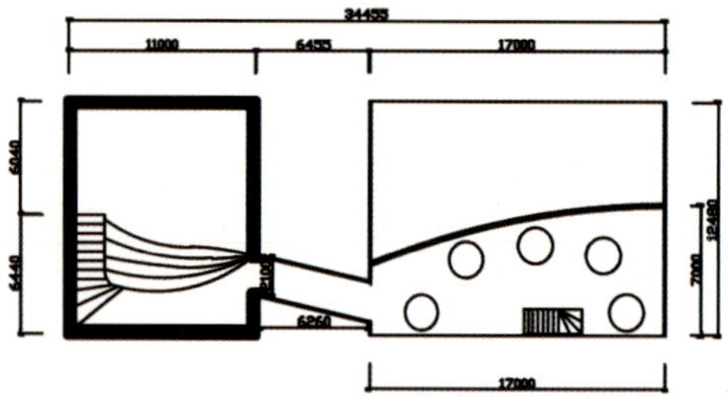

2F

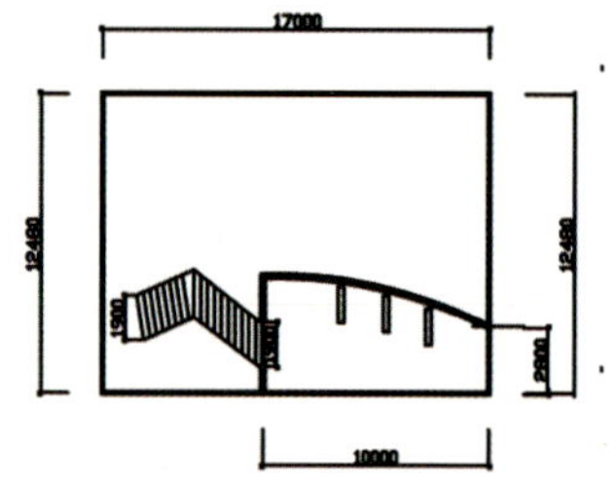

3F

3F 그래픽이 섞인 그림자놀이를 통해 즐거움을 느낀다.

3F Feel joy through making shadow pictures with mixed graphics.

2F 각각의 나무 방음재로 구성된 실린더 공간에서 음악놀이를 체험한다.

2F Experience music play in each cylinder space composed with wooden soundproofing material.

1F 바닥의 압력을 통해 나타나는 그래픽을 밟으며 여러 가지 INTERACTING한 놀이를 체험한다.

1F Step on the graphics appeared through fundal pressure and experience various interacting plays.

Interaction design

빛이 인도하는 곳으로 나가게 되면 디지털화 된 공간을 마주하며 눈에 탁 트인 넓은 INTERACTION WALL과 하나가 된 객석과 피아노계단이 한 눈에 들어오게 된다. 천정에 고인 물이 외부자연의 리듬에 따라 시시각각 변하여 바닥에 비쳐 자연의 음악을 시각적으로 환상적이게 느낄 수 있다.

When people go outside where light glimmers, they face digitalized space and take in a grand view of auditorium that became one with unhampered wide INTERACTION WALL and piano stairway. They get to feel the music of nature visually and fantastically as stagnant water in ceiling changes from moment to moment to the rhythm of outside nature.

다양한 음악 놀이 공간이 1층, 2층, 3층에 거쳐있고, 넓은 1층 공간에서 전체적인 흐름을 볼 수 있다. 공간이 통유리로 되어 있어 반투명적이지만 외부를 바라볼 수 있다.

Various music play spaces are on the 1st, 2nd, and 3rd floor, and people can see the overall flow in a wide space on the 1st floor. It is translucent as the space is made with glass but can view the outside.

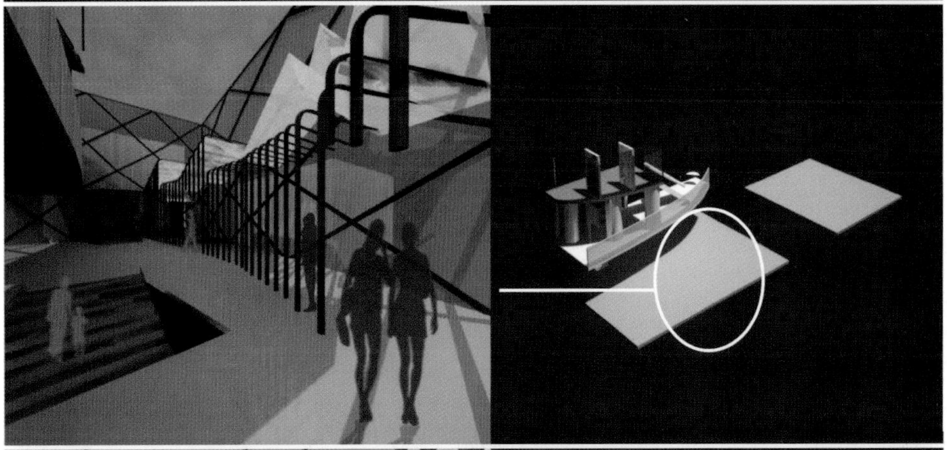

놀이공간에서 감상공간으로 이동하는 좁고 긴 세 가지의 통로에서 벽에서 나타나는 여러가지 효과와 놀이로 다른 공간에 온 듯 한 착각과 함께 음악의 흐름과 같은 자연스러운 공간 이동을 경험한다.

Various effects and games that appear from three narrow and long passageways from a play space to appreciation space make people to delude themselves as if they came to a different space and to experience the free space movement like flow of music

외부 철골구조가 내부로 인입된 모양의 반복적인 파이프 통로 구간에서 휴식을 취하거나 음악을 감상한다.

People will rest or appreciate music in a repetitive pipe passageway section of an inserted-shape of exterior steel construction.

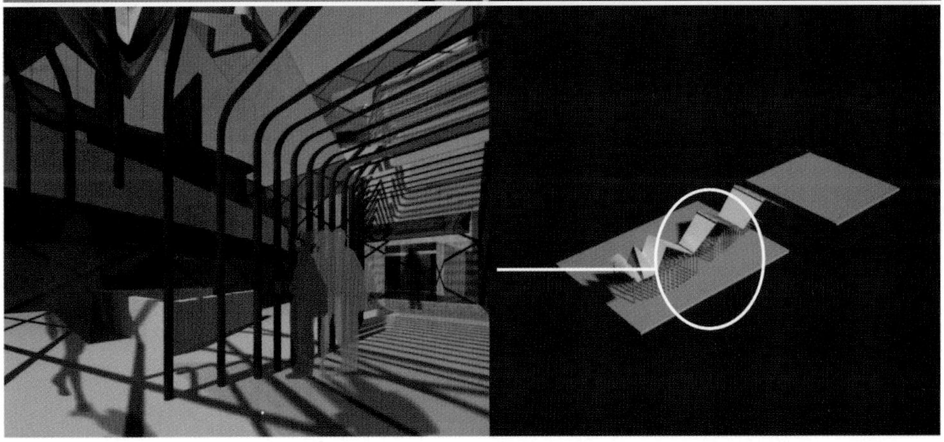

Composer process

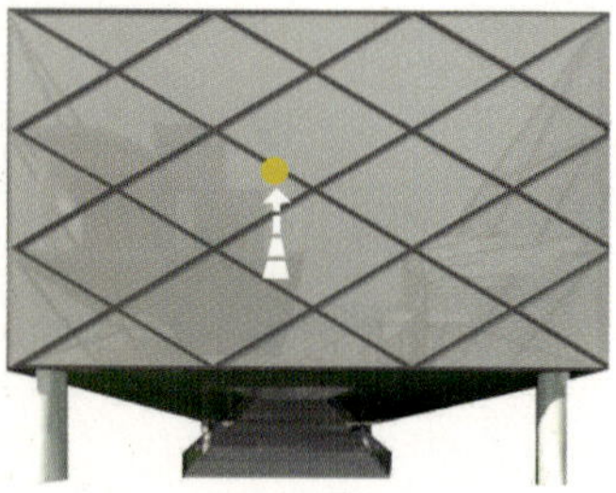

Low & slow
도입부

건물로 들어서면서 가장 먼저 만나게되는 출입구에서 내부가 한눈에 보이지 않는 답답함과 함께 재개발의 기대에 찬 이촌동 주민들의 모습을 연상시킨다.

The entrance that people get to encounter as soon as they enter the building reminds stuffiness of not being able to see the inside at a look and image of Ichon-dong residents who are in anticipation for the redevelopment.

Medium & stringendo
중반부 – 1

좁고 긴 통로를 마주하며 세 갈레의 길 중 1관에서 2관으로 이동하는 통로에서 다른 공간에 온 듯 한 착각과 함께 음악의 흐름과 같은 자연스러운 공간이동을 경험한다.

People will experience the natural space movement as the flow of music along with delusion as if they came to a different space in a passageway from 1st building to 2nd building among 3 parted passageways while facing narrow and long passageway.

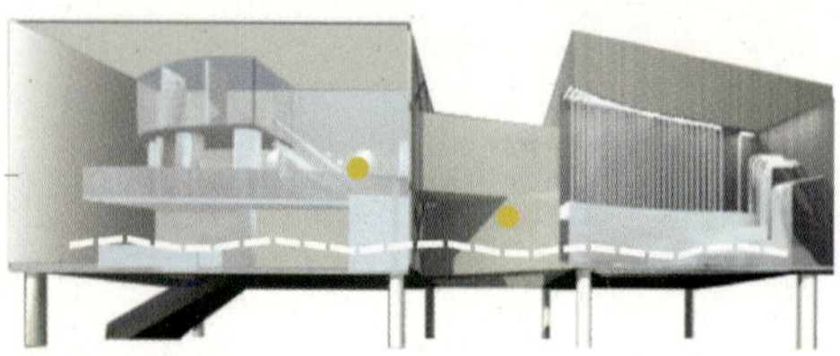

Medium & fast
중반부 – 2

빛이 인도하는 곳으로 나가자 디지털화 된 공간을 마주하며 눈에 탁트인 넓은 INTERACTION한 벽면과 하나가 된 객석과 피아노계단이 한 눈에 들어오면서 환상적인 느낌을 자아낸다.

When they go outside where light leads them to, they get to face the digitalized space and fantastical feeling is evoked while the piano stairway that became one with unhampered wide INTERACTION WALL comes into view at a glance.

Medium & fast
중반부 – 3

기대감 고조 – 아날로그적인 소소한 느낌의 음악적 풍경이 맞이한다.

Upsurge of expectancy - People will be greeted by a musical scenery of analog trivial feeling.

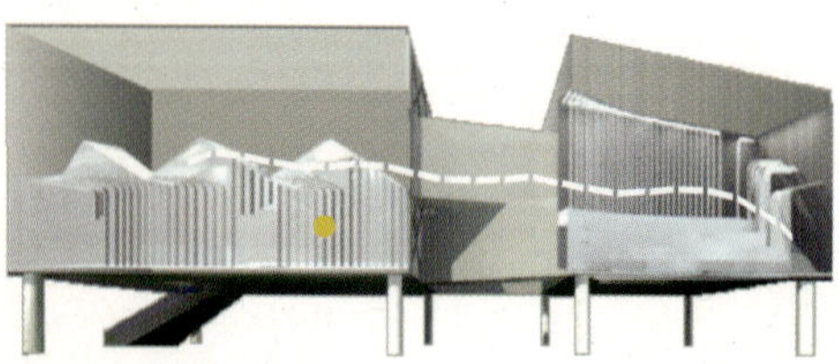

High & fast
클라이막스

피아노계단과 객석이 이어진 공간을 따라 올라가면서 커다란 Mass 의 공중에 떠있는 듯한 위태로운 인터렉션 공간을 발견하며 하이라이트에 도달한다.

People will reach the highlight while discovering a dangerous interaction space that feels like floating in the air of large mass.

Low & ritardando
후반부

반복적인 음악감상공간이 나온다.

A repetitive music appreciation space appears.

Thinktank

이촌2동 주민들의 꿈과 희망을 키워주는 도서관
Ichon 2-dong library where it gives people to dream and hope

유영주 김경희 | Yu, young-ju Kim, gyoung-hee

Background

이촌 2동은 단군이래 최대 민간개발사업으로 주목받은 용산국제업무지구 개발사업이 6년 만에 청산절차를 밟게 되었다. 그로 인해 현재 주민들은 정신적으로 경제적으로 엄청난 피해를 입은 상태이다. 피해를 입은 대부분의 사람들은 삶의 희망과 의욕을 잃어버렸다. 주민들의 이러한 상태를 반영하듯 지역 분위기 또한 굉장히 어둡다.
그렇기 때문에 이들이 희망을 되찾을 수 있도록 도움을 주는 공간을 마련해 주고자 한다. 그래서 선택한 공간이 물이 있는 도서관이다. 물이란, 편안함과 고요함, 자연 속에 있는 듯한 느낌을 주기도 하며 향수를 불러 일으키기도 한다.
도서관이란, 남녀노소 모두가 이용이 가능한 곳이고, 누구나 부담을 가지지 않고 편하게 올 수 있는 곳이다. 이 두 가지 요소를 섞으면 지역주민들이 이곳에 와서 어지러워진 마음과 머리를 편하게 식히기도 하고, 서로 대화를 통해 좀 더 가까워지기도 하며, 다시 한번 일어날 수 있는 용기와 희망을 얻을 것이라 기대한다.

Yongsan International Business District development in E-Chon-2-dong, receiving attention with the largest private development projects since Dangun era, finally went into liquidation after six years of inauguration. In order of that, residents there are currently damaged both mentally and economically. Most of the injured inhabitants lost the hope and enthusiasm of their lives. Reflecting these states of those people, the atmosphere of the region is also very dark.
Therefore, now being made is the arrangement of a space to help them find some hope,

Concept

THINKTANK (think + tank)
물탱크+도서관=생각하는 탱크
물탱크에서 생각을 하다

물을 채운 물탱크
미래의 꿈으로 채워질 물탱크
곧 밝은 미래로 바뀔 지역을 의미

현재 사이트에는 사용하고 있는 물탱크가 있다.
이 점을 고려하여, 물이라는 요소를 선택했고,
도서관이라는 공간도 생각하게 됐다.
결국 물이 있는 도서관으로 정하게 되었다.
물이란, 자연의 일부이고, 그로 인해 편안함과
안정감을 주기도 한다.
그리고 도서관이란 모두에게 접근하기 쉽고,
안락함을 주는 곳이기도 하다.
그렇기 때문에 이 공간을 생각하게 되었고,
콘셉트를 thinktank라고 정하게 되었다.

THINKTANK (think+tank)
tank+Library=thinking tank
think in the thinktank

A tank filled with water
A tank to be filled with the future dreams,
meaning the area soon to transform
into a bright future

There is a tank currently used in the site.
Considering the fact, the element of water and
the space of a library came up in mind. Water which
is a part of nature gives some comfort and
the stability. Furthermore, it is also a place cozy
as well as easily accessible to all.
Thus, the idea of such a theme is chosen and
affectionately called "Think tank."

Site

서울시 용산구 이촌2동 396_주민센터 옆 대림아파트 테니스코트+주차장 공간
지역특성: 한강대교와 원효대교 사이 한강변을 따라 동서로 길게 위치
면적: 1.22㎢(구의 5.57%)
주택: 3,556동(단독 474, 연립 580, 아파트 2,502)
인구: 4,389세대, 10,518명(남 5,043/여 5,475)
–개발사업의 실패로 인해 지역 거주자도 많이 감소하고, 지역의 분위기마저 어둡다.
–상가의 개수가 200여 개에서 90여 개로 줄었다.
이것은 그만큼 유동인구가 줄었음을 의미한다.
–바로 옆에 주민센터가 있지만, 조금 더 활동적이고 쉴 수 있는 공간이 필요하다.

주변의 번화한 지역들 옆에 위치한 이촌 2동은 아직까지
남아있는 시위의 흔적으로, 분위기마저 다소 어둡다.
거기에 노후화된 건물도 그 분위기에 한몫을 한다.
그로 인해 맘 놓고 쉴 곳 없는 삭막한 공간에 보다
활동적이고 생기 넘치고 안락함을 갖춘 공간을
구성하여 지역 분위기를 바꿀 수 있도록 계획한다.

Ichon-2-dong, Yongsan-gu, Seoul,
Daelim apartment next to the community center
396_Parking space
Regional Characteristics: Bridge between
Han River Bridge and along Wonhyo

Long east-west location
Size: 1.22 km^2 (5.57% sphere)
House: 3,556 East (Single 474, Alliance 580,
Apartment 2502)
Population: 4,389 households, 10,518 people

-As a result of the failure of development project, the local residents has decreased and the local atmosphere has become dark.
-The number of the stores there is reduced by 90 from 200, which means the decrease of the floating population.
-Right next is the community center, but it still needs more active and comfortable spaces to take some rest.

E-chon-2-dong, located right next to the populated areas, is dark because of the remaining graces of protests. The aging buildings play a big part in such an atmosphere. Therefore, the plan is made to turn the dark atmosphere into the lively and animated space with comfort.

Design process

Cartoon area

Magazine area

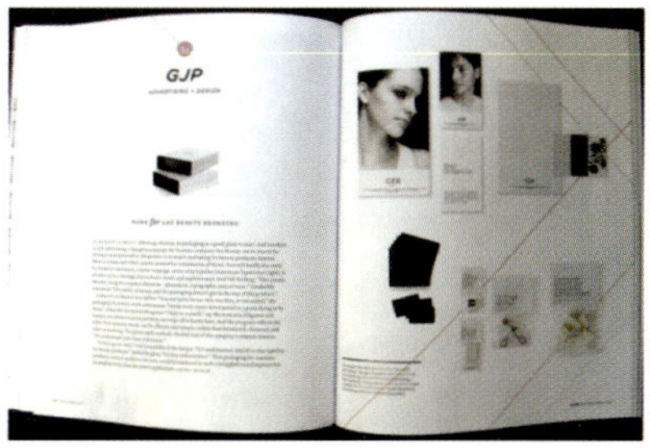

Novel area

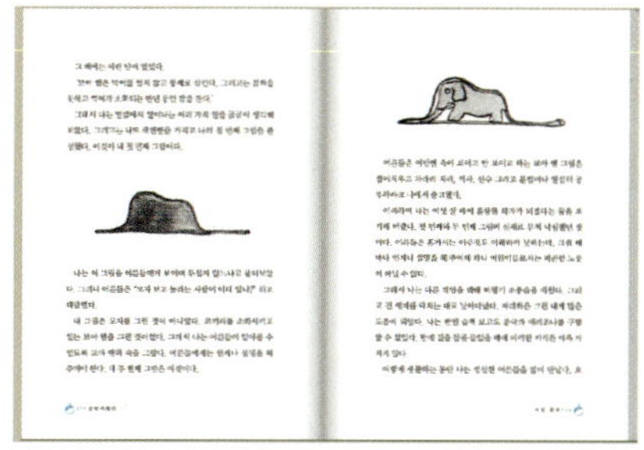

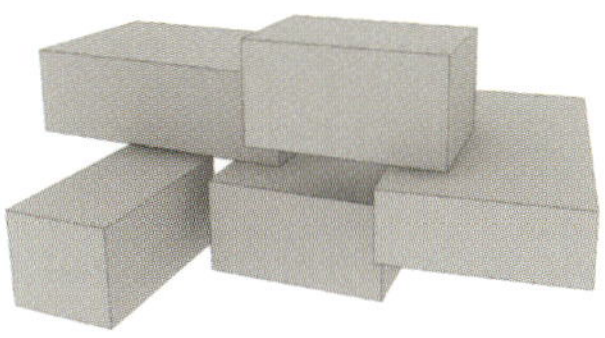

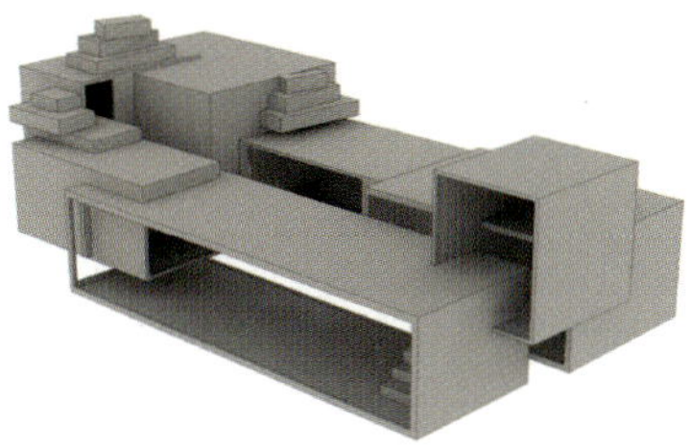

크기가 다른 상자들을 불규칙하게 배치하여 아이들이 그것을 자유자재로 움직일 수 있도록 하였다.

The boxes with different sizes are arranged so that children can move those around freely.

잡지 공간은 나무 상자들을 겹쳐 놓아 작은 방처럼 공간을 나누었다

Magazine space is divided into some spaces like separate small rooms, stacked with some wooden boxes..

나무로 된 계단 하나하나에 상자를 두었고, 그 안엔 조명과 작은 책상을 놓아 책을 읽을 수 있도록 하였다.

There is a box on the each of the wooden stairs, and the lights and the little tables are in the box .

Water tank

GLASS WATER TANK

유리 바닥 밑에 물탱크를 두었다.

There's a water tank in under the glass floor.

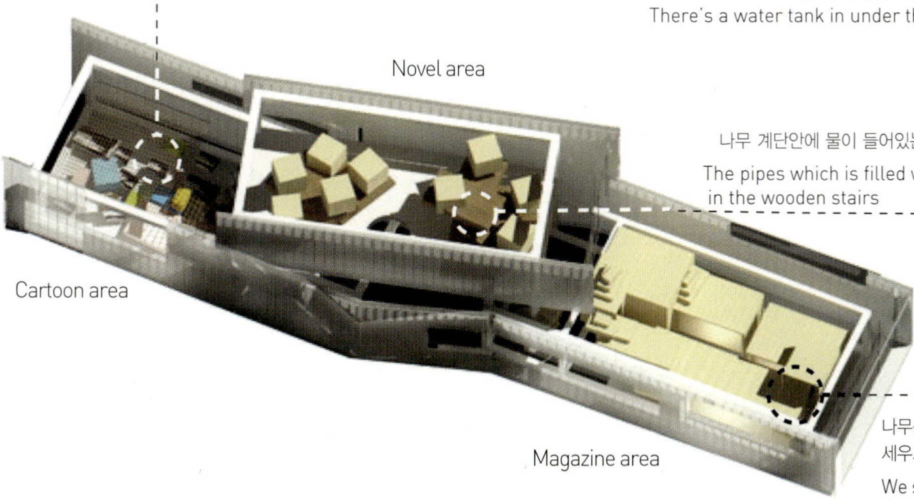

WOOD PIPE

나무 계단안에 물이 들어있는 파이프가 있다.

The pipes which is filled with water are in the wooden stairs

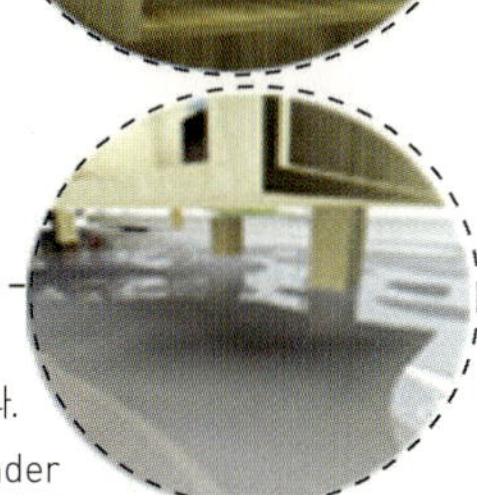

WOOD WATER TANK

나무상자 밑에 기둥을 세우고 물이 흐르게 했다.

We set up a pillar under the wooden box to let water run.

겹쳐진 나무 상자는 책상이나 의자가 되기도 한다.

The wooden box which is overlapped can become a desk or chair.

공간 안에는 카페도 있고, 독서활동을 편히 할 수 있게 되어있다.

In this space there is a cafeteria, in order to read books in relaxedatmosphere.

PLAN_F1

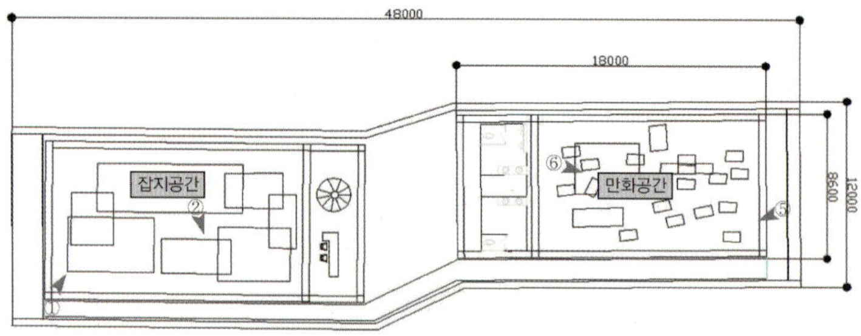

PLAN_F2

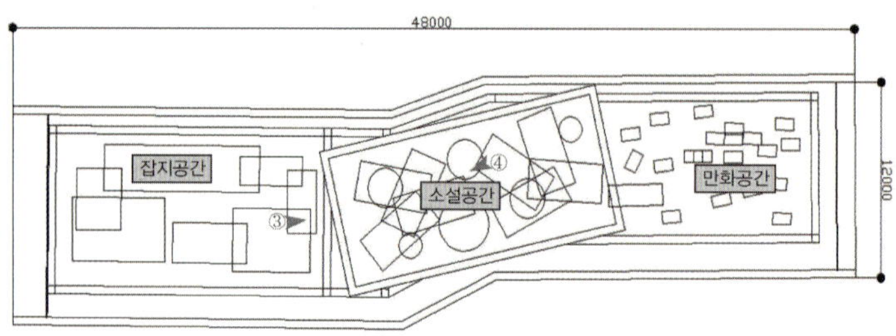

크기가 다른 상자들은 책상이 되고 의자가 되며 아이들 스스로 책을 읽을 수 있는 공간을 만들게 된다.

The boxes with different sizes can become chairs as well as tables, and children create their own space for reading.

공중에 그물을 설치하여 보다 자유로운 곳에서 책을 읽을 수 있는 환경을 만들어 주고자 하였다.

The nets are installed up in the air for more freedom of reading.

나무상자들을 이용하여 2층 공간인 소설 공간과 연결되게 하였다.

Using these wooden boxes it is possible for this space to be connected to Novel space, located in the second floor.

소설 공간의 바닥 부분에 천장과 마찬가지로 웅덩이를 만들었고 그 주위에는 의자들을 두어 여러 사람이 대화를 할 수 있는 공간도 만들었다.

Just like the ceiling, there are some puddles on the bottom of the novel space, and the space for the conversation between a lot of people is also made there in the middle of the rounding chairs.

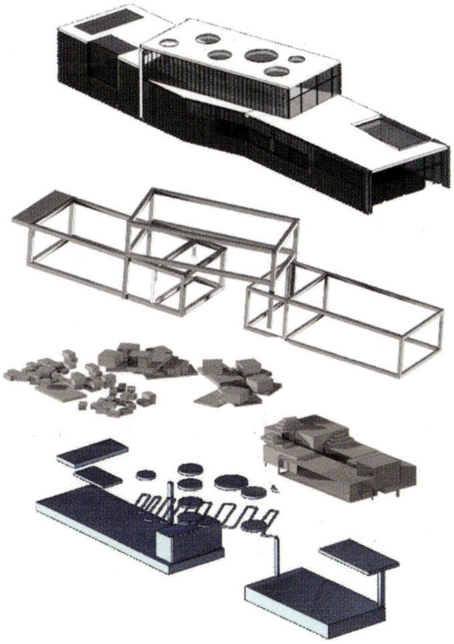

천장에는 물웅덩이가 있으며 이것은 실내의 디자인 역할과 빛의 양을 조절하는 역할,
내부에 있는 잔디에 빛을 쐬게 하는 역할이 있다.

There are puddles on the ceiling and this works as a design effect as well as the light-controlling device or the lighting device to the grass inside.

프레임은 공간을 세 개로 나눠줌.
프레임안에 물이 이동하는 파이프가 들어있으며 기둥역할과 물 공급 역할을 한다.

Frame divides one space into three zone. Water runs inside the pipes in the frame, which work as pillars and the water-suppliers.

2층에는 소설 공간이 있다.
1층에는 만화, 잡지 공간, 대여실, 화장실이 있다.
만화와 잡지 공간 외에
공간에는 잔디를 깔아 자연의 느낌을 주었다.
미끄럼틀과 계단 등으로 모든 공간은 이동이 용이하다.

There is a novel space on the second floor.
On the first floor is the cartoon and magazine space, library and toilet. Comics and magazines the grass covers all the other spaces, giving the feel of nature. The slides and the stairs makes it easy to move from spaces to spaces.

지하에는 물탱크, 기계실, 주차장이 있다.

There are water tanks, engineering room, and parking lot on the basement.

대여소와 소설 공간으로 올라가는 나선형 계단

The staircase leading to a rental shop and a novel space is spiral.

소설 공간 나무 상자 내부

A novel space the inside of wooden box.

流 이촌동이 흐르다 Ichondong flows

김시은 이명섭 | Kim, si-eun Lee, myeong-sub

Background

어떻게 하면 도시에 에너지를 불러 일으킬 수 있을까? 이촌동은 재개발 과정에서 피해를 받고 유보중인 지역이다. 이러한 지역에 에너지를 불어 일으키고자 한다. 성공적인 지역 재활성을 위해선 소수에 의한 거대계획이 아닌 주민들의 애정과 관심이 필요하다. 따라서 주민들이 외부에서 쉽게 접근하고 편하게 즐길 수 있는 공간을 계획해 정신적으로 물리적으로 새로운 재충전과 소통할 수 있는 공간을 제안한다. 이 공간 속에 또 하나의 이촌동이 만들어진다.

What can we do to increase the energy of a city? Ichondong is an area that has been damaged in the process of redevelopment, and it remains damaged. There is a need to increase the energy of this city. A successful city's reactivation requires no enormous plans, but love and care from its residents. Thus, it is recommended to plan an easily accessible outdoor area where residents are capable of comfortably communicating and participating in mental and physical recreation to recharge.

Site

서울특별시 용산구 이촌2동 위치
〈이촌2동 일반현황〉
면 적 : 1.22㎢(구의 5.57%)
인 구 : 4,389세대 10,518명 (남 5,043명, 여 5,475명)
주 택 : 3,556동 (단독 474, 연립 580, 아파트 2,502)
한강대교와 원효대교 사이 한강변을 따라 동서로 길게 위치 용산 역세권 및 한강로의 배후지로서 개발 기대지역 노후서민 아파트 등의 재건축 시행으로 고층 아파트화 되는 지역

Ichon 2-dong Yongsan-gu Seoul Special City location
←Ichon 2-dong General Information→
Area : 1.22km^2(5.57% of the Gu)
Population : 4,389 Households with a population of 10,518 people (Male: 5,043; Female: 5,475)
Housing : 3,556 Units (detached houses 474, row-houses 580, apartments 2,502)
Located along the Han Riverside between the Han River Bridge and the Won Ho Bridge running east to west
A development expected regional due to being hinterland of Yongsan station area and Hangangno
Rebuild low-income apartment to become high-income apartment area

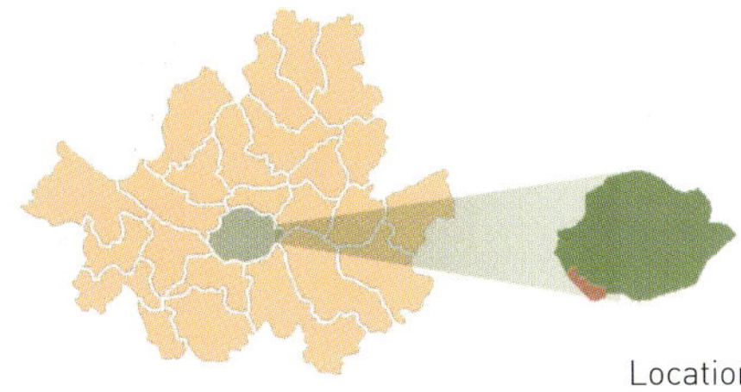

Location

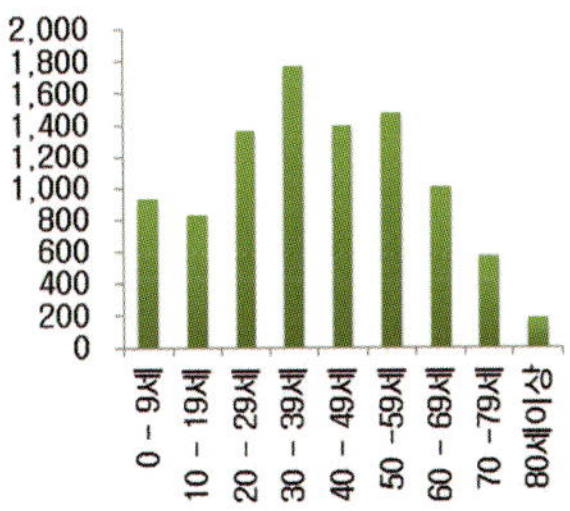

Population by Age

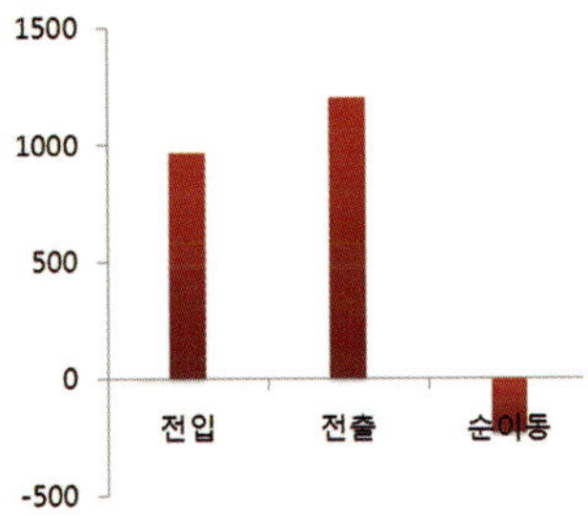

The number of people moving

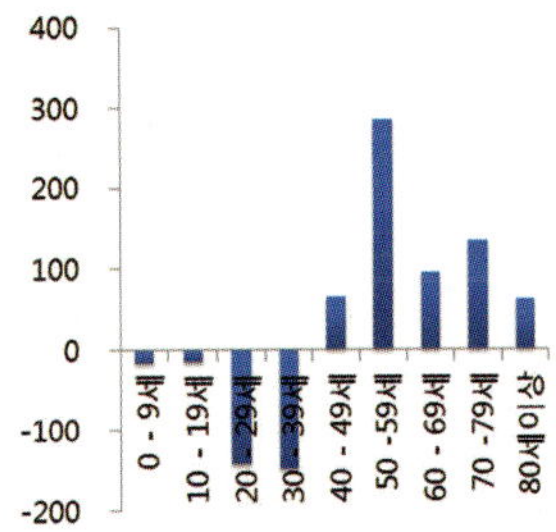

Non-over-year increase since the period

전입보다 전출인구가 많고 연령을 보자면 30대 인구가 가장 많이 줄어 들고 있다. 현재 30대 인구가 많지만 이대로 계속 가다 가는 이촌2동의 젊은층이 많이 사라질 것으로 보인다. 이들을 고려한 문화공간을 만든다.

Population is moving out more rapidly than moving-in. People in their 30s have decreased the most.

Target site

서울특별시 용산구 이촌2동 396
면적 : 848m^2

Address : 396-Ichon 2-dong, Yongsan-gu, Seoul location
Area : 848m^2

Accessibility

Proposal

주민들을 자연스럽게 유도하고 공간들을 서로 연결하여 소통을 이끌어낸다.
Naturally lead residents to communicate through interconnected spaces..

쓰이지 않고 있는 잠재된 공간을 설정하여 주민들에게 제공한다.
Take advantage of the unused potential space and make it accessible to residents.

Concept

流; 흐름 흐르다 Flow

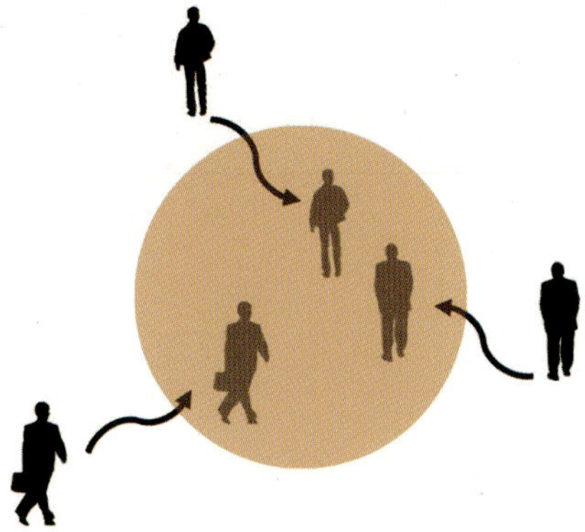

사람들이 자연스럽게 흘러들어와 큰 흐름이 되고 그 흐름들이 모인 이 공간은 또 다른 이촌동이 된다. 이 공간속에서 이촌동이 담겨 흐르다.

People naturally flow into a large current, and the place where that current gathers will become another Ichon-dong. Ichondong flows inside this space.

Motive

물의 흐름 The Flow Of Water

물은 한가지 방향, 한가지 모양으로 흐르는 것이 아니라 어떤 변수에 따라 각기 다르게 흐른다.
그 흐름들은 각자의 길을 가다가 어느 순간 다시 하나로 합쳐지기도 한다.
이처럼 사람들도 각기 다른 흐름으로 움직이며 그런 흐름들이 자연스럽게 공간과 연결되도록 한다.

Water does not flow in one direction or in one shape; it flows differently depending on certain variables. The individual flows follow their own paths and then at some point come together as one. Like this people move and flow on different courses and naturally connect with the space.

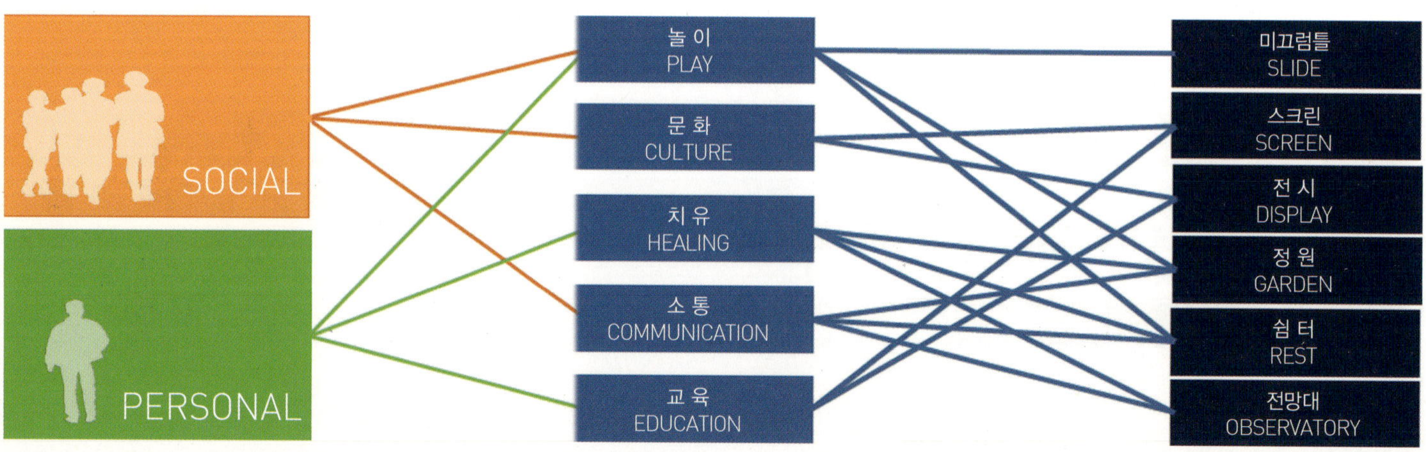

Background

각기 다른 흐름의 선을 표현
Different flow lines' expression

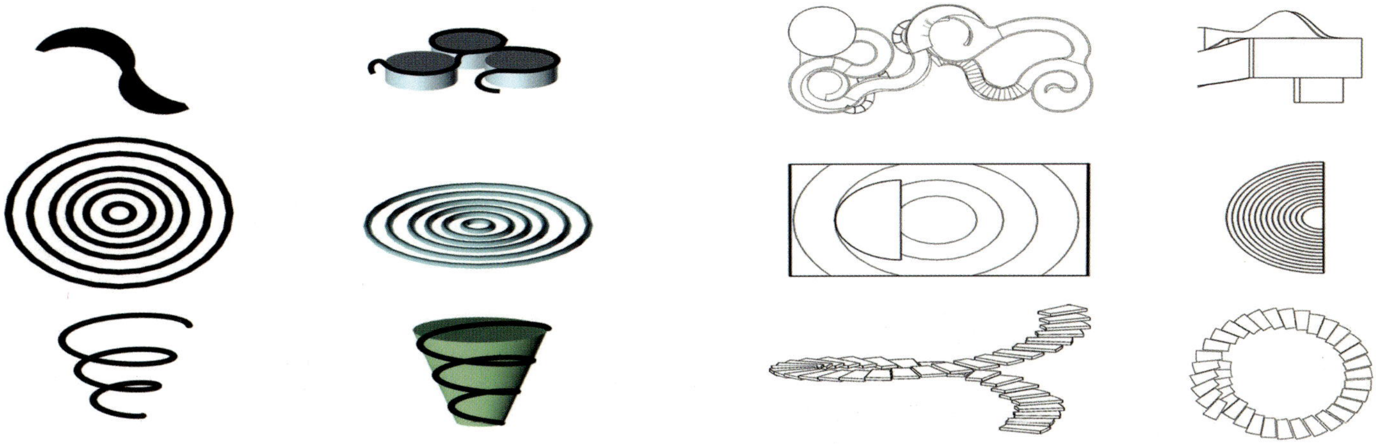

Mass study

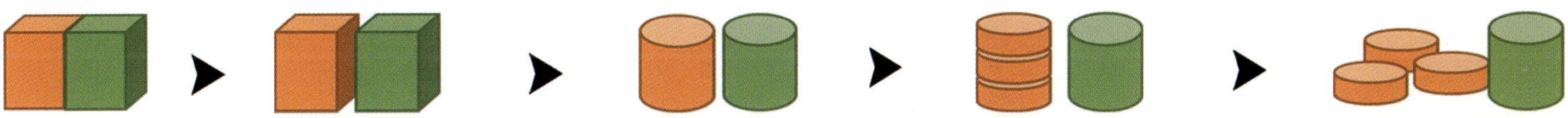

MASS	DIVISION	TRANSFORMATION	REPETITION	MOVEMENT
하나의 공간.	필요한 공간을 구획한다.	목적에 맞게 변형시킨다.	세분화하여 다시 나눈다.	조건에 맞게 이동 후 배치한다.

Elevation

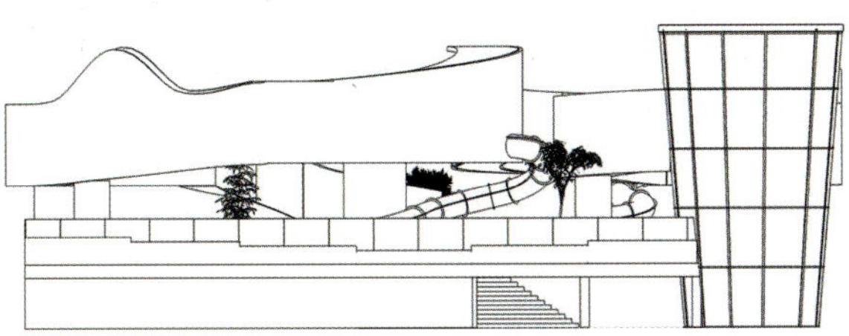

FRONT VIEW

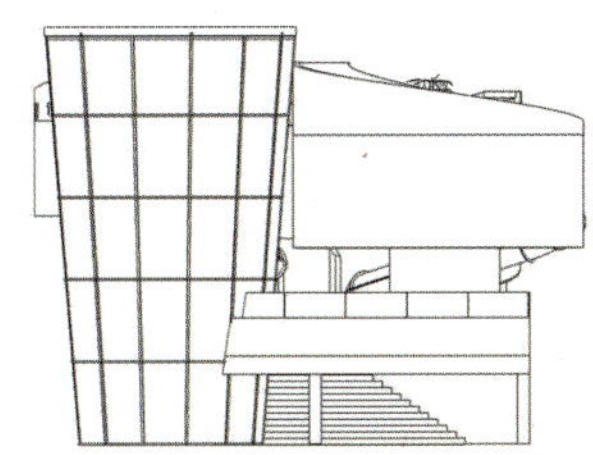

L-SIDE VIEW

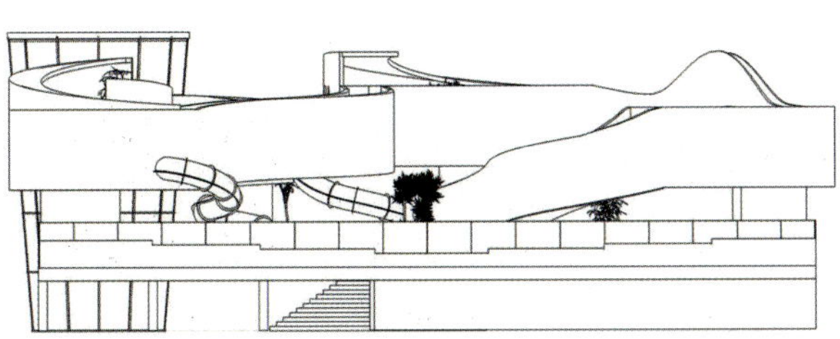

REAR VIEW

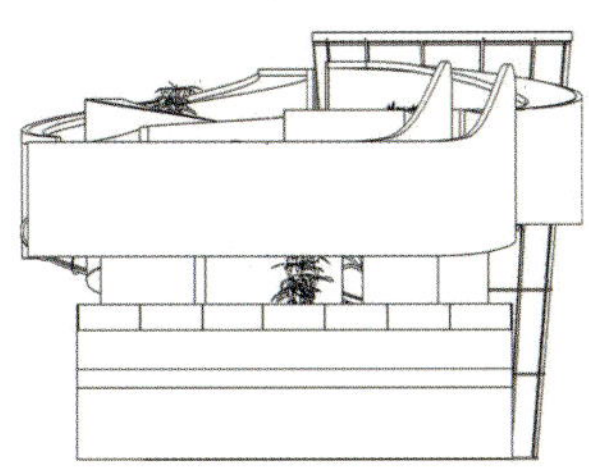

R-SIDE VIEW

이촌동을 이끌어갈 주민들을 위해서 쉽게 즐기고 모일 수 있는 문화시설을 제안한다. 기다란 통로를 돌아다니며 바닷속을 스크린을 통해 간접 체험 할 수 있다. 안으로 들어 갈수록 점점 깊이를 느낄 수 있으며 다양한 즐거움을 느낄 수 있다. 디지털세대에 사는 우리들을 만족시키기 위해서 스크린은 터치에 민감하게 반응하여 우리의 오감을 만족시켜준다. 통로의 중간 중간에는 쉬어 갈 수 있는 세 개의 공간이 있으며 그 공간들은 점점 공간 속에 자연스럽게 하나로 스며들고 감싸진다. 그리고 그 안들은 주민들간의 소통을 이끌어 낸다.

We suggest culture facilities that allow residents who lead Ichondong enjoy and gather easily. They can experience indirectly while wandering along the long passage through a screen inside the sea. The more we get inside the deeper and more enjoyable we can feel. To satisfy us who live in the digital generation they create touch screen with high sensitive responsibility to satisfy our feel. There are three resting spaces located on the passage and those spaces will gradually and naturally harmonize into one space. And inside it residents will get to communicate.

첫번째 마주하게 되는 공간은 천장은 모두 오픈되어 있다. 중심에 의자를 배치해 공간이 분리되는 느낌을 주고 S자형으로 디자인해서 앞뒤 어디든 자유롭게 앉을 수 있게하였다.

The first space we see-the ceiling will be wide opened. At the centera chair will be placed to give the feeling of a separated space and it is designed in S-shape so that residents can sit on the front or back wherever they want.

두번째 공간은 첫번째 공간보다 더 공간에 스며드는 느낌을 받을 수 있다. 중심에 나무를 배치하였고 벽 가장자리에 위치한 의자에 앉으면 앞에 앉은 사람들과 자연스럽게 눈이 마주치게 된다.

The second space will give more feeling of space harmonization. A wood chair will be place at the center and if people sit down on the chair at the edge of the wall they will naturally see each other.

세번째 공간은 의자를 없애고 바닥에 잔디를 깔아 어디든지 앉을 수 있게 하여 공간 속에 자연스럽게 스며들게 하였다. 천장을 많이 감싸 깊이감을 더 주었다.

In the third space there will be no chair and the lawn would be paved so that residents can sit down anywhere and can be connected naturally inside the space. The ceiling will be widely covered to give us the feeling of depth.

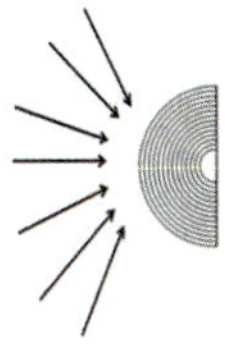

전망대로 올라가는 입구와 놀이공간으로 가는 입구로 나눠져 있다. 놀이공간으로 이어지는 입구는 반원으로 어느 방향에서든지 쉽게 접근할 수 있도록 하였다.

Entrance to observatory and entrance to play space is separate. Entrance leading to the play space is a half circle so that it can be approached easily from any direction.

Concrete

통로 중간에 있는 미끄럼틀을 타고 내려 오면 계단 형식의 쉼터로 이어진다.쉽터는 계단 형식으로 되어있고 앉아서 일어서서 자유롭게 쉴 수 있다. 또한 밑으로 내려 올수록 중앙으로 모여지게 하였다.

Going down by the slide location in the middle of the passage you will reach the shelter in the type of stairs. The stairs type allows you to rest free whether you stand or sit. The stairs get narrower as you going down.

Slide: Plastic
Shelter: Grass
Handrail: Grass

나선형 계단을 올라서 꼭대기에 르면 아파트 틈사이로 한강이 보인다. 길다란 통로의 끝과도 연결되며 주민들을 위한 소통의 공간으로서 역할도 한다. 천장의 조명은 커다란 물방울들이 떨어지는 것 같은 착각을 불러 일으킨다.

If we go up the spiral stairs and stand on top we can see Han River through the gaps between apartments. The long passage is connected end-to-end to perform the role of the space for residents' communication. The light on the ceiling gives the illusion of large drop dropping.

Time control

이은비 김윤애 채지연 | Lee, eun-bi Kim, Youn-ae Chae, ji-yeon

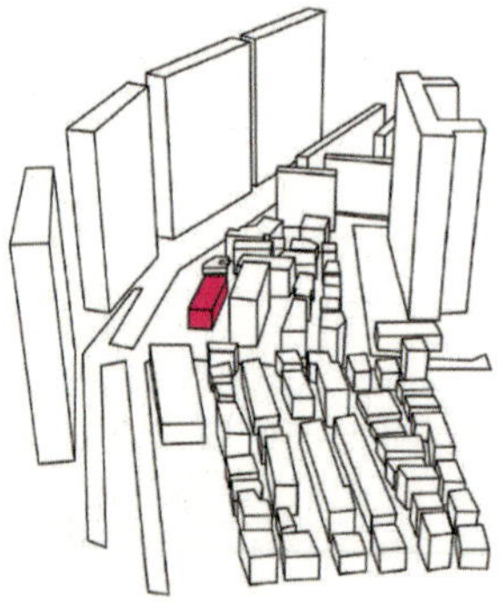

Site

서울시 용산구 서부이촌동 대림아파트 옆 주차공간

Ichon-dong, Yongsan-gu, Seoul, Daelim Apartment West Side parking space

용산구 서부이촌동 주민들은 재개발 문제로 힘들어 하고 있다. 삭막해진 서부이촌동에 새로운 활력을 불어넣기 위한 'Re-Energizing' 공간을 만들었다. 과학의 발전으로 편리하고 용이한 생활을 할 수 있는 지금, 과거보다 더 삭막해진 삶을 살아 가고 있는 서부이촌동 주민들에게 과학 발전의 소중함을 일깨워주는 동시에 이 공간으로 인해 동네 주민간의 화합이 이루어지도록 한다.

Yongsan-dong Ichon western residents are difficult to re-development issues. Now you are able to be useful for the development of. Science that made the "Re-Energizing 'space and an easy life for insufflate a new vitality in the western part Ichon-dong became abandoned and became lonely than the past at the same time bring home to us the importance of scientific development are living life, in the western part Ichon-dong residents, for this space, harmony between the residents of the neighborhood to be performed.

대림아파트 주변에 아이들이 안전하게 놀 수 있는 공간이 없다는 점과 아이들로 인해 밝아질 이촌동의 미래를 그리며 공간을 구성했다. 재개발로 힘든 주민들 사이에서도 순수하고 밝은 영혼을 지닌 아이들의 웃음소리로 가득 찰 놀이공간을 기대한다.

We organized space considering how there is no space around Daelim Apartment complex for children to place safely and by picturing the future of Ichon-dong that will brighten along with children. The play space which will be filled with laughter of children of pure and bright spirit even among with residents who are having hard time with redevelopment is expected.

현대에 맞게 빽빽이 서있는 아파트들과 도로, 숨바쁘고 어지럽게 돌아가는 현대 사이에 과거 산업시대로 돌아간 듯 한 건물. 삐걱삐걱 기름칠이 부족한 철들의 마찰소리와 투박한 볼트와 너트는 공포감을 조성한다. 하지만 안에서 흘러나오는 아름다운 오르골 소리는 우리를 네버랜드로 데려다 줄 것만 같다. 빠르게 느리게 급하게 천천히 흘러가는 시간이 다른 공간들에서 우리는 시간의 지배자가 된다. 내가 원하는 시간대로 TIME CONTROL!

A building that seems to have gone back to the industrial society of the past from modern society of packed apartments and streets and frenetic busy lives. Creaky friction sound of ungreased steels and rough bolt and nut create a sensation of fear. However, the beautiful orgel sound that comes from inside feel as though they will take us to the Neverland. We become ruler of the time in spaces of different lapse of time that flows rapidly, slowly, urgently, or leisurely. I CONTROL TIME as I want!

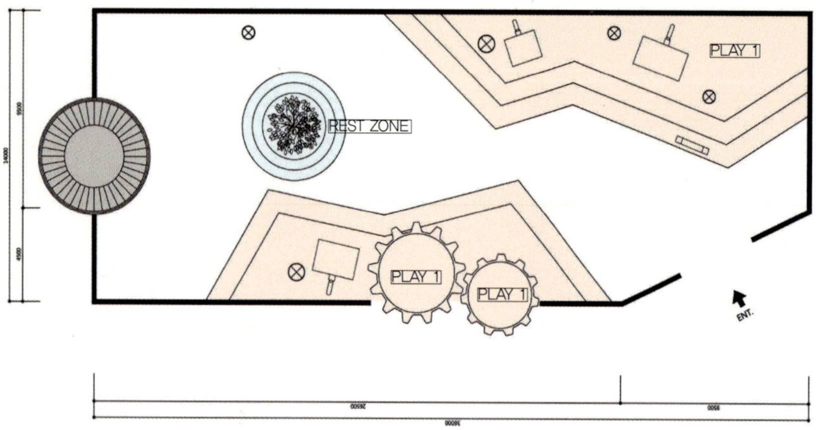

1st Floor Plan

PLAY 1

입구와 가장 가까운 놀이공간으로, 넓은 단을 두어 계단의 역할을 하는 것과 동시에 하나의 놀이공간이 되며 오르골 놀이를 즐길 수 있다. 움직이는 톱니바퀴는 호기심을 유발시킨다.

A place space that is located closest from the entrance performs a role of stairways with wide steps, and at the same time, it becomes one play space where people get to enjoy the orgel game. The moving saw-toothed wheel arouses curiosity.

REST

휴식공간으로, 가운데에 사과나무를 심어 과학원리인 '만류인력의 법칙'을 체험할 수 있다. 나무에서 떨어지는 사과를 직접 봄으로써 원리를 더욱 쉽게 이해할 수 있도록 한다.

A resting place with planted apple tree in the middle allows people to experience a scientific principle, 'the laws of gravitation.' Seeing an apple falling from the tree in person helps people to understand the principle more easily.

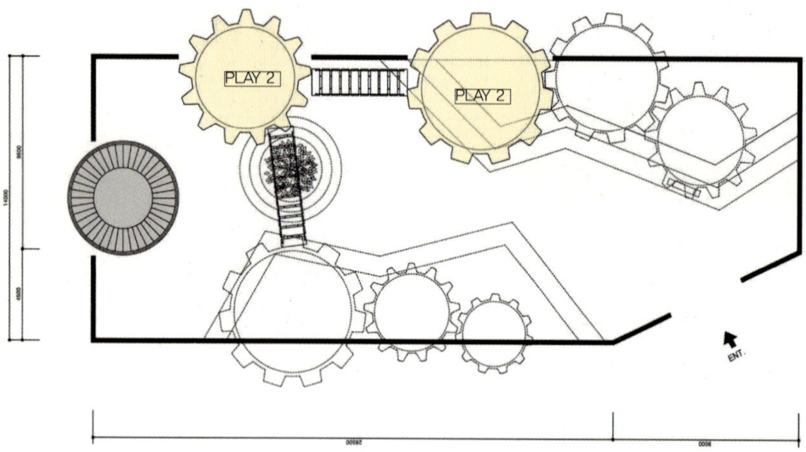

2nd Floor Plan

PLAY 2

PLAY 1의 오르골에 나는 소리를 들으면서 천장에 매달린 추를 잡아당겨 음의 높낮이를 맞추고 조절하는 놀이다.

PLAY 1 is a game of tuning and controlling high and low of sound by pulling weights hanging from the ceiling while listening to the sound from an orgel.

EXPERIENCE

'만류인력의 법칙' 을 몸으로 직접 체험하는 공간이다. 피사의 사탑의 꼭대기에 올라가는 동안 법칙에 대한 호기심을 유발시키며, 탑 위에서는 탄성이 있는 공을 아래로 던질 수 있다. 또한 이 탑은 이 지역의 상징이 된다.

A space of experiencing 'the laws of gravitation' in person. Going up to the top of the Leaning Tower of Pisa arouses curiosity about the law, and can throw an elastic ball down from the top. Also, the top becomes a symbol of this region.

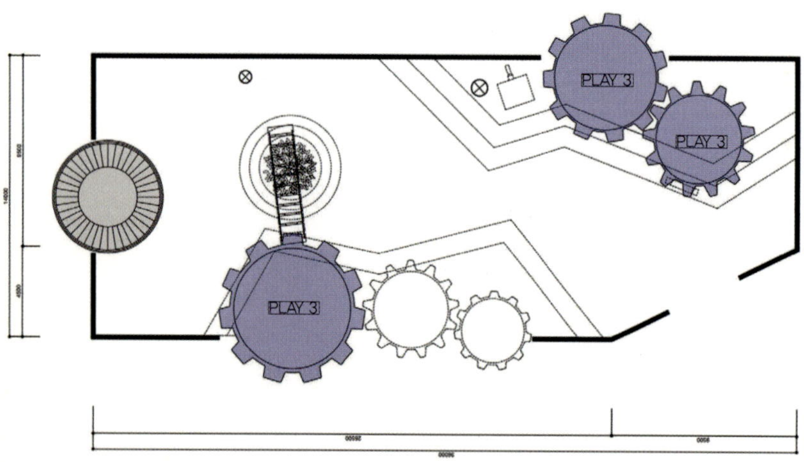

3rd Floor Plan

PLAY 3

놀이공간의 마지막 단계. 오르골과 추의 화음이 이루어지면 안에 있는 톱니들이 움직이면서 시공간이 변한다. 마치 타임머신을 타고 과거나 미래로 가는 듯 한 기분을 느끼게해준다.

The final stage of the play space. When harmony of an orgel and weight is composed, the sawteeth inside move and space and time changes. It makes people to feel as though they went back to the past or future in a time machine.

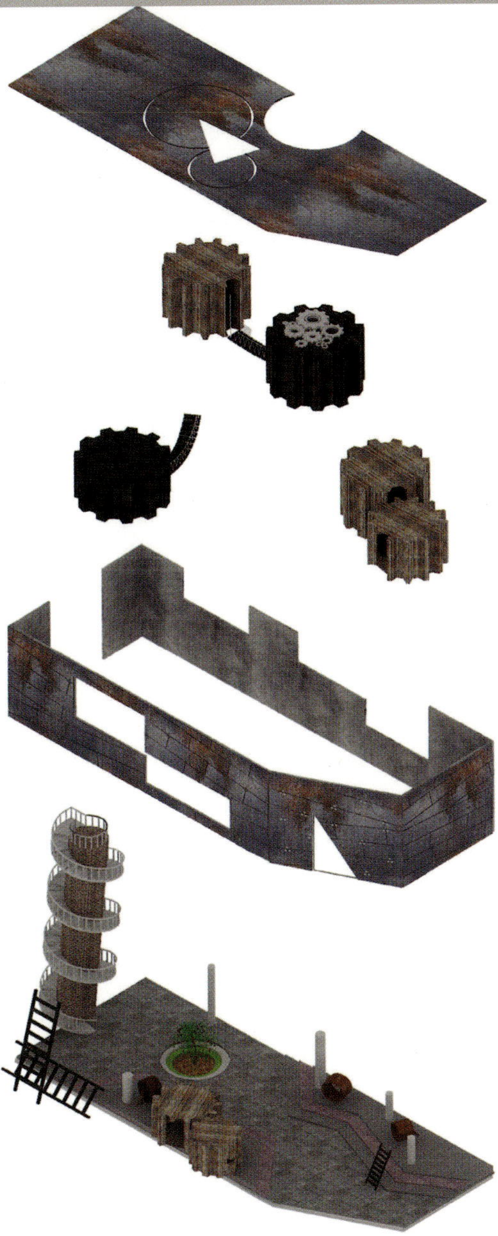

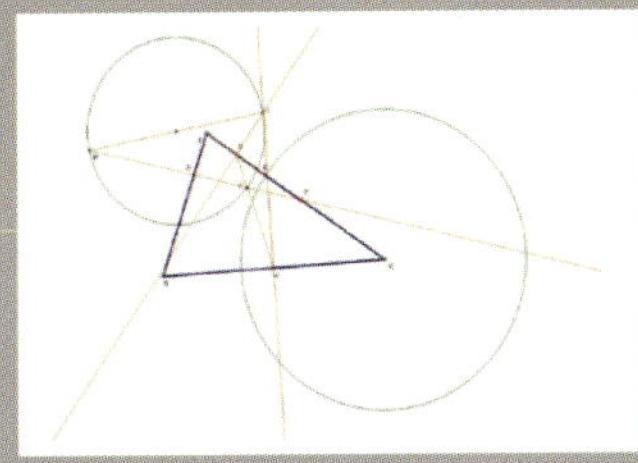

작도법을 이용하여 천장의 창을 뚫었다. 중첩되어 있는 커다란 삼각형과 원의 모습을 보며 아이들은 자기도 모르게 익숙해지는 작도법을 배울 수 있다.

A skylight window was bored with geometric construction. Children can learn the geometric construction method that they become accustomed to unknowingly by looking at a large triangle and circle that are piled up onto one another.

톱니바퀴의 특징은 맞물리며 돌아간다는 것이다. 이런 맞물리면서 돌아가는 톱니바퀴의 성질을 가져와 맞물리며 돌아가는 공간, 즉 움직이는 공간을 만들었다. 외부에서나 내부에서나 돌아가는 톱니바퀴는 신기하게 느껴진다.

A distinct feature of the saw-toothed wheel is that they mesh and turn. Such property of the saw-toothed wheel of turning in mesh is used to create a space that interlocks and turns, meaning a kinetic space.
The spinning saw-toothed wheel inside and outside feels marvelous.

건물외벽은 콘크리트벽으로 이루어져있다. 콘크리트벽 위에 녹슨 쇠로 마감처리 했다. 겉과 안이 다른 반전의 매력을 느낄 수 있다

Building facades are composed with concrete walls. The concrete wall is finished with rusted iron. People can feel the attraction of opposites of different inside and outside.

과학법칙 중 뉴턴의 만류인력의 법칙을 아이들이 알기쉽게 시각적으로 보여주기 위한 요소이다. 놀다가 휴식을 취하면서 가만히 있던 사과가 떨어지는 원리를 눈으로 직접 보고 느낄 수 있다.

An element to visibly display the Newton's law of universal gravitation among scientific laws to children more easily. Children can see and feel the principle of apples falling down in person while they play games and take rest.

Section

맞물리는 톱니바퀴들 마다 서로 다른 놀이들이 있어 그 공간들을 오가며 서로 다른 재미를 느낄 수 있다.

Children can feel different interest from each other when they come and go to these spaces as there are different games in each interlocking saw-toothed wheel.

Material plan

① 

콘크리트는 벽 뿐만 아니라 바닥 등 여러곳에 쓰이며 현재 가장 흔한 재료라고 할 수있다. 아이들이 낯설어 할 수도 있는 폐자재들 사이에서 익숙한 콘크리트가 색다른 느낌을 준다.

Concrete is not only used on walls but in many other places as well, and can be called as the most common material. Familiar concrete in between waste resources that children can be unfamiliar with gives a different feeling.

② 

나무폐자재를 사용하여 건축폐자재들이 환경오염의 주범인이 아니라 다시 쓸 수 있는 자원으로 주목받고 있다.

Construction wastes are receiving attention as the reusable resource, not the main culprit of environmental pollution by using wooden waste resources.

③ 

건물의 외피,외벽의 재료는 녹슨 쇠이다.
TIME CONTROL에 맞춰 산업혁명시대로 돌아간 듯한 현실감을 주기위해 당시 사용되었던 증기기관차의 낡은 외피 이미지를 가져왔다.

Material for the outer cover, outer wall of building is rusted iron. Old outer cover of a steam locomotive that was used at the time to give a sense of reality of returning back to the industrial revolutionary age according to the TIME CONTROL.

④ 

벽돌은 산업혁명 때 쉽게 볼 수 있는 재료중 하나다. 그 중에서도 으스스한 내부를 완화시켜주는 밝은 색의 벽돌을 사용하였다.

Brick is one of the materials that were easily seen during the industrial revolutionary age. Among them, bright colored bricks that mitigate the chilly interior are used.

⑤ 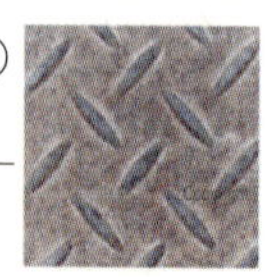

공장에서 많이 쓰던 재료를 사용하여 산업시대 공장의 느낌을 재현하려고 노력하였다

We tried to reenact the feeling of factories in the period of industrialization by using frequently used materials in factories.

⑥ 

7개의 톱니바퀴중에서 가장 큰 2개의 톱니바퀴는 진한 낡은 나무이다. 산업혁명 당시 공장에서 사용하던 공장 폐자재 나무를 가져왔다.

The biggest 2 saw-toothed wheels among 7 saw-toothed wheels are dark and old woods. The factory waste wooden materials that were used in factories at the time of industrial revolution were collected.

천장에 작도법을 적용한 창이 답답함을 해소시키며 채광에 도움을 준다.
The geometrical construction method-applied windows on the ceiling removes stuffiness and helps the lighting.

Music and the appearance of swing bridge

Apple tree by the law of gravitation Stream

밤에 본 건물의 외관. The appearance of this building at night

Among the gears are connected to each other, I experience scientific principles

Play and want to play freely in the space jump Enjoy the laws of science

조경찬 석진주 이혜원 | Jo, kyoung-chan Seok, jin-ju Lee, hye-won

Background

재개발의 추진으로 생존을 위한 반대와 삶의 여유를 위한 찬성으로 나뉘었고 함께 지냈던 주민들은 대화가 줄어들게 되며 점점 폐쇄적으로 변하게 되었다. 아이들 또한 어른들의 분위기로 물들어가게되었다. 변해가는 아이들의 마음에 에너지를 불어넣기 위해 에너지를 이용한 놀이터를 만들어 빛을 잃어가는 서부이촌동을 RE-Energizing 하고자한다.

Due to redevelopment, residents were divided into two groups. one group objected to redevelopment for survival and the other group supported redevelopment for their leisurely life. residents became closed, reducing their conversation. children also became closed following the mood of adults.
To inspire energy to changing children, by establishing playground using energy, we would like to re-energize west Ichondong which is losing their light.

Site

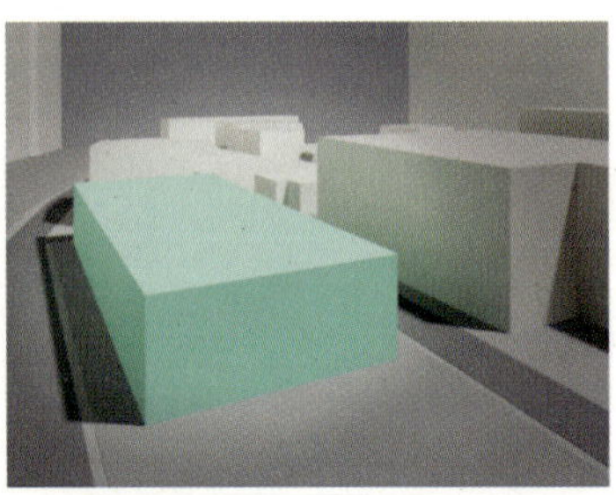

서울 특별시 용산구 이촌2동 대림아파트 옆 주차장.
West Ichon 2-dong, Yongsan-gu, Seoul

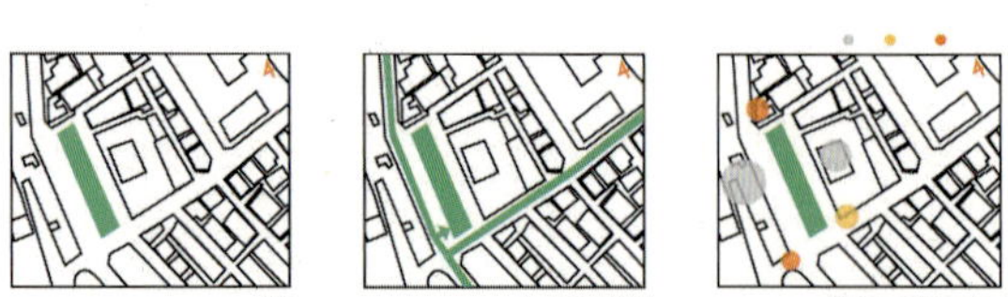

Site Aceessibility Perspective view

Needs

주변의 도로가 인접하여 소음에 취약하고 복잡하여 아이들이 자유롭게 놀수있는 놀이공간이 부족.
노인과 부모들이 아이들과 함께 쉴 수 있는 공간 부족.

The road of circumference is contiguous and it is weak in sound arresting and it is complicated and the children to be free the playing space which is glow possibility the tribe.
The spatial tribe who is the possibility the old people and the parents resting with the children together.

어울림

EO-UL-LIM : Harmony

Taget

서부 이촌동의 분위기로 인해 폐쇄적이게 되어가는 아이들을 타겟으로 잡아 재미와 에너지를 동시에 얻을수 있는 놀이터를 건설 하려한다.

By targeting children who became exclusive because of atmosphere of west Ichondong, we aim to build a playground where children can feel joy and energy at the same time.

Concept

어우러지다 [Together] + 울림 [Echo] = EO 어 UL 울 LIM 림

아이들을 통해 주민들을 한데 어우르고 에코 공간을 통해 서부 이촌동을 밝히고자 한다.
갈등이 된 주민들을 한데 어우르다와 용도인 에코의 동음이의 어 울림을 더해 "서부 이촌동의 주민들을 서로 어울러 에너지를 퍼지게 하다" 란 의미로 사용 하였다.

Through children, we want to gather residents together and through eco-space, we want to brighten west Ichondong.
By combining meaning of 'make residents who suffer from conflict put together' and 'echo' which is a homonym of eco, we used a term 'harmony' to mean 'make the residents join together' so they can spread energy.

MOTIVE

Scenario

자극
[Stimulate]

확산
[Diffusion]

고요
[Silence]

자극 = 아이
[Stimulate = Kid]

압전놀이
[Piezo-electric play]

LED스케치북
[LED sketchbook]

Design process

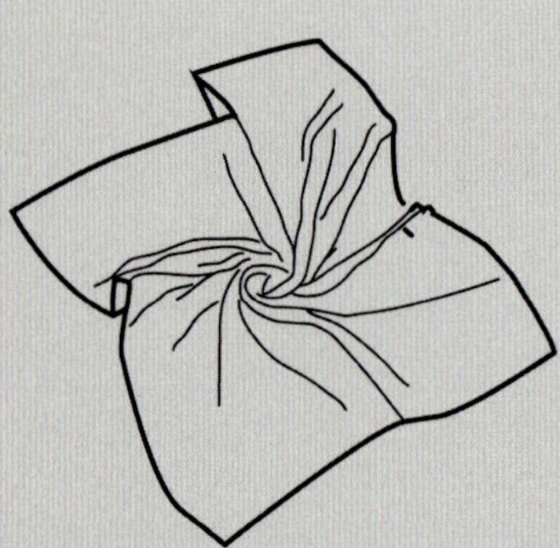

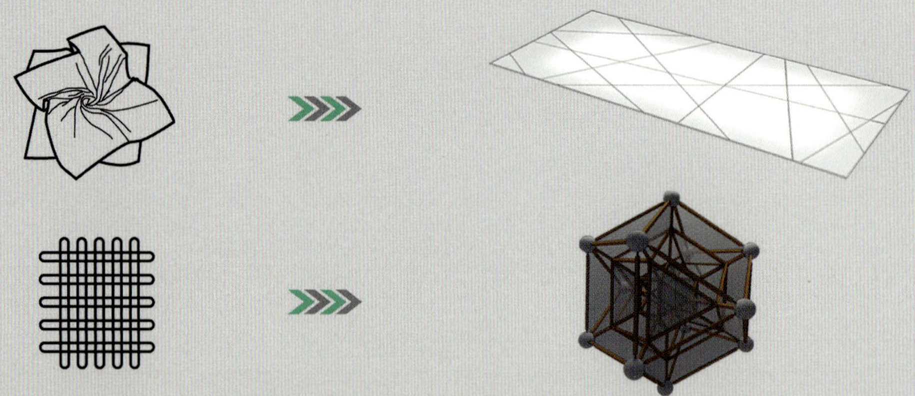

어우르다란 컨셉을 "실을 엮어 하나의 천을 만드는 직조방법" 과 같은의미로 사용 하여 천의 모습을 확대해 교차되는 패턴의 모습과 천의 짜여진 모습을 입체화한 구체로 표현하였다.

By using the concept "join together" as a metaphor for sentence that "a weaving method to make a cloth by weaving threads",
we expressed image of crossed pattern that we can see when the cloth is magnified and a sphere which represents image of woven cloth in three dimension.

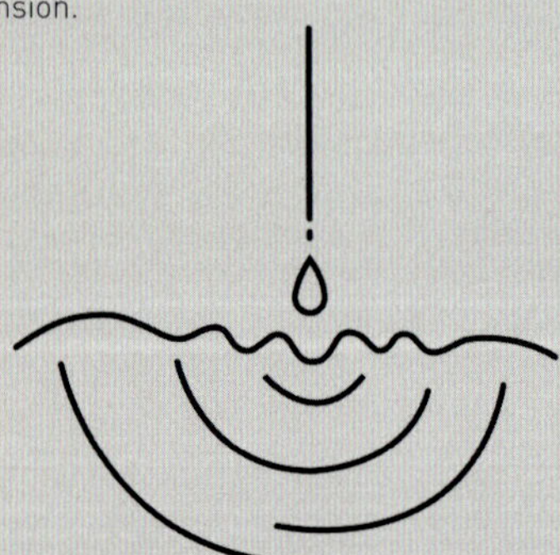

물이 떨어지는 모습을 형상화하여 물방울 모양은 해먹, 수면에 형태는 매스의 분할, 퍼지는 효과는 공간 분할에 이용하였고, 물이 떨어지는 형태를 단면적으로 보아 천이 일렁이는 모습으로 나타내었다.

By embodying the image of falling water drop, we made a hammock in water drop shape. we also used shape of the surface in partition of space and wave of water in partition of mass.
By seeing shape of falling water drop fragmentarily, we expressed it as swaying cloth.

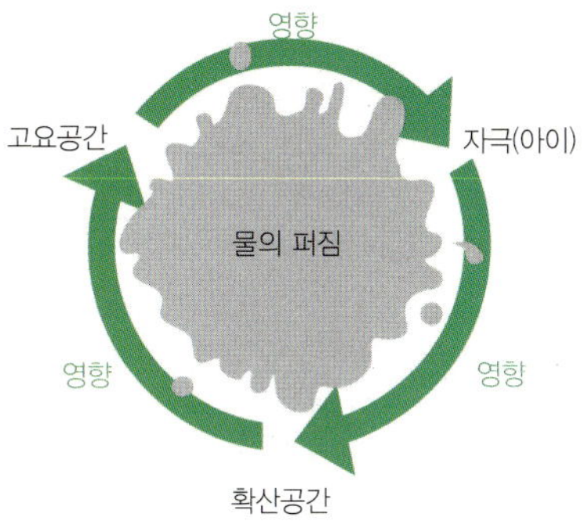

아이란 자극이 공간으로 들어가게 되면 입구부분의 압전센서를 통해 물이 퍼지는 듯한 스크린 효과가 나타나며 에너지가 생성된다. 생성된 에너지는 모아져 고요의 공간으로 전달되며 LED 스크린 스케치북에 사용된다.
아이들은 확산의 공간에서 구체를 굴리거나 던지면서 놀이를 하게 되면 구체에 에너지가 모아져 조명으로 사용된다. 에너지를 소모한 아이들은 고요의 공간으로 이동하여 해먹에서 휴식을 하며 재충전을 한다.

When stimulation(a child) is entered into space, through piezo-electric sensor which located at the entrance, Screen effect occurs and energy is generated. generated energy is gathered, delivered to space of silence and used in Screen Sketchbook.
when children roll or throw a sphere in the place of spread, the energy is focused and used as lighting. children who use up their energy move to the space of silence, take a rest and recharge their energy.

Exploded view

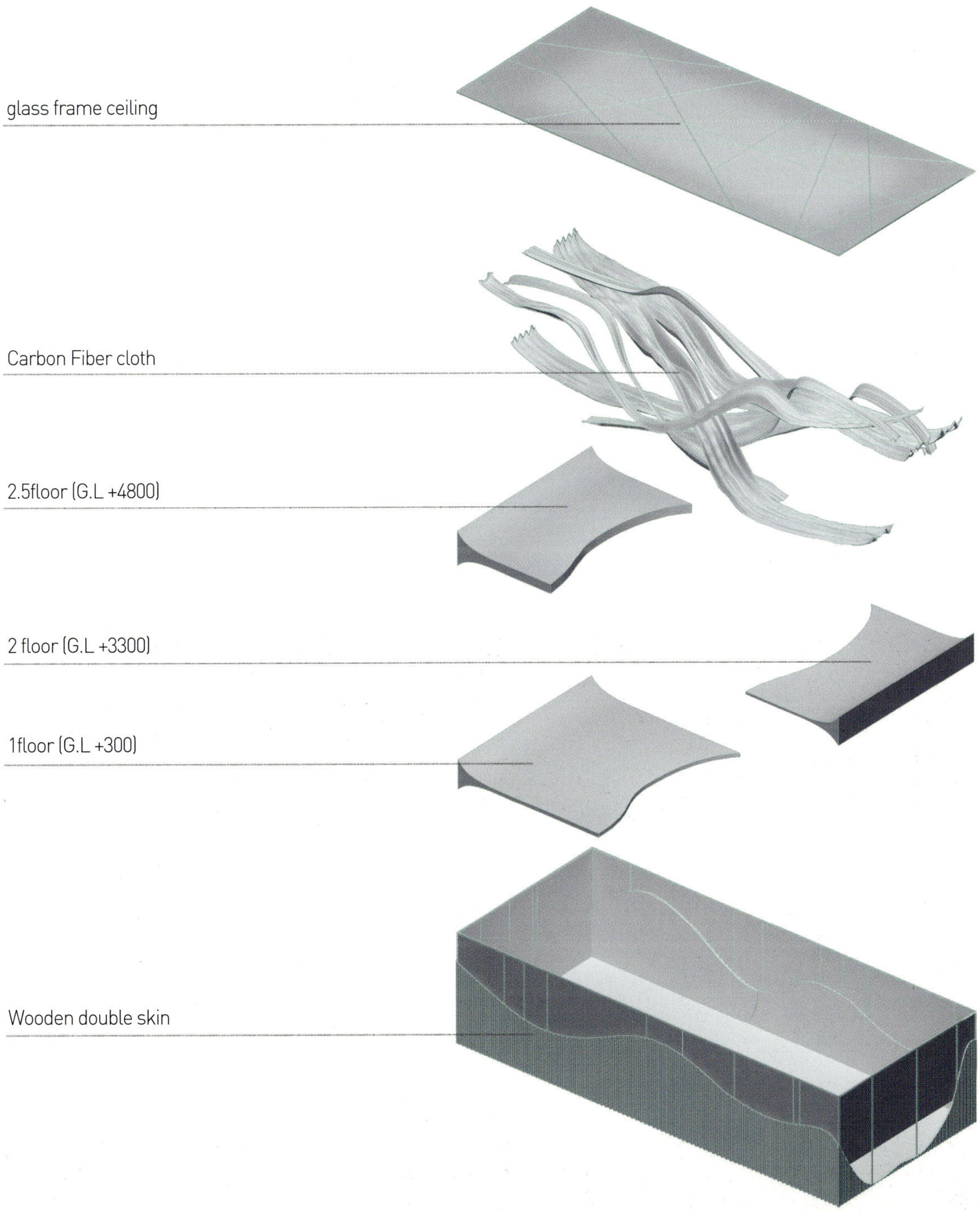

'갈등이된 서부이촌동의 주민들을 서로 어울러 에너지를 퍼지게 하다' 란 의미를 담은 공간으로 낮에는 프레임에 의해 보여지는 매스속에서 아이들이 천을 타거나 가지고 놀게 되며, 압전센서를 이용한 에너지를 생성하여 사용 할 수 있고 또한 휴식을 할 수 있는 에너지 놀이터이다.
밤에는 에너지를 퍼짐을 보여주고자 매스에 불을 밝혀 낮과는 다른 외부를 느낄 수 있게 하였다.

As a place that covers a meaning that 'residents of western Ichon dong who suffer from conflict join together and spread energy', in the afternoon, in mass which is shown by frame, children can play together with cloth, use energy generated by piezo-electric sensor and take a rest.
In the evening, to show spread of energy, by lighting up mass, you can feel different atmosphere of outside in comparison to afternoon's atmosphere.

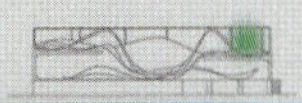

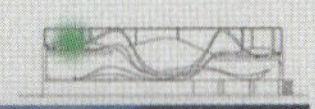

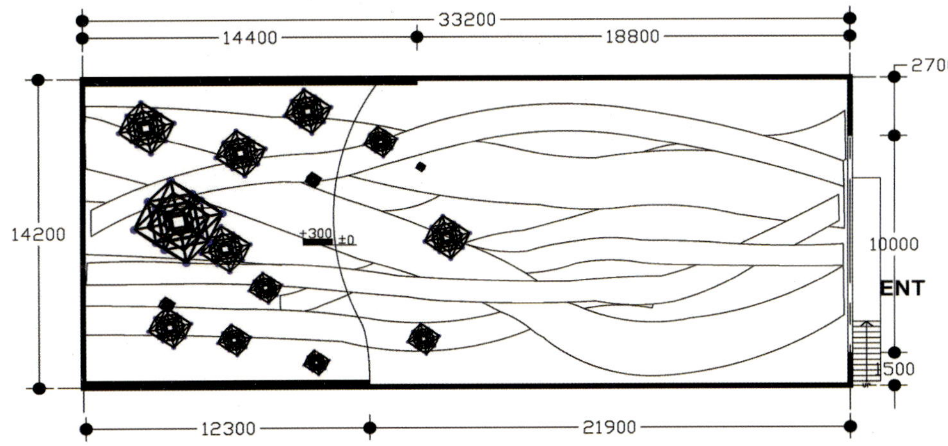
33200
14400
18800
2700
14200
10000
+300
±0
ENT
1500
12300
21900

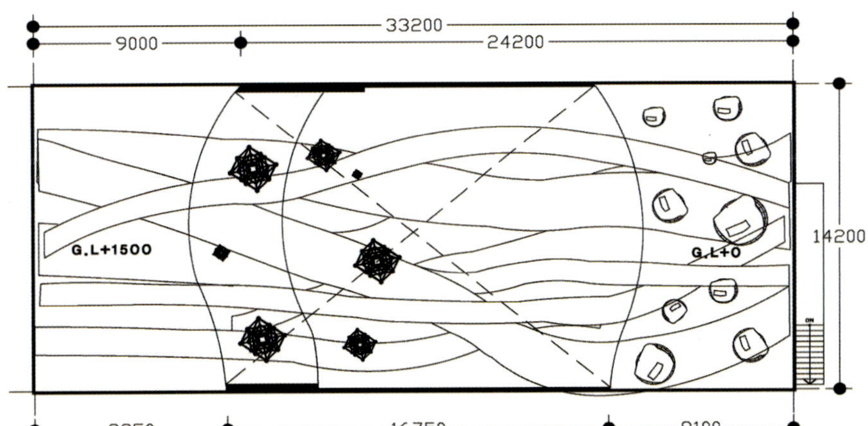
33200
9000
24200
G.L+1500
G.L+0
14200
8350
16750
8100

FOLLOW ME

사람을 부르는 공간

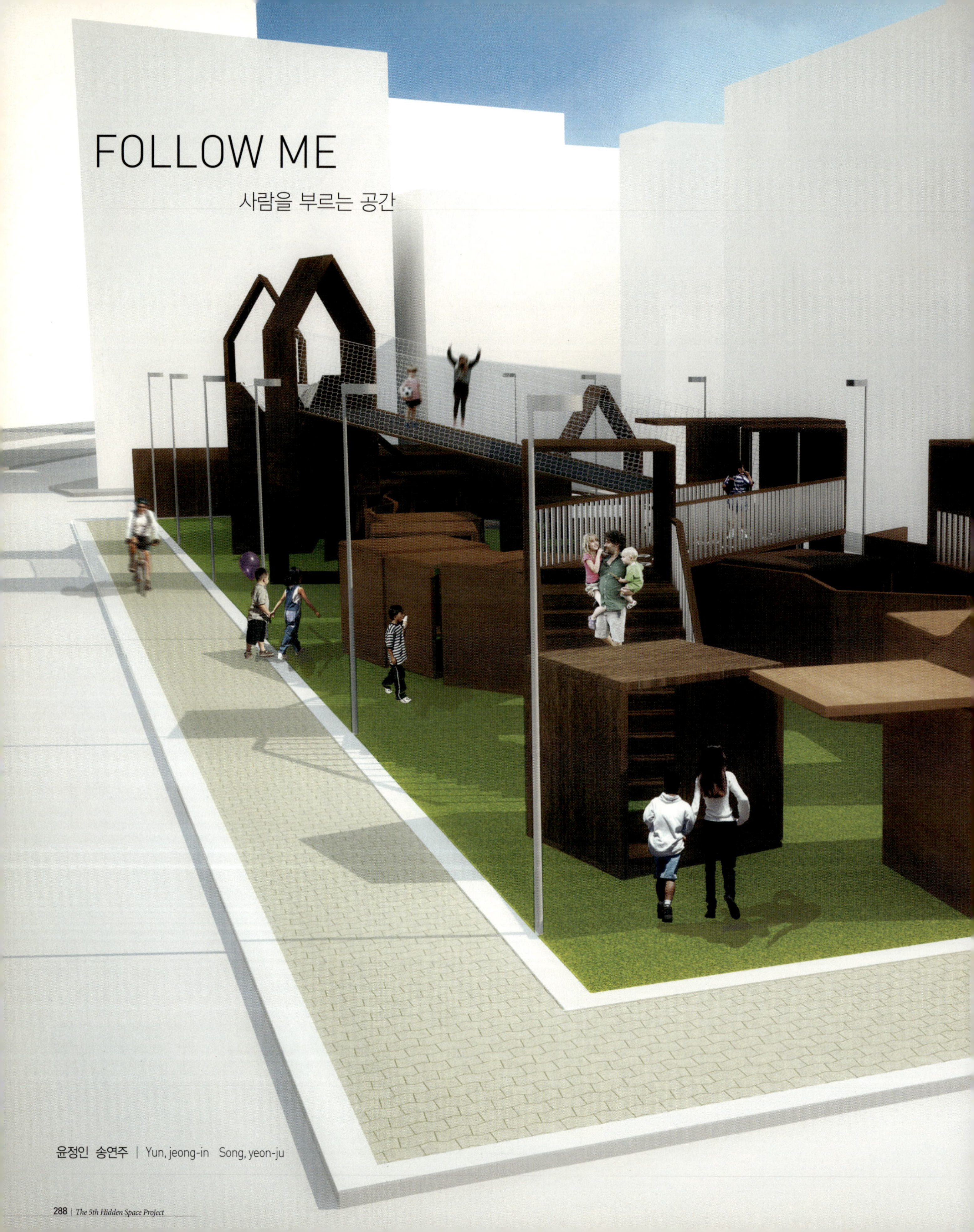

윤정인 송연주 | Yun, jeong-in Song, yeon-ju

Background

재개발 속에서 피해를 입어 복잡 미묘한 분위기, 침체되어 있는 분위기를 가지고 있는 지역에 활기를 불어 넣을 수 있는 공간을 제공함으로써 지역 전체에 Re-energising을 유도한다. 침체된 분위기가 될 수 밖에 없었던 원인을 주민들간의 의사소통 문제에서 찾고 그 문제를 해결하기 위해 모두가 쉴 수 있고 즐길 수 있는 PLAYSCAPE로 공간의 기능을 정하였다. 복잡한 문제들을 풀어낼 때일수록 단순하게 생각하면 문제가 생각보다 쉽게 해결될 수 있다는 생각으로, 최대한 쉽고, 단순하게 공간에 접근한다.

Induces re-energising across the region of indefinable and depressed atmosphere from damage by redevelopment by providing space that can stimulate vitality. The function of space was set into the PLAYSCAPE where everybody can rest and enjoy to find and resolve communication problems between residents that had to be the cause of the depressed atmosphere. With a belief that complex problems can be resolved more easily when they are approached with simple perspective, the space was approached simple and easy as much as possible.

Target

용산구 서부이촌동 이촌2동 주민들은 갑작스럽게 무산된 재개발 때문에, 감정이 고조 되어 있는 상태이고, 주민들 간에 이해관계 또한 달라 갈등을 조금씩 겪고 있다. 분위기 때문에 주민들끼리의 소통 또한 잦은 편이 아니며 지역에 대한 애정도도 떨어진다. 따라서 침체되어 있는 이 지역에 분위기에 변화를 줄 수 있는 시발점을 제공한다면 충분히 쉽게 주민들의 태도도 변화할 수 있을 것이라고 예측한다.

Because of the redevelopment was canceled suddenly, community of Yongsan-gu Seobuichon-dong has been struggling to solve the conflict generated from the different interests between residents. Therefore, they do not communicate often or care about community. If PLAYSCAPE can act as starting point, we are expecting that attitude of the residents can be changed easily.

Site

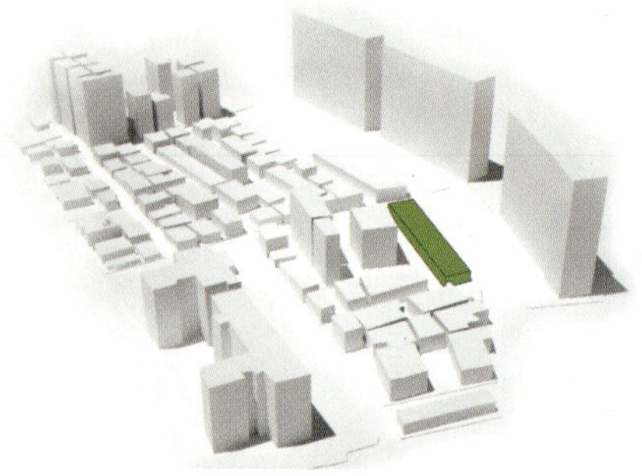

용산구 서부이촌동 이촌2동
주민센터 옆 대림아파트 주차장 공간

드림허브 프로젝트가 진행되었던 용산구 서부 이촌동은, 비싼 땅값, 취약한 자본구조, 무리한 사업부지 확장 등 내부적인 모순이 주된 원인으로 개발이 청산되었다. 드림허브 프로젝트의 무산으로 전체적인 동네의 분위기가 많이 침체되어 있다. 지리적으로는 가까운 곳에 한강과 원효대교, 강변북로 또 용산역이 위치하여 굉장히 우수한 위치를 선점하고 있다. 주민들이 모일 수 있는 공간을 마련함에 따라 상호소통이 가능해진다면 충분히 다시 활기를 되찾을 수 있는 공간이다.

The Yongsan western Ichon-dong dream hub project has been progress, the development has been liquidated high land prices, a weak capital structure, such as the expansion of excessive office, an internal inconsistency is the main cause. Atmosphere of the town overall is depressed many ended in failure of the Dream hub project. Yongsan Station Wonhyo Bridge and Han River, also Riverside North Road is located, you are ahead of the very excellent position closer geographically. By providing a space capable of residents gather, is a space which can take heart again sufficiently mutual communication becomes possible.

Purpose

기존 아파트에 함께 있는 놀이터가 가지고 있는 기본적인 놀이 공간과 함께 기존 공간보다는 좀더 다양한 놀 거리를 제공하고, 호기심을 가지고 한번 쯤은 방문하고 싶은 공간, 사람들에게 쉽게 다가가는 느낌을 가진 공간을 목적으로 가진다.
공간 자체의 움직임은 많지 않고 정적이지만, 공간의 특성상 공간이 이동성에 힘을 실어주기 때문에, 공간 안을 자유롭게 이동하는 사람으로 인해 공간 안에 생기와 활력이 생긴다.

More diverse recreation devices were provided rather than the existing space along with basic place space that the playground had in existing apartments, and intends to create space that people wish to visit at one time or another with curiosity and space that approaches people easily.
There are not many movement of space itself and is static, but as space empowers the mobility due to the nature of space, vitality and energy are created in space by people who freely move around.

Space SWOT

STRENGTH

놀이 공간이 주는 흥미로 인해 삭막함이 감소한다.
주민센터와의 근접성으로 주민센터와 프로그래밍과 연동해서 공간 활용이 가능하다.
공간과 사람간의 의사소통이 최대화된다.

desolation is reduced because of the interest that space of play give.
In conjunction with the programming in the Community Service Center in proximity with the Community Service Center, Effective use of space is possible. Communicating with people and space is maximized.

WEAKNESS

분위기에 민감하여 지속가능한 업그레이드가 늘 필요하다.
기후(눈, 비)에 따른 공간 사용의 어려움이 있다.

Sustainable upgrading is always needed as it is sensitive to the atmosphere.
There is difficulty in space usage by climate(snow, rain).

OPPORTUNITY

서부이촌동이라고 하면 가장 먼저 떠올리는 건물이 될 수 있다.
새로운 장소에 대한 필요를 충족시킨다.

It can be a building that people think of first when they hear Seobuichon-dong.
Necessity for a new place will be fulfilled.

THREAT

유지 관리 비용이 지속적으로 든다.
다양한 이용자 계층이 점차 어린이만으로 축소될 수 있다.

Continuous cost for maintenance will be required.
Various class of users can be gradually reduced down to children.

Process 1

Concept : follow me!

따라가다, 뒤를 잇다라는 뜻을 가지고있는 Follow는 근래에 와서 SNS의 활성화로 친구를 맺다라는 뜻으로 많이 사용되고 있다. 아무 관계가 없는 것들끼리의 관계의 성립의 의미로 사용되고 있는 Follow는 단순히 사람과 사람 사이에서만 적용되는 개념이 아니고 공간과 사람, 공간과 공간사이에도 적용시킬 수 있다. 사용자가 이동함에 따라 공간 간에 전혀 없던 관계가 생겨나고 그러한 관계들이 모여서 공간이 완성된다. 공간의 연결들로 더 큰 개념의 공간을 완성을 시키는 주체는 사람이다. 공간이 자신을 따라 오라고 말하는 듯한 Follow me는 공간과 사람의 상호소통 또한 원활하게 계속되고 있다는 것을 의미한다.

Follow meaning 'to go after or to follow' is often used with a meaning of 'to make friends' recently with vitalization of SNS. Follow, which is used as a meaning of to form relationships with people or things with no relation, is not a concept that is just applied between people, but can be applied between space and people or space and space as well. Completely new relationships are created between spaces as users move around, and space become created when these relationships get together. The main agent who completes space of a greater concept through connected spaces is a person. 'Follow me' which represents space telling people to follow signifies that communication between space and people is continuing on smoothly.

Motive

공간을 생각하면 우리가 가장 먼저 쉽게 떠올리는 네모난 박스 형태의 큐브에서 형태적 모티브를 따왔고 공간간의 관계를 통해 자연스럽게 풀어지는 큐브의 여러가지 형태의 모양을 통해 다양한 모습의 공간을 구성하였다.

A morphological motive came from a rectangular box-shaped cube that comes into our head first when we think about space, and organized space of various shapes through various forms that naturally become untangled through communication between spaces.

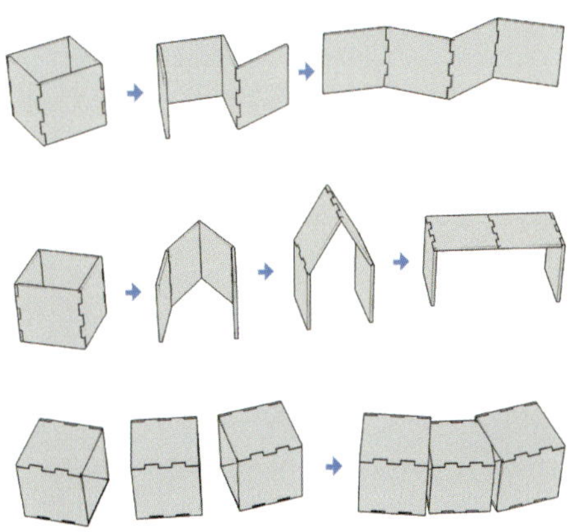

Plan

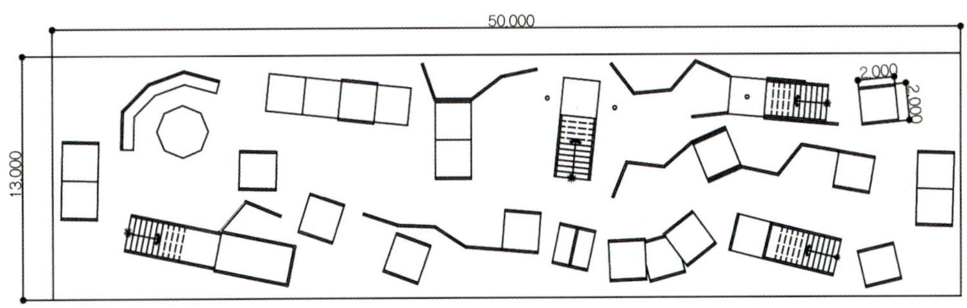

1st FLOOR PLAN

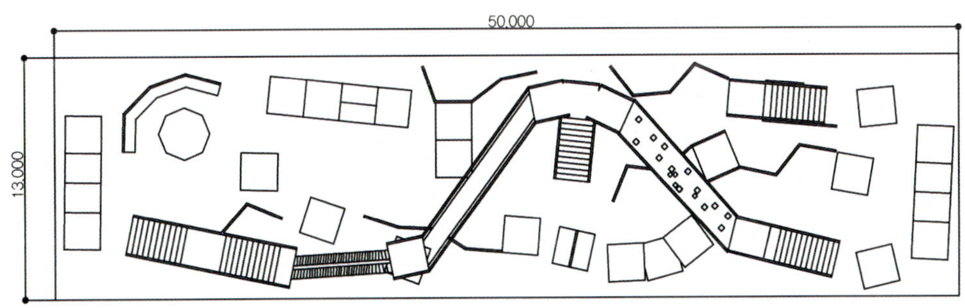

3rd FLOOR PLAN

Process 2

point of contact occurs

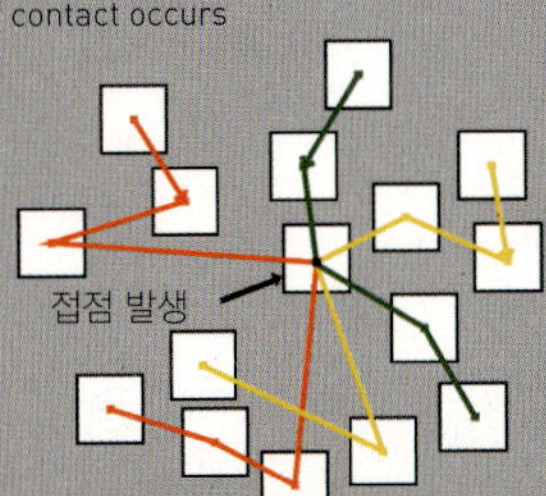

Space that builds a relationship

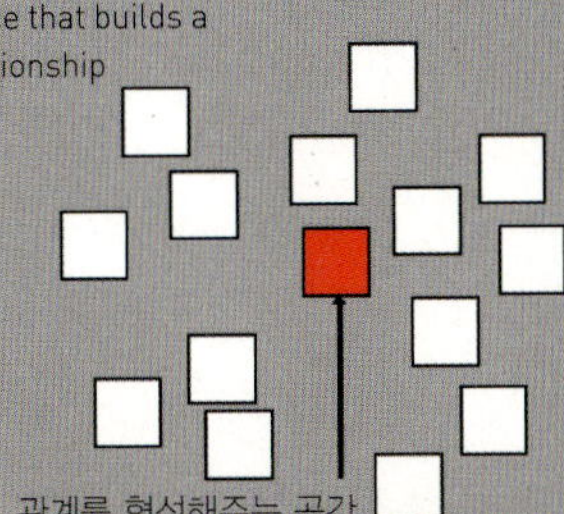

Relationship occurs between the space

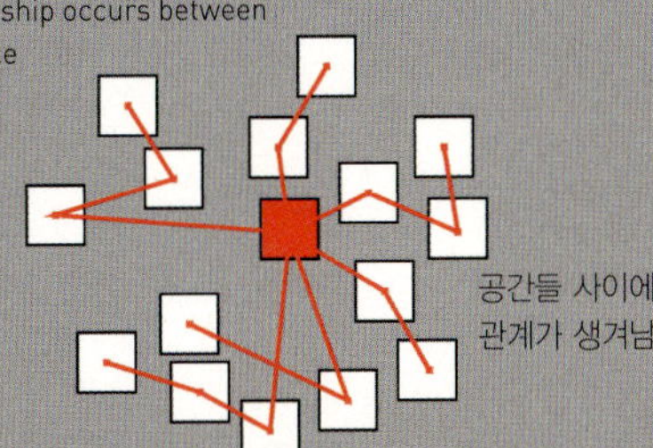

Materials

Wood 목재

Cherry
담황색을 띄고 있고 색이 진하여 중후한 느낌을 준다. 재질은 단단하고 탄력 있다.

Gives stately ambience with its dark lemon yellow color. Quality of the lumber is solid and has elasticity.

Mahogany
적갈색을 띄고 있고 건조가 쉽고 내구성이 좋으며, 가공성이 매우 좋다.

It is reddish brown, easy to dry, very durable, and excellent machinability.

Walnut
진한 갈색을 띄고 있고 강도가 매우 높은 편이긴 하지만, 작업하기에 아주 용이한 나무이다

It is dark brown and has high intensity, but very easy to work with.

Artificial turf elastic mat

폐타이어 고무매트와 비교하여 색감이 다르다. 보통 수명이 약 5~7년으로 이후에는 상부의 인조잔디만 새로이 교체하면 되므로 유지보수 비용을 약 50%이상 절감된다. 중금속 또는 어떠한 발암물질도 검출되지 않는 안전한 제품이고 불에 강한 난연처리로 아이들의 안전을 먼저 생각한 국내 최고의 탄성율을 가진 매트이다.

Color is different compare to the rubber mat made out of waste tires. They usually last about 5~7 years and can decrease approximately 50% of the maintenance fee since the upper artificial turf only needs to be replaced. A safe product with no detected heavy metal or carcinogen and it is a mat with the best elasticity in the country that considered safety of children above all else through fire treatment.

Advantages and Disadvantages

- 장점
재료의 가공 및 시공이 쉽고 나뭇결 수직방향에 대한 강도가 크다. 전기와 열에 대한 전도성이 없으며 적당한 정도의 흡수성이 있다.

- advantages
Process of the material and construction are easy and solidity for vertical direction is intense. There is no thermal and electrical conductivity and has just enough absorptiveness.

- 단점
건습과 열에 약하여 재료의 변형이 크다.
내구성이 약하다.

-Disadvantages
Great transformation of material as it is vulnerable to dryness and heat.
Not durable.

단점 보완
To complement

- 가을이나 겨울에 벌채한 것을 사용한다.
- 수령은 장목기의 나무를 선택한다.
- 곧은결을 사용한다.
- 도장을 알맞게 한다.

- Use trees that were cut down in fall or winter.
- Select long timbers.
- Use straight grains.
- Paint appropriately.

1. 바닥엔 그림도 그릴 수 있고 오른쪽 위에는 〈Follow my step road〉가 있다.
이 road는 춤을 못 추더라도 발판의 위치에 따라 몸을 움직일 수 있다.

Can draw pictures on the bottom
and there is ←Follow my step road→ in upper right.
People can move there body according to the position
of stepping stones even if they are not good at dancing.

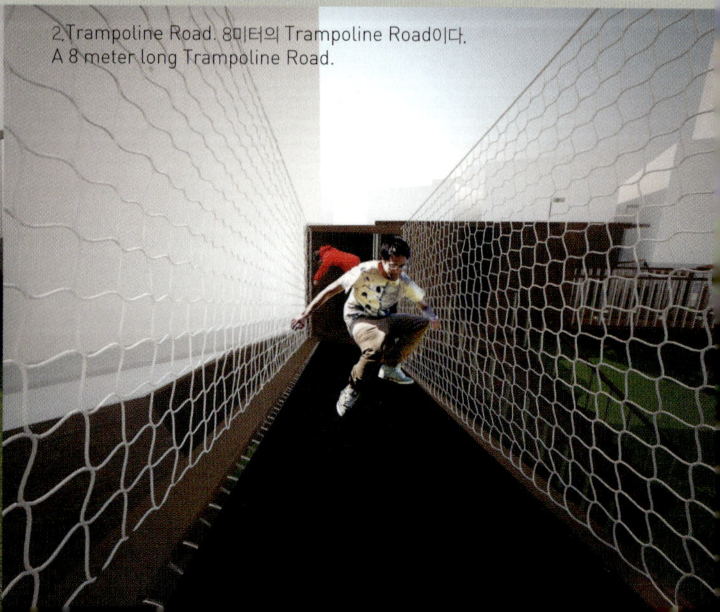

2.Trampoline Road. 8미터의 Trampoline Road이다.
A 8 meter long Trampoline Road.

Course Suggestion

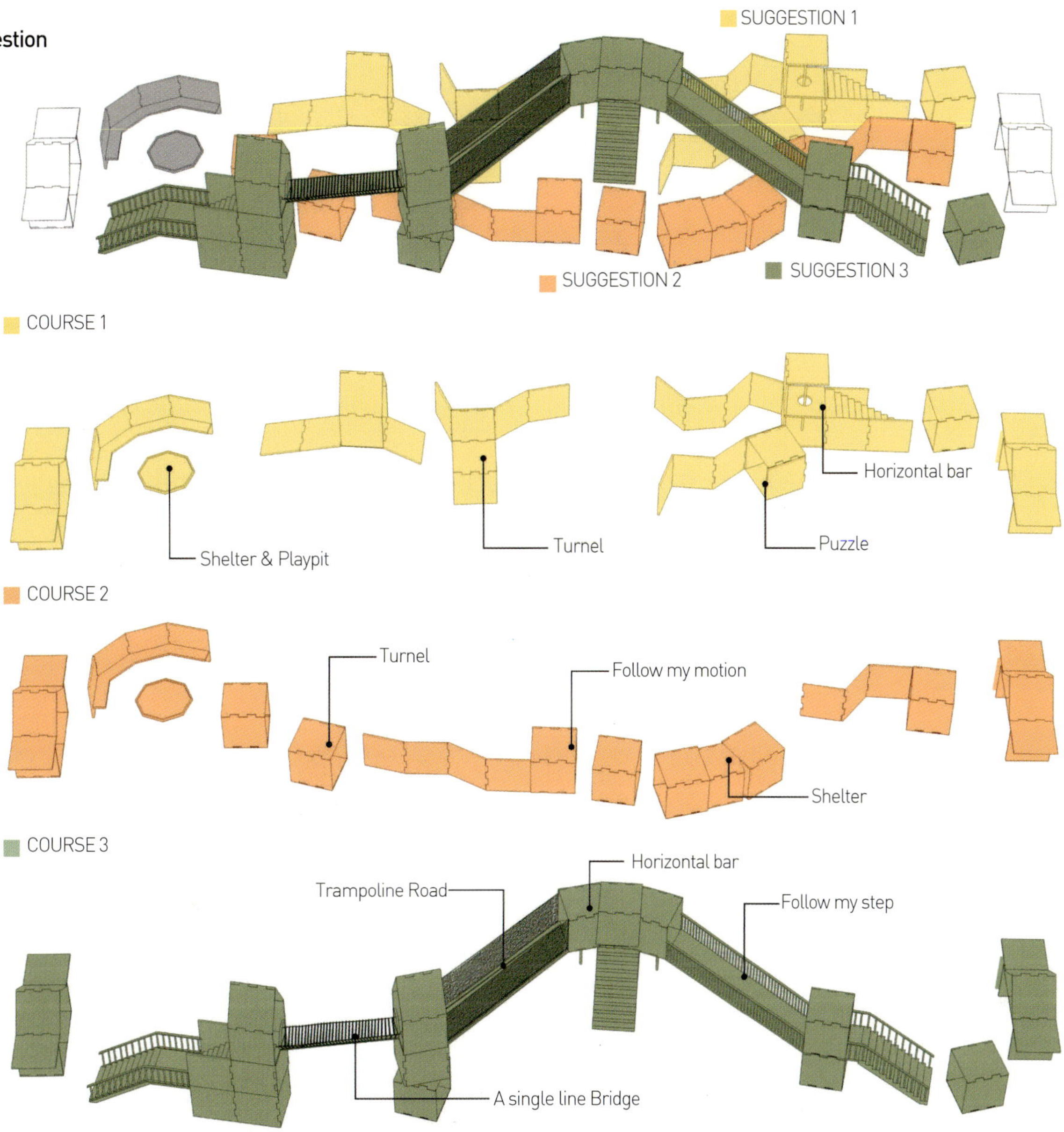

3.〈벽면 퍼즐 맞추기〉, 〈봉타기〉 등 공간 안에 놀이가 존재한다.
There are games as ←Putting the wall puzzle together→ and ←To climb up pea[illegible] inside.

4. 낮은 미끄럼틀과 모래 놀이터가 존재한다.
There are low slides and sand playgound.

김미소 이혜경 | Kim, mi-so Lee, hye-kyeong

Assimilate With Nature

Background

서부이촌동은 용산국제업무지구 개발사업 백지화로 인해 낡은 건물과 시위의 흔적 속에서 침체된 분위기에 있는 동네이다. 주변에는 녹지공간과 아이들이 뛰어놀 수 있는 놀이공간이 부족하다. 이런 점을 보완하여 우리는 아이들이 자연을 쉽게 접하며 그 속에서 뛰어놀 수 있는 공간을 만들고자 한다.
녹지환경이 부족한 현 상황을 보완하여 아이들이 자연에 동화되고 자연에서 뛰놀며 스스로만의 동화를 만들어가는 공간이다.

Seobuichon-dong is the village in depression mood with the old buildings and the traces of protest due to the invalidation of development work in Yongsan international business district.
In the vicinity, the green spaces and the play spaces for children to play out is insufficient. Supplementing this matter, we wish to create the spaces so that children can access the nature easily and can play about in the nature.
Supplementing the current situation of the insufficient green spaces, it is the place where children assimilate with the nature, creating their own fairy tales by frolicking in the nature.

Site

주소 : 서울특별시 용산구 서부이촌동 이촌2동
주민센터 옆 대림아파트 주차장 공간
면적 : 약 650㎡ (약 196.625 평)
용도 : 1층 주차공간, 2층 테니스장

Address:
Seobuichon-dong, Yongsan-gu,Seoul Daelim apartment parking space next to the community center
Area: about 650 M^2
Application: the first floor of the parking area, two tennis courts layer

Target

타겟 : 4세~초등학생 어린이
Target: 4-year-old to elementary school children

Concept

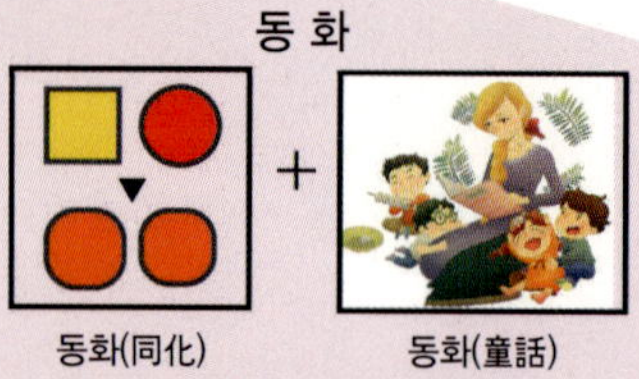

'같이 화합한다' 라는 뜻의 동화(同化)와 어린이를 위 하여 동심을 바탕으로 지은 이야기를 뜻하는 동화(童話) 라는 두가지 의미를 가지고 있다. 우리는 부족한 녹지환 경을 개선하여 자연에 동화(同化)되고 아이들이 뛰어놀며 스스로 이야기를 만들어나가 자신만의 동화(童話)를 완성해내는 체험공간을 표현하고자 한다.

It has a dual meaning. One of the meanings is "assimilation" which is defined as 'harmonizing together' and the other meaning is "fairy tale" which refers to the story that is created, based on the innocence of children. We wish to express the experience spaces by assimilating with the nature to improve the insufficient green spaces and by depicting the children who frolic in the nature and who create their fairy tales, completing the fairy tale of their own.

Section

1st FLOOR PLAN

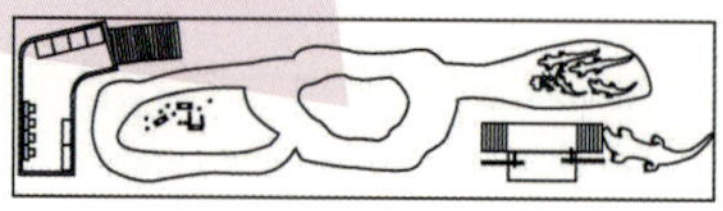

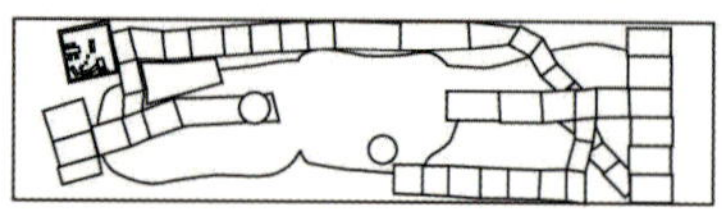

2nd FLOOR PLAN

Process

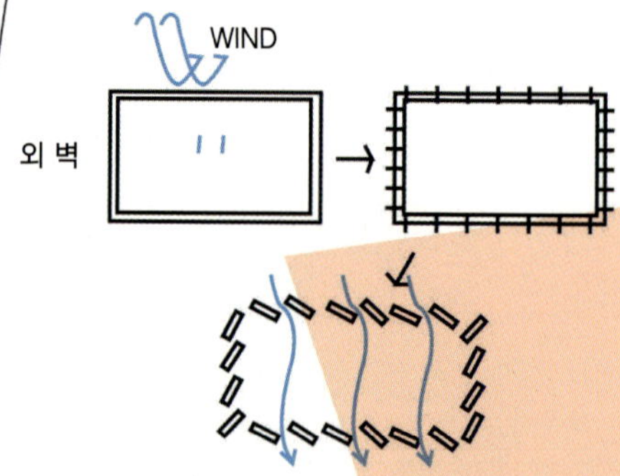

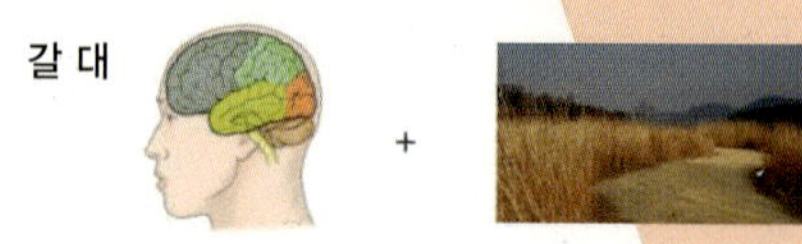

아이들은 자신의 키보다 높은 갈대숲을 헤치고 나간다. 어떤 공간이 나올지 모르는 상태에서 헤치고 나갔을 때 생기는 임의의 공간에서 만들어지는 행동으로 스스로의 이야기를 만들어 나가며 상상력을 키우게 된다.

The children wedge their way through the reeds higher than their height. Wedging own's way without knowing what kind of spaces will come up nurtures imaginations, creating their own story by the actions taken in the arbitrary spaces.

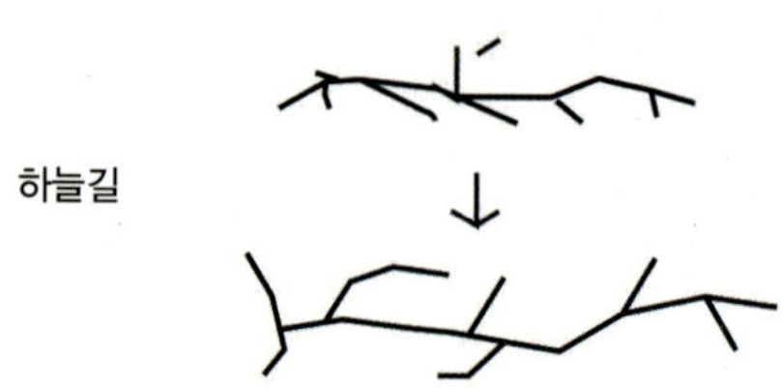

2nd floor program

물에 직접 닿지않아 젖지않고 물위를 걸어다니는듯한 느낌을 받으며 놀 수 있는 공간이다.

One do not get wet as no direct contact occurs with the water and can play in the space with the feeling of walking on the water.

→

이상한 나라의 앨리스에서 앨리스의 몸이 커지고 작아지는 부분을 체험할 수 있는 공간이다.

One can experience the part where the body of Alice from the Alice in wonderland expands and contracts in this space.

→

→

다양한 레벨차이의 길에 하늘의 모습이 그대로 반사되어 하늘 위에서 구름 위를 걷는 느낌을 주는 공간이다.

The space gives the feeling of walking on the cloud as the appearance of the sky is directly reflected from the streets of various differences in the level.

트램펄린으로 피터팬에서 하늘을 나는 장면을 연출한 공간이다.

It is the space directing the scene where Peter Pan flies in the sky by a trampoline.

→

1st floor program

시계소리, 제한 시간마다 악어의 입이 열리고 닫힘에 따라 건너가는 미션을 가진 공간이다.

It is a place that has a mission cross according to aligator's mouth open and close per each time.

→

동화 속 공간을 체험하기 전에 분장을 통해 진짜 주인공이 된 듯한 느낌을 받을 수 있게 도와주는 공간이다.

It is the space where one can feel as if they really become the hero by putting on make-up before experiencing the spaces of fairy tales.

→

→

빔으로 인해 비춰지는 영상에 따라 다양한 놀이를 할 수 있는 가변적인 공간이다.

It is the changeable space where one can enjoy various games according to the images projected by beam.

동화 피터팬에서 나온 악어를 극대화시켜서 어려움을 헤쳐 나가는 공간이다.

It is the space where one can win through hardships by maximizing the alligator in the Peter Pan fairy tale.

→

넝쿨을 타고 매달려 높은 곳과 낮은 곳을 자유롭게 왔다 갔다 하며 놀 수 있는 공간이다.

It is the space where one can freely come and go to high and low places and play, dangling on the vine.

→

빔 영상에 따라 다른 배경,다른 느낌으로 오르락 내리락 할 수 있는 가변적인 공간이다.

It is the changeable space where one can rise and fall with different background and different feeling according to the beam images.

→

Story

용케도 여기까지 왔군!
CHANG!
CHANG
WOW~
윽! 이럴수가!
내가 지다니...
이번엔 어딜까?
여긴 어디지~?
라푼젤이 우리를 도와주고 있어~ 저기 라푼젤 머리 잡고 올라가야겠다
얼른 도망가자!
이번엔 또 어딜까?
라푼젤 고마워!
hahahahahaha~!!
둘은 행복하게 살았습니다~

박종창 박소망 최경천 | Park, jong-chang Park, so-mang Choi, kyung-chun

Flap space inside

공간속에서 펄럭이다.

Concept

펄럭이다 – 깃발이 바람이라는 힘의 의해 펄럭이는 모습 그리고 서부이촌동 주민들이 고난과 역경이라는 힘의 의해 흔들리지만 깃발이 봉하나를 잡고 버티며 펄럭이듯이 서부이촌동 주민들도 희망이라는 봉을 잡은채 버티며 살아가는 모습이 유사하여 이 컨셉을 선정하였다. 깃발은 하나하나 각기 다른 모습으로 펄럭인다 사람들은 각기 다른생각과 다른 모습을 가지고 있다. 사람들이 모여 서로 화합한다면 용산 서부이촌동은 훨씬 살기좋은 동네로 탈바꿈 할 것이다.

Fluttering - The look of a flag flurttering due to the force called the wind. This concept was set since residents of the Seobu Ichon-dong lead their life while withstanding while holding on the rod of hope since Ichon-dong who sway due to hardship and turmoil. This neighborhood will become a very livable, nicer neighborhood if people gather together to form harmony.

Background

용산 서부이촌동 용산역 과 한강사이에 자리잡은 서부이촌동은 용산국제업무지구 개발사업으로 큰 기대를 한몸에 받았다 땅값이 치솟았으나 최근 용산국제업무지구 개발사업 무산으로 약 7년간 고통받아온 서부이촌동 주민들은 이제서야 무산된 개발사업의 벗어났지만 그 피해는 무지막대하다. 주변에 매말라버린 상권과 개발사업 속에서 다투어온 주민들간의 불화 겉으로 보기에도 지저분하고 유령동네가 되어버린 서부이촌동은 그곳의 어른들을 보며 커가는 아이들, 어른들을 사회생활에 지치고 아이들은 주차공간의 부족으로 자동차들에게 골목을 빼앗겼고 작은 아파트 놀이터에서 놀고있다.

Seobu Ichon-dong where is located between Yongsan station and Han River got huge expectation about development project, called 'Yongsan International Business District'. Naturally the land value of this area increased a lot. However, the residents of Seobu Ichon-dong have been troubled with Yongsan International Business District's foundering development project for 7 years. They would seem to escape that situation but they have had damages yet. Market, relationships with neighborhood and community, etc. all about these things were deteriorated. Now, the Seobu Ichon-dong is like a ghost town no one lives.

Site

용산 서부이촌동 전방의 한강이 흐르고 있고 주변과의 높낮이 차가 거의 없다. 근처 주택가와 아파트들은 쉬운 접근이 가능하며, 용산역에서 버스 이용시 약 17분이 소요되고, 아파트 앞 이촌한강공원 에서도 도보로 이용한 접근이 용이하다.

Han river flows in front of the Yongsan Seobu Ichon-dong, and there is almost no difference with the surrounding area in terms of the height. It is readily accessible due to the nearby residential area and apartments. When taking a bus from the Yongsan Station, it takes about 17 minutes, and it is also possible to approach it easily from the Ichon Han River Park located in front of the apartment.

Metarial

EPDM은 고무 재질로써 요즘한창 아이들 놀이터 바닥에 모래 대신 사용되는 재질이다. 신장률이 450%나 되어 균열 진동에도 대응력이 뛰어나 아이들이 넘어지고 부딪혔을 때 다치는 일이 거의 일어나지 않는다. 자외선이나 외부에 노출된 상태에서도 초기물성을 평균20년이상 유지하는 뛰어난 내구성을 가지고 있다.

EPDM is a rubber material that it is used often for the playground for children these days, replacing sand. Since its expansion coefficient reaches up to 450%, it is superb against crevice and vibration. As such, children seldom get hurt when they fall down or clash. its endurance level is superb to the point that its initial physical properties can be maintained for 20 years on average even when exposed to ultraviolet rays and to the outside.

Design program

형태적인 해석 – 말그대로에 깃발이 펄럭이는 모습을 표현
의미적인 해석 – 깃발이 바람이라는 외부적 힘의 의해 변형 되는 모습을 표현

– 매스표현
– 공간표현

Formative interpretation – Expression of a fluttering flag as the word itself indicates
Significance interpretation – Expression of how a flag change due to the external force, called the wind

– mass expression
– space expression

형태적인 해석 – 말그대로에 깃발이 펄럭이는 모습을 표현

- Expression of a fluttering flag as the word itself indicates

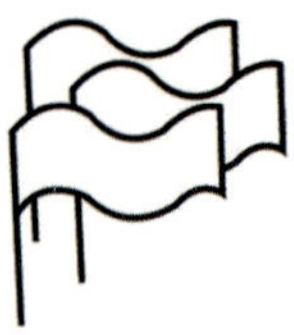

서로 각기 다른 깃발의 펄럭이는 모습

Sight of different flags fluttering

각기 다른 깃발들의 형태를 추출

Extraction of the flag forms that are different from each other

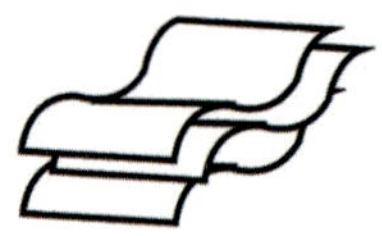

형태를 매스화

Development of the form into mass

KEYWORD

의미적인 해석 – 깃발이 바람이라는 외부적 힘의 의해 변형 되는 모습을 표현

Semantic interpretation - that the flag in the wind like being deformed by external force representing the

Distortion

자유롭게 형성된 곡선의 거울은 물체를 왜곡시킨다 아이들은 거울 앞에서 왜곡된 모습을 보고 호기심을 느껴며 창의성을 길러준다

Mirror of the freely formed curved line distorts object. Children feel sense of curiosity while looking at the distorted look in front of the mirror and their creativity is cultivated.

Rotation

외부적인 힘의 의해 공간이 회전하며 아이들에게 호기심 심어주고 회전하는 벽을 보고 만지며 공간성을 기르게한다.

Space circulates due to external force. Children's sense of curiosity is stimulated, and their spatiality is cultivated while looking at and touching the circulating wall.

Wobble

공중에 매달린 짐볼에 호기심을 심어주고 아이들이 직접 매달려 흔들리는 짐볼 위에서 운동감각과 균형성을 기르게한다.

Children's sense of curiosity towards the gym ball that is hanging from the ceiling is stimulated, and children's physical is cultivated while on the gym ball that is moving by hanging onto it.

Stretch

외부적인 힘의 의해 벽이 고무줄 처럼 늘어남과 동시에 출입구가 될 수 있으며 늘어나는 벽은 아이들에게 창의력과 공간성을 기르게 한다.

The wall stretches out like a rubber due to the external force and it serve as an entrance at the same time. Thus, this helps to cultivate sense of creativity and spatiality in children.

Space program

주차공간의 부족으로 아이들의 주된 공간이였던 골목이 주차장으로 변질, 앞으로 계획될 공간에 아이들의 주된 놀이공간을 형성한다. 깃발이 외부적인 힘인 바람을 만나 펄럭이듯이 공간이 아이들이라는 힘으로 인해 사용되므로써 아이들은 공간 속 에서 뛰고 만지고 부딪히고 하며 그 행위 자체에서 에너지가 형성되고 그 에너지는 공간 속과 이 공간을 사용하는 사람들에게 형태적 의미적 변형을 일으키게 한다.

Streets that were the playground for the children turned into parking lot due to the lack of parking space. Formation of the playground for children in the space that will be planned going forth. Just as a flag flutters as it meets the external force, called the wind, ensure that the space will be used by the power called the children so that the children can run around, touch and clash amidst the space.

These acts in turn form energy, and that energy leads to the formative and significance level transmutation in this space and in case of the people who use this space.

Distortion

Rotation

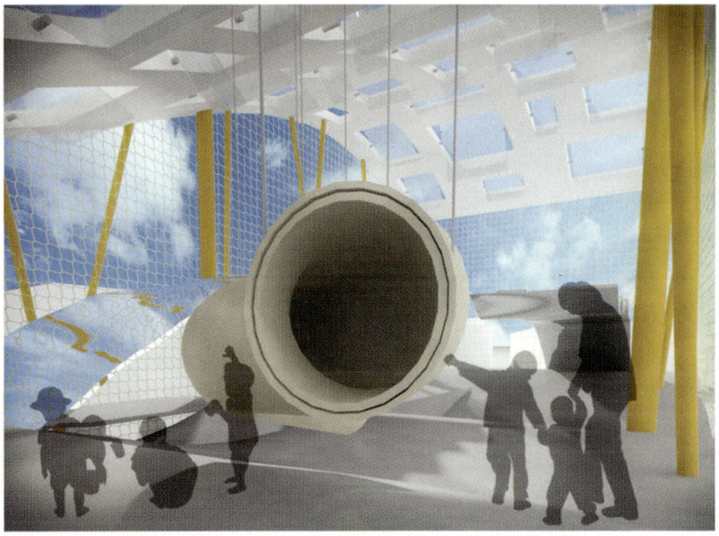

Wobble

Stretch

Plan

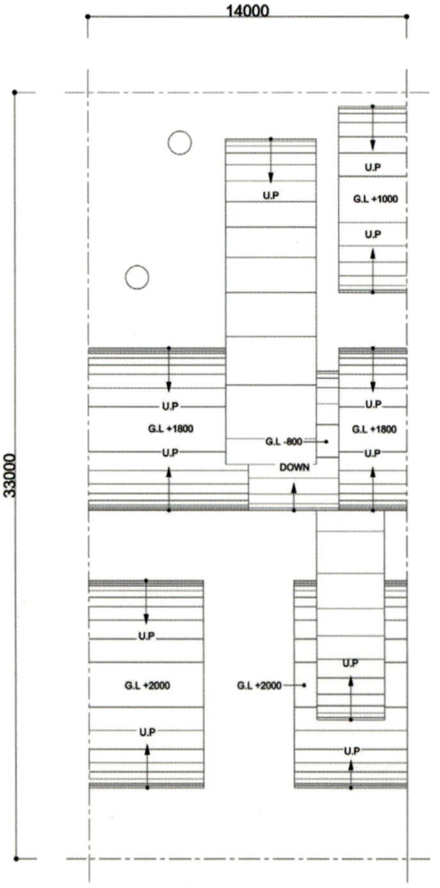

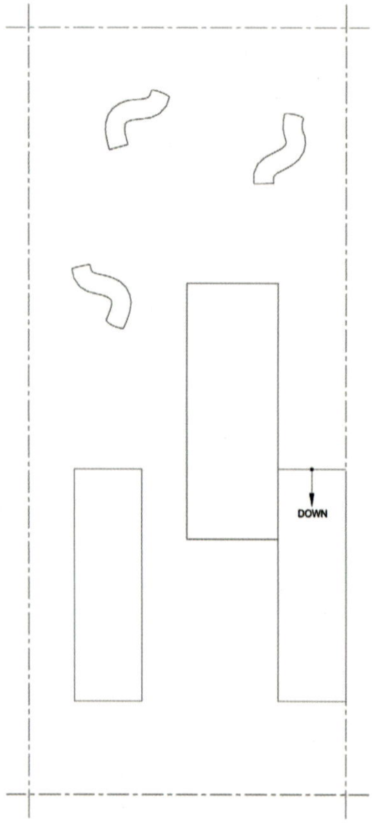

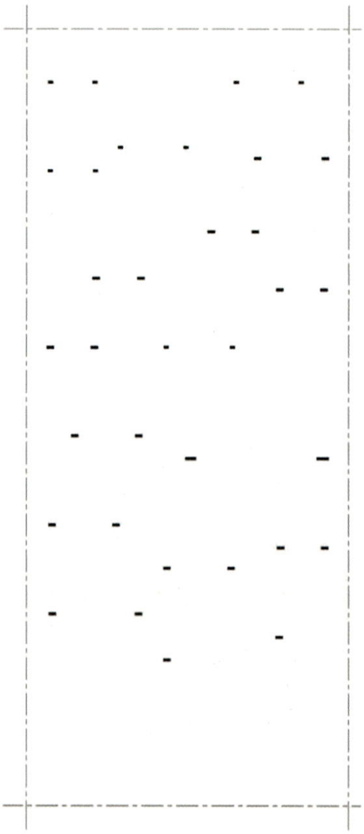

Elevation

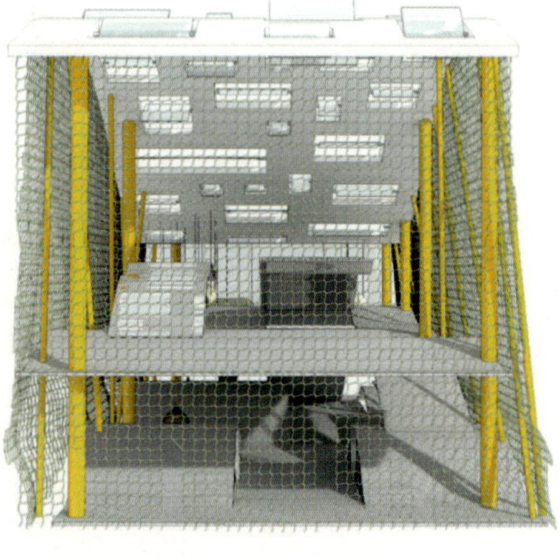

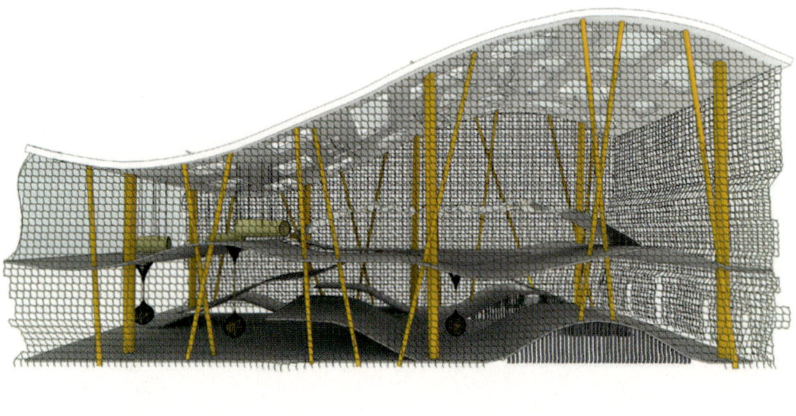

Axonometric

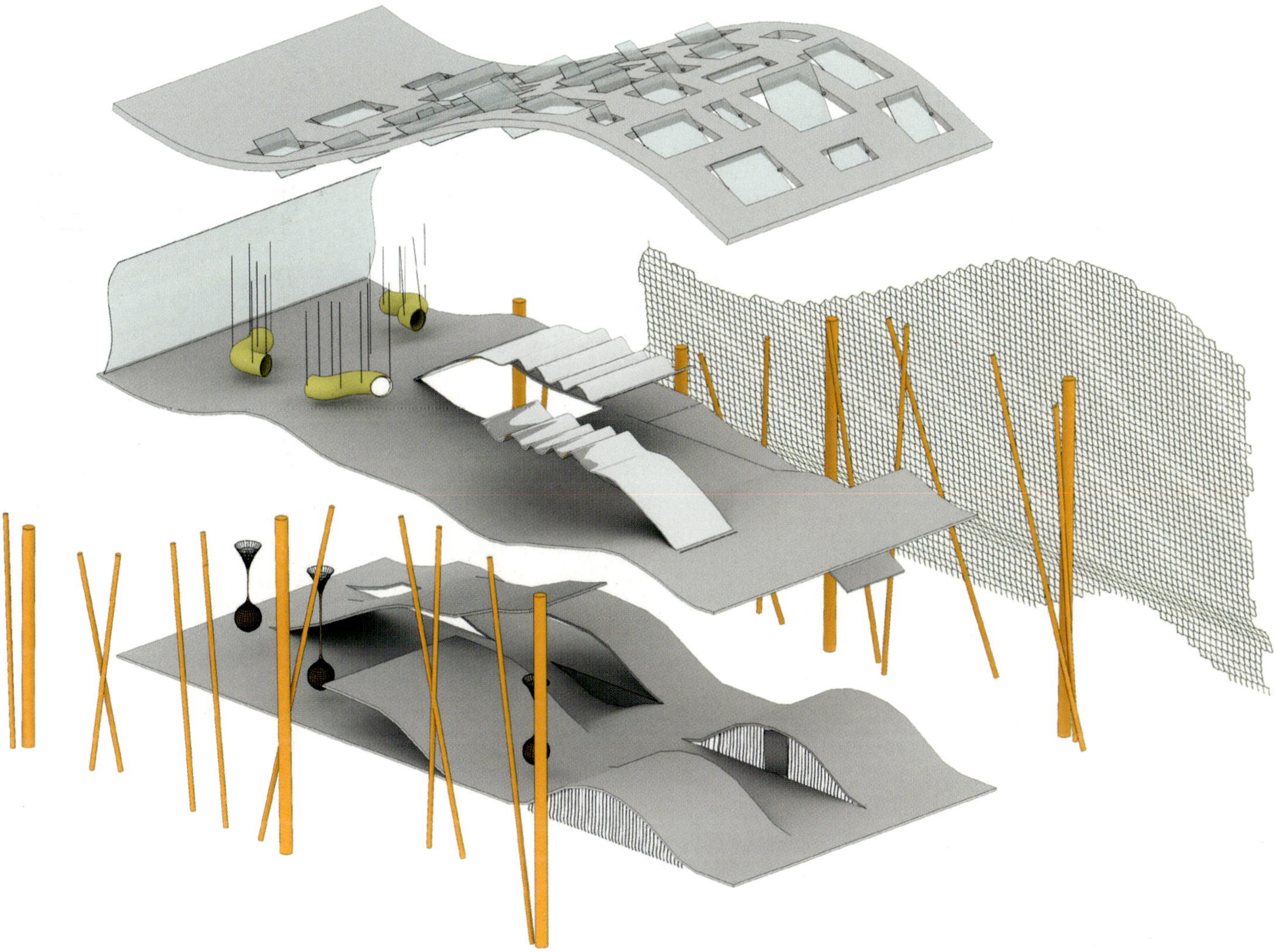

Inside image

전채희 최수진 한어진 | Jun, chae-hee Choi, su-jin Han, eo-jin

Freedom in the frame

A playground responding to people's moves and ideas, changeable and easily approachable

Site analysis

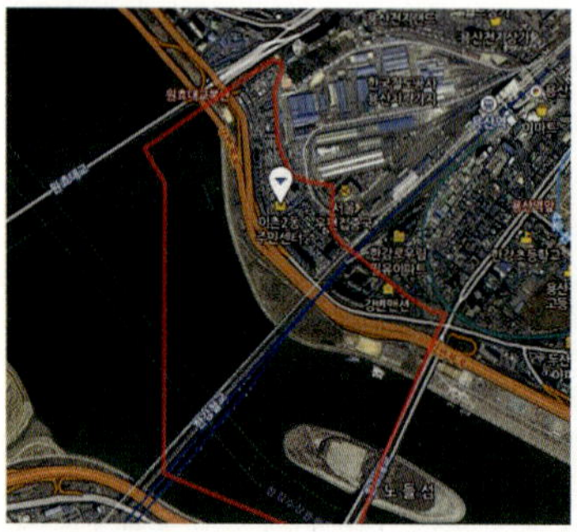

위치 :서울시 용산구 이촌2동 주민센터 앞 (대림아파트 주차장)
면적 : 약 848m² (256평)

서울시는 용산 국제업무지구 계획 무산으로 인한 주민들이 겪을 피해를 줄이기 위해 도시 개발 사업을 해제하였다. 이것은 건전한 부동산 시장 형성을 통해 주택거래를 활성화 시키기 위한 방안이었다.

Seoul has withdrawn a city development project to reduce side effects of a cancelled plan for Yongsan international business district on residents. It was a way to revitalize real estate transaction through forming fair real estate market.

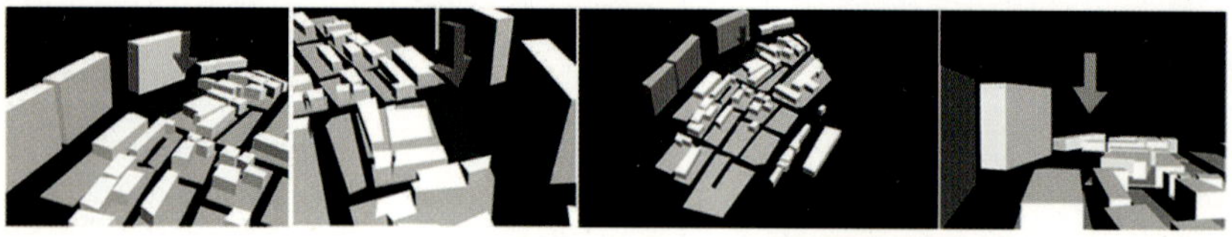

서울시는 주거환경이 노후된 구역에 대하여 지역재생을 위해 적극 나서기로하였다. 그 중 대표적인곳이 이 서부이촌동이고, 주변 지역에 비해 낙후되고 침체된 분위기인 이곳의 주민들이 편안한 마음으로 쉽게 찾고 즐길 수 있는 놀이공간을 만들고자 한다. 주민들 뿐만 아니라 더 많은 사람들이 함께 어울릴 수 있는 즐거운 공간이 필요하다.

Seoul aims to improve abandoned areas with deteriorated accommodation. One of representative areas is Seobuichon-dong more abandoned and depressing. Seoul sets a goal to create an enjoyable and approachable playground for Seobuichon-dong residents so that they take a break at the playground, while relaxed. An enjoyable place is required for not only residents but also more people to hang out together.

Background

스트레스, 인간소외화, 급속하고 과밀한 정보화 속에 지친 현대인들은 도시생활로 인해 삶의 여유를 점점 잃어가고 있다. 또한 디지털 기기의 등장과 발전으로 어른들은 물론 한참 뛰어놀아야 할 나이의 아이들까지 활동적인 모습을 찾아보기 힘들게 되었다. 따라서 도시 공간에서 잠시나마 몸과 마음을 치유할 수 있는 곳을 찾게되고 삶의 질을 향상 시키기 위해 레저활동과 사회적 참여 등 다양한 활동을 하게 된다.

Modern people who suffer stress, human alienation and, rapid and intense informatization have forgotten how to enjoy life due to city life. Kids supposed to play outdoors as well as adults are hardly committed to outdoor activities due to the appearance and development of digital devices. Thus, they seek a place to take a rest and participate in various activities such as leisure activities or social gatherings to improve the quality of life.

이곳에서 사람들의 움직임과 생각이 반응하며 변화해 가는, 쉽고 편안하게 다가갈 수 있는 '놀이공간' 을 제안해 보고자 한다.

'A playground' responding to people's moves and idea, changeable and easily approachable is suggested.

여기서 공원은 도시의 삭막한 공간 속에서 살아가는 사람들에 삶의 고단함을 달래고 휴식을 취할 수 있는 일상 속에서 자연스럽게 이용할 수 있는 소중한 문화공간이 된다. 활동적이며 즐거운 놀이를 찾아 보던 중 요즘 젊은층에서 유행하는 스케이트 보드를 떠올리게 되었고, 보드장 보다는 길이나 공터에서 위험하게 타거나 연습하는 사람들을 종종 볼 수 있었다. 갖춰진 보드장을 찾아보기 힘든 요즘, 일상에 지친 주민들에게 여유롭고 편안하게 찾을 수 있는 놀이공간을 제공하며 그 안에서 활기를 되찾게 해주고자 한다.

Here, parks can be a precious cultural space for people suffering tough city life to be relaxed and rest during daily life and they are approachable. During a research on active and fun activities, skateboarding getting trendy among young generation came up. People who skateboard or practice to skateboard on the street or empty lots rather than skateboard parks have often been seen and it is dangerous. Since it is hard to find a decent skateboard park, a playground was suggested to revitalize people's life so that residents tired of daily life can rest and enjoy at the playground.

Design motive

서부이촌동은 용산국제업무지구 개발사업 백지화로 인해 낡은 건물과 시위의 흔적 속에서침체된 분위기에 있는 동네이다 서부이촌동은 용산국제업무지구이다. 개발사업이다.

Plaza refers to a vacant lot a number of people can gather, which means a space many people share their ideas. Communication and emotion between people forms a space.

관심, 선호, 흥미, 바램, 확신, 필요 → SPACE형성 → 행동,행위 유발

Attention, preference, interests, wishes, conviction, needs → space formation → Behaviors, causing behavior.

쉴 틈 없이 반복되는, 삭막하고 여유가 없는 주민들의 바쁜 생활모습

Residents' constantly repetitive, heartless and busy daily life

다양한 크기의 톱니바퀴는 각자의 방향으로 끊임없이 회전한다.

Tooth wheels with different sizes are consistently spinning round towards each direction.

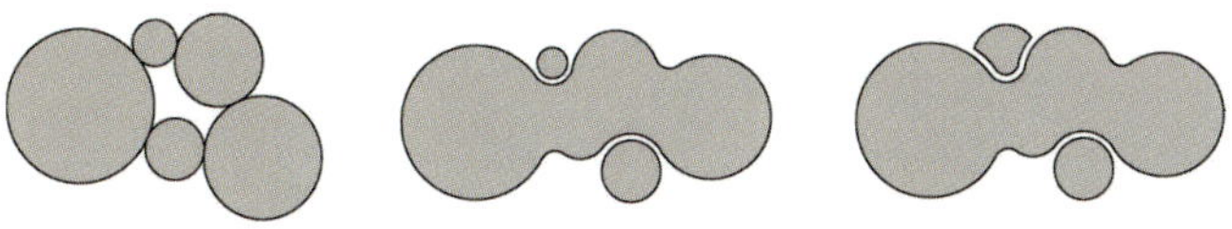

그들이 만나 공존하며 에너지를 나눔으로써 활력이 넘치는 공간으로 재탄생

They meet, coexist and share energy, which recreates a space full of energy.

다양한 크기의 톱니바퀴가 서로 맞물리고, 각자 다른 방향으로 끊임없이 회전한다. 이것은 쉴 틈 없이 반복되며 삭막하고 여유가없이 한 동네 안에서 열심히 생활하고 있는 서부이촌동 주민들의 생활모습과 닮았다. 쉴 틈 없이 반복되며 삭막하고 여유가 없는 주민들의 바쁜 생활은 톱니바퀴가 돌아가는 모습과 오버랩 된다. 다르게 생각해보면 톱니바퀴는 맞물려서 무언가를 움직일 수 있는 원동력이 되어주고 주민들도 마찬가지로 각자의 위치에서 공존하며 에너지를 나눔으로써 서부이촌동에 활력을 불어 넣어 줄 것이다.

Tooth wheels with different sizes are spinning around and consistently rotating toward different directions. It is repetitive and similar to life of Seobuichon-dong residents engaged in repetitive and heartless life in a hurry at the same town. Tooth wheels spinning around remind of the residents' life. In other words, tooth wheels spinning around become motive power to move something and residents will also revitalize Seobuichon-dong by sharing energy and coexisting as they are.

FUNNY

SOAR

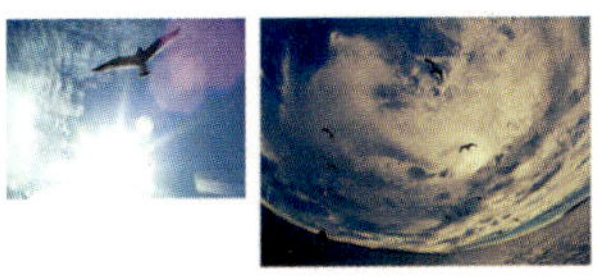

BOARD

COMMUNICATION

Client ndeeds

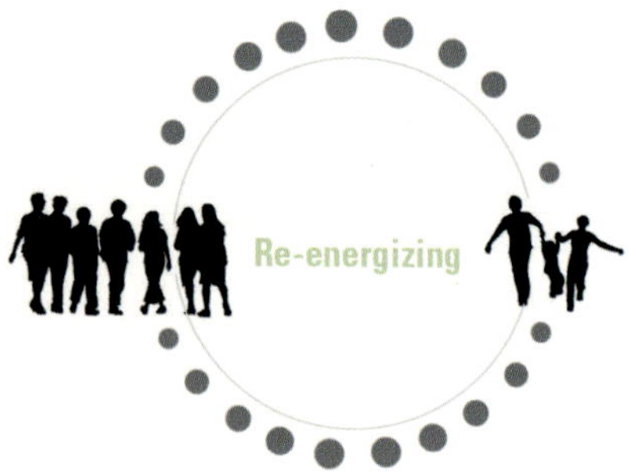

서부이촌동은 주위에 비해 상대적으로 낙후된 지역이며 분위기도 매우 침체되어있다. 대부분의 주민들이 안고있는 경제적 어려움과 부족한 여가공간은 열심히 살고 있는 주민들에게 육체적, 정신적 피로를 준다. 조금 더 나은 생활을 위해 열심히 살고 있는 서부이촌동의 주민들이지만 상대적으로 침체되고 낙후된 분위기인 서부이촌동에서 주민 대부분이 안고있는 경제적 어려움과 부족한 여가시간, 공간으로 인해 육체적으로는 물론 정신적인 피로까지 쌓여있다. 모든 주민들이 쉽게 이용할 수 있으며 시간과 비용의 최소화로 부담없이 찾을 수 있는 언제나 열려있는 놀이공간이 필요하다.

Seobuichon-dong is relatively more lagged than other areas and highly stagnant. Financial problems most of the residents suffer and insufficient leisure spaces stress the resident out physically and mentally.
Although Seobuichon-dong residents live life to the full for better life, they suffer physical and mental stress because they have financial problems with insufficient leisure time and spaces, living relatively stagnant and abandoned area, Seobuichon-dong. An open space all residents use easily with less time and cost is necessary and it should be always open.

Design process

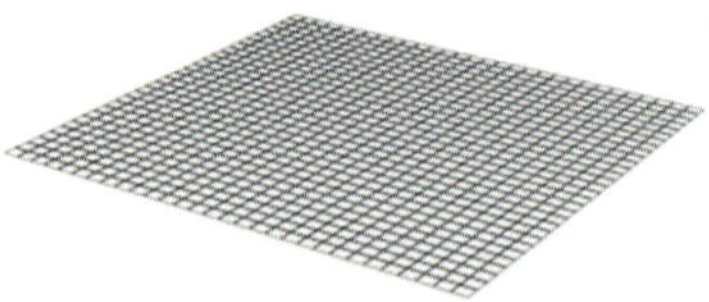

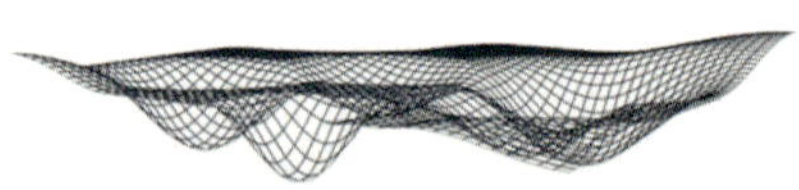

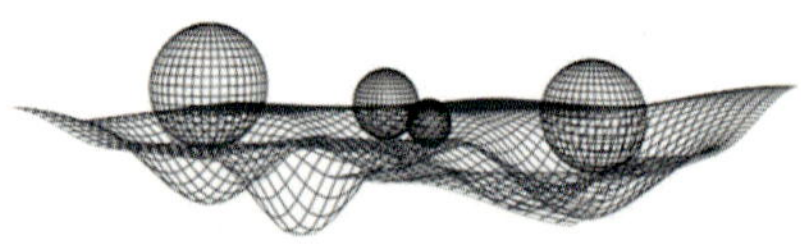

외관에서 느껴지는 견고함은 내부의 유동성을 담는 형태를 취하고 있다. 내부로 진입할수록 느껴지는 리듬성과 자유로운 형태는 이용하는 사람들의 체험하지 못한 경험을 하게 된다.

The solid appearance includes liquidity of the inside. As people approach the center of the playground, people experience rhythm and see liberating forms they haven't experienced.

Concept

Wheels - rotation
Speed - dynamic

Wind

바퀴가 회전하는 모습과 보드를 타며 즐기는 스피드를 생각했을때 공통적으로 '바람' 이 떠올랐다. 바람은 공기가 있는 곳이면 어디든 존재할 수 있으며 구체적으로 눈에 보이는 형태는 없다. 이러한 바람의 특성을 살린 채 이용할 수 있는 놀이공간을 생각해보았다. 딱딱하고 무거운 일상에서 벗어나 자유를 느낄 수 있는 공간.
〈FREEDOM IN THE FRAME – 틀 안의 자유〉를 공간에 담아내고싶었다.

Both spinning tooth wheels and speed of skateboarding reminded of 'wind'. A space for skateboarding was shown as a sort of winds rather than structures we often see.

주민들의 모습을 형상화 한 톱니바퀴와 스케이트보드를 타며 느낄수 있는 스피드, 바람을 이용해 공간을 풀어나가려고 한다.

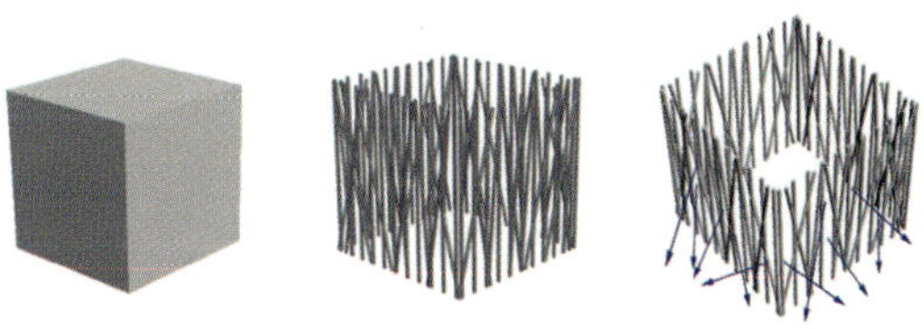

꽉 막힌 외벽보다 바람이 잘 통하고, 자연 그대로의 바람을 느낄 수 있는 벽을 구상해 보았고, 우리가 평소 흔히 볼 수 있는 스케이트보드의 구조물 대신 보드를 즐길 수 있는 바닥공간을 바람의 종류로 표현하려한다.

West dong Yongsan International Business District development project is to negate Due to traces of an old building in protest and West dong Yongsan International Business District development project is to negate Due to traces of an old building in protest and West dong Yongsan District development project is to negate Due to traces of an old building in protest and West dong Yongsan

KEYWORD

Mobility

Openness

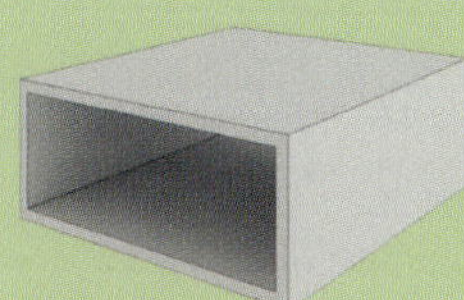

Freedom in the frame

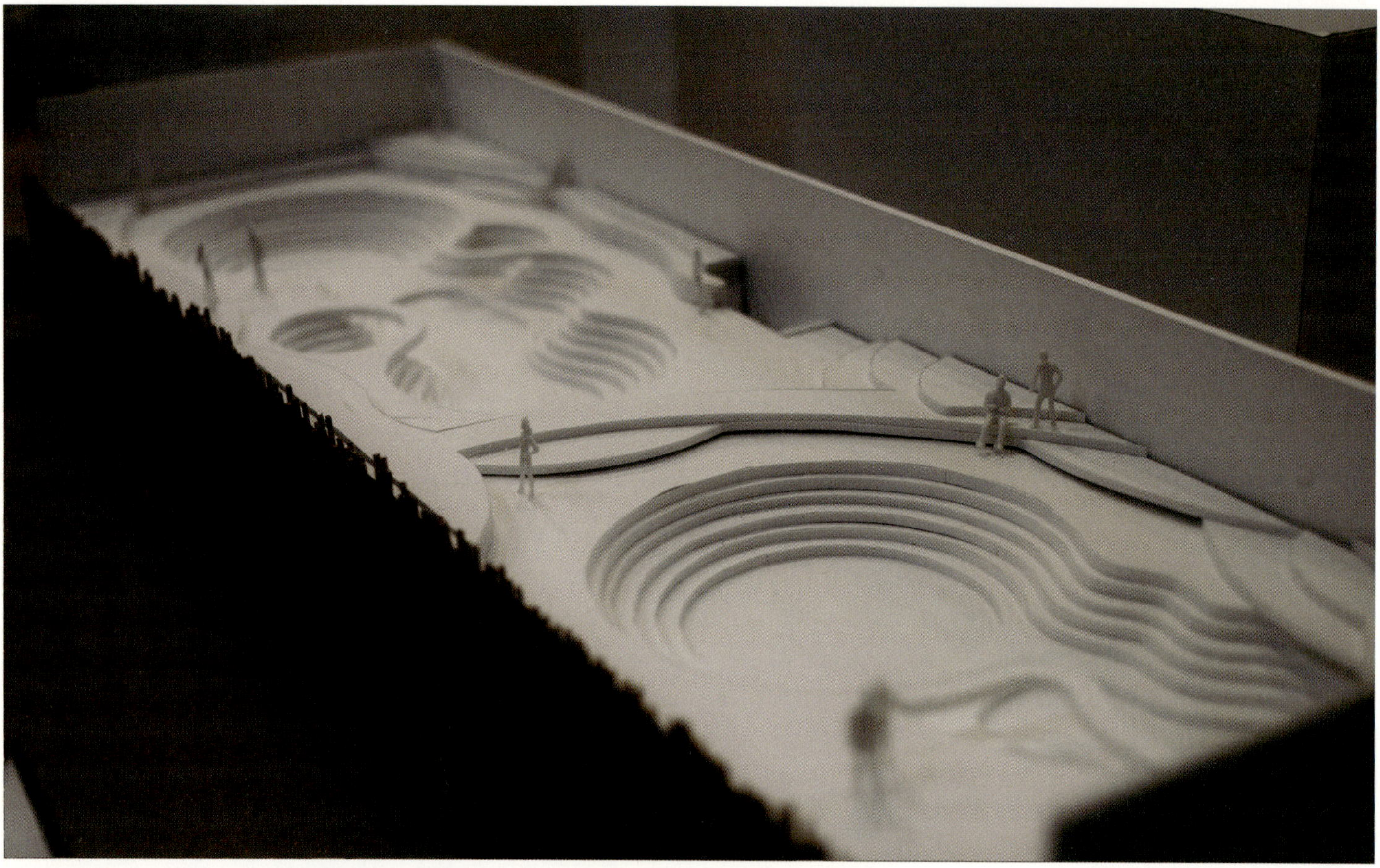

천장에서 쏟아들어오는 빛은 볼륨감을 형성하는 동시에 공간을 구획하게 하는 중요한 요소가 된다.

Light from the ceiling both forms massing and becomes a significant element to divide the space at once.

유기적인 형태의 바닥은 공간의 부드러운 움직임을 포함하며 유동적이고 연속하는 공간의 감성을 가지고 있다.

West dong Yongsan International Business District development project is to negate Due to traces of an old building in protest and West dong Yongsan International Business District develo

공간으로 유입되는 대기의 흐름은 유연한 동선인 동시에 공간 확장성 매개체가 된다. 공간 전체를 흐르는 공간의 연속성과 유동성을 의미한다.

The stream of the air towards the space is not only a flexible traffic line but also a medium to expand the space, which means liquidity and continuity of the space.

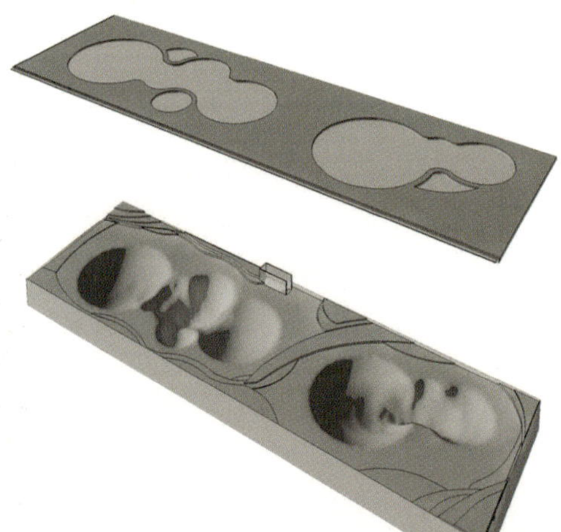

Floor plan

●	산들바람	산들산들 불어오는 바람 넓고 완만한 경사로 보드를 탈때 마치 산들바람이 불어오는 듯 한 느낌을 준다.
●	황소바람	좁은 골목으로 세차게 부는 바람 넓은 -> 좁은 공간으로 바람이 세게 몰아치듯이 출발점에서부터 안쪽으로 들어갈수록 살짝 좁아지다가, 곧 넓은 구조가 나타난다.
●	용수바람	작은 회오리바람 황소바람과는 조금 다르게 공간 넓이의 변화는 없으며 동그랗고 좁은듯 한 모양이 작은 회오리 바람을 연상시킨다.
●	왜 바 람	이리저리 방향 없이 부는 바람 중간 크기의 구조 두개가 합쳐있어 여러 방향으로 자유롭게 즐길 수 있다.

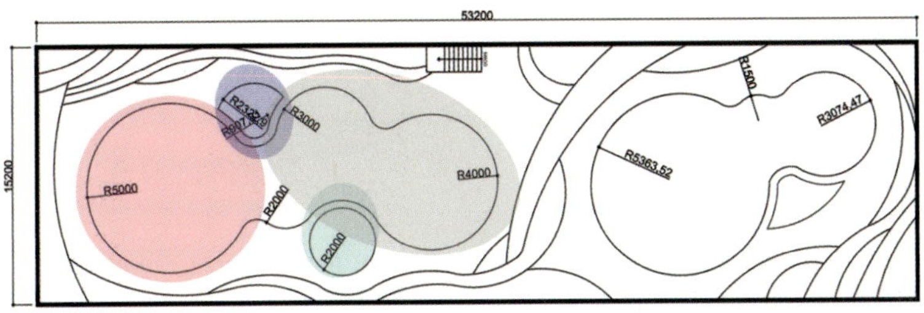

2nd FLOOR

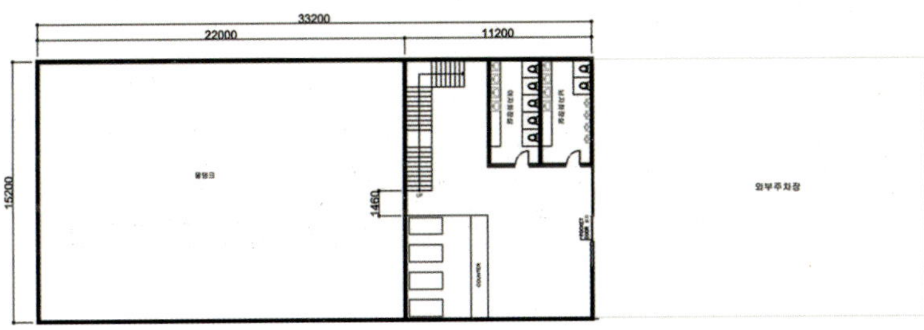

1st FLOOR

Breeze
Slight air
When skateboarding on a wide and gentle slope, it feels like a light breeze is blowing.

Big blow
A strong wind on a narrow alley
As a heavy wind is blowing from wide to narrow space, it is getting a bit of narrowing towards the inside from the entrance. Then a wide structure appears soon.

Whirlwind
A small dust devil
Unlike big blow, there is no difference of space size. The round and narrow shape reminds of a small dust devil.

Fickle wind
Changeable wind
As two medium structures are combined, it can be enjoyed in many directions.

장혜선 이유리 | Jang, hye-sun Lee, yu-ri

Folding Space

Music Hall & Studio

Background

Play ! Join ! Show ! 적극적으로 참여하여 체험하고 즐길 수 있는 공간.
The place where can joy and participate actively

개방되고 열린 공간 속에서 인간들을 적극적으로 참여하게 만드는 21세기의 새로운 패러다임에 비추어 수동적 형태를 벗어나 관람자가 주인이 되어 적극적으로 참여하는 능동적 개념으로의 전환이라는 명재를 가지고 이 프로젝트는 시작된다. 용도상으로는 불명확한 공간의 발생을 억제하고 사용자에 의해 자유롭게 조합되고 구성됨을 의도한다.

Reflect the 21century's new paradigm which make people participate actively in open space, this project will be begin with the subject that converting to the active concept from passive form. In terms of use, control the causing ambiguous space and be organized unrestrictedly by user.

Site analysis

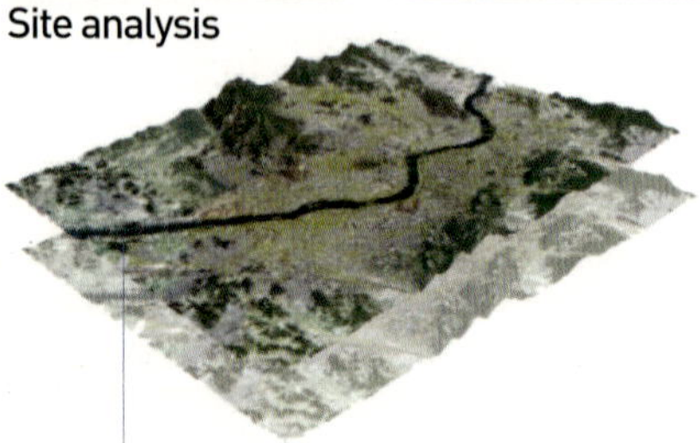

용산구 서부 이촌2동 대림아파트 주차장
2-dong Ichon, Western Daelim apartment parking lot

Population	Area	Traffic	Housing	Use
10,518명	740㎡	지하철 용산역	3556동	1층:주차장 2층:테니스장

용산국제개발 무산으로 분위기가 다소 침체되어있고 낙후된 시설로 인해 녹지 환경이 부족하고 주거이주, 노인층이 대부분이다.

Because of the failure of Yongsan international development, the atmosphere is depressed. Also, green area is insufficient due to the old facilities and there are a lot of residential migrations and the elderly.

NEEDS

OPEN CLOSE

녹지환경 Green environment
음료자판기 Beverage vending machines
ATM
편의시설 Amenities
감성공간 Emotional space

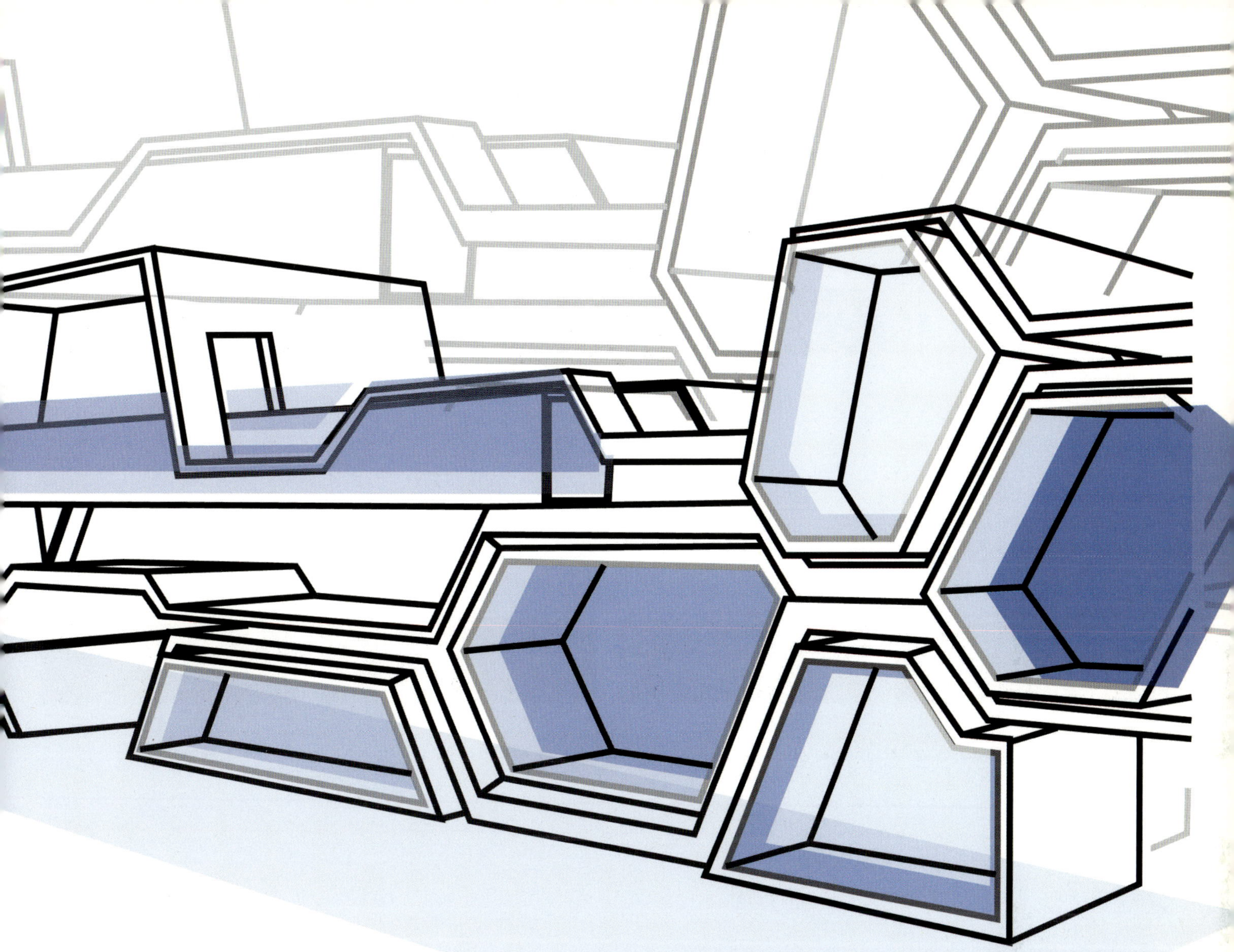

Concept FOLD 음악의 선율에 따른공간. The space of the music melody

Ⅰ. 두 겹으로 겹쳐지거나 구부러진 현상
Ⅱ. 종이가 반으로 접혀졌을 때 접혀진 부분.

I. Condition which is covered with two layers or bended
II .Folded part when a paper is folded by half

FOLD

변환성
Transform

능동적
Active

놀이성
Play

누구나
Anyone

Design direction

일상 속에서 무심히 놓쳐버린 음악적 영감들을 되살릴 수 있는 공간. 자기 안에 잠재되어 있는 음악성을 최대한 끄집어 낼 수 있는 공간. 담담한 희망을 갖고 들어설 수 있는 뮤직 스튜디오를 목표로 한다.

We are trying to set up music studio where we can revive musical inspiration, bring out potential musicality, and enter with small wish.

Design key word

- Be folded
- Freely, unrestrictedly
- Melody of Music

Design motive

소멸의 재탄생
rebirth of extinction

악보의 흐름
flow of musical note

한강 고수부지 공연장
performing place in Han river park

침묵이 없는 공간
the place without silence

Process

Concept design

Step 1_folding space

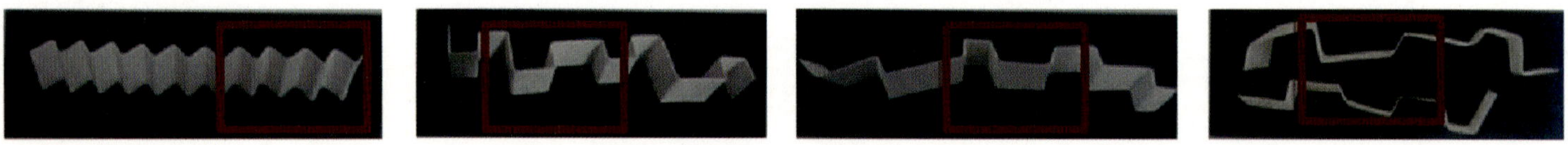

편편한 종이는 그 자체로도 서 있을 수 없다. 그러나 약간의 접힘이 있을 때 자체의 힘에 견줄 수 없는 견고함을 지닌 채 공간을 만들기 시작한다. 종이를 규칙적 혹은 불규칙적으로도 접어보는 FOLD의 공간형성방법으로 제안한다.

Flat pater cannot stand itself. However, when it has a little fold, it starts to make space with stability cannot compare with its power. I suggest FOLD space formation method which is folding a paper regularly or irregularly.

Step 2_ 재구성과 변형 Reconstruction and transformation

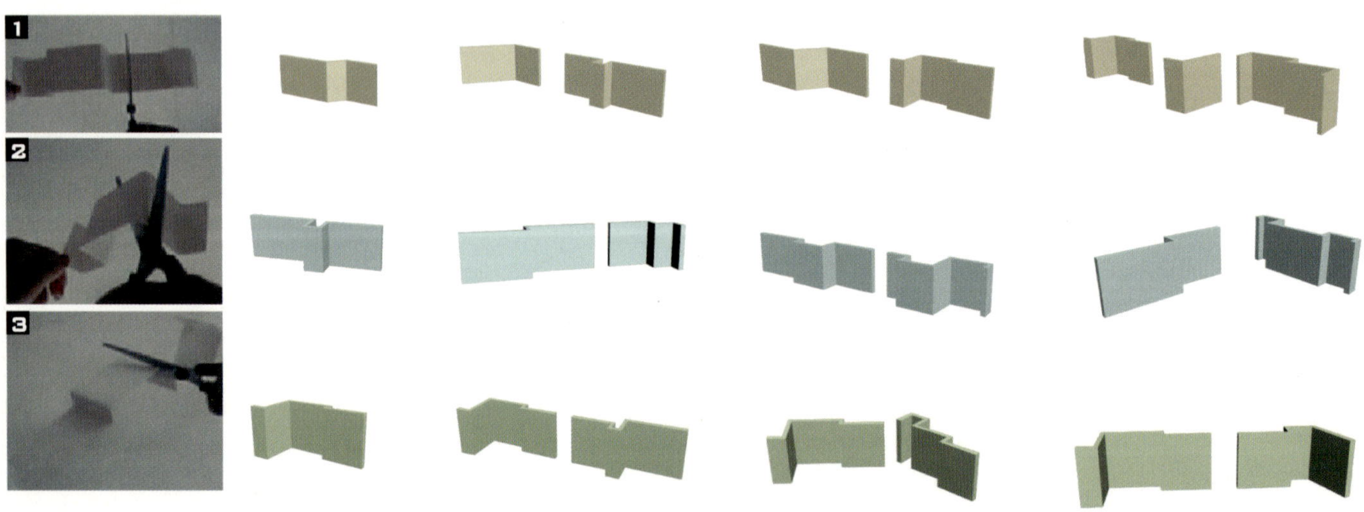

활동 및 흐름을 위한 가장 뛰어난 확장성과 이동성을 제시하기 위해 frame은 필요에 따라 다양한 형태로 조합 변형되어 리듬있는 외관을 형성한다.
Frame forms rhythmical shape by combining or transforming as necessary due to suggest perfect expandability and mobility for activity and flow.

Step 3_ 접힌 벽체와 인간의 대면 Walls collapsed and the human face

벽체의 접힘을 통한 다양한 변화는 인간으로 하여금 자연스러운 동선을 유도한다. 접힌 벽체는 공간을 형성한다.
Various changes through the folding of wall induce people to natural flow of human traffic. Folded wall makes space.

Space design

Step 1_ 음표의 흐름 Note the flow

악보의 음표와 음표의 연결을 통해 공간레벨을 구획.

The notes and the notes of the score level through the connection compartment space.

Step 1_ 선율의 기승전결 The level of melody

1악장 INTRO

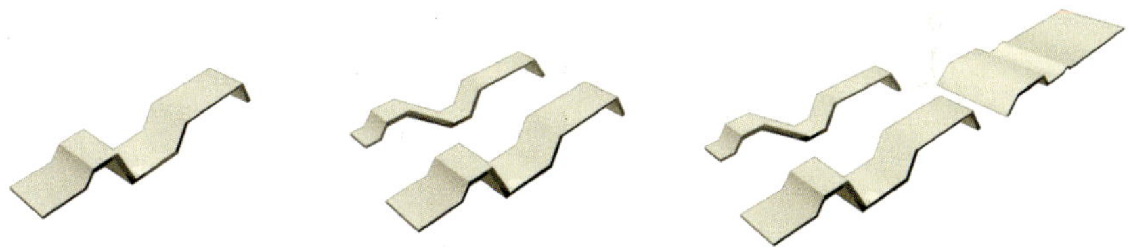

선율의 도입 – 잔잔함, 감미로운, 조용한, 소근거림
The introduction of the melody - still, sweet, quiet, in whispers

2악장 BRIDGE

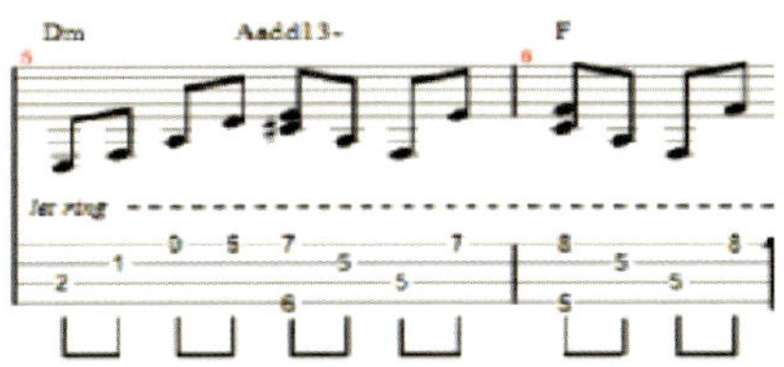

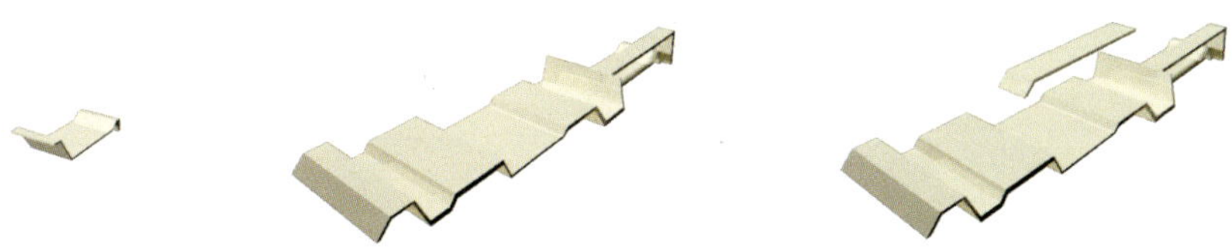

선율의 진보 – 인트로와 클라이맥스 연결부,기대감 상승, 도약부, 브릿지
Advances in tune - intro and climax connections, expectations rise, hopper, Bridge

3악장 CHORUS

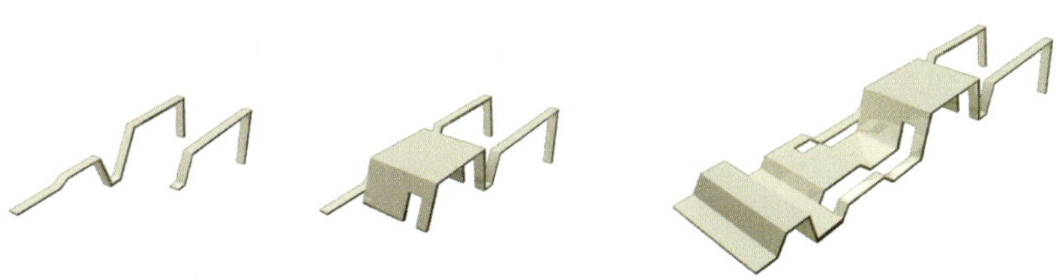

선율의 클라이맥스 – 경쾌함, 요동침, 힘찬리듬, 소프라노, 떨림
The climax of the melody - the lightness, John lay, energetic rhythms, soprano, tremors

Plan

1F

1. Piano studio
2. Band studio
3. The seats
4. Stage

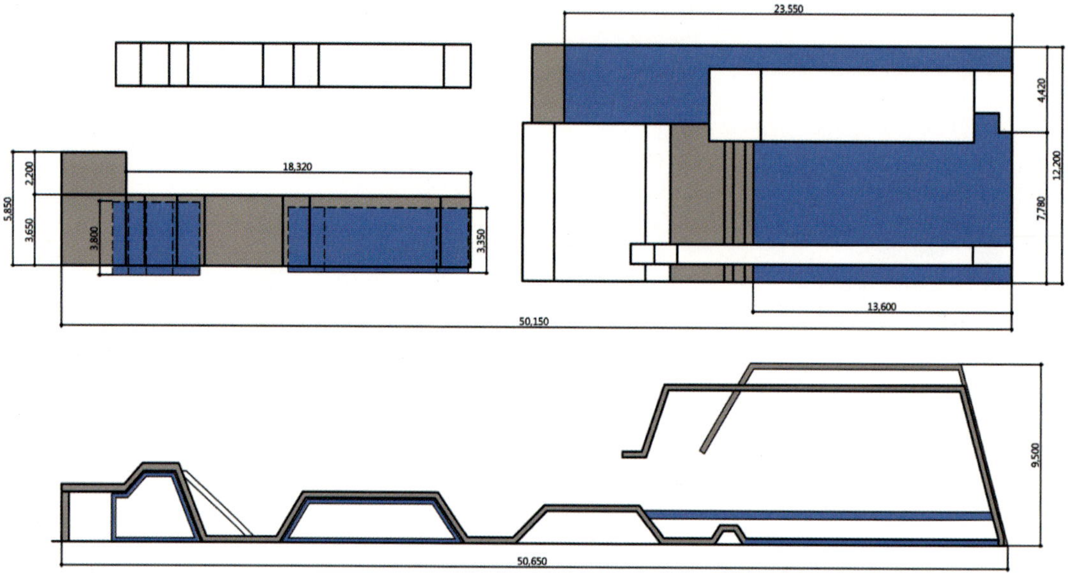

2F

5. Piano studio
6. Guitar studio
7. Music zoon
8. Stage

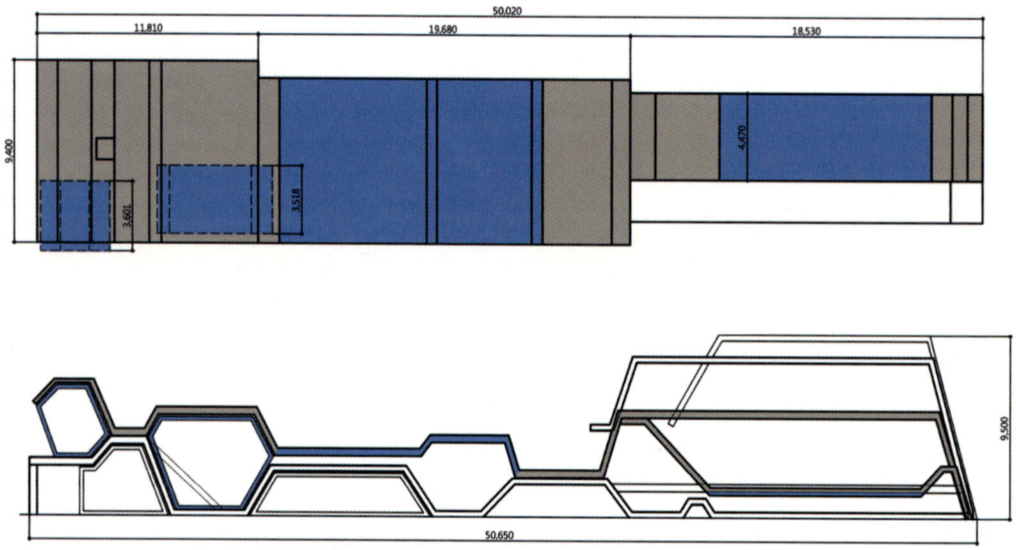

3F

9. Music studio
10. The seats bridge

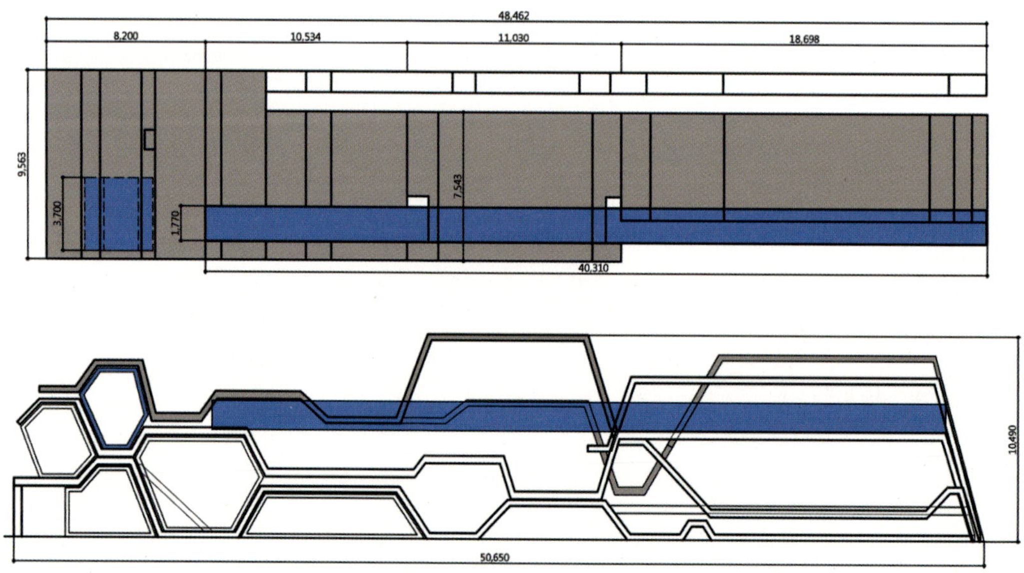

Elevation

음악 선율의 흐름에 따른 형태로 공간이 나뉘어져 있는 Music Hall & Studio.

Music to the flow of the melody iAs divided into space in the form of Music Hall & Studio.

개방되고 열린 공간 속에서 연주자와 관람객 모두가 주인공이 되며 이들의 적극적인 참여를 유도한다.

In an open and open space hero of all performers and audience, and leads to their active participation.

오픈 스페이스의 공연장에서는 관람객과 연주자, 사용자들에 의해 공간이 자유롭게 구성되어지며 변화하게 된다.

The venue, the audience and the performers open space, free space, configured by the user and is being changed.

이웃들과 함께 자기 안의 잠재되어 있는 음악성을 마음껏 펼칠 수 있는 뮤직 스튜디오.

Potential in the neighborhood of his musicality with fully expandable music studio.

폴딩된 형태에서 자연스러운 동선을 유도하게 된다.

Fold induction in the form of the natural copper.

각각의 공간 속에서 이들은 자유롭게 보고, 듣고, 즐기며 희망을 품게 된다.

Freely in space, they each see, hear, and enjoy the wish to bear.

interesting links

Site 서울특별시 용산구 이촌 2동 주민센터 옆 대림아파트 주차 + 테니스장

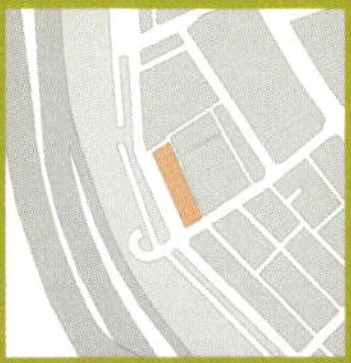

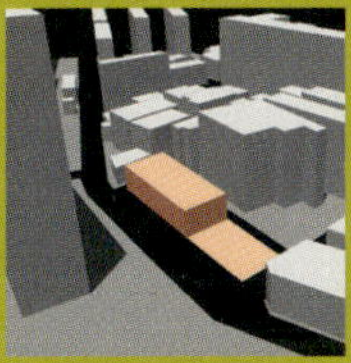

가로로 긴 직사각형 형태의 대지이며, 사면에 도로가 있고 테니스장과 주차장 · 물탱크의 용도로 사용되고 있다.
바로 옆에 이촌 2동 주민센터가 있고 근처에 지하철 1호선 용산 역이 있어 철도 교통이 편리하지만 유동인구는 많지 않다.
주변에 어린이 놀이공간이나 노인복지센터 등 휴식 공간이 부족하고 안전시설 역시 부족하며 주택간의 좁은 거리로 인해 소음이나 프라이버시 문제, 채광 등의 문제가 있다. 또한 낙후된 건물들과 건물 외벽의 고압선 정리가 미흡한 곳이 많으며 개발사업의 찬반으로 인한 내부 갈등이 심하고 물질적 또는 정신적으로 지치고 피폐해진 주민들의 분위기가 어둡고 무겁다.

Horizontally long rectangular shape site and the slope side is road, being used for Tennis courts, parking lot and water tank. Right next is Ichon 2 dong community service center, Metro line 1 Yongsan station is nearby located so that reason railway transportation is good but there ara few floating population.
There is lack of resting space and security facilities as like Kid playing place, elder welfare center also has some noise problem, privacy issue, natural light problem because the narrowness of between houses.
Most of them are outdated buildings, outside power cable need to be cleanup, and internal conflict is severe due to pros and cons for development, the resident are physically or mentally tired and the overall atmosphere is dark and heavy.

Background

'용산개발' 사실상 무산

민간출자사 29곳중 11곳 반대
코레일 주도 사업추진 못해

서부이촌동은 대한민국 건국 이래 가장 큰 규모의 개발 사업에 포함되었다. 하지만 주민들의 경제권이 묶이면서 개발의 더딘 진행과 죽어가는 상권들로 인해 찬반논란이 커다란 갈등으로 자리 잡았고, 생활은 더욱 어려워졌다. 이것은 곧 지역의 침체와 무거운 분위기로 이어졌고, 개발이 무기한 연장 · 중단 위기에 처하면서 상황은 더욱 악화되었다.

Seobuichon dong had been included in largest development project since foundation of Korea. but tardy development and declining commerce makes sluggish economy of this area. residents conflict is ongoing and the life is getting harder. this led to heavy and depressing atmosphere. development is indefinitely suspended or extended and the situation is exacerbated.

SWOT analysis

STRENGTH

- 한강으로의 접근이 빠르고 쉬운 편이다
- 서울의 중심부에 위치(용산구)
- 한강대교와 원효대교를 통해 강남으로의 접근이 편리
- 용산 역과 가까이에 위치, 철도교통망 발달

-Quick and Easy access to the Han river
-Located in center of of Seoul(Yongsan-gu)
-Easy access to Gangnam through wonhyo bridge or hangang bridge.
-Located near by Yongsan Station, good railway transportation

WEAKNESS

- 부족한 주차공간, 좁은 도로에 의한 위험성
- 주민들의 복지/문화공간 부족
- 고층의 판상형 아파트로 인한 한강경관 훼손
- 밀집 되어있는 주택들로 인한 채광 부족
- 침체되어있는 지역 분위기
- 약해진 상권

-Short of parking space, risk from narrow road
-Short of residents welfare/culture space
-River view interruption due to high-rise building
-Dense house area hinder of natural light
-Depressing area atmosphere
-Declining commercial zone

OPPORTUNITY

- 낙후된 대지의 효율적인 활용
- 지역부가가치 상승
- 주민들의 문화, 경제적 활성화
- 개발 기대 지역

-Efficient use of underdeveloped land
-Elevating local value
-Activate the residents culture/economy
-Expecting regional development

THREAT

- 주민센터와 겹칠 수 있다
- 주민들의 경제 격차 심화
- 교통망 확충 계획의 차질
- 개발 실행의 연기 또는 중단

-Overlapping possibility with the community center
-Deepening the residents economic disparity
-Delay of transportation expansion plan
-Stop or delay of development

Keyword | motive

잇다, 연결하다.

개발사업이 무기한 연기 · 지연됨에 따라 주민들의 찬반논쟁 등 갈등이 더 심해졌고, 외부와의 또는 주민들 서로간의 소통이 단절되면서 점점 더 삭막하고 냉담한 분위기가 조성되었다. 이런 상황에서 주민들에게 활력을 줄 수 있는 공간으로 이웃간의 소통을 이루어 줄 수 있는, 연결고리의 역할을 하는 공간이 필요하다고 생각했다.

Delayed development project makes severe residents conflict, communication disconnection between residents, apathetic and heartless atmosphere. In this situation, we thought that there need some vitality giving space which can communicate and rest between neighbors.

interesting links

서부이촌동 주민들의 다양한 취미 · 흥미 · 취향 등 다양성을 표현하고 공유하며, 화합시키는 RE-ENERGISING SPACE

West dong Yongsan International Business District development project is to negate Due to traces of an development project is to negate Due to traces of an

Concept | interesting links

새로운 공간으로의 LINKS

공간에서 또 다른 공간으로의 이동, 예상하지 못했던 방향으로의 이동 등을 통해서 이전과 다른 새로운 공간으로의 연결을 제공한다.

LINKS OF NEW SPACE

provide the new born space through movement space to other space or move to unexpected direction.

과정이 흥미로운 LINKS

단순히 걷고 뛰는 이동에서 공간 자체를 움직여 이동하는 과정을 보다 흥미롭게 하고, 시각적으로 즐거움을 제공할 수 있으며 공간 안에서 일어나는 과정들로 흥미를 느낄 수 있게 한다

THIS EXCITING COURSE LINKS

Simply moving by walk and run to moving the space itself, makes more interesting process and give visual pleasure, can feel more exciting from the process.

자유로운 LINKS

연결에서 선택이라는 권한을 부여함으로써 정해진 길이 아닌 선택적인 연결을 가능하게 하고, 지친 서부이촌동 주민들에게 여유와 활력을 되찾을 수 있는 공간을 제공한다.

FREE LINKS

Simply moving by walk and run to moving the space itself, makes more interesting process and give visual pleasure, can feel more exciting from the process.

Space program | interesting links

실내공원 Indoor park

날씨나 외부조건에 영향을 받지 않는 실내공원 + 옥상정원

Indoor park which is unaffected by weather or out-side condition+ Roof garden

전시공간 Exhibition area

테마에 따라 다양한 주제로 작품을 전시할 수 있는 공간

Wide range of topics displaying space on the theme.

놀이공간 playground

아이들을 위한 안전하고 보다 흥미로운 놀이 공간

Safe and exciting play place for children.

Disign process | links

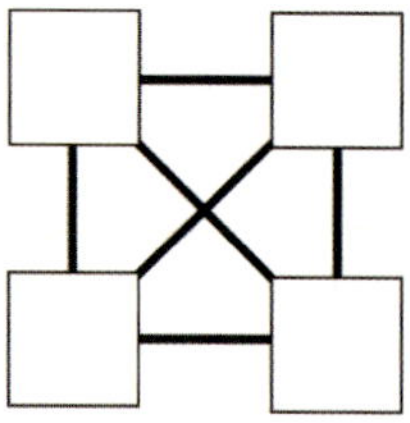

수평적인 연결
가로 · 세로 · 대각선 등
평면 안에서 일어나는 모든 방향의 이동

Horizontal connection
all direction movement of Horizontal, vertical, diagonal

수직적인 연결
높낮이가 서로 다른 입면 안에서의 이동

Vertical connection
moving af different elevation

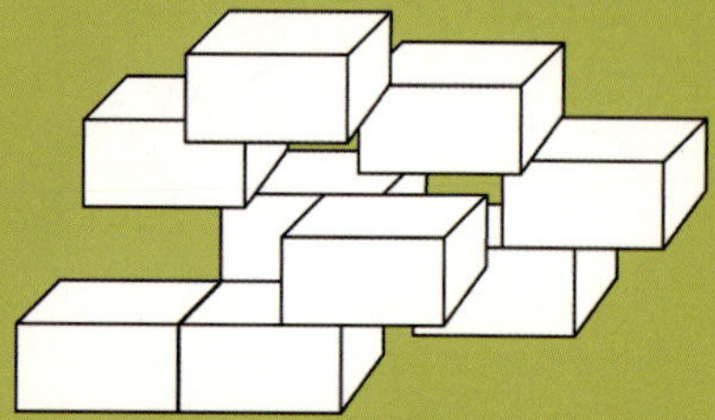

복합적인 연결
방향과 높낮이가 자유롭고 다양한
여러가지의 입체적인 구조

Multiple connections
direction or height free structure, diverse dimensional structure

Mess process | interesting links

외벽의 이동을 이용한 공간의 변화 (폐쇄적▷개방적/합침 · 분리)
Spatial change from outer wall movement(closed→open /combining,separating)

다양한 방향으로의 이동 (의외성)
Moving in different directions (unexpected)

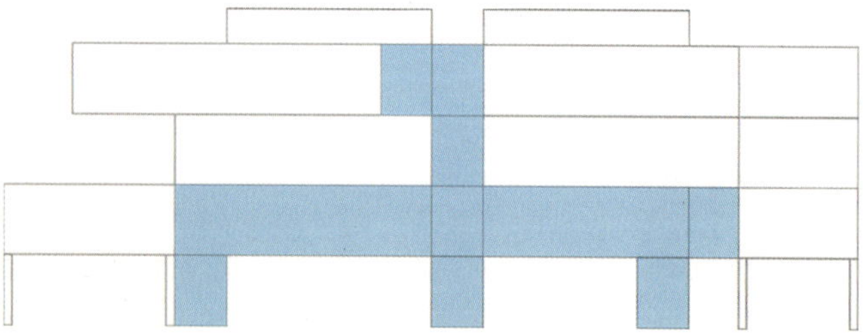

유리로 이루어진 공간을 통해 단순한 미로 형태를 구성하여 질리지 않게 함
Help not feel boredom constructing a glass formed simple maze space

움직임을 감지하는 센서 벽을 이용해 이동하는 모습을 볼 수 있게 함
See the moving figure using a motion detection sensor wall

직육면체의 공간을 움직여 자유로운 이동과 선택적인 연결을 가능하게 함
moving the cuboid space enables free movement and optional connection

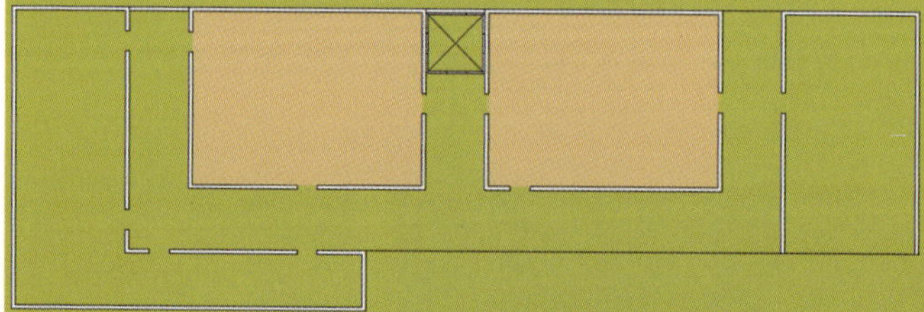

지친 서부이촌동 주민들을 위해 하늘을 올려다 볼 수 있는 휴식 공간 제공
Provide a relaxing space to see the sky for the weary seobuichon dong residents.

Floor plan

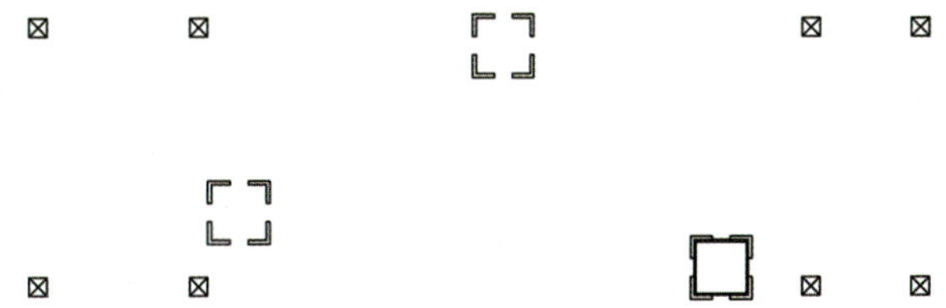

건물의 1층은 공간들 사이를 자유롭게 이동하기위한 탑승장과 기둥만으로 이루어져있다. 기둥 사이의 공간들은 주차공간의 용도로 쓰일 수 있으며, 유리로 된 셔틀을 통해 건물의 안을 들여다 볼 수 있다.

First floor consists of freely movable boarding space and poles. the space between poles can use parking area, can see the building inside through glass formed shuttle.

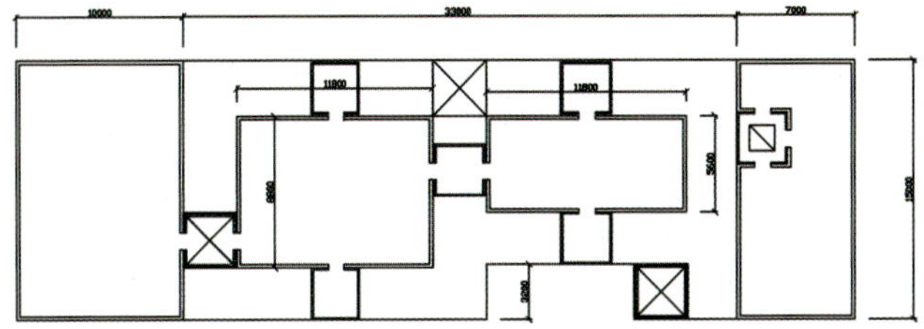

건물의 2층은 양쪽 끝에 직사각형 형태의 공간이 있으며, 1층에서 올라온 박스가 미로처럼 만들어진 셔틀을 돌아다니면서 자유로운 이동과 선택적인 연결을 할 수 있게 한다.

At the both ends of the second floor has rectangular shape space, the box from the first floor go around the shuttle as like maze and help the free movement and optional connection.

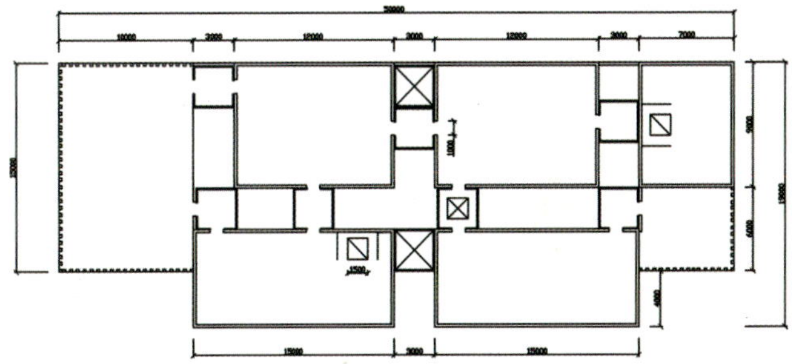

건물의 3층은 도로 쪽을 향해 튀어나온 켄틀레버 구조의 공간과 실내 공원의 개념을 가진 공간으로 되어있다. 도로 쪽으로 만들어진 켄틀레버 구조의 공간은 공중에 떠 있는 기분을 줌과 동시에 시각적으로 흥미로움을 느끼게 한다. 또한, 3층은 위로 높게 이어지는 실내 공원이 시작되는 층이다. 심적으로 지치고 힘든 주민들이 잠시 하늘을 보며 여유를 갖고 편안함을 느낄 수 있게 구성 될 것이다.

The third floor of building has protruding the cantilever structure space on the road side and inside park concept space. cantilever structure space on the road side help floating mood in the air also give the visual interesting. And the third floor is starting point of upside leading inside park. Mentally tired residents can feel the relaxed time with seeing the sky.

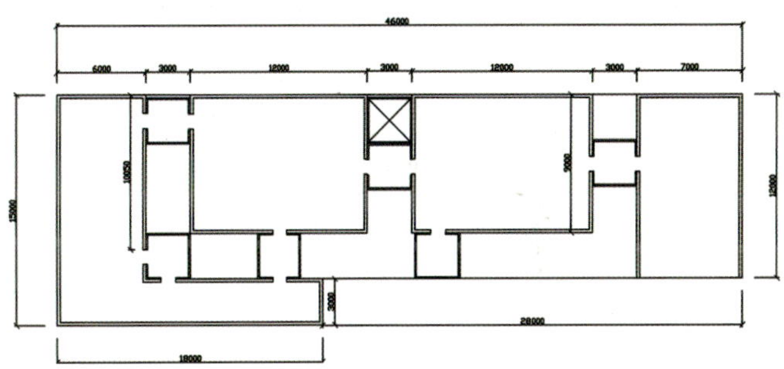

건물의 4층은 3층부터 이어지는 실내공원과 L자 모형의 공간으로 이루어져있다. 2 · 3층의 옥상 정원을 볼 수 있는 위치이며, 공간의 이동이 수직적으로 최고 높이에 이르는 공간이다.

The forth floor of the building consists of inside park led from the third floor and L shape space. where can be seen the 2·3 floor roof garden and vertically maximum height space.

김홍미 한아름 | Kim, hong-mee Han, ah-reum

Active Cycle

Background

용산 역세권 및 한강로의 배후지로서 개발 기대지역으로, 노후 서민 아파트 등의 재건축 시행으로 고층아파트화 되는 지역으로 재개발되는 지역이다. 그속에서 재개발과 개발무산등 부정적인 사건과 이슈들로 지역 분위기는 삭막하다. 이러한 지역에서 이곳 주민들을 위한 공간 즉 자녀들과 함께 시간을 보낼 수 있는 공간, 아이들이 신나게 뛰어놀 수 있는 공간 등 소통이 가능한 공간은 협소하기만하다. 이 문제점을 해결할 수 있는 공간을 계획함으로써 지역의 분위기를 반전시키고자 한다.

As a subway influential area of Yongsan as well as hinterland of Hangangno, the area which was filled with deteriorated apartment houses for the low incomer is now expected to be reconstructed into skyscraper complex. But the problem we want to focus on is dreary local atmosphere due to negative incidents and issues such as redevelopment and abandonment.

It is pretty natural that lack of place for the residents to spend an hour with their children and for kids to play around is unavoidable. In order to solve these problems, we arranged some space to freshen up the local atmosphere.

Energy playscape

Production space

– 운동에너지가 전기에너지로 바뀌는 공간.
아이들이 뛰어놀면서 전기를 만들어내는 공간이다.

The place where conversion between kinetic energy and potential energy occurs.
Power production occurs while kids play around.

Consumption space

– 생산된 에너지를 소비하는 공간으로써 휴식이 이루어지는 공간이다. 이 때는 에너지놀이터의 기능뿐만 아니라 공원의 기능이 추가 된다.

Consume the power mentioned-above, thereby Relaxation is performed. In other words, it is functioned as an Energy Playground as well as a Park.

Site analysis

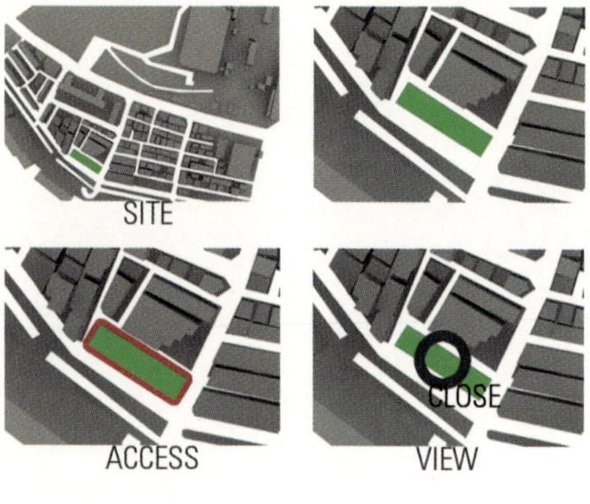

기존 테니스장으로 운영되던 이 대지는 주변 주차공간 부족으로 1층은 주차장으로 이용되고 있다. 대지로 접근성이 용이하지만 주변에 고층아파트와 주택단지들로 인해 경관이 좋지 않다.

The place where used as a tennis court before, its first floor has been used for a parking lot due to lack of parking places. Accessibility to the ground is fine but its landscape is disturbed due to skyscrapers and a housing complex.

Design concept

순환 자전거 사이클 회전 오토바이 주기

CIRCULATION BICYCLE CYCLE ROTATION MOTORCYCLE PERIOD

쳇바퀴를 모티브로 하였다. 사람들이 쳇바퀴에서 달리면서 에너지를 생산하고, 생산된 에너지로 빛을 만들어 공간을 비추고, 빛이 생김으로서 소통하는 공간이 생기게 된다.
이러한 형태의 반복적인 순환은 이 공간을 삭막한 분위기를 반전시켜 역동적이고 활기찬 공간으로 바꿔준다.

The motive of this system is a treadmill. People produce energy by running in a treadmill, and brighten the space with the produced energy. Finally lightened place leads to the area where people can communicate with each other. This type of repetitive circulation can change dreary atmosphere into vibrant air.

User analysis

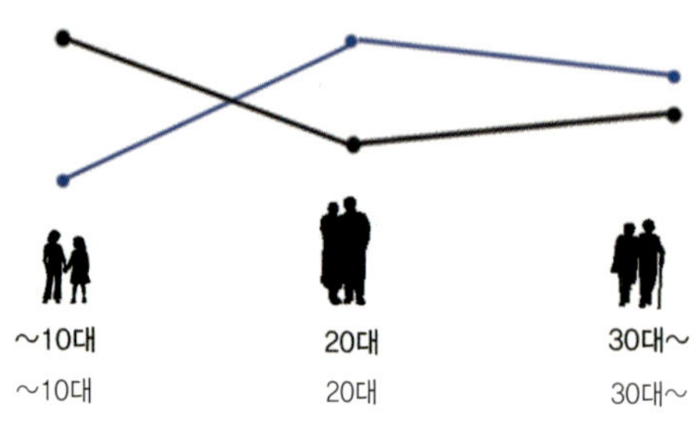

시간에 따른 이용변화
Use change over time

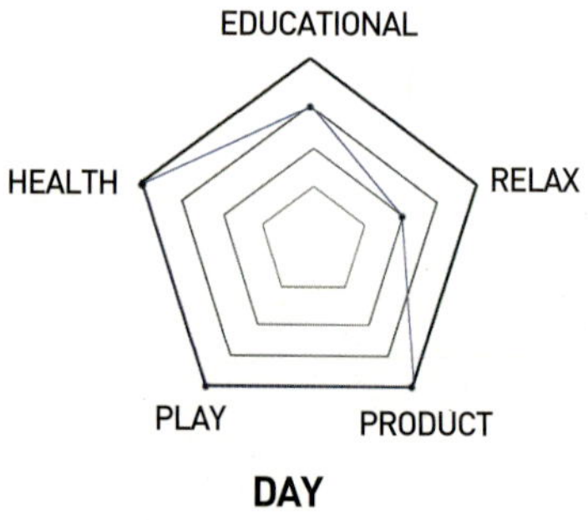

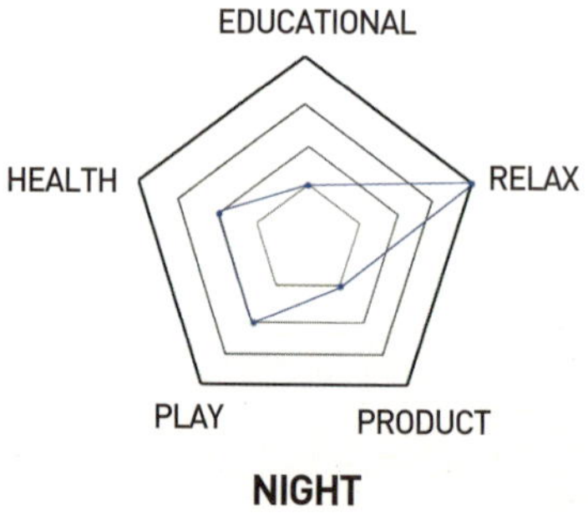

Process

공간에 CYCLE이라는 컨셉으로 활동적이고 역동적이며, 소통이 이루어지는 공간을 계획하려고 한다.

We are modeling a dynamic and vigorous space where human interaction is accompanied under the name of 'CYCLE of SPACE'.

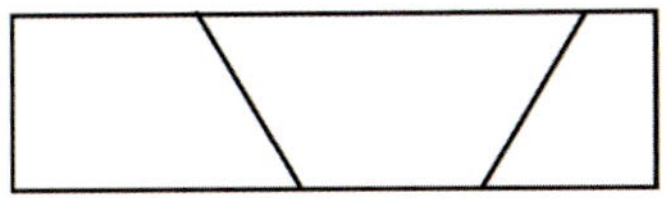

사선을 이용해 공간을 분할한다.

Divide up the space by using diagonal line.

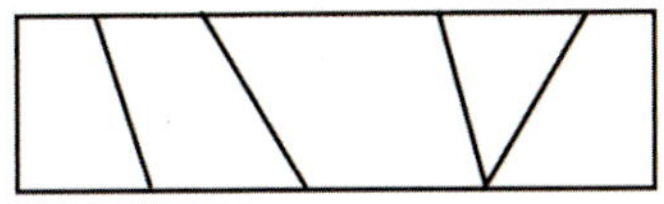

분할된 공간에 사선을 추가하여 경사를 두어 공간을 역동적으로 구성한다.

Draw a diagonal line again on the divided spaces so that a slope would make it more dynamic.

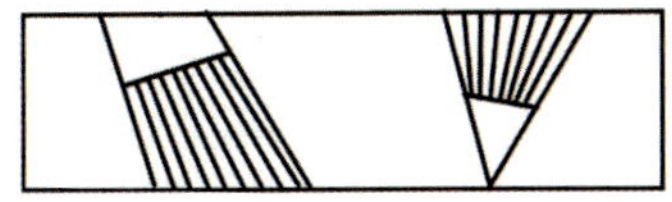

두개의 경사면 각각에 계단을 두어 좀 더 역동적이고 다양한 활동이 이루어질 수 있게 구성한다.

Place some stairs to the two inclined planes to make it more active, able to perform various activities.

Exonometric view

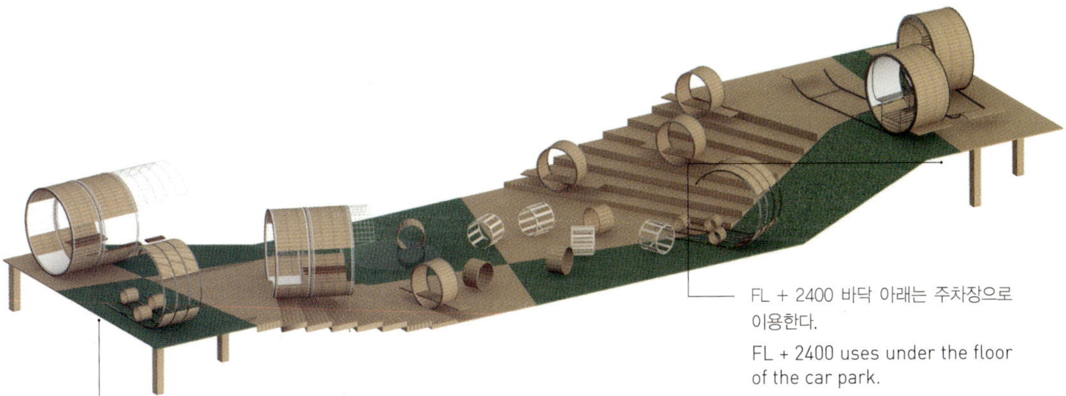

FL + 1600 바닥 아래로 아이들이 자유롭게 출입이 가능하며 이 아래에서 굴다리를 통해 내부로 진입이 가능하다.

FL + 1600 kids down to the floor below is freely available and accessible through the inside passage to enter the cylinder can be.

FL + 2400 바닥 아래는 주차장으로 이용한다.

FL + 2400 uses under the floor of the car park.

Section

Floor plan

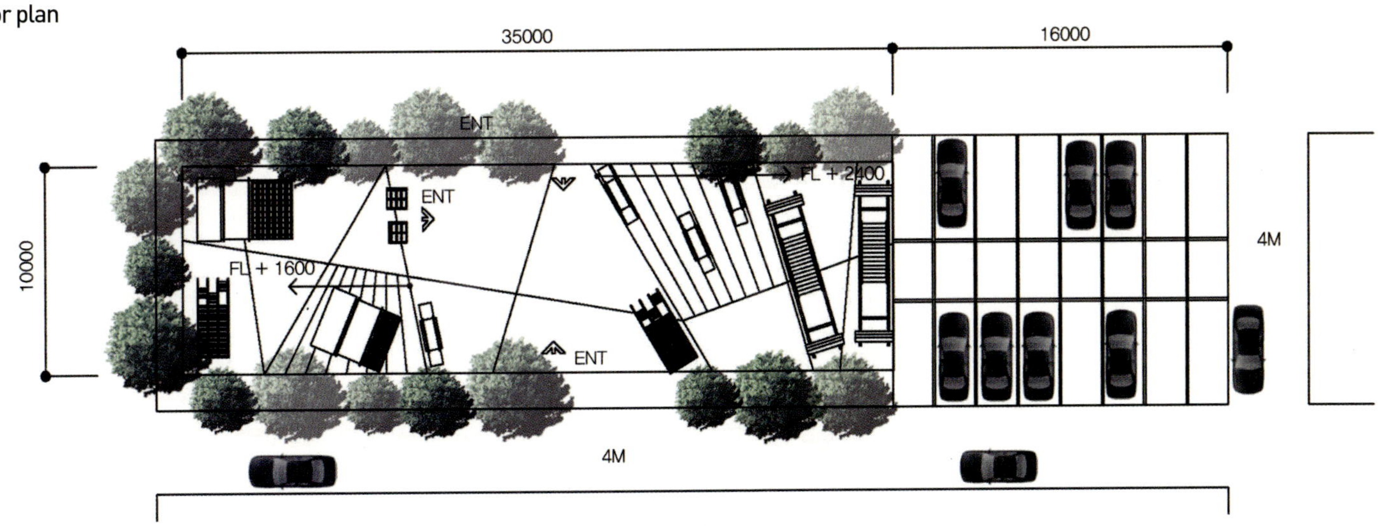

Unit detail

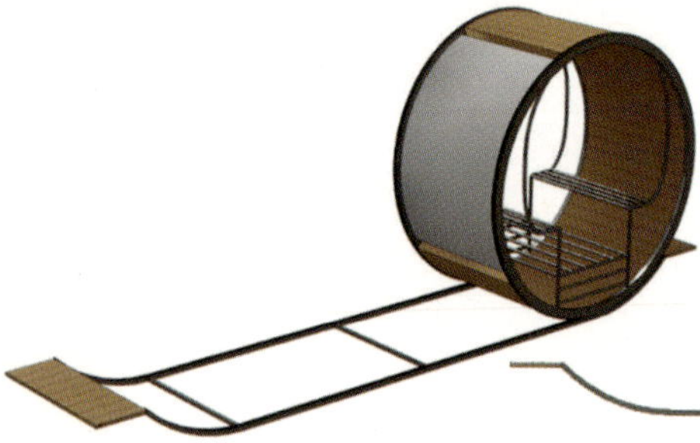

TYPE A

휴식+놀이+전기생산
Rest + play + electricity production

사람이 레일 위에 있는 원통안에 들어가서 자유롭게 굴림으로써 전기에너지를 생산해낸다. 원통을 돌리면서 내부의 형태가 바뀌게 됨으로써 다양한 활동을 하게끔 한다. 또한 레일 끝은 벤치의 기능도 하여 전기생산뿐만 아니라 휴식의 기능도 있다.

A person freely run in the treadmill which is located on the rail so as to produce electric energy. While running in the cylinder, changing of its shape make it possible to perform various activities. In addition the edge of the rail functions as not only a device producing electricity but also a bench offering a rest.

TYPE B

휴식+놀이+전기생산
Rest + play + electricity production

벤치의 기능을 하는 이 개체는 고정되어 있는 의자가 아닌 내부 레일에서 자유롭게 움직일 수 있다. 다양한 형태의 벤치로 만들 수 있다. 또한 굴릴 때 전기가 만들어 진다.

This part which can function as a bench is not a stationary chair. It can move freely on the inner rail. It can be changed to many forms of benches and also it makes electricity while rolling over.

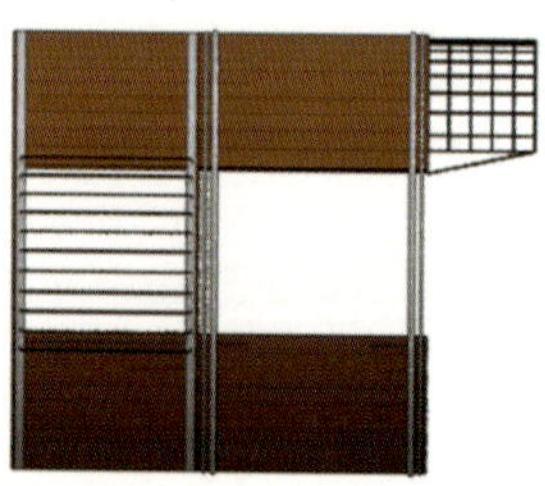

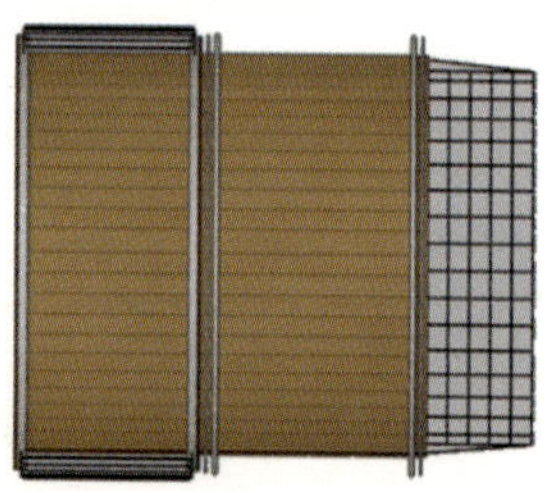

TYPE C

휴식+운동+전기생산
Rest + exercise + electricity production

레일형 원통 + 벤치형 원통

두개의 개체가 합쳐져 있는 형태로 레일에서 사람이 달리기 운동을 통해서 전기에너지를 생산해 내고, 벤치형 원통안에서 사람은 쉴 수 있다. 운동과 휴식이 함께 이루어지는 개체이다. 레일에서 달리는 사람과 휴식하는 사람 사이에서는 소통이 이루어진다.

A rail tpye cylinder + A bench type cylinder

This is an object which is put two part together. It generate electricity when a person run on the rail and the other person can take a rest in the bench type cylinder. These two operation is accomplished at one time. It means that a person who runs and the other one who take a rest can communicate with each other.

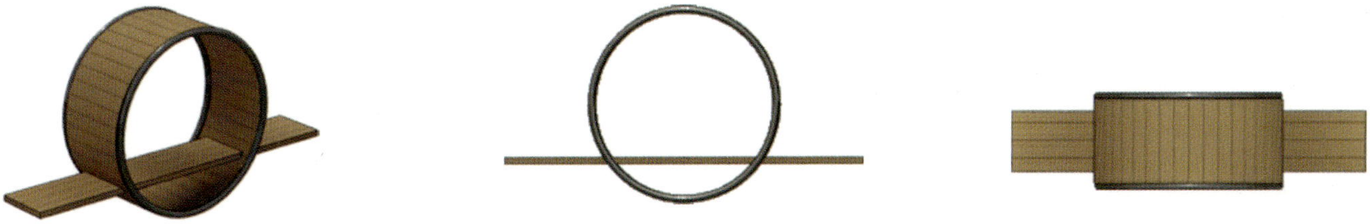

TYPE D

휴식
Rest

휴식이 이루어지는 벤치형 원통이다.

A bench type cylinder which offers a place to take a rest.

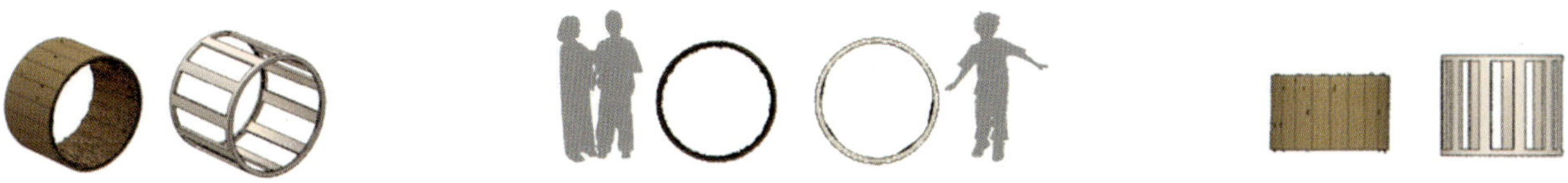

TYPE E

휴식 + 놀이
Rest + Play

아이들이 공간의 제약없이 자유롭게 굴릴 수 있는 원통이다. 또 한 굴리는 기능 말고도 고정시켜 벤치의 역할도 한다.

This is a cylinder which kids can roll over freely without any restrictions. Moreover it functions as a stationary chair.

Day view

물이 있는 이 공간은 사람들이 쳇바퀴를 통해 만든 에너지로 수증기가 형성되어 새로운 공간이 생긴다. 이 공간은 아이들이 자유롭게 작은 쳇바퀴를 굴리고 놀 수 있어야해서 열린 공간이다. 공간의 제약없이 자유롭게 새로운 공간을 만들어 낼 수 있다.

The area which contains water create a new place where vapor is produced by the energy people make while running in the treadmill. This space needs to be opened so that children can play freely with little treadmill. It can create new place without any restriction.

굴다리를 통해 자유롭게 아이들이 왔다갔다 하면서 놀 수있다. 메인입구가 딱 정해진 것이 아니라 다양한 통로를 통해 진입이 가능하다. 아이들은 굴다리 안/밖에서 타고 놀 수 있는 기능이 있다.

Kids can play by coming and going through a land bridge freely. There is no main entrance, but it is able to enter through many paths. Kids can play around on the land bridge.

쳇바퀴 안에서 사람들이 달리는 행위를 함으로서 전기 에너지가 생산되어진다. 이때 달리는 사람과 벤치에 앉아있는 사람과의 소통이 이루어질 수 있다.

By performing an act of running in the treadmill, people can make electricity. At this point, a person who runs and the other one who take a rest can communicate with each other.

Night view

밤이 되면 사람들이 운동의 목적이 아닌 휴식의 목적으로 전 연령대가 즐길 수 있는 공원의 역할을 한다. 낮 동안 모아진 전기로 저녁이 되면 개체마다 빛이 들어와 주변을 밝게 비춘다. 빛으로 다른 공간감을 느낄 수 있다.

When it comes to night, people can enjoy the park not for the exercising purpose but for the purpose of relaxation. And each objects emit light to brighten surrounding area by using electricity stored during a daytime. The light produce other space senses.

노승희 임정연 | No,seung-hee Lim, jeong-yeon

소리항해
voyage of sound

Site analysis

서울특별시 용산구 이촌 2동 주민센터 옆 대림아파트 주차장 공간. 용산구 서부이촌동 일대는 용산국제업무지구 사업의 여파로 주민들의 갈등이 생기면서 지역 분위기가 침체되고 우울해졌고 주민들의 소통과 교류가 단절되었다. 지역적으로는 근처에 한강과 고속도로가 있으며 용산역과도 가까워 교통이 편리하다.

Parking area in Daelim Apartment complex next to the Office of Seoul-si, Yongsan-gu, Ichon2-dong. The local atmosphere of Seobuichon-dong, Yongsan-gu area became depressed as conflict between residents was created in the aftermath of the Yongsan International Zone project, and communication and exchange between residents were cut off. Regionally, the transportation is convenient as it is located near Han River and freeway as well as the Yongsan Station.

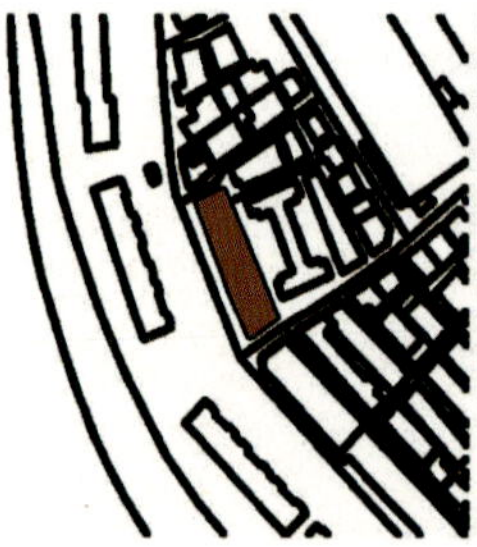

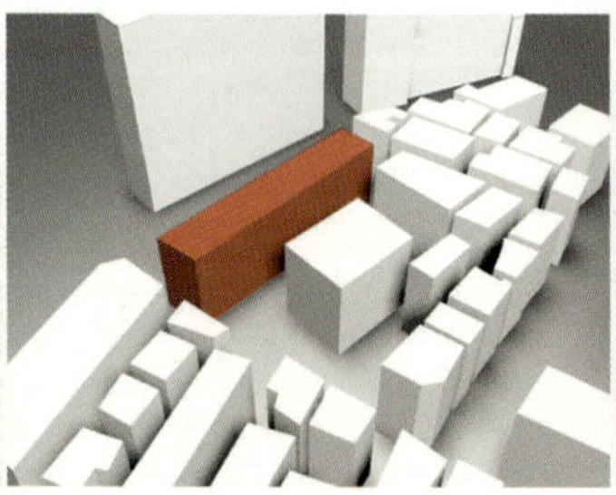

Background

이촌동에는 다른 동네와 다를 것 없이 사각형의 매스로 이루어진 천편일률적인 주택들이 빽빽하게 자리하고 있다. 그리고 근처에 한강이 있는데 이 점을 고려해 '배' 라는 공간을 생각했고 침체 된 지역에 주택들과 전혀 어울리지 않는 '배' 가 놓여진다면 어떨까 고민해보았다.

지루한 매스들 사이에 '배' 라는 이질적인 공간이 생기면서 사람들의 호기심과 모험심을 자극하는 흥미로운 공간을 제안할 수 있다는 생각을 했다. 침체되어 멈춰있는 배 안에서 활동적인 놀이를 하면서 활기넘치는 놀이공간을 계획하여 리에너자이징 프로젝트를 진행했다.

Just like other areas, monotonous housings composed with quadrangle mass are densely situated in Ichon-dong. Also, the Han River is nearby that we came up with a space, called 'ship,' with this point in view, and considered about how it would be like when a 'ship,'which is mismatched with housings in a depressed region, is placed there.

We thought it was a way to suggest an interesting space that stimulates curiosity and adventurous spirit of people by creating a disparate space, called 'ship,' in between tedious masses. We progressed the re-energizing project by planning a lively play space of playing active games inside an inactive stationary ship.

Proposal

서부이촌동 주민들이 자신들이 처한 상황과 비슷한 이야기 를 가진 공간에서 현재 처한 상황과 대비되는 즐거운 놀이 활동을 하면서 서로의 마음의 벽을 허물고 조금이나마 소α통할 수 있게 한다. 동시에 아이들과 함께 즐겨찾을 수 있는 공간이 되게 해서 가족들이 즐거운 시간을 보내도록 한다.

Residents of Seobuichon-dong will get to communicate somewhat by tearing down each other's barriers while playing fun activities which contrast with their current situations in space with a story similar to their situations. At the same time, it will be made into a favorite space for families to visit with their children for them to have fun and make precious memories.

Target

주요 타겟은 엄마, 아빠와 같이 놀러오는 유치원생부터 초등학생까지의 아이들이다.

The main target is children from kindergarten to elementary school who will come with their mother and father.

Story

#1. 한적하게 바다를 순항하고 있는 배.

A boat, quietly cruising across the ocean.

#2. 한 순간의 실수로 인해 배는 암초에 부딪히게 되고

The boat bumps into a reef with a moment's inattention and

#3. 암초에 부딪힌 배는 표류하게 되는데

the boat crashed in a reef got adrift

#4. 선원들은 보수파와 진보파로 갈라져 갈등을 빚는다.

that the crew come into conflict with each other by dividing into conservatives and progressives.

#5. 하지만 시간이 지날수록 상황은 더 열악해지고

However, the situation becomes poorer as time goes by

#6. 결국 선원들은 힘을 합쳐 생존방법을 모색한다.

and in the end, the sailors are working together to search for survival methods.

Concept

'난파된 배' 는 실패한 용산개발지구라는 부정적인 의미를 가지고 있다. 또는 두 동강 난 배는 현재 서부이촌동 주민들이 두 갈래로 갈라져 갈등을 빚고 있는 모습을 나타내기도 한다. 이러한 '난파된 배' 라는 부정적인 요소를 역으로 생각해 놀이 공간을 계획했다. '난파된 배' 안에서 선원들이 구조 요청하는 행위를 소리와 물을 이용한 놀이활동으로 승화시켜 공간을 구성했고 항해하는 배에서 물을 이용해 소리를 만들어내는 모습을 통해 '소리항해' 라는 컨셉을 잡았다.

The 'wrecked ship' has a negative meaning of a failed Yongsan development division. Or the ship broken into two represents an image of how residents of Seobuichon-dong are currently divided into two who came into conflict with each other. The play space was designed by considering such negative element as the 'wrecked ship' into a station.

The space was composed by sublimating a behavior of sailors requesting for help inside a 'wrecked ship' into a play activity using sound and water, and established a concept of the 'sound voyage' through images of creating the sound using water inside a cruising ship.

Plan

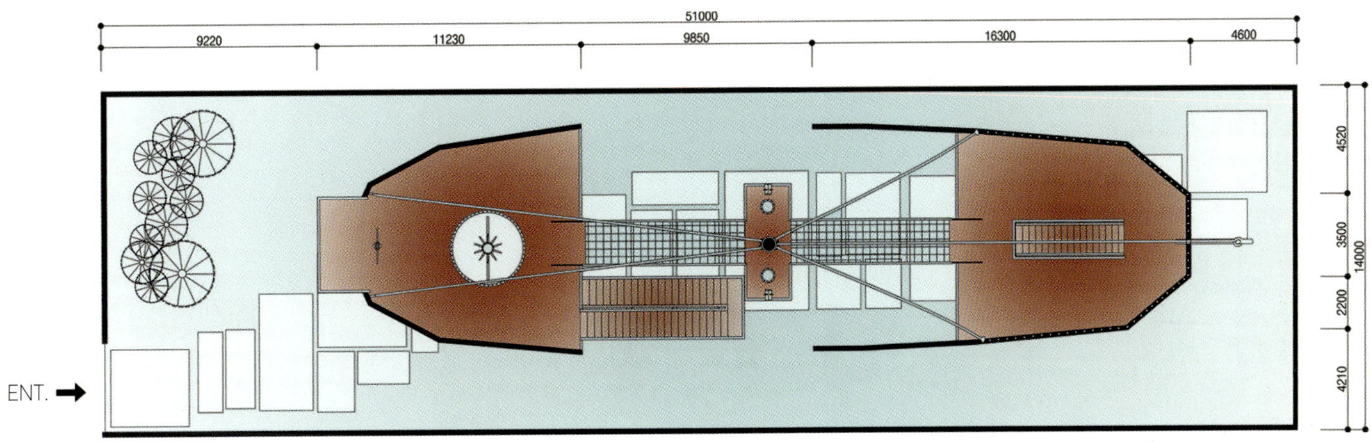

Elevation

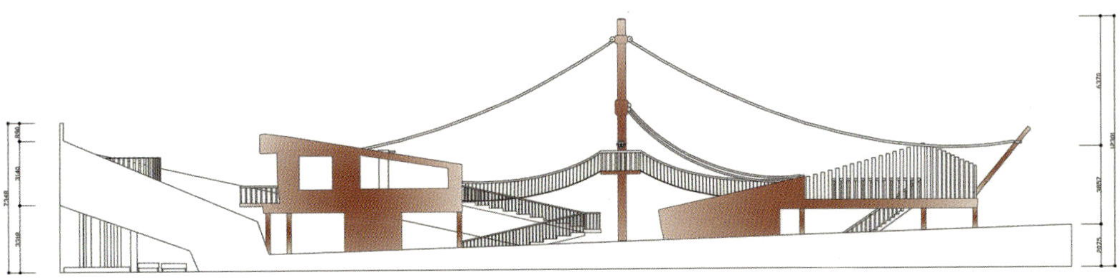

SIDE VIEW

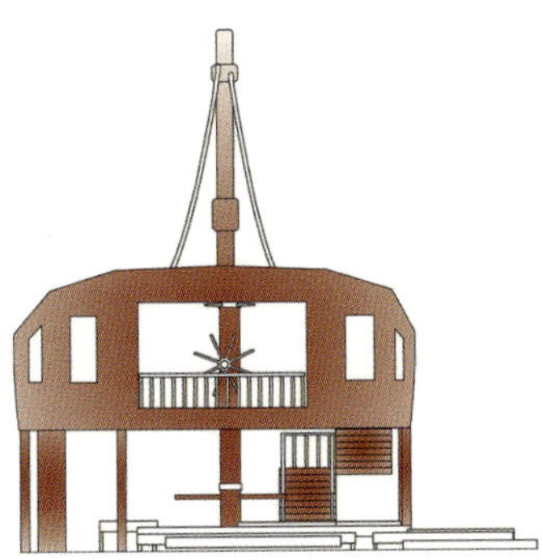

FRONT VIEW

REAR VIEW

Space programing

바닥에 깔린 높낮이가 다른 유리데크가 동선을 유도하고 물 위를 걷는 듯 한 느낌이 들게 한다.

The glass deck on the floor of uneven height induces the flow and makes people to feel as though they are walking on water.

기둥의 손잡이를 돌리면서 친구와의 협동심을 기를 수 있다.

They will develop cooperative spirit with friends by turning the knob on column.

일층에서 이어진 기둥이 회전하면서 물이 사방으로 뿜어져나온다.

Water spurt out in all directions as the column connected to the first floor rotates.

그물다리를 건너면 멀리 내다볼 수 있는 망원경이 있어서 아이들의 모험심을 자극한다.

A telescope that helps to see the far distance across the netted bridge stimulates adventurous spirit of children.

유리관 실로폰을 치면서 아이들이 소리와 물을 동시에 이용한 놀이를 즐긴다.

Children are joining water and sound at the same time with playing a glass tube xylophone.

임대홍 주민희 김아란 | Lim, dae-hong Ju, min-hee Kim, a-ran

逸脫

: 일상을 깨뜨려 일탈하다.
To break the daily deviation.

Site analysis

서울특별시 용산구 서부이촌동 396번지 대림아파트 주차장 396 dong, Yongsan–gu, Seoul
Parking lot 396 of Daelim apartment in Yongsan-gu, Seobuichon-dong, number 396 in Seoul.

현재 서부이촌동은 재개발이라는 상황에 묶여있어 개인의 재산권을 침해받고 있는 지역이다. 그리고 주민들의 계속된 반대로 재개발은 재개발대로 점점 늦어지고, 그로인해 부동산 거래는 계속 중지되어서 도시는 황량한 모습으로 약 8년동안 그대로 방치되어 있다.

Currently, Seobuichon-dong is a region where individual property rights are being violated as it is tied down with a situation of so-called redevelopment. Also, the redevelopment is delayed with continuous opposition of residents, and as a result, the real estate transaction stopped unceasingly that the city of a forlorn look has been neglected for approximately 8 years.

Background

서부이촌동 주민들은 재개발의 반대를 원해왔다. 그러나 실제로 재개발이 무산되어 원하는대로 됐지만 좋아할 수 없는 상황에 놓이게 되었다. 이렇게 힘들고 지친 사람들은 그런 자신의 상황에서 벗어나고싶은 '일탈' 을 꿈꾼다.

Seobuichon-dong residents wanted to oppose the redevelopment. However, they were placed in a situation that they cannot be happy with even though the redevelopment foundered as residents wanted.

Design concept

"일상을 깨뜨림으로써 일탈하다."
"Daliy deviation is cracking."

요즘 사람들은 쳇바퀴처럼 반복된 일상에 지쳐 있다. 서부이촌동 주민들은 여기에 재개발이라는 상황이 겹쳐져 현재 이촌동은 정체되있다. 그로인해 사람들은 점점 활기를 잃어가고 도시는 생기가 사라져간다. 우리는 주민들의 반복적인 일상을 깨뜨려 그들의 마음 속 일탈을 이루어 도시와 주민에게 생기를 불어 넣어주려고 한다.

People are tired of repetitive everyday life these days. Residents and Seobuichon-dong are currently stagnated with a mix of redevelopment here. As a result, people gradually lost vitality and liveliness of a city disappeared. We intend to
infuse life into the city and people by breaking repetitive daily life of residents and to attain deviation in people's mind.

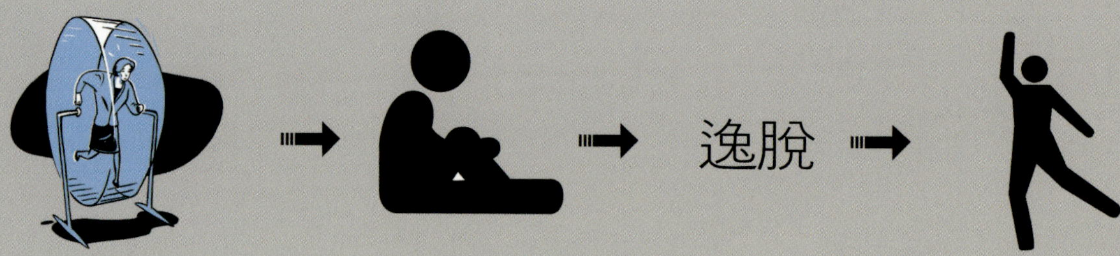

Keyword

"일상을 깨뜨림으로써 일탈하다."
"Daliy deviation is cracking."

Deviation	Crime	Strike	Break
Daily	Immoral	FLY	Crime
Escape	Unreality	Rest	Violence

Daily Break

일상(Daily)을 깨뜨리다.(Break)

Break routine.

BREAK
1. 건물을 깨뜨리다.
   Break building.
2. 생각을 깨다.
   I think the break.

Design process

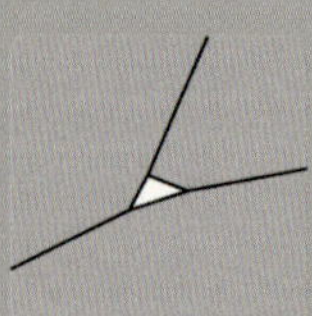

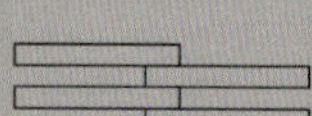

쌓다 Amass

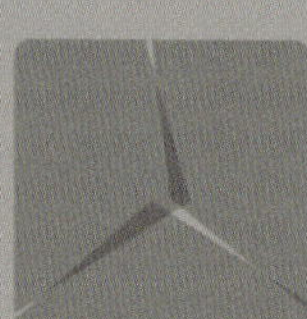

깨지다 Break

틀에박힌 쳇바퀴같은 일상
We created new space with fixed daily life

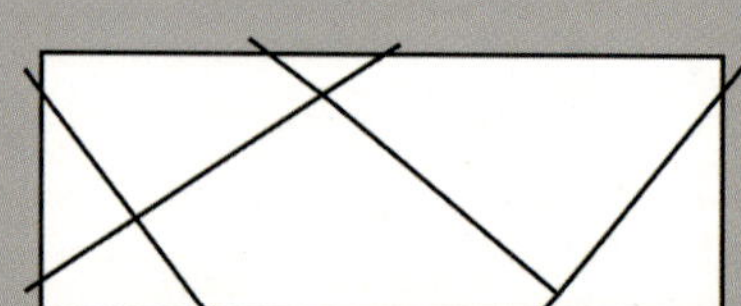

깨지기 시작한 일상
daily life that began to shatter

Space program

수영장 Swimming pool

1층 입구와 탈의실을 제외한 나머지 전체적인 공간을 물로 채워서 마치 해변가에 있는 듯한 느낌을 주었다.

We gave the beach-like feeling by filling entire space with water except the 1st floor entrance and dressing room.

제일 고층인 3층 수영장에 설치되있는 미끄럼틀을 타고 아래로 빠르게 내려갈때 사용자는 마치 날아가는 느낌을 받는다. 이러하듯 평소엔 느낄 수 없는 날아가는 느낌을 주게해 일탈을 느끼게 한다.

When a user goes down rapidly on the slide installed in a swimmin pool on the highest 3rd floor, they feel as though they are flying. People will experience deviation through this flying feeling that they cannot feel in everyday life.

이 공간은 깊은 물로 이루어져 있어서 스킨스쿠버로 사용한다. 깊은 물 속에서 자유롭게 헤엄칠 땐 마치 내가 중력에 상관없이 날아가는 듯한 느낌을 받는다. 이렇듯 평소엔 느껴보지 못하는 날아가는 듯한 느낌을 주게하여 사용자로 하여금 일탈하게 하는 공간이다.

The space is used by skin scuba diving as it is composed with deep water. When I swim freely in deep water, I feel like flying regardless of gravity. As explained, it is a space that allows users to deviate by giving flying-like sense that they cannot feel usually. People are living with a lot of stress as they are trapped inside repetitive daily life.

고정관념을 깨뜨림으로써 더 자유로운 공간을 만들 수 있게 한다.
More free to get rid of stereotypes was able to create space.

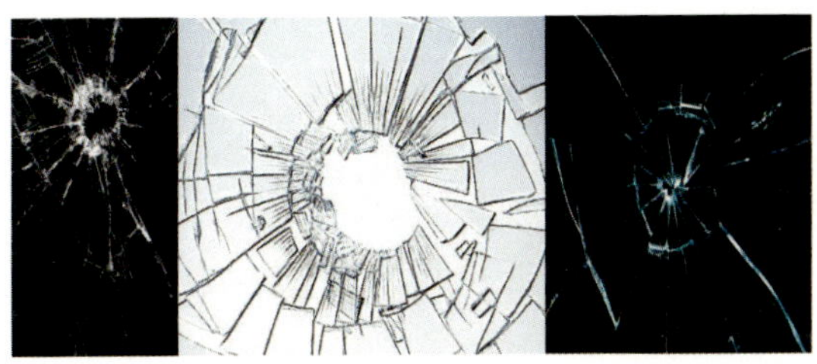

사람들은 반복적인 일상에 갇혀 많은 스트레스를 안고 살아가고있다. 그들은 항상 그런 일상에서 벗어나고 싶은 일탈(逸脫)을 꿈꾼다. 우리는 건물이 쳇바퀴같은 일상이라고 보고 그것을 깨뜨림으로써 깨진 일상에서 일탈을 하게 한다.

hey always dream about deviating from such everyday life. We saw building as a tread-wheel-like daily life and created deviation from shattered everyday life by breaking it. More free space was created by breaking fixed ideas. More free space was created by breaking fixed ideas.

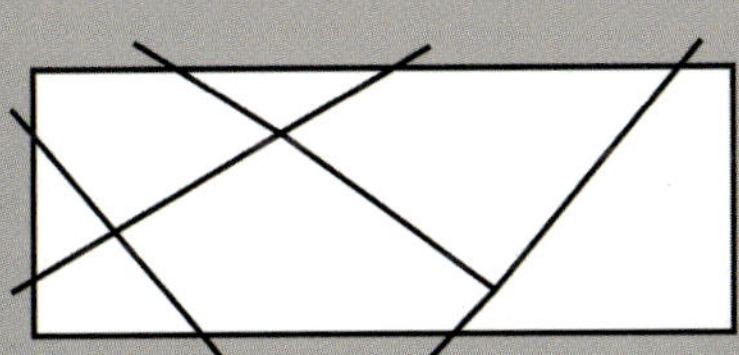

완전히 깨진 일상
Completely broken everyday

깨진 일상 조각들이 모여 새로운 공간을 만듬
Broken pieces are gathered daily Makes new space.

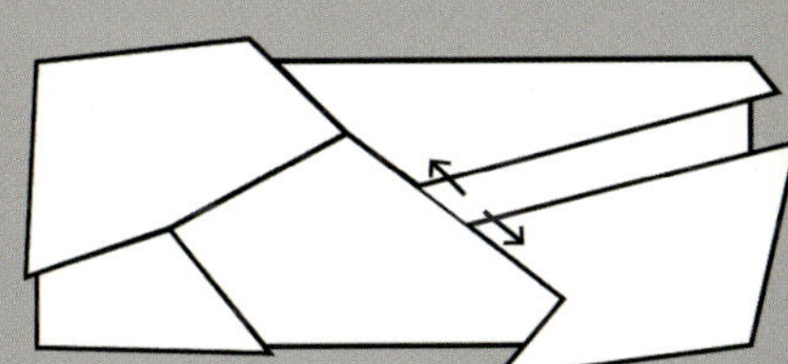

완전히 깨진 일상
Completely broken everyday

Plan

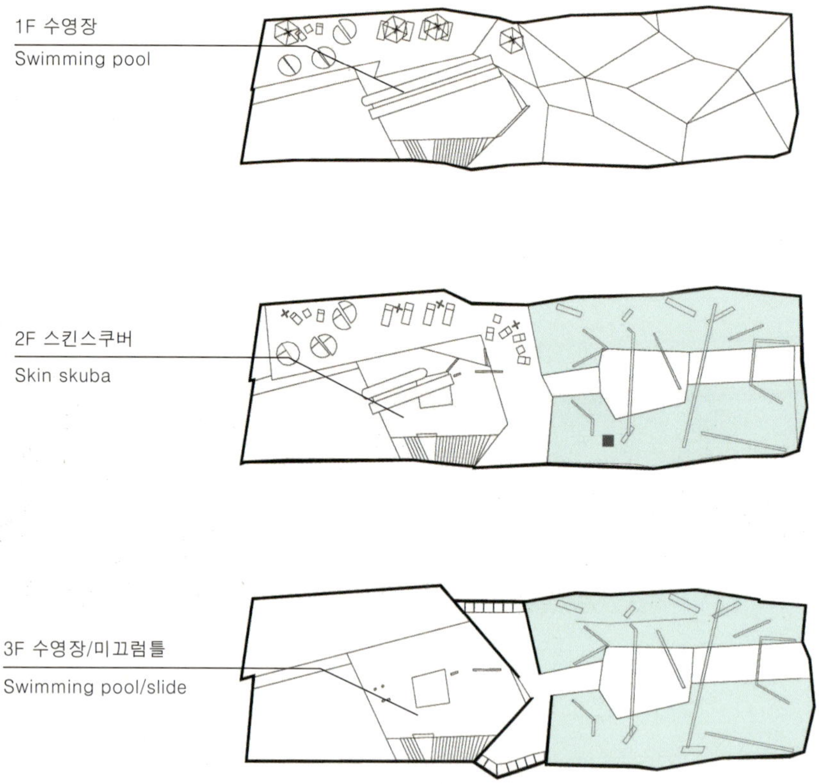

ISOMETRIC

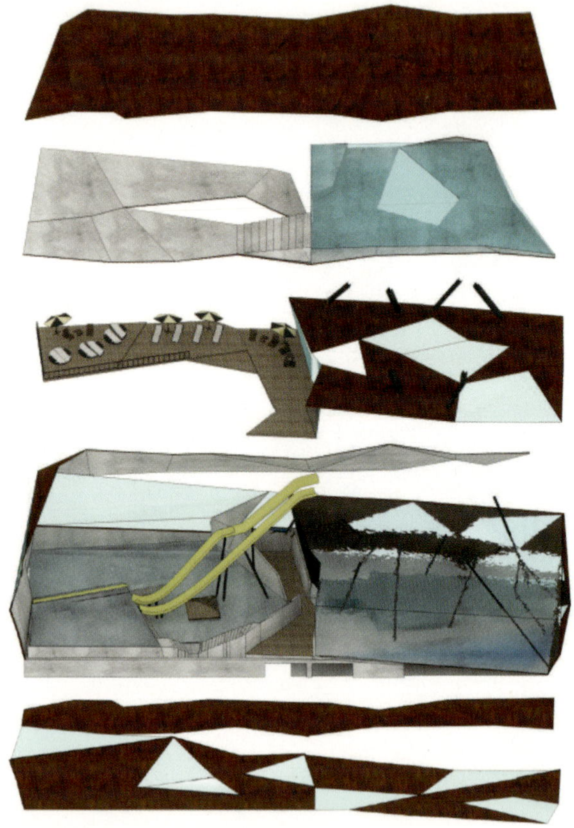

외관 모습 : 크게 깨진 파편들로 이루어진 외관 모습 지나가는 행인들의 호기심 유발을 위해 가려진 공간과 오픈된 공간을 만들었다. 오픈된 공간은 투명한 유리로 가려진 공간은 코르텐 철판을 이용했다.

Exterior appearance : composed with large pieces of shattered chips. We created hidden space and open space to arouse curiosity from passerby. Transparent glass for open space and cor-ten steel was used on hidden space.

3층 수영장을 올라가는 부분 : 일부로 공간을 넓게 만들었는데 이곳에서 사람들은 이동로로 사용하기도 하지만, 편히 쉴 수 있는 편한 공간으로도 이용하게 된다.

Section of going up to the swimming pool on 3rd floor : The space was created largely on purpose and people use it as a movement route, but it will be used as as place where people can rest comfortably.

Material

외관에서부터 코르텐 소재를 사용해서 녹이 슨 느낌을 주었다. 1층은 입구와 탈의실을 제외한 나머지 부분은 다 물로 채워져 있어 어느 구역을 수영장이라고 구분짓지 않고 어디에서든 물을 느낄 수 있다.

Gave rusted feeling with cor-ten steel. The rest of the 1st floor except the entrance and dressing room is filled with water that no specific region is classified as a swimming pool but can feel and enjoy water anywhere. We used cor-ten material from the exterior to give the rusted feeling.

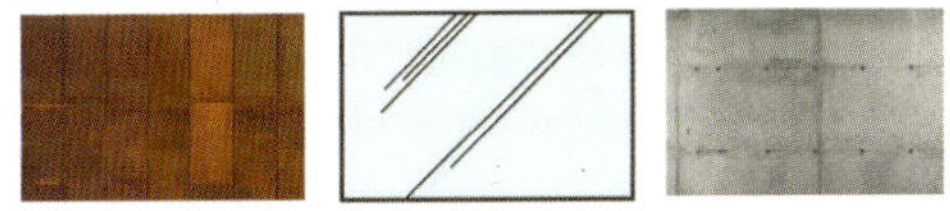

2층은 폭이 넓고 깊은 수영장이 있는데 여기는 스킨스쿠버 공간으로 이용한다. 이 수영장은 1층 입구까지 내려가는데, 여기에서 스킨스쿠버를 이용하는 사람과 입구로 들어오는 사람이 서로 소통할 수 있게 한다.

There is wide and deep swimming pool on the 2nd floor and this is used for skin scuba diving. This swimming pool goes down to the 1st floor entrance, and people who skin scuba dive and people who enter can communicate with each other.

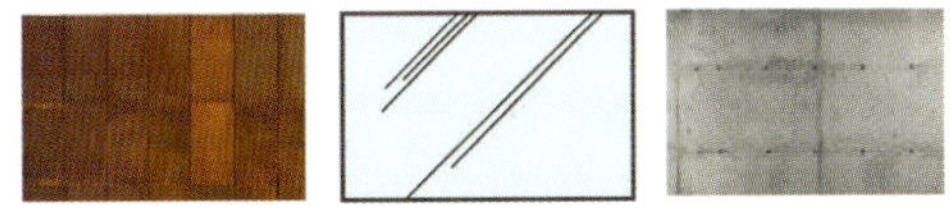

3층은 미끄럼틀 수영장 공간이다. 그곳이 설치된 2개의 미끄럼틀을 타고 내려가면 1층까지 내려가게 된다. 이 미끄럼틀은 끝이 올라가 있어서 마치 날아가는 듯한 느낌을 줄 수 있다.

The 3rd floor is a slide swimming pool area. When users go down on 2 installed slides, they get to go down to the 1st floor. Slightly tilted upward end of this slide gives that flying-like feeling.

2층 쉼터 부분 : 나무재질로 마감을 했다. 이부분은 뒷편이 투명하게 오픈되 있어서 선텐을 할 수 있고, 아랫층을 볼 수 있게 려있어서 1층 수영장을 볼 수 있고, 3층 미끄럼틀을 타고 내려오는 사람들을 볼 수 있게 했다.

Resting area on the 2nd floor : finished with wood. People can suntan here as the back is opened transparently, can see the swimming pool on 1st floor as it is opened to make the lower floor visible, and can see people coming down the slide from 3rd floor.

1층 수영장 부분 :1층은 거의 모든 부분에 물이 깔려있어서 사람들은 마치 해변가에 있는 듯한 느낌을 들게 한다.

Swimming pool on the 1st floor : People will get to feel as if they are on the beach as nearly every part of the 1st floor with covered with water.

S L I D I N G

PUZZLE

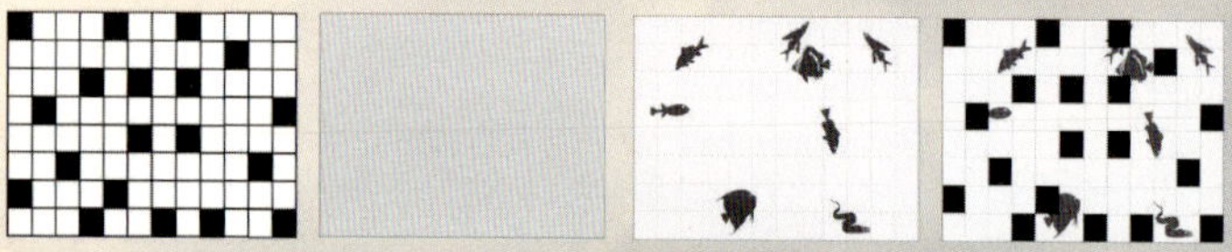

SOLID (Rails + Walls)GLASS AQUARIUM SLIDING PUZZLE

사람들은 퍼즐을 직접적으로 보고 만지고 이동한다.
동시에 그들은 어항 속에 있는 작은 생명과 소통한다.
그들은 공간의 흐름속에 이끌린 것처럼 건물안으로 들어간다.

People go to see, touch directly with the puzzle.
At the same time, they should communicate with life in a small fishing port in. As if they are headed in the flow of space into the building entered.

정호진 김해조 | Jeong, ho-jin Kim, hoe-jo

Project background

상대적으로 연령이 지긋하신 어르신들이 많이 거주하고 있고, 용산 국제업무지구로 인해 정신적, 육체적 피해를 입은 주민들에게 생동감 넘치며 긴장을 덜어주기 위해 우리는 수족관이라는 공간을 생각했다. 수족관은 자연 가습 효과, 두뇌발달, IQ, EQ,정신적인 안정에 효과적이다 .인간이 자연과 공생한다는 친환경적 교육적 효과를 동시에 이미지화 할 수 있다.

There are many elderly relative living in older, and because of the Yongsan Business District residents to be mentally and physically harmed.
We have to give them a lively tension to ease We thought the aquarium. The effect of the aquarium is a natural humidifying effect, brain development, IQ, EQ, emotional stability. It can have a greener image symbiosis of nature and the human.

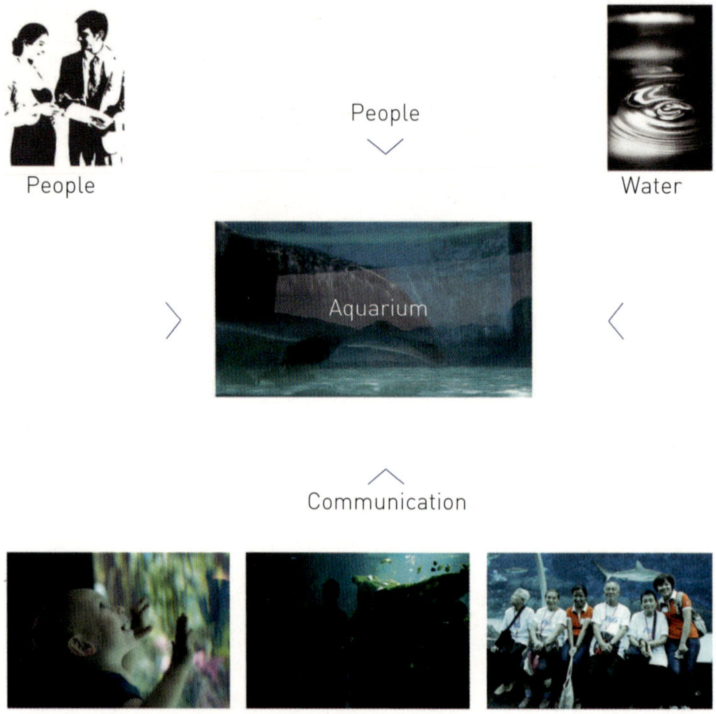

Purpose

자연과 사람의 조화를 실현하기 위해 우리는 도심지에 숨을 불어넣고 사람과 자연 문화가 어우러지는 매개체를 만들었다.

To realize the harmony of nature and man, we have to breathe into the downtown and Media culture where people and nature are harmonized is made.

Site analysis

용산국제업무지구 바람이 불면서 서부 이촌동 주민들은 극심한 고통에 시달려왔다. 집이 경매로 넘어가 한강다리에 올라가 흐르는 강물을 바라봤다는 노인. 오순도순 살던 주민들이 용산사업을 둘러싸고 입장이 갈리면서 반목이 나날이 심해지고 있다. 7년간 재산권을 행사하지 못한 주민들은 물론이고 개발사업과 함께 상권이 죽어버려 수입이 급감한 서부이촌동

Yongsan International Business District development news to residents of the West ichondong Residents extreme distress. Move on to the auction house, an old man trying to suicide. Position of the repeated conflicts between residents and developers. Seven years the residents not to exercise property rights. Because of the start of the development project, Reduction in the income of traders

서울특별시 용산구 이촌 2동 주민센터 옆 테니스장

2-dong, Yongsan-gu, Seoul, Ichon tennis courts next to the community center

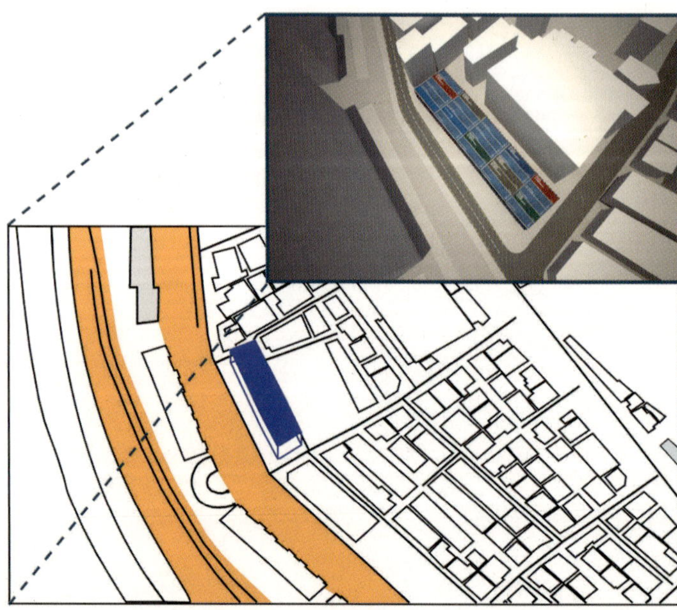

Design concept

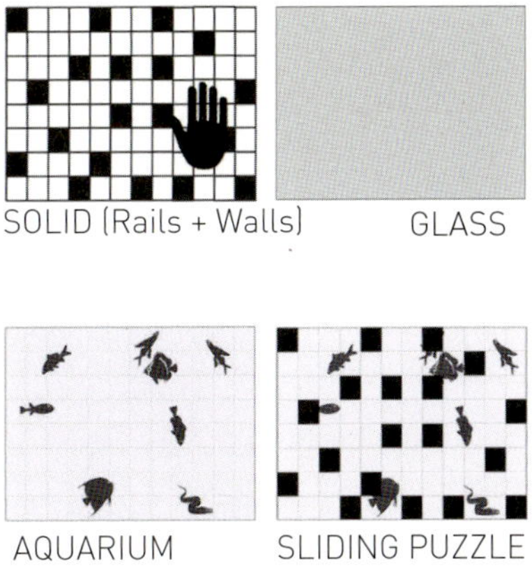

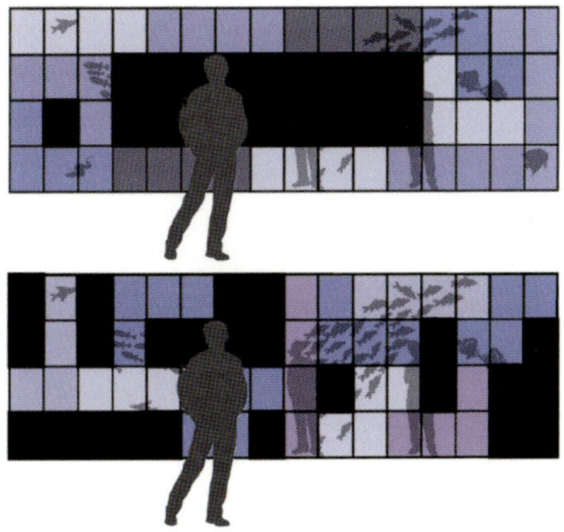

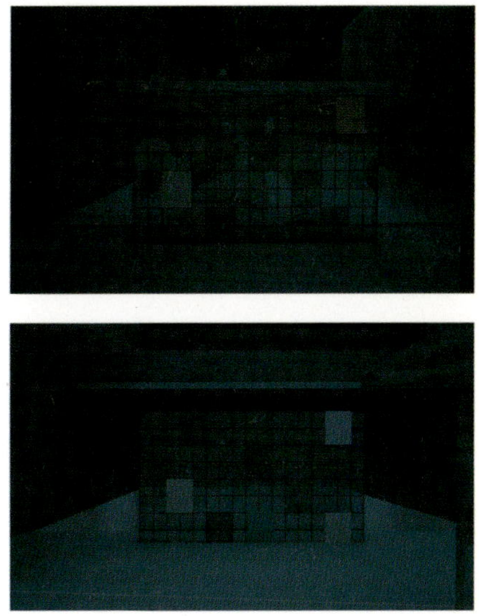

Space program

주민들이 퍼즐을 직접 보고 만지고 이동시키면서 공간이 변화하고 빛이 드나들며, 빛과 어항속 작은 생명과 소통한다.

People go to see, touch directly with the puzzle. while Space is changing and the light coming in, light shall communicate with small life in fishing port.

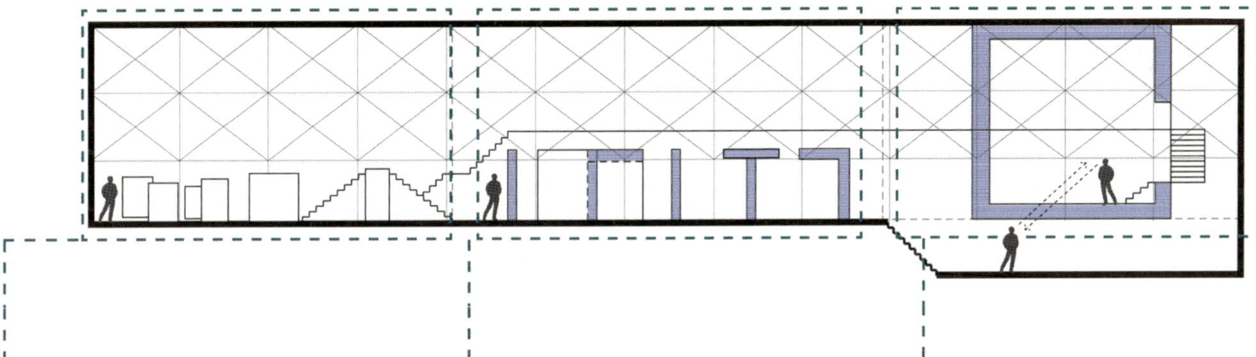

ZONE 1
the avant-garde space

입구 및 전위 공간에 입장을 하게 되면 이제 외부에서 들어오는 빛,소리가 차단되게 된다. 사용자들은 떠 있거나 불규칙적으로 배열된 퍼즐 조각을 보고 방향을 유도 받게 된다. 방향을 유도 받게되면 앞에 사람의 시야를 차단하여 긴장감을 갖게 하는 계단이 나오게 된다. 계단 뒤에서 나오는 은은한 푸른 불빛과 시야가 차단 되었기 때문에 사용자들에게 궁금증을 유발하게 된다.

If the position of the inlet and potential space, the light and the sound coming from outside is to be blocked. Users can induce to get directions while they see randomly arranged and floation puzzle pieces. If you receive orientation induced, Stairs comes out Blocking the view of the front man and having a tension. Because Soft blue light comes from behind the stairs and Vision is blocked, Users will have questions.

ZONE 2
the maze space

도입부에 전면 시야를 차단하는 슬라이딩 퍼즐이 나오게 된다. 어항으로 된 슬라이딩 퍼즐을 움직이면서 빛이 움직이게 되고 수족관 뒤가 조금씩 보이기 시작한다. 여기서 사용자들은 호기심과 궁금증을 갖게 된다. 슬라이딩 퍼즐을 지나 퍼즐이 미로처럼 배열되어 있고 Void, Solid, transparency, translucence, omok, convex 로 되어 있어 사용자들에게 재미를 준다. 어항중에서 윗부분에 유리관을 올려놓은 물탱크는 그 안에서 사용자들이 물이 흐르는 소리를 들을 수 있고 또 다른 개인의 공간으로서 역할을 한다. 끝 부분에 슬라이딩 퍼즐이 한 번 더 나오게 되어 궁금증을 유발시키고 퍼즐을 움직이면서 다음 공간의 재미와 호기심을 느끼게 한다.

in the front of space Sliding puzzle appeared blocking all view. Moving the sliding puzzle made aquarium, The light moves and it Gradually begins to see the appearance of the aquarium. Here The user will have the curiosity and questions. Passing over a sliding puzzle, puzzle is arranged like a maze, Void, because it been to Solid, transparency, translucence, omok, convex, Users can feel the fun. the glass tube is placed on top of water tank in the aquarium, User can hear the sound of running water in the tank water, and it serves as a private space. while the sliding puzzle comes out at the end, People is moving puzzle with questions and they feel the fun and curiosity in the following space.

ZONE 3
the floating aquarium

슬라이딩 퍼즐을 지나서 떠 있는 수족관 아래로 들어오게 되면 안개속에서 떠 있는 수족관을 본다. 그 위로 보이는 사람들과 상호작용을 하게 되면서 위에서 아래를 내려다 볼 때 물과 안개 속에서 사람들이 있는 것 처럼 보이고 아래서 위를 볼 때 물 위에 사람들이 떠 있어 보여 초현실적인 분위기를 연출한다.

passing the sliding puzzle, when it comes to down, see the aquarium in the fog. interacting with people When you look down in the top, people seems to be in Water and mist Directed to surreal atmosphere.

Floor plan

floor B1

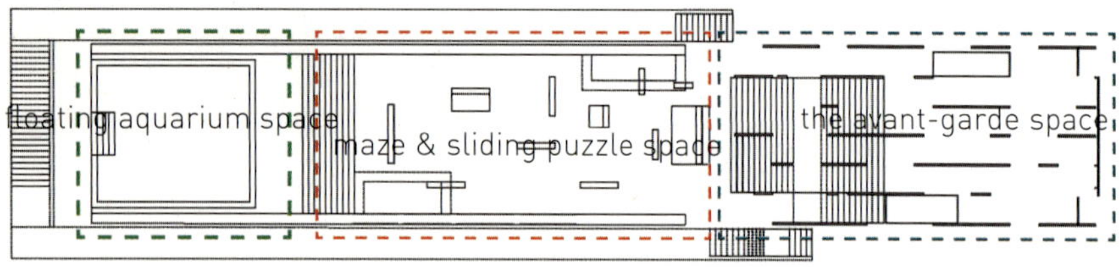

floor 1

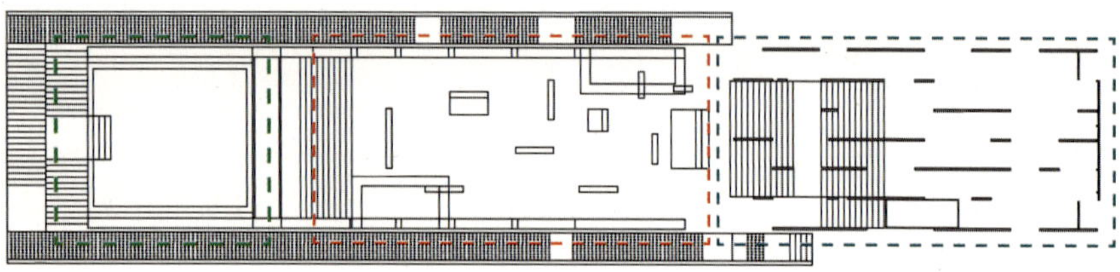

floor 2

Under Space

전위 공간의 밑 지하 공간이다. 지하 공간으로 진입하게 되면 1층에 있던 잔잔한 물이 보이게 되면서 물속에 잠긴듯한 초현실적인 분위기를 연출한다. 1층에서 지하를 내려다 볼 때는 물 속에서 사람들이 걸어다니는 분위기를 연출한다.

Underground space is a underneath space. When entering the underground space, still water on the first floor was seen like submerged and a surreal atmosphere. looking down in the first floor to the basement, It seems that people are walking around in the water.

The Avant-Garde Space

건물의 진입 공간으로 빛과 소리, 시야가 차단되면서 긴장감을 유발하게 되고 다음 공간에 대한 호기심을 갖게 된다.

While entering the space of the building blocked a light and the sound and the vision, people caused by tension, had a curiosity about the next space.

Maze & sliding puzzle space

사용자들이 슬라이딩 퍼즐을 움직이면서 재미를 주고 투명, 반투명, 오목, 볼록, 비어 있고 막혀있는 사각형의 어항이 불규칙하게 배열되어 미로처럼 펼쳐져 있는 공간을 지나가게 된다.

While users move the sliding puzzle, they feel funny. The transparent, translucent, concave, convex, empty, fill-in squares glass tube is randomly arranged and space passes as a maze.

Floating aquarium space

건물의 맨 끝 부분으로 떠 있는 수족관이 있다. 사용자들은 아래로 진입하게 되고 바닥에는 안개가 잔잔하게 있다. 위에서 아래를 내려다보면 물 속에 잠긴 듯한 분위기를 연출하고 아래서 위를 내려다 보면 수족관 안에서 사람들이 움직이는 듯한 분위기를 연출하여 초현실적인 분위기를 연출한다.

There is a aquarium floating at the top end of the building. The user will have to enter down, and fog is calm at the bottom. if you look down from above then look like submerged. if you look at under the above, people move in aquarium as a surreal atmosphere.

The Avant-Garde Space

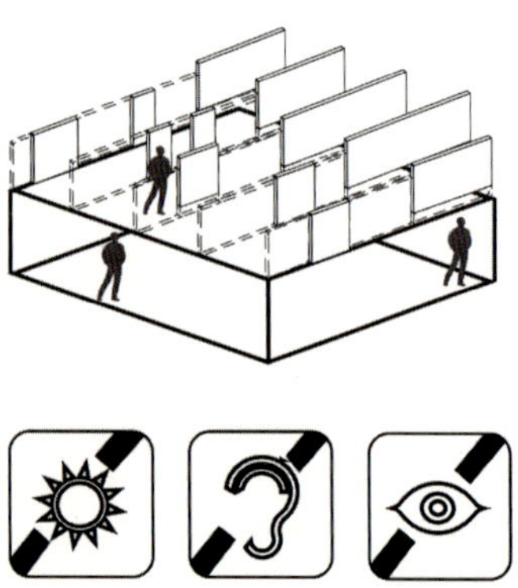

Maze & Sliding Puzzle Space

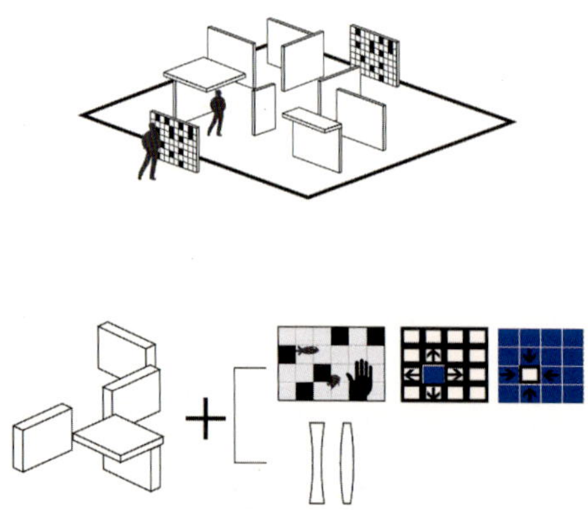

Floating Aquarium Space

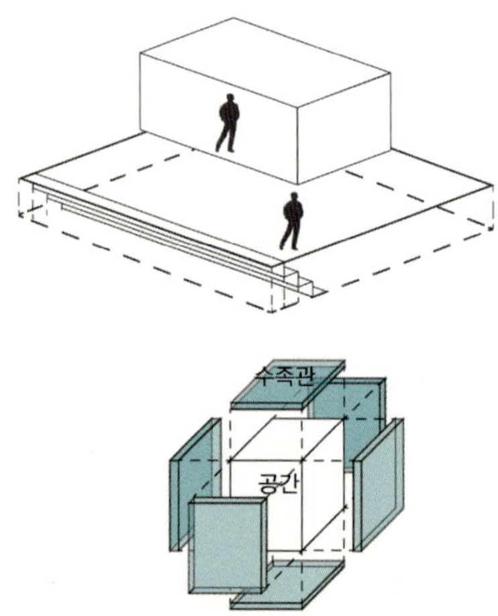

Exploded view

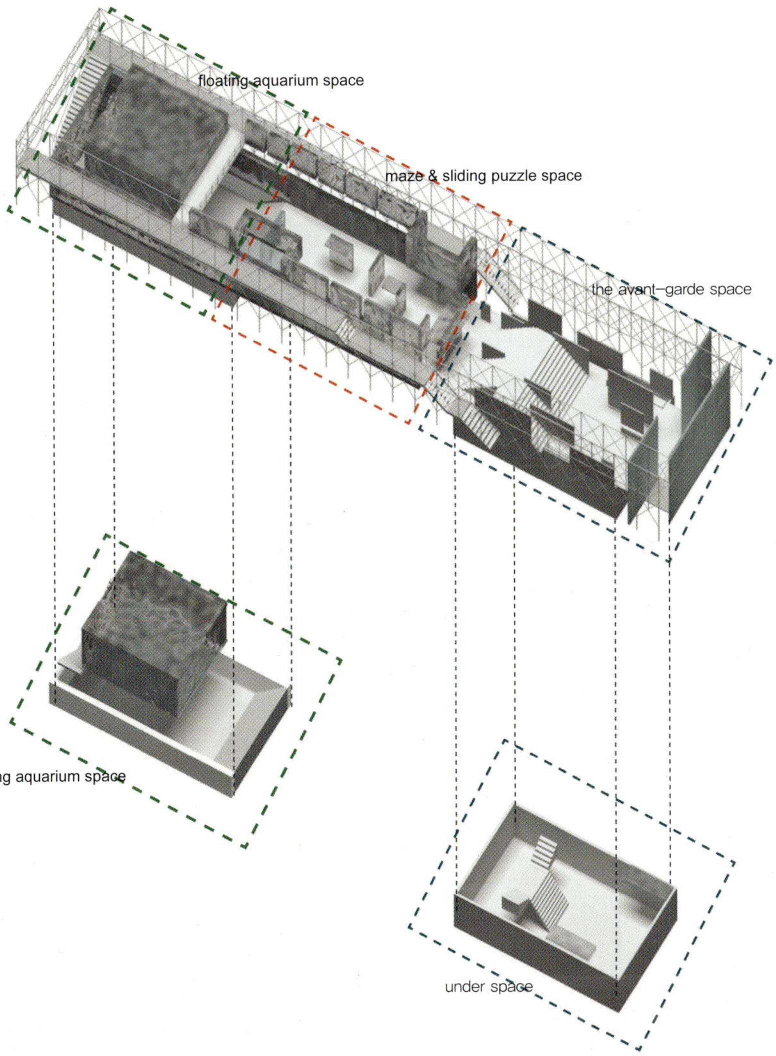

최승민 여운비 | Choi, seung-min Yeo, woon-bi

EQ_III

Play With Noise & Vibration

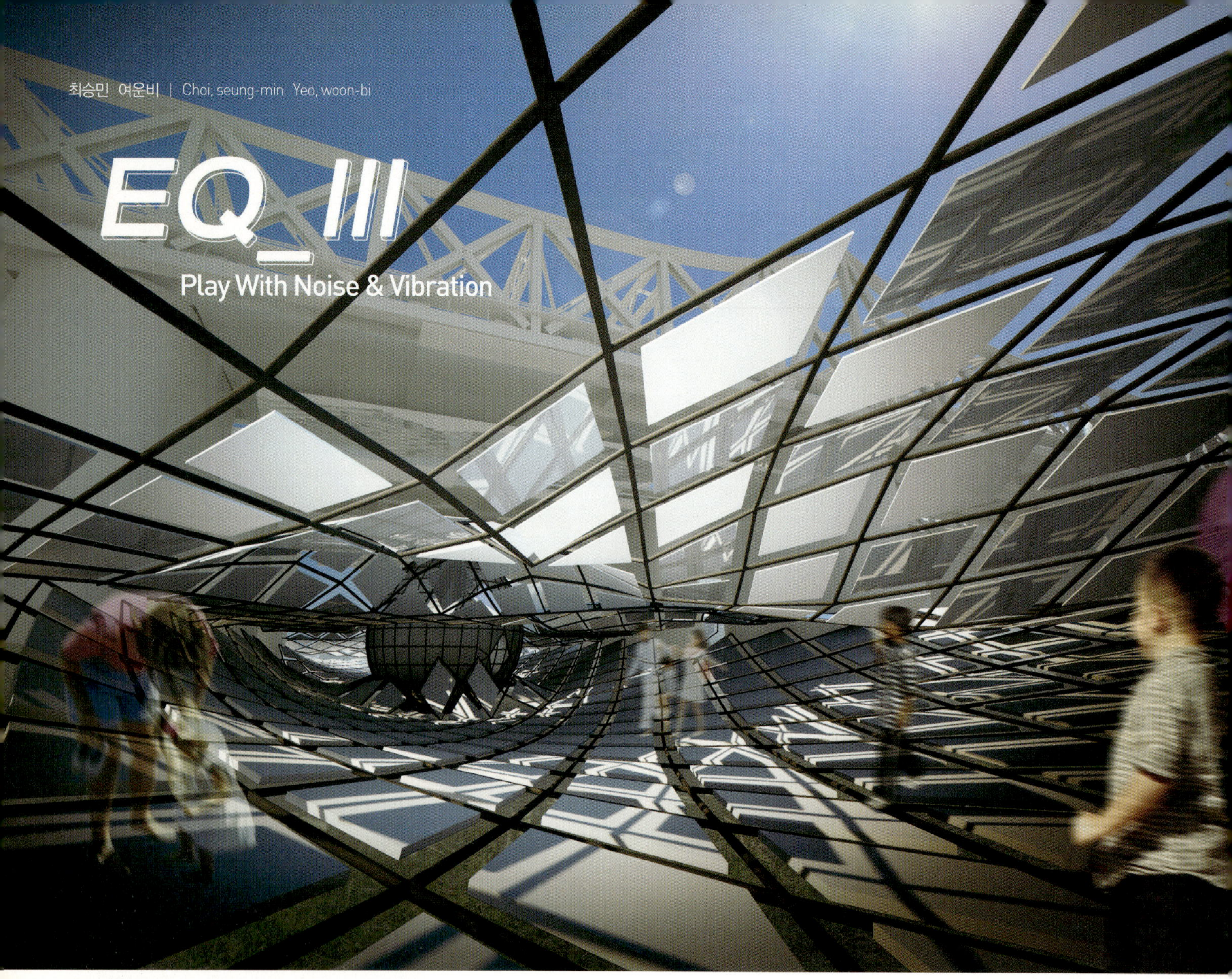

Site

이촌 1동 한강 시민 공원에는 체육 및 여가 시설이 11개소나 위치하고 있다. 그러나 이촌 2동 한강 시민 공원에는 단 하나의 여가 시설도 위치하지 않는다. 무엇이 문제일까? 두 지역의 빈부 격차가 주민들의 여가생활까지 차이를 보인다면, 분명 큰 사회적 문제이다. 이 지역에 가장 필요한 것은 무엇일까. 이촌 2동 주민들의 활력을 되찾아줄 그 무언가가 필요하다. 우리는 '놀이' 라는 단순한 즐거움을 목표로 하는 행위에서 그 해답을 찾아본다.

Location :
Echon-dong, Yongsan-gu,
Seoul, Korea

There are 11 facilities for sports and leisure at Han River Park Ichon 1-dong, while there is nothing at Ichon 2-dong. If the gap between the rich and poor between two regions brought the gap between the leisure lives of residents, it is certainly a big social problem. What is the most necessary thing to regain Ichon 2-dong residents' vigor? We found the solution in 'play' an act whose purpose is just pleasure.

Background

재개발 사업 계획에 따른 갈등, 갈증

용산 국제 업무지구는 철도청 부지와 서부 이촌동 지역을 통합 개발하여, 용산역 주변을 국제 업무기능을 갖춘 서울의 부도심으로서의 위상확보와 동시에 대한민국을 대표하는 명품 수변도시로 조성하여 국제경쟁력을 강화하겠다는 목표를 가진 사업계획이 실패로 돌아서면서 서부 이촌동의 많은 문제점들이 야기되고 있다.

Conflict due to redevelopment project plan

Many problems of Western Ichon are arisen because of the failure of International Business District of Yongsan's plan; developing the land of National Railroad Administration and Western Ichon to enable to dealing International Business functions as a sub center of Seoul.

Solution

비포장 노면, 그늘, 습기 등의 낮은 쾌적성은 관리로 개선될 수 있다.
하지만 철도 차량의 이동으로 인해 발생하는 소음과 진동은 개선이 어려운 문제다.
굳이 피해야 할 필요가 있을까 ?
재해석에 의한 즐거움이 될 수 없을까 ?
우리의 발이 되어주는 고마운 교통수단에 대해 미간의 찡그림 대신, 새로운 이해와 즐거움으로 승화시키기 위해 '소음과 진동의 놀이화' 라는 해답을 찾는다.

It can be improved by management for uncomfortableness because of unpaved road, shade, and humidity.
However, it is difficult to make better for the noise and vibration because of the frequent movement of railcar. But, do we have to avoid that problem?
Is it impossible to make it enjoy by reinterpretation?
We have solution for it, 'Make noise and vibration as a play', instead of displeasure.

Keyword

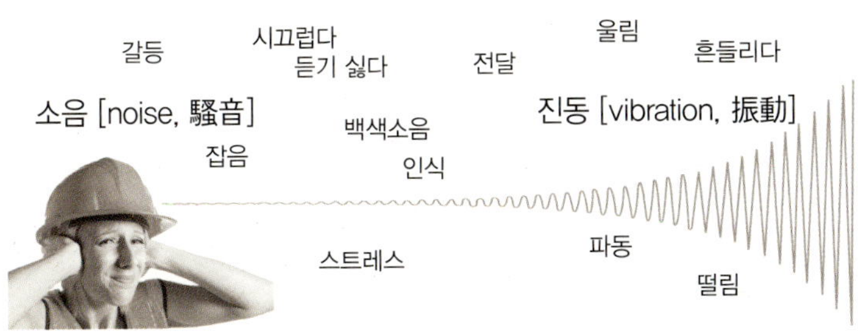

Motive

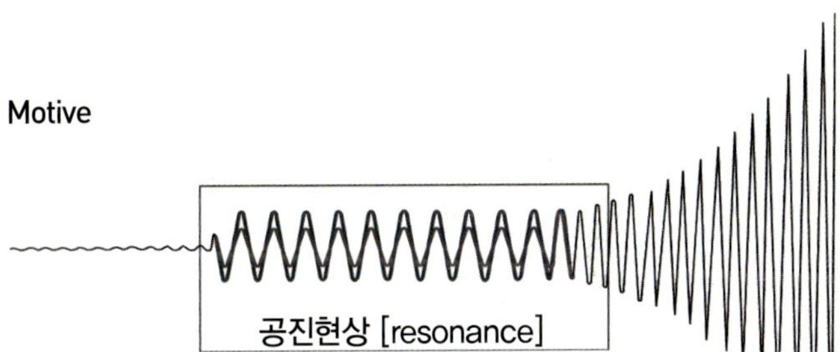

특정 진동수를 가진 물체가 같은 진동수의 힘이 외부에서 가해질 때, 진폭이 커지면서 에너지가 증가하는 현상
A phenomenon that an object that has specific frequency receives the same frequency energy, the amplitude of the object is broaden.

Concept

EQ = equalizer

개개의 소리 에너지는 고유의 주파수를 가지고 있다.
소리들의 특성을 조율하고 불필요한 저음이나 고음을 컨트롤 함으로써 음의 파장을 조절하여 필요한 소리로 '재창조' 하는 장치.
이러한 이퀄라이져의 특성을 공간에 대입하여, 소음과 진동의 수준에 의한 3가지의 공간 구성으로 이 지역을 기피의 대상으로만 여기던 사용자들에게 즐거움의 공간으로 '재창조' 된다.
나아가, 소리 에너지 (소음, 진동) 는 운동 에너지 (놀이) 로 변환이 되고 운동 에너지는 다시 소리 에너지로 변환이 됨을 되풀이하며, 서로 Act 와 Re_act 가 되어 Interaction (상호 작용) 의 공간이 된다.

EQ = equalizer

Each sound has natural frequency. Apply the feature of equalizer to the site; tuning the features and wave length of sound by controlling the unnecessary high or low-pitched tone to reinvent as a necessary sound. It will make this place "reinvention" as a place of joy to the users who used to consider this place unpleasant. Furthermore, sound energy-noise and vibration- convert to the kinetic energy-play- and kinetic energy converts to sound energy again. Repeating this act and react procedure, it became a place of interaction.

III = Three

사이트를 채우고 있는 하나의 소음과 진동을 개개의 음역대로 분리하여, 각각 다른 소음과 진동의 수준을 갖는 3가지의 공간으로 구성한다.
익사이팅한 놀이로 소음과 진동의 주체가 되어 온몸으로 즐기는 Exiting_zone, 놀이를 즐기며 소음과 진동에 대한 이해와 학습이 가능한 Playing_zone, 증폭된 소음과 진동이 가득한 놀이 공간 속 그 요소들이 철저하게 절제된 공간인 Relaxing_zone으로 나뉘게 된다.
굳이 아라비아 숫자 '3' 이 아닌 'Ⅲ' 로 3가지의 공간을 표현함으로써, 3개의 철교가 가르는 사이트의 특성을 내포한다.

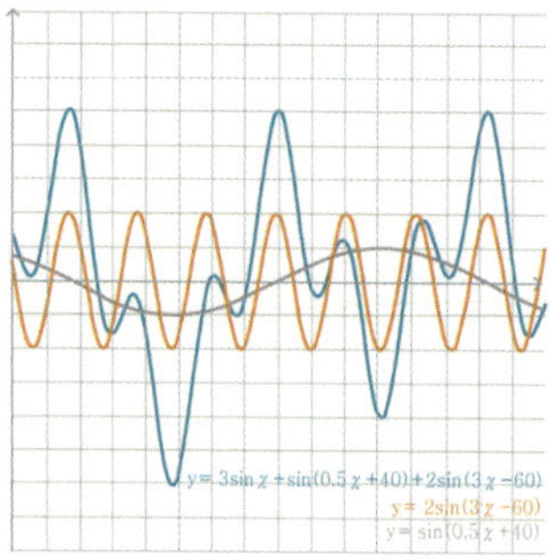

III = Three

It composes three spaces which has different level of noise and vibrations each: Exciting-zone, making noisy and vibration personally and enjoying that by playing excitement things, Playing_zone, understanding and learning about the noisy and vibration while enjoying play, Relaxing_zone, the place there is anechoic and no noise among the play zones where its noise and vibration is amplified. Designate these three zones by Romanize 'Ⅲ' instead of '3'. It implies the feature of the rail of the site.

EQ_III

더 이상 소음과 진동은 기피가 아닌 오히려 기다림의 대상이 되고, 남녀노소 즐길 수 있는 놀이로 활력을 되찾는다.
사용자의 의해 컨트롤 되는 공간.
주체가 되는 능동적 공간.
그것이 우리가 추구하는 진정한 EQ_Ⅲ의 공간이 될 것이다.

EQ_III

Now, the noise and vibration become a thing that people love to do, and enjoyable play regardless of age or gender. And the site regains its vigor. This is the place of 'EQ_Ⅲ' what we pursue: users can control and take an active part in.

Programming

Exciting_zone

커다란 투명 비닐 공 안에 들어가 땅위 또는 언덕 아래로 굴리는 스포츠.
소음과 진동에 의한 피해를 받는 입장에서 벗어나, 능동적으로 소음과 진동의 주체 즉, 음원이 되어 놀이를 즐김으로써 사용자들에게 활력을 불어넣는 공간이 된다.

Zonning

sports that go inside of big transparent vinyl ball and rolling it on the ground or down the hill. Users are not the victim of noise and vibration anymore. They take an active part in noise and vibrations, as sound source, so they enjoy the play that the place makes people regain their vigor.

Playing_zone

클라드니 도형이란 어떤 판 위에 건조한 모래를 올려 놓고 그 판에 소리를 입력했을 때, 판 위의 모래가 주파수에 따라 서로 다른 무늬를 만들어 내는 것을 말한다.
이러한 클라드니 장치를 철교 하부에 다운 라이트를 설치한 후 layer 형태로 매달아 그림자를 형성시킨다. 철도차량의 이동이나 놀이의 소음으로 인한 주파수 간섭에 의해 변형되는 형태를 보고 느낄 수 있다.

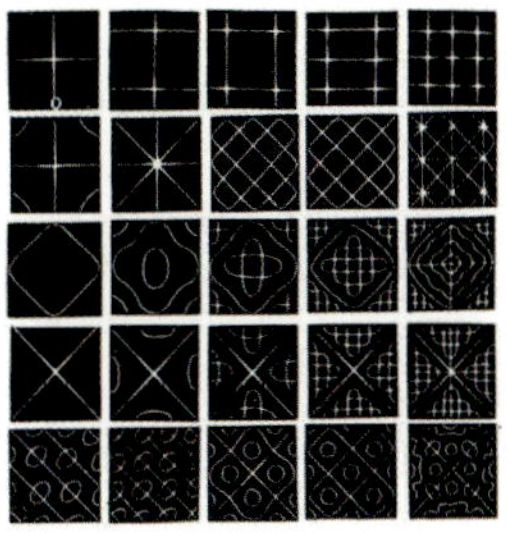

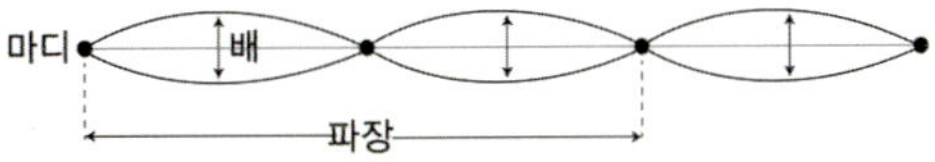

Chladni' figures

Chladni' figures a technique to show the various modes of vibration of a rigid surface. A piece of metal whose surface was lightly covered with sand reached resonance, the vibration causes the sand to move. Therefore, different patterns formed depending on frequencies. Install down-light underneath the rail-road bridge, and hang Chladni' equipment as layer form to make shadow. People can see the figures' various forms transformed by frequency interference that is because of the noisy by movement of railcar or plays.

Relaxing_zone

바닥, 천장 및 주위의 벽이 모두 흡음재로 내장되고, 음향의 반사를 극력 작아지도록 설계된 방.
유일하게 MASS감을 갖는 이 공간은 외부에서 발생하는 소음과 진동을 차단함과 동시에, 내부에서 발생하는 모든 소음과 진동조차 그대로 외부로 빠져 나가는 구조를 이용하여 소음과 진동이 최소치인 상태로 만들어, 소음과 진동을 이용한 이 놀이 공간과 극적인 대조를 이루는 공간이 된다.

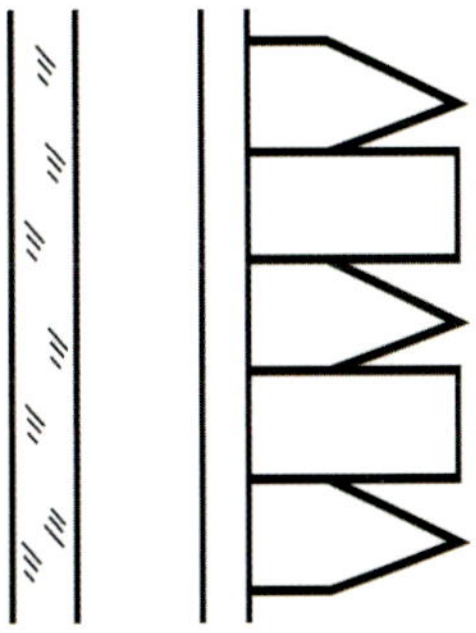

Anechoic room

The only room which has dignity, will be built in sound-absorbing materials to every wall includes the floor, ceiling and designed to minimize acoustic reflection. This place will shut off the noise and vibration from outside. At the same time, using the structure which can make the noise and vibration, even it generated from inside, make the state of the noise and vibration minimum. This space will be a place which is a dramatically contrast from the play zone.

Process

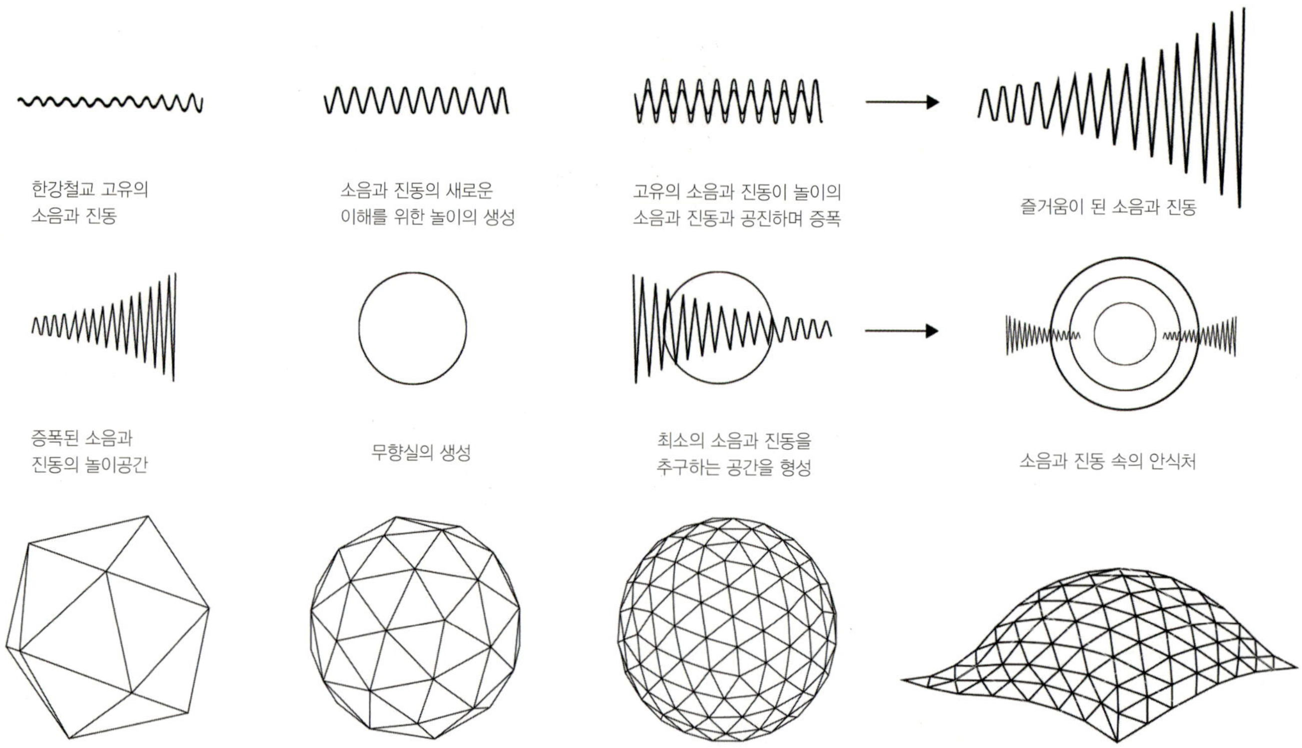

기둥없이 넓은 볼륨을 갖는 공간을 보다 안정적인 구조로 만들기 위하여, 지오데식 돔을 펼쳐 공간을 구성

Material

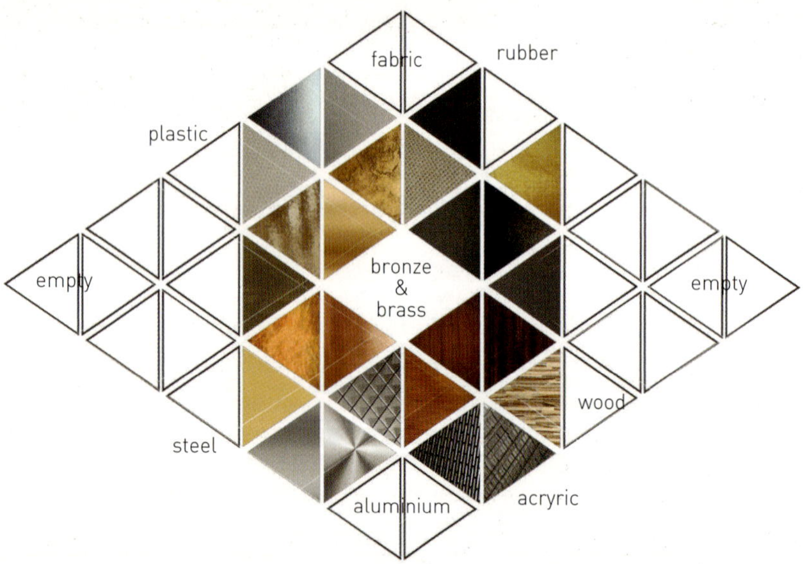

Units are various materials ranged from softness to hardness. They make different sound and vibrations.

각 유닛들은 연질에서 경질까지 이르는 다양한 물성으로 제각각 다른 소음과 진동을 발생시킨다.

Plan

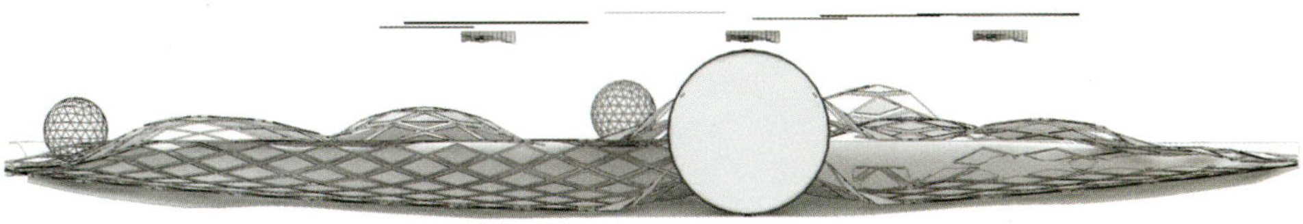

Sectional view

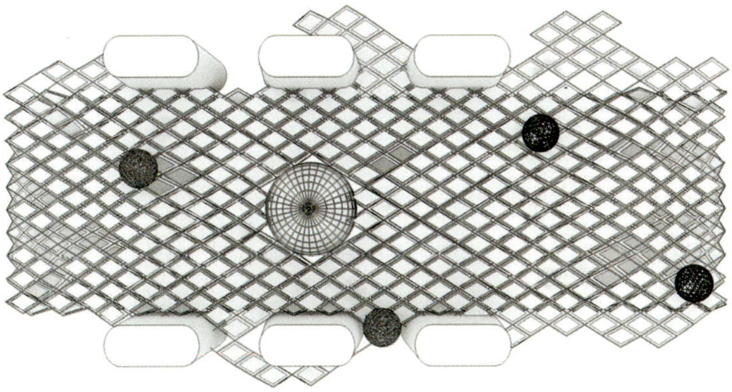

[Floor plan]

Exploded view

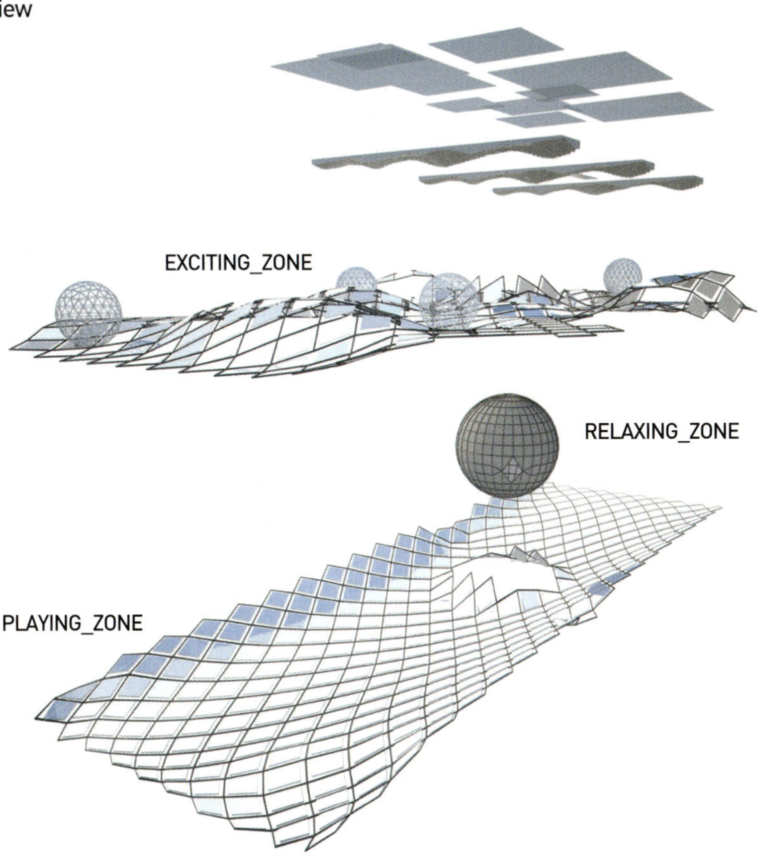

Silver Lining

구름 속 희망과 놀다.

이호림 이성경 | Lee, ho-rim Lee, seong-kyung

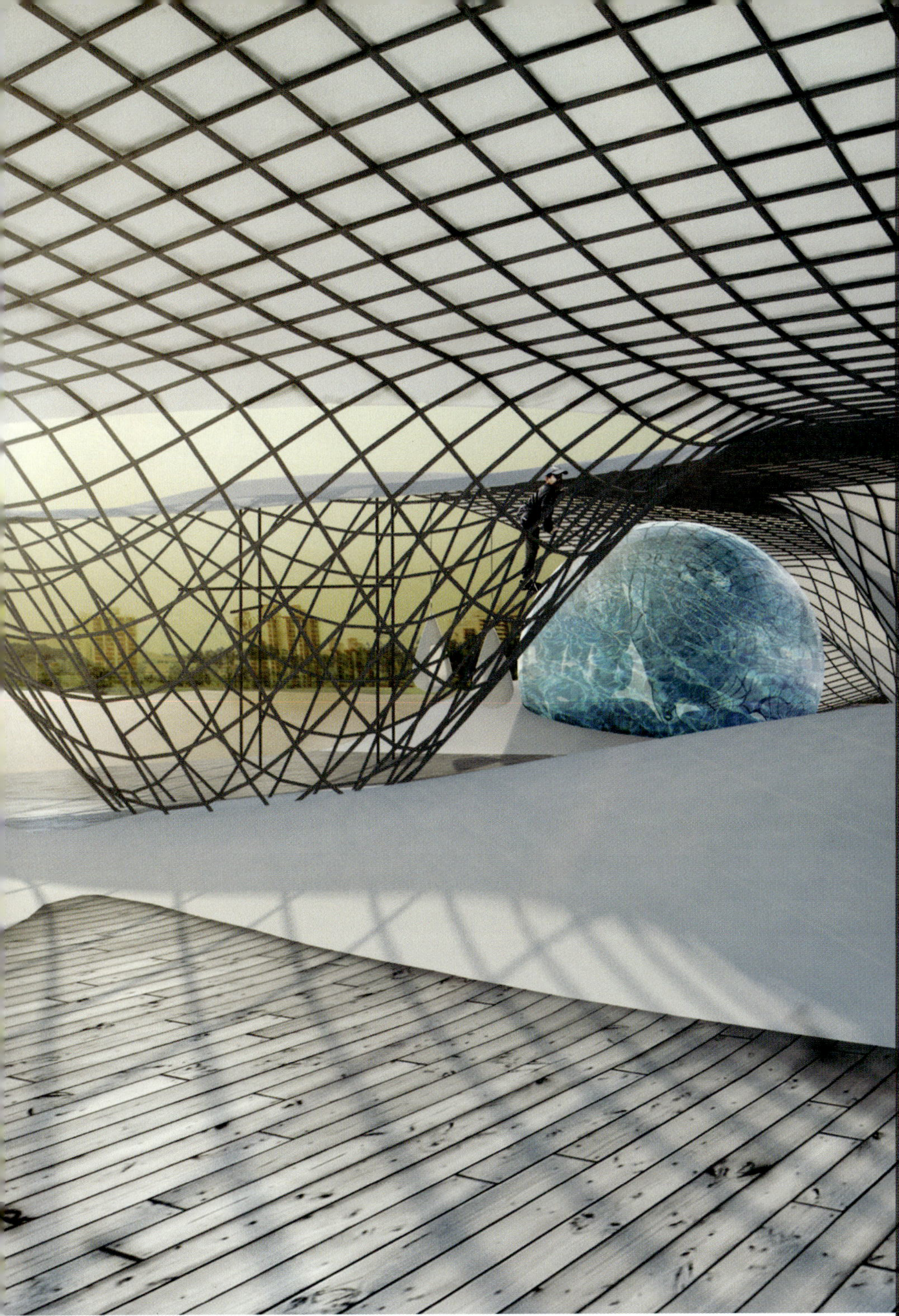

Background

서부이촌동은 용산국제업무지구

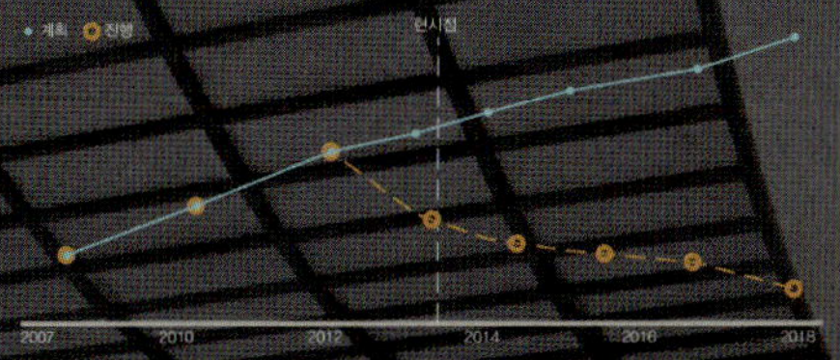

동부이촌동과 서부이촌동은 한강대교와 원효대교 사이 한강변을 따라 동서로 길게 위치한 용산 역세권 및 한강로의 배후지로서 2007년 재개발 지역으로 선정되어 많은 사람들이 밝은 미래를 꿈꾸며 용산구 서부이촌동 이촌2동으로 많은 세대 가구들이 이주하게 되었다. 심지어 이들중에는 빚을 지면서까지 이주하는 사람도 있었다. 하지만 생각과는 다르게 재개 발이 무산 되고 용산구 서부이촌동 이촌2동 주민들은 한순간 에 희망을 잃어버렸다. 우려하던 집값은 폭락하고 이사오면서 생긴 이자도 감당하기도 힘든 상황이다. 심지어 2300여 가구 중 절반 이상이 용산개발 보상을 기대하고 평균 3억 4000만원 가량의 금융권 대출을 받아 생활비와 학자금 등으로 쓰고있는 상태이다. 하지만 이러한 상황에서도 용산시는 정부가 나서서 주민들을 지원하지 않는한 시 자체 예산으로는 도울 여력이 없 다는 입장 이다. 그들도 맞는게 재개발의 투자 가치를 보고 얻은 빚을 시가 나서서 구제하는 것은 적절치 않다고 보는 것이다. 이제 그 누구 도 이들편을 들어줄 사람이 없다. 이자는 쌓여서 빚은 늘어나고 집값은 폭락하는 이러한 시점에서 무기 력하고 지친 주민들은 삶에서 느낄수있는 진정한 행복, 소소한 감정, 잠시나마 현실을 잊게 만드는 그러한 공간이 있다면 어떨까?

Dongbuichon-dong and Seobuichon-dong are the hinderland to Yongsan station area and Han River located from east to west along the banks of the river in between Han River Bridge and Won Hyo Bridge that many families moved into Yongsan-gu Seobuichon-dong and Ichon2-dong with a hope for brighter future as it was selected as the 2007 re-development district. However, Seobuichon-dong and Ichon2-dong residents lost hope as the redevelopment fell through. The housing price plunged drastically and it was a difficult situation for people to cope with debt. How would it be like if these lethargic and exhausted residents can have place where they can feel true happiness, emotions, and where they can forget reality for a moment?

Purpose

이미 서부이촌동 재개발은 무산이되었고 주변 집값은 폭락되었다. 심지어 주민들의 빚은 점점 산더미처럼 늘어가고있다. 더 이상 잃을게 없고 너무 힘든상황에 처한 이들에게 진정 필요한 공간은 무엇일까? 사실 너무나도 힘든 이들에게는 가장 필요한 것은 아파트재개발 일것이다. 하지만 우리는 공간개념에서 생각을 멈추지 않고 공간에서 일어날 수있는 스토리들을 상상해보았다. 자신의 아이들이 노는 모습을 볼수있는곳, 주민들이 모여 일상속에서 진정한 행복을 느낄수 있는곳, 지친 현실에서 한박자 쉬어 갈수있는곳, 다시 일어설수 있게 희망 씨앗을 얻을수 있는곳, 우리는 서부이촌동에 다시 에너지와 희망을 심어주려고 한다.

Redevelopment of Seobuichon-dong became foundered and house prices nearby declined drastically. In fact, what these people in difficult situations need the most would be the redevelopment of apartments. However, we did not just stop at the space concept, but imagined about stories that can happen in space. A place where people can see their children having a delightful time, where residents gather together to feel true happiness in trivial everyday life, place where they can take a break in exhausted reality, where they can get seeds of happiness to stand up again, we are trying to plant energy and hope again in Seobuichon-dong.

Site analysis

위치 : 용산구 서부이촌동 주민센터 옆 대림아파트 주차공간
크기 : 53000x15000 = 약 240평
- 남쪽으로 높은 아파트가 위치해 조망권이 좋지 않다.
- 대중교통 이용이 불편하다.
- 주변상권이 없어서 외부 사람들의 발길이 적다.
- 주민센터의 옥상공원의 좋은시설(연결이 가능하다면 더 넓고 서로 협조하는 공간 가능)
- 근처에 청소년 공부방과, 주민센터의 여가시설(헬스장,컴퓨터교실)등이 위치함으로서 주변공간에 힘이 생김
- 비교적 타 지역에 비해서 재개발 지연으로 인해 같은 문제를 갖고 있어 취향이 비슷하고 동질감이 높다.

Location: Yongsan-dong Ichon 2 Daelim apartment parking space next to the community center
Size: 53000x15000 = about 240 ft
Maximum parking space: external 16 available , internal 14 available
Height of the surrounding buildings : Daelim apartment (66m), community center (12m), real estate (9m)
Surrounding buildings Distance : Daelim apartment (30m), community center (5m), real estate (7.5m)
Surrounding population: 10,518 people 4,389 households (male 5,043 people, female 5,475 people)
Around the house: 3556 dong (Single 474, Alliance 580, Apartment 2502)

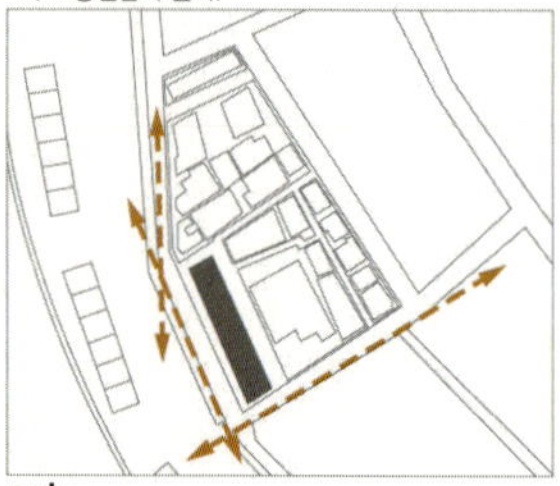
axis

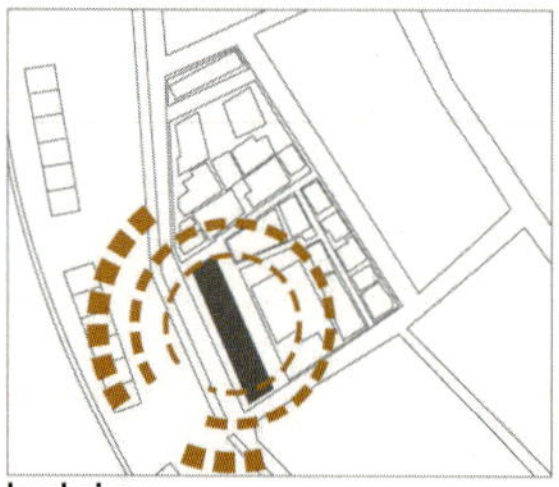
bed view

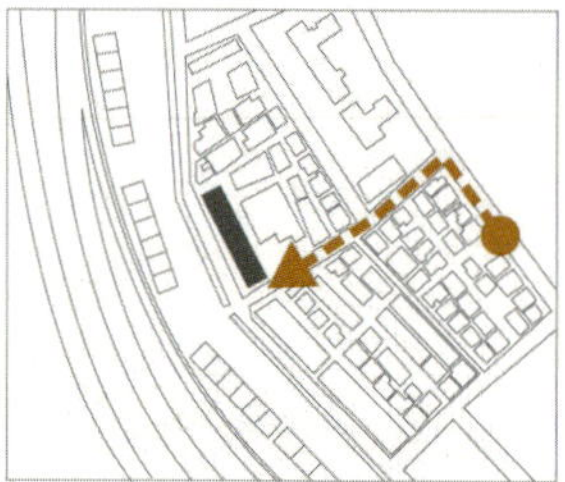
bus, subway

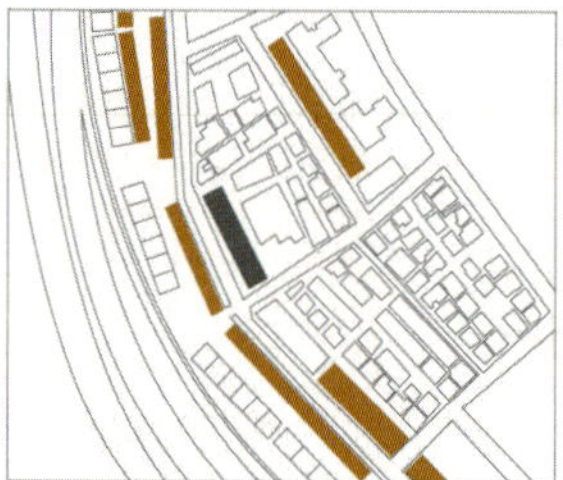
parking space

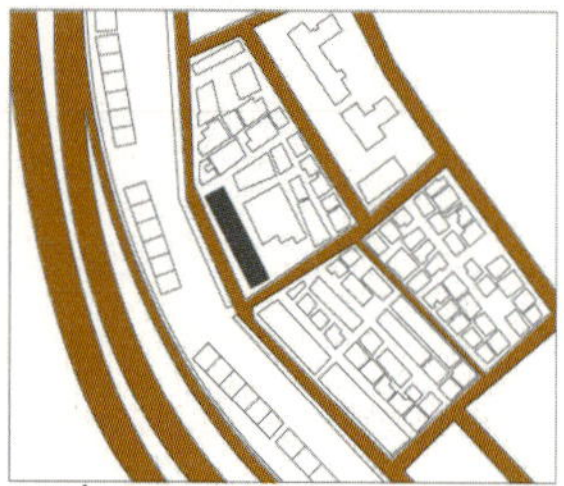
road

남쪽으로 있는 대림아파트의 주차공간이자 주민들의 여가를 (테니스)을 즐길 수 있는 공간이다. 하지만 테니스는 커녕 들어가지도 못하게 문을 잠궈놓았다. 주변 주차공간은 턱없이 부족하여 길가에 주차를 하여 도대체 이공간이 뭐하는 공간 인지 모를정도로 공간을 낭비하고 있었다. 남쪽으론 높은아파 트가있어 조망권이 좋지않고 북쪽으론 주민센터가 있는데 옥상에는 정원이 아주 잘되어있다. 동쪽으로는 부동산등이 있고 서쪽으론 재활용수거하는 곳이있다. 사방이 도로라 접근성은 좋지만 위험할수도 있다. 대중교통으로는 3분거리에 버스는 있지만 두개의 노선 밖에 없고 지하철은 한참 떨어져있어 외부인들이 접근하기 힘들고 불편하다.

A parking space of Daelim apartment on the south and a place where residents can enjoy leisure facilities. However, the door is locked. The nearby parking space was wasting space. The viewing are is not very pleasant as there are tall apartments to the south and there is a community service center to the north that there is a very nice garden prepared on the roof. The accessibility is excellent with streets on every side but can be also dangerous. For public transportation, there is a bus station, but of just 2 routes, about 3 minutes away and subway station is in a far distant that it is difficult and incovenient for outsiders to come.

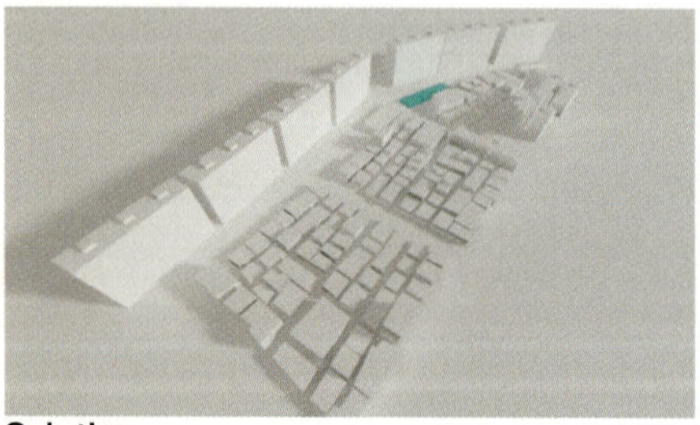

Solution

[Re-Energising space]
사람들의 발길이 잦고 거부감이 없는 외부 주차장 건물에 테니스장과 물탱크가 위치해 있다. 하지만 제 역할을 하지 못하고 있음으로 이 지역의 문제점을 찾아 새로운 PLAY SPACE을 제안하려고 한다. 주민들에게 버림 받은 dead space가 아닌 남녀노소 누구나 쉽게 찾고 즐길 수 있는 이지역에 활력소가 될 수 있는 공간.

A tennis court and water tank are located on the exterior parking structure where people frequently visit that they are not repulsed by. However, we intended to suggest a new PLAY SPACE by finding problems of this region as it is not performing its role adequately. A space that can become a tonic of this region that everybody, men and women of all ages, can visit and enjoy, not an abandoned dead space.

SWOT analysis

S
- 사방이 도로라서 답답하지 않다 (소음문제에 유리)
- 평소에 다니던 주차장이라는 특징 때문에 사람들의 발길이 잦다 (거부감이 없는 위치)
- 주변에 풀이 우거지고 높은 나무가 많아 방음효과를 주며 그나마 삭막하지 않다
- There is not frustrated because on all sides road (there is good noise problems)
- This often comes from people. Because it is the usual features that my old garage.(no reluctance position)
- There is not dreary Because there are many trees around

W
- 지상에 물탱크가 자리하고 있어 공간을 낭비한다.
- 남쪽으로 높은 아파트가 위치해 조망권이 좋지 않다.
- 대중교통 이용이 불편하다.
- 주변상권이 없어서 외부 사람들의 발길이 적다.

- The water tank is located on the ground so wasting space.
- high Apartment located in the south so there is not good view
- Public transportation is inconvenient.
- People of outside is not visit because there is no commercial low.

O
- 주민센터의 옥상공원의 좋은시설(연결이 가능하다면 더 넓고 서로 협조하는 공간 가능)
- 근처에 청소년 공부방과, 주민센터의 여가시설(헬스장,컴퓨터교실)등이 위치함으로서 주변공간에 힘이 생김
- 비교적 타 지역에 비해서 재개발 지연으로 인해 같은 문제를 갖고 있어 취향이 비슷하고 동질감이 높다.
- 테니스장이 제 역활을 하지 못함으로써 데드 스페이스가 지속되고 있다.

- The roof of the community center of the park with good facilities (If it is possible to connect a wider space can cooperate with each other)
- Near the youth study room and a recreation of the community center (gym, computer classes) are located in the surrounding space by causing the power
- Chemistry is relatively high compared to other regions. Because of the same problems.
- Tennis court is dead space .

T
- 계속된 재개발 지연으로 인해 주민들이 심리적으로 민감한 상황이다.
- 부채가 많은 주민들이 갖고 있는 경제적 부담

- Due to continued delays in the redevelopment, residents sensitive for psychologically
- Due to Increased debt, residents sensitive for economic

Target

Concept keyword

Concept process

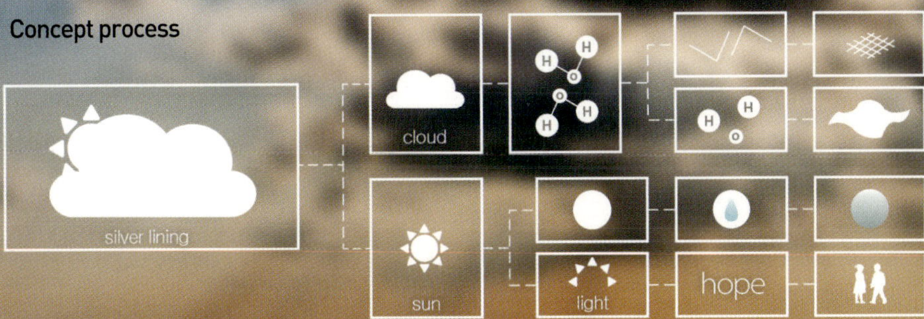

구름의 분자를 연결해주는 연결매체에서 프레임을 떠올렸다.
Recalled frame from a media that connects cloud molecules.

구름의 분자들의 성질을 이용해 막구조를 이용하려고한다.
Intend to use a membrane with properties of cloud molecule.

태양을 지상으로 가져와 물을 채워 의미는 같지만 성격이 다른 태양을 연출하려고 한다.
We are trying to express sun with a same meaning but of different properties by bringing sun to the ground and filling up with water.

자라나는 아이들이 뛰어 노는 것 자체가 지쳐 있는 사람에게 활력을 준다.
Growing children running around and playing itself become light and hope and gives vitality to exhausted people.

Mass process

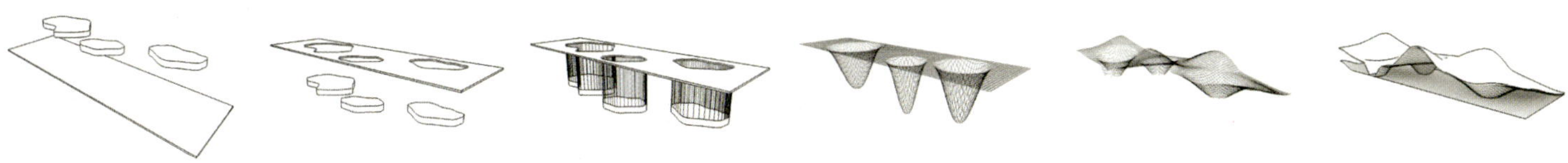

Design process

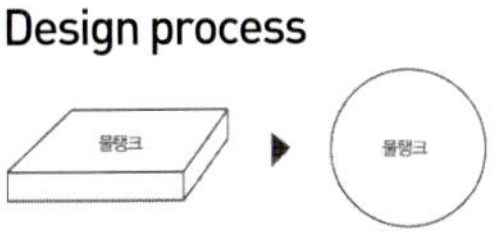

지상에 큼지막하게 자리를 차지하고 있던 물탱크에 컨셉을 적용해 가장 군더더기 없는 구의 형태로 물탱크를 바꿔보았다.

Changed the water tank into a sphere shape with no superfluousness by applying a concept to the water tank that was occupying large space on the ground.

지상에 주차장이 있어 사람과 차량의 동선이겹쳐 위험하다. 그래서 주차장을 지하로 내려 안전과 넓은 공간을 동시에 확보하였다.

Parking space on the ground is dangerous as the flow of human and car traffic overlap. Therefore, safety and wide open space are secured by taking down parking space to the basement.

놀이공간과 휴식공간, 운동공간이 따로 나 누어져 있지 않고 세 공간을 융합해 역동 적이고 생동감 있는 공간을 구성해 보았다.

Play space, resting space, and exercise are not divided and created dynamic and lively space by fusing the three space together.

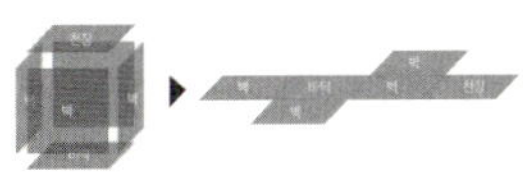

공간의 가장 기본적인 구성요소인 바닥, 천장, 벽을 구분하지 않는 하나되는 공간을 만들어 외부와 내부의 경계가 없는 공간연출

Created one space without classifying most basic components, the floor, ceiling, and wall to create space without boundaries between inside and outside.

공간에 수증기라는 요소를 사용해 오감으로 느끼면서 구름속으로 들어가는 듯한 느낌을 주게 된다.

Gives a feeling of going inside the cloud while feeling five senses by using vapor in the air.

Material

step1 H_2O

액체인 물은 분자간의 인력이 별로 없어 자유로우며, 기체인 수증기의 분자들은 매우 자유로워 날아다닌다. 컨셉을 적용해 액체상태의 물은 투과성이 좋은 유리구 형태의 물탱크를 이용해 오브제 역할과 실용성 디자인하였고 여름에 수증기를 이용한 또다른 공간을 연출하려고한다.

The liquid, water, is free as there are not much gravitation between molecules, and molecules of gas, the vapor, is very free that they are floating around in the air. An objet role and practicality of water in liquid state was designed with high penetrability and glass sphere shaped water tank, and another different space will be designed with vapor in summer.

step2 frame

'틀' 이라는 뜻으로 공간에 작용하는 여러가지 외력을 안전하게 지탱할 목적으로 부재를 접합하여 만든 골조. 프레임의 외력을 버텨내는 힘을 이용해 2층의 구조를 지탱할 뿐 만아니라 구름사다리 또는 철봉 등의 역할도 할 수 있도록 연출하였다.

A framework created by grafting materials with a purpose to support various external forces safely that affect space with a meaning of 'frame.' A structure of 2 floors is not just supported with a force that endures external force of the frame, but is designed to play the role of cloud ladder or pull-up bar.

step3 Membrane structure

천막지의 인장 강도를 높이고 수명을 연장시켜 공간, 구조에도 자유롭게 활용할수 있도록 개량된 구조. 막의 Tension감을 이용해 PLAY 공간을 창출하였다. 트램펄린 역할을 하기도 하고 미끄럼틀 또는 막을 이용한 CACOON을 이용해 아늑한 혼자만의 공간을 연출하였다.

A reformed structure to create free movement in space and structure by increasing the tensile strength of tent fabrics and by expanding the lifespan. The PLAY are was created with tension of curtains. It works as a trampoline and created cozy space with CACOON with slide or curtain that people can enjoy alone.

Axonometric view

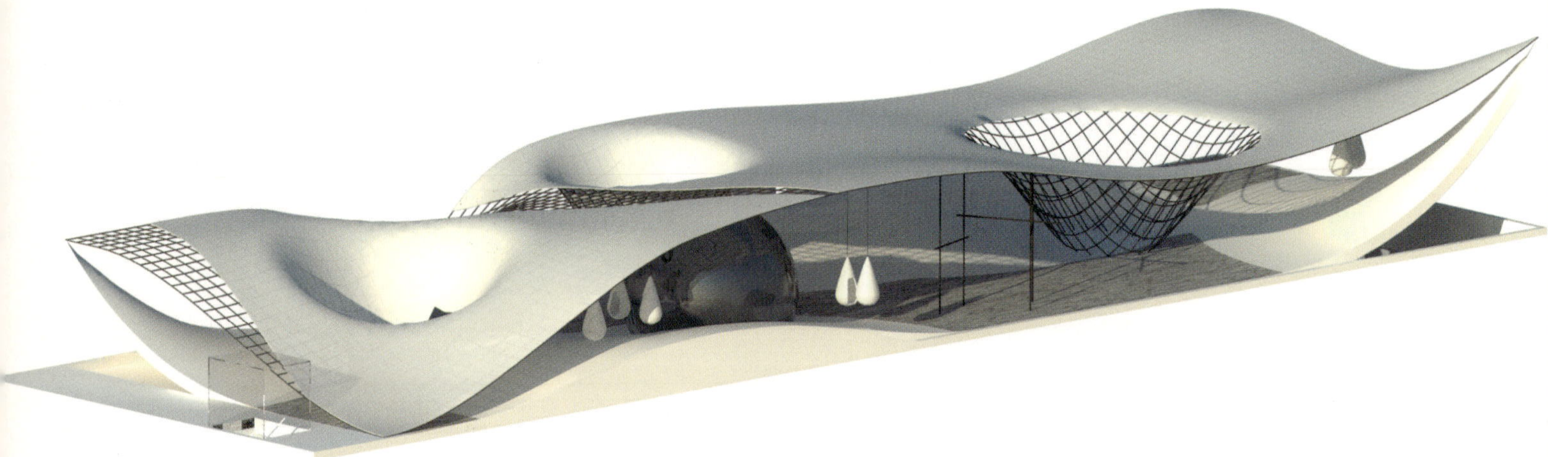

exploded view

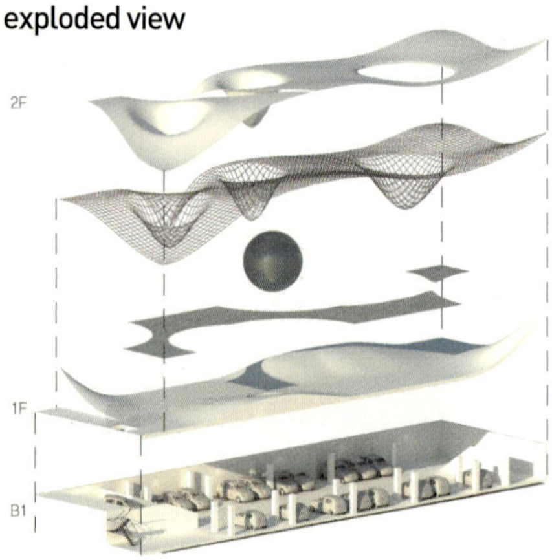

5

천장이자 벽이 되는 프레임에 막으로 되어있는 텐션 감있는 카쿤이 매달려 있어 안에 들어가서 쉴수있고 또 어린이들은 들어가서 놀수있다. 밑에는 트램블린재질인 공기막구조로 되어있어 안전하다.

There is an elastic cacoon made out of curtain on the ceiling and wall frame where anybody can go inside to rest or children can go inside to play. The bottom is made out of air-barrier structure, a material used for trampoline, that it is very safe for use.

6

주차공간이다. 1층에 있는 물탱크가 지하까지 내려와 아침에는 햇빛을 전달하고 밤에는 주차장의 빛이 물탱크를 통해 1층으로 나가 지하와 1층을 연결해주는 매개체 역활을 한다.

A parking space. Water tank on the 1st floor goes all the way down to the basement that it works as a medium that connects the 1st floor with basement as it delivers sunlight in the morning and as light penetrates the 1st floor through water tank at night.

1

1층에서 2층으로 올라가는 공간이다. 프레임으로 되어 있어 이용객 이 프레임을 잡고 2층으로 올라가게 된다.

A space of going up to the 2nd floor from 1st floor. It is made with frames that users can hold onto when they go upto the 2nd floor.

2

2층에 올라왔을때 보이는 광경이다. 프레임으로 되어있는 골조를 막이 둘러 쌓여있다. 가운데 홀이 3군대 뚫려있는데 올라갔다 내려갔다 할수있고 또 미끄럼틀을 탈수도있다.

A scenery that users get to enjoy when they go up to the 2nd floor. Curtain is wrapped around this framework made out of frames. There are 3 holes in the middle for people to go up and down or to go down the slide.

3

구멍이 뚤린곳 안쪽 모습이다. 2층으로 올라갈수 있는데 프레임으로 되어있는곳 보다 좀 더 안전하게 올라가고 내려올수 있다.

It is an interior image where holes are perforated. People can go up to the 2nd floor and it is safer than where it is made out of frames.

4

세개의 홀이 만나는 곳이자 물탱크가 보이는 메인공간이다. 앞에 철봉이 있어 사람들이 운동하고 쿠션감있는 바닥에 앉아 쉴수도 있다.

A place where 3 holes meet and a main space where the water tank can be seen. There is a pull-up bar that people can exercise and rest on the cushiony floor.

1

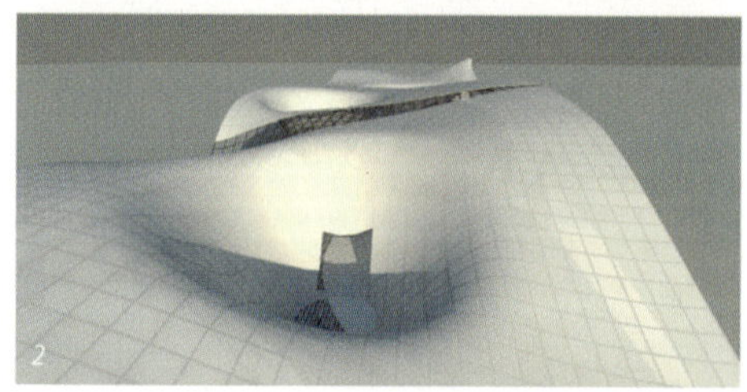
2

3

6

5

4

floor plan

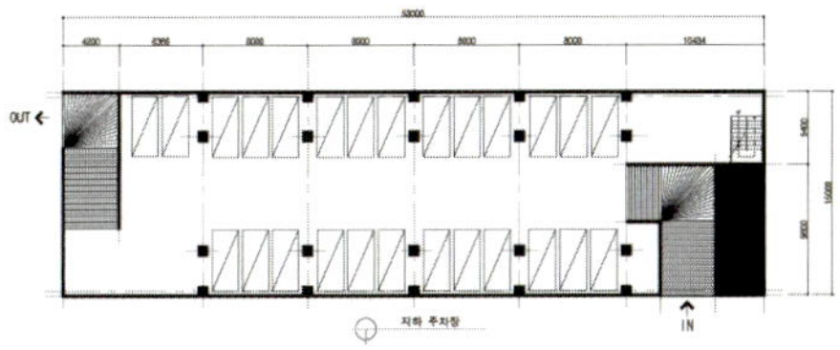
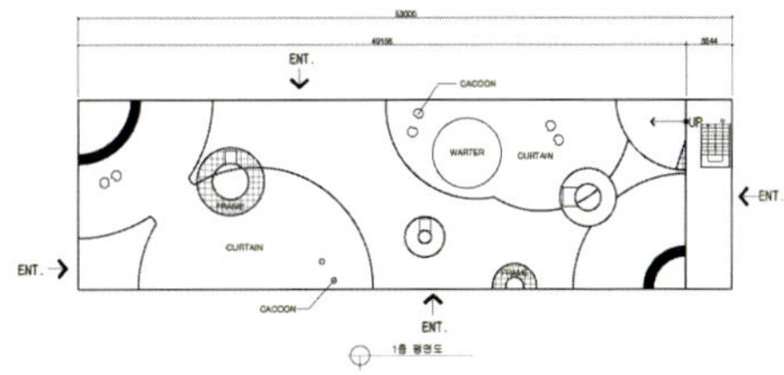
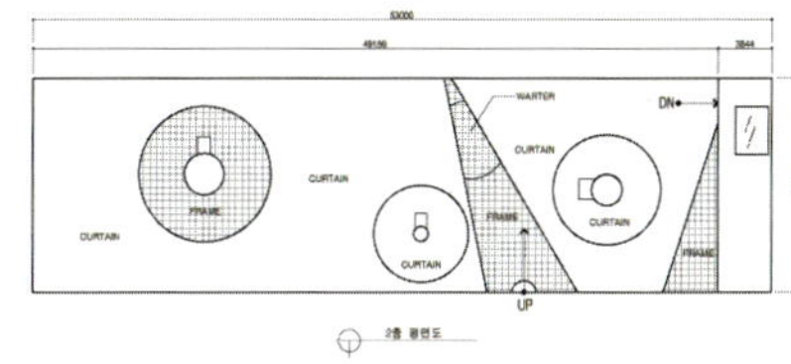

summer view

더운 여름 날 지친 일상을 떠나 휴가를 가기에는 짧은 연휴, 평소와는 다르게 구름형태의 공간 안에 수증기를 이용해 가족들과의 시원한 연휴를 보낼 수 있는 매력적인 공간. 멀리 휴양지를 찾아 떠나지 않아도 집 앞에서 가족들과 즐길 수 있는 곳 주민들이 여름을 기다리게 하는 공간을 서부이촌동 이촌2동 주민들에게 제안해 본다.

Holidays too short for families to take a break from exhausted everyday life and go on vacation on hot summer days. We are suggesting place where families can enjoy right in front of their houses and a place that makes residents to long for summer to Seobuichon-dong Ichon-2dong residents.

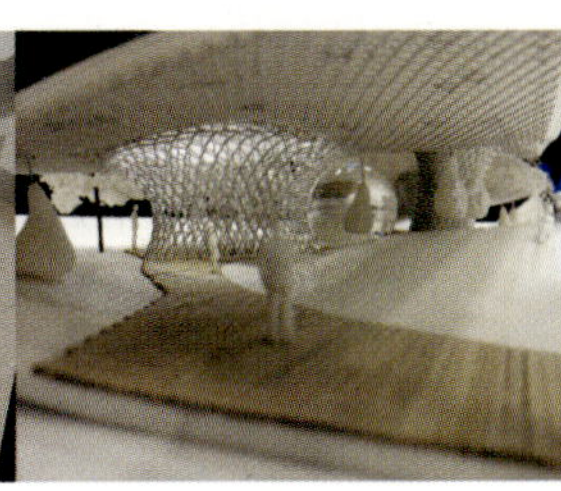

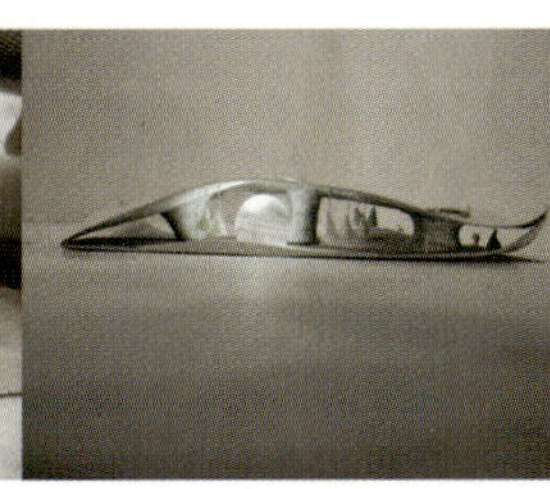

Hiding in the Book

숨어 있는 책과의 관계를 만들어주는 책방을 통해 세상을 둘러보기 위한 방식은 하나씩 넓어져 간다.

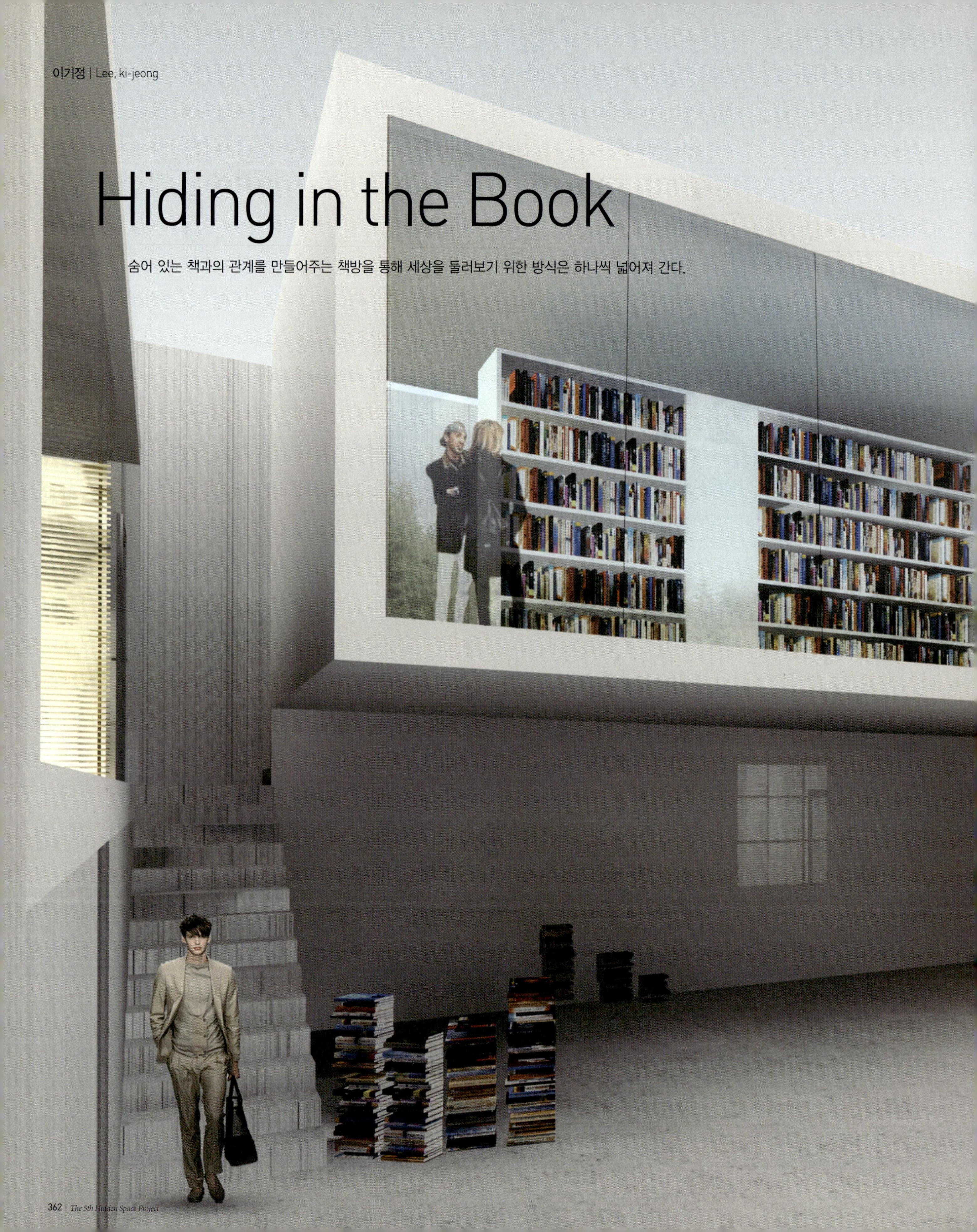

Background

헌책방은 우리가 살고있고 살아가고 있는 생활속에서 삶을 같이하는 보내며 생기있게 해주는 곳이다. 옛부터 헌책방은 봇물처럼 쏟아져 나오는 서적들을 헌책방을 거쳐 손에 손으로 전달되어 지식의 매개체 역할을 해온 곳이 바로 헌책방이다. 헌책방에 숨어있는 책을 찾아가는 길은 미로처럼 꼬불꼬불 복잡하며 그 길에서 이동감과 공간들의 연속 흐름을 느끼고 자연스런 시퀀스가 생성되었다. 미로처럼 쌓인 책들속에서 새책방에서 찾을 수 없는 책들을 찾을 수 있고 그 책을 통해 소통과 교감을 나눌 수 있었다. 헌책방을 찾는 사람들은 책더미 속에서 자신만의 '보물' 을 찾으며 이곳에서 긴시간을 보내며 자신만에 휴식을 보낸다. 쓰윽 훑어보고 그칠 게 아니라, 찬찬히 찾아보면 뜻밖의 좋은 책을 책방 한 구석에서 찾을 수 있다. 보물처럼 고른 책을 들고 앉아 하나씩 곱씹어보는 재미도 만만치 않다. 이러한 헌책방이 배다리에서 위기에 처해있다. 배다리에는 산업화도로계획으로 40여개나 되던 헌책방들이 현재는 6~7개만이 그 명맥을 이어가고 있다. 배다리 헌책방 골목의 부활로 배다리의 재생을 꾀하고 있다.

Used book stores are the place that stay with vivid memory of people in their lives. In the past, used book stores have been serving as a role of delivering all different types of books to customers. Searching for books through used book stores is definitely complicated that created natural string of sequences in the continuum of movement and space. In the used book stores, it is feasible to find books that cannot be found at regular book stores sharing and communicating thoughts through used books. People visiting used book stores find their own 'treasure' in the stack of books spending long time to relax themselves there. Not running eyes on used books but carefully looking for them makes it possible to find valuable books in the corner. It is still fun to read every page of the books that are as valuable as treasure. These used book stores are now at risk on the pontoon bridge. On the pontoon bridge, there used to be about 40 used book stores in the past. However, there are only about 6 to 7 used book stores due to the industrialized road plans. It is currently a hope to restore the pontoon bridge with the survival of used book stores.

Site analysis

인천시 동구 금창동 배다리의 헌책방 거리는 인천에서 태어나고 자란 어른들에게 정감어린 착창시절의 옛 향수를 떠 올리게 하는 곳이다. 해마다 새 학기가 되면 이 헌책방 거리는 참고서나 교과서를 구입하기 위래서 책방들을 둘러보는 학생들과 헌책을 이곳으로 팔러 나온 학생들로 늘 붐볐다. 하지만 빽빽하게 들어선 헌책방들이 이제는 몇 집만 남아 명맥을 잊고 있다. 예전에 이 지역까지 바닷물이 드나들고 이 일대에 배를 댈 수 있도록 만든 다리가 있다고 해서 붙여진 이름 배다리는 동인천과 제물포를 이어주는 곳으로 인천에서 뿌리를 내린 사람이라면 배다리를 모르는 사람이 없을 정도로 인천의 역사와 함께 잘 알려졌으며, 아직까지 인천를 대표할 수 있는 소박함이 그대로 남아 있는 곳이다. 배다리의 골목을 시작으로 골목에서 또 다른 길로 이어지는 미로처럼 작은 길에서 느끼는 넓고 깊은 삶의 흔적을 엿볼 수 있으며 요즘처럼 높고 크고 빠른 것만을 발전이라고 우선시하는현재와 상반되게 느리고 자연스럽고 조금은 촌스럽지만 마음의 여유를 갖고 살아가는 또 다른 아름다움과 정겨움을 생각해 보게 한다.

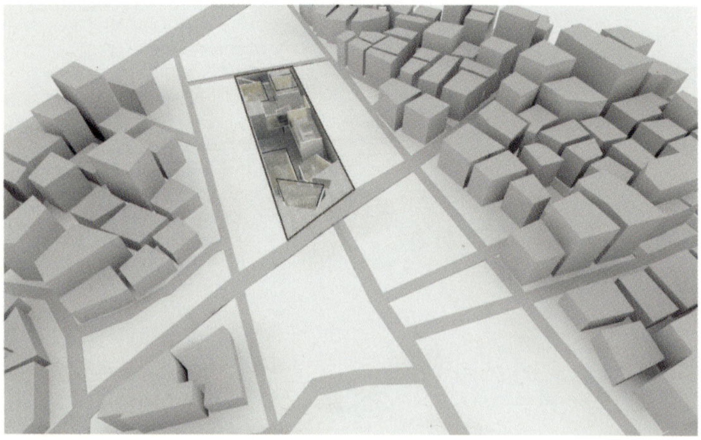

Used book stores street on the pontoon bridge at Keungchang-dong, Dong-gu, in Incheon-si is the place that brings back the reminiscence in the days of creation to childrens that were born and raised in Incheon. In the upcoming semester every year, used book store street is full of students that look around used books to purchase textbook or reference books or to sell used books to others. However, only several used book stores are left in business unlike the past. The pontoon bridge was named after how sea water was flowing up to the area with the bridge that ships were brought. Hence, the pontoon bridge connects the eastern Incheon and Jemulpo that everyone raised in Incheon knew in the history of Incheon. This place is still left with the image of reminiscence that represents Incheon. Traces of lives are left on the narrow road that is connected to another route that starts from the alley of the pontoon bridge. Unlike the current period that people regard something high and fast as the one related to the development, this place brings people an opportunity to relax their mind, enjoy the tranquil environment in somewhat old but natural manner, and think of different beauty and affection.

Concept

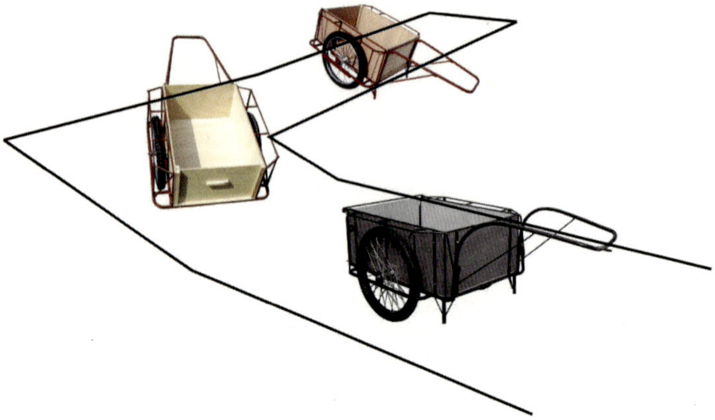

갈 대

꼬불꼬불 복잡하고 미로처럼 이어지는 길에서 사람들은 만나며 마주치며 소통을 하며 놀이를 하는 곳으로 우리에 삶을 함께하던 공간이다.

Alley

In the complicated and crooked road, people saw others communicating and sharing thoughts with them that always stayed in their lives.

리어카

헌책방은 길거리에서 헌책을 담고 다니던 리어카를 통해 사람들이 모여 들고 그 곳에서 자리를 잡고 커뮤니케이션을 하며 형성되어 지역에 녹아 들었다.

Handcart

Used books are the place where people gathered with hard-cart that carried used books and communicated with each other.

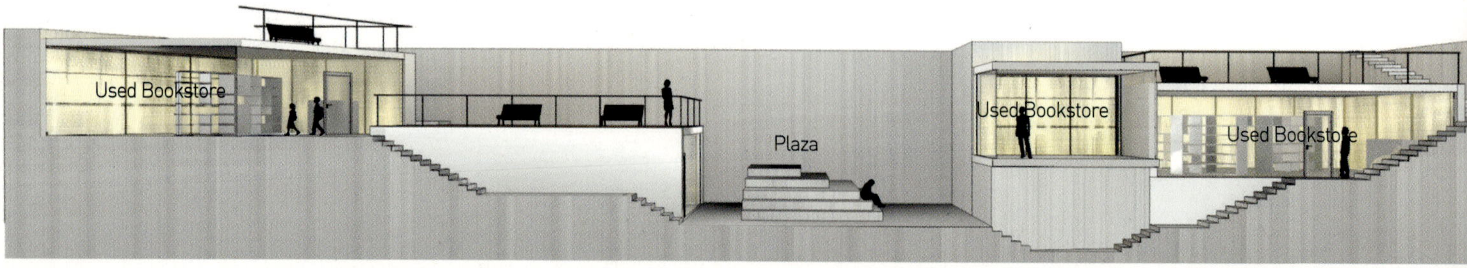

Purpose

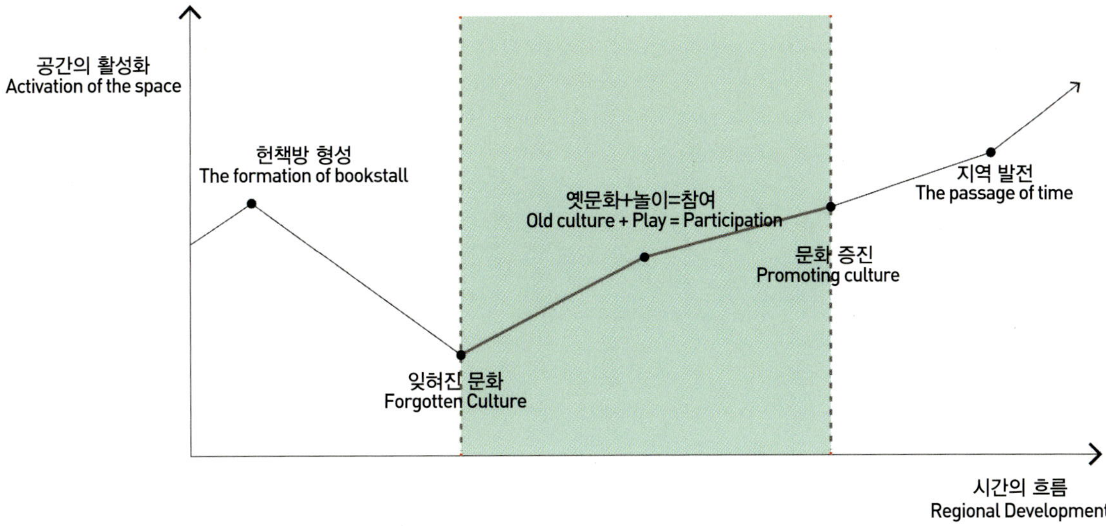

과거 Past	해방 이후 피난민들이 배다리 지역에 집중하면서 배다리 일대는 서민들이 책을 사고 팔던 헌책방으로 형성.	Refugees came to the pontoon bridge area after Korea was liberated that people purchased and sold books at used book stores near the area.
현재 Present	없어져가는 헌책방 문화 공간에 '놀이'를 결합해 복합 문화 공간으로 발전시켜 지역 주민들의 참여를 유도.	Play' was combined in the used book cultural space that was nearly lost in the area promoting residents to participate in it by developing it as a complex cultural space.
미래 Future	배다리 지역의 옛문화인 헌책방이 활성화 되어 지역주민들의 삶의 질이 높아지며 지역발전에 힘씀.	Used book stores, as a traditional culture in the pontoon bridge area, were promoted improving the quality of lives of residents to supports the local development.

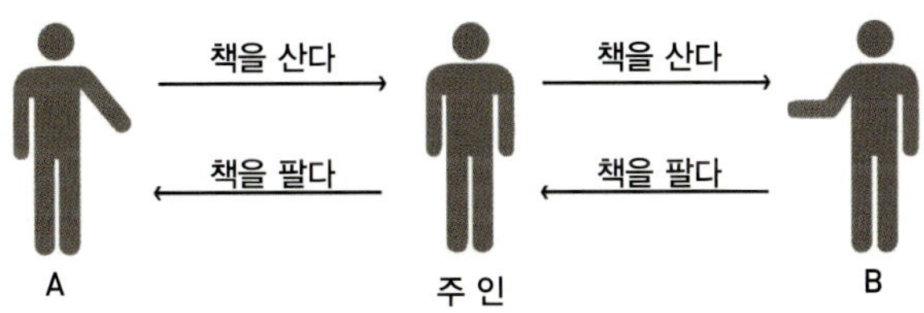

헌책방의 문화는 단순히 책을 사는 것만이 아닌 책을 팔고 그 곳에서 소통하며 서로의 상호 작용을 한다. 그로 인해 커뮤니케션이 생겨나고 헌책방 주인을 거지치 않고 판매자와 소비자가 바로만나는 책방 장터가 생성된다.

The culture of used book store is not merely to purchase books but to sell books, communicate with people, and exchange thoughts with one another. Therefore, they end up communicating with each other and creating the book store market that sellers and consumers meet without going through the owner of used book stores.

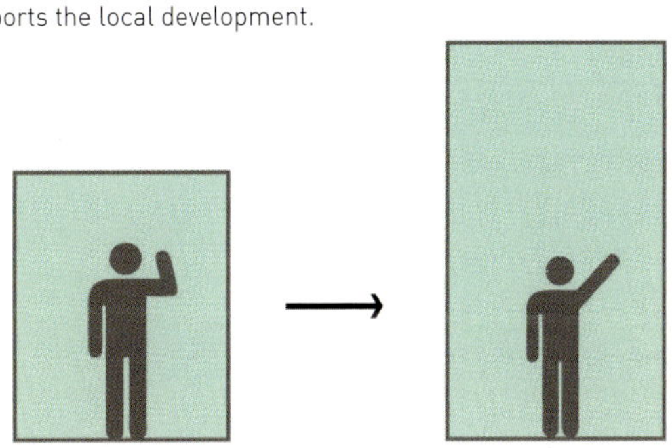

헌책방 이용자들이 휴먼스케일의 헌책방과 골목을 지나 초월적 스케일에 광장으로 공간과 공간을 넘나들며 헌책방 골목이 재미와 낭만이 충족한 문화임을 공감하고 헌책방 거리라는 한계를 넘어 영혼의 안식처가 될 수 있다.

Used book store users tend to enjoy the transcendental scale of plaza beyond the scope of used book store alley in the human scale sharing their sympathy for how it is the place for fulfilling the fun and romance and also serves as a place to relax by overcoming the limitation of used book store street.

Roof Plan

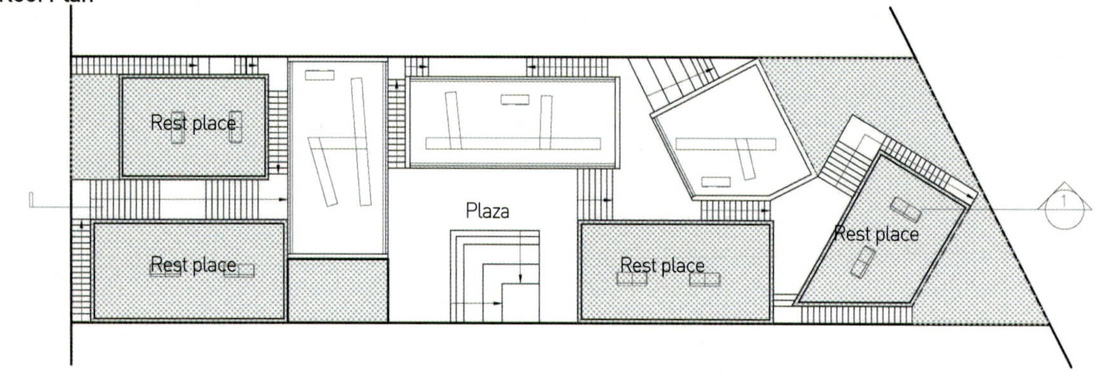

Elevation

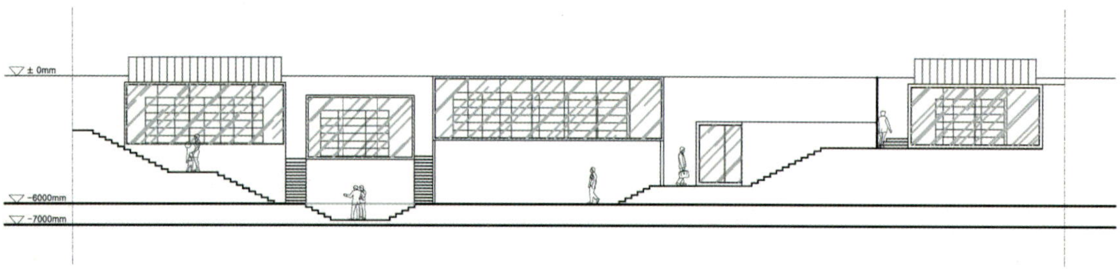

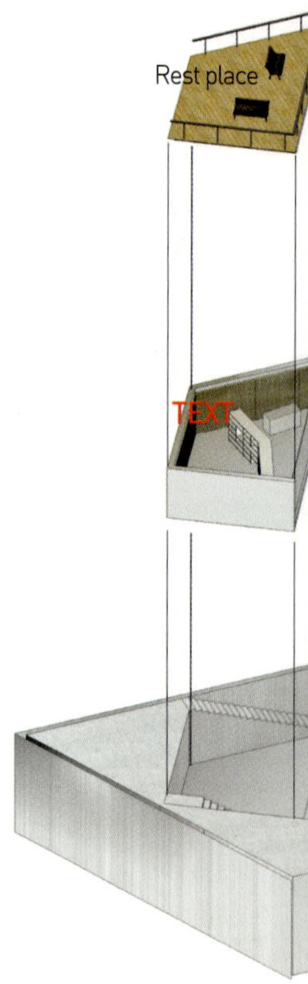

헌책방은 미로처럼 쌓인 책들속에서 새책방에서 찾을 수 없는 책들을 찾을 수 있다. 헌책방을 찾는 사람들은 책더미 속에서 자신만의 '보물' 을 찾으며 이곳에서 긴시간을 보낸다. 쓰윽 훑어보고 그칠게 아니라. 찬찬히 찾아보면 뜻밖의 좋은 책을 책방 한 구석에서 찾을 수 있다. 보물처럼 고른 책을 들고 앉아 하나씩 곱씹어보는 재미도 만만치 않을 성 싶다.

Used book stores provide people an opportunity to find books that cannot be found in the new book store through the stack of books. People visiting the used book store try to find their own 'treasury' in the stack of books spending a long time there. Not running eyes on used books but carefully looking for them makes it possible to find valuable books in the corner. It is still fun to read every page of the books that are as valuable as treasure.

Nowadays, so called circulating readers that purchased a new book and then sells it to purchase another book is in increasing trend. Money is not the only one that is circulating. Books shall be circulating to serve the appropriate role.

Exploded view

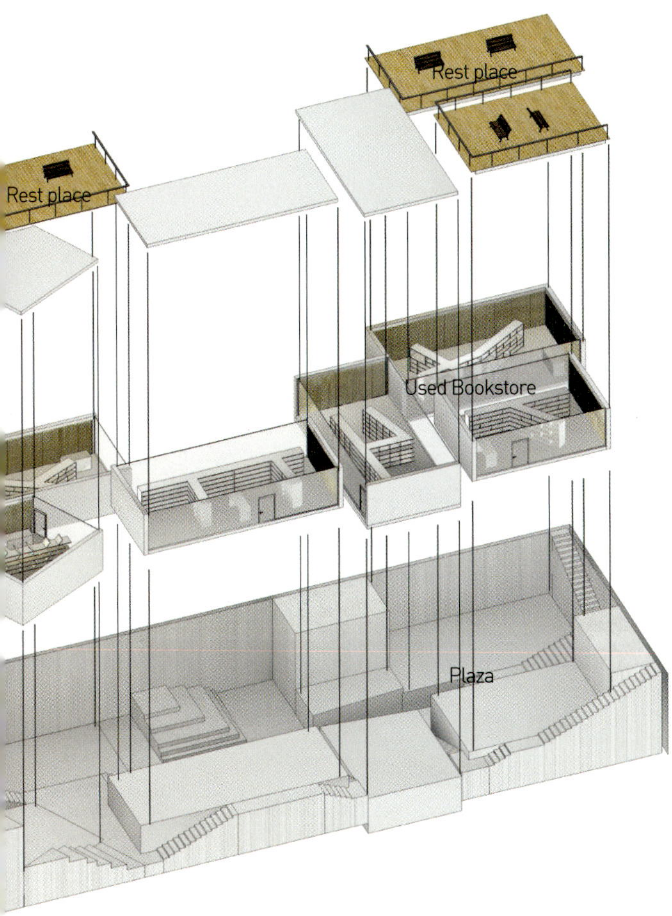

Used books are becoming new these days. They are flexible and fun. People see less of obstacles and have more opportunity. When [illegible] used book stores, people might catch an unexpected luck. They can find a valuable book to have insight, listen to music, a[illegible] a rest. What has happened t[illegible]sed book store?

Used book stores are the way to protect the history in an effort to acquire out-of-print books, evaluate works, and cultivate rare and old books. In the perspective of recycling the resources, used books have much value. Programs for the local society are being operated in this place.

2014
INTERNATIONAL DESIGN EXCHANGE PROJECT

The 5th Hidden Space Project

RE-ENERGISING SPACE

The Cass London Metropolitan University
The Sir John Cass Faculty of Art, Architecture and Design
Interior Design BA (Hons)
UK

Hunan University
Dept. of Industrial Design, in school of Design
CHN

Dongyang Mirae University
Dept. of Interior Design, in school of Design
Rep. of KOREA

19 June -
4 July 2010
DEU
HUN
IAS
IDN
ITA
Intl.
Architectural
Showcase
Presented by foreign
embassies and cultural
institutes in collaboration with The British Council and
The Architecture Foundation for the London Festival
of Architecture 2010

JPN
KOR
LVA
Living Bridges
A UK and Korea collaboration
26. Korean Cultural Centre and London Metropolitan University
31
16 Goodge Street, London E1 7TP
22 June – 27 June
10am – 5pm
Private View: 24 June, 6–8pm
An exhibition of work by 25 students from three collaborative universities; London Metropolitan, Kyung Won and Dong Yang. The project brief focused on urban and community issues shared by the UK and Korea and asked the students to create a 'living bridge' structure that is both an architectural metaphor for community support and a space for the communities to enjoy, integrate and communicate. Each student has built a model and will show the structure, its location and context.
The work from this project has been compiled in a book that will be published alongside the exhibition. Living Bridges has given us a valuable insight into how young creative minds work, and how they deal with physical evidence and community challenges.
Image: Night Visual Designer, 2010. © Anelika Ghouralal

2010
INTERNATIONAL DESIGN EXCHANGE PROJECT

The 1st Hidden Space Project

LIVING BRIDGES

The Cass London Metropolitan University
Dongyang Mirae University
Kyungwon University

Symposium Hidden Spaces :
Wednesday 27th June 10.30am-3.30pm
Gensler Architects
Aldgate House
33 Aldgate High Street

2012
INTERNATIONAL DESIGN EXCHANGE PROJECT

The 3rd Hidden Space Project

HIGH STREET

The Cass London Metropolitan University
Dongyang Mirae University
Kyungwon University

FHHS2010 - *The 1st Hidden Space Project* **"LIVING BRIDGES"**

FHHS2011 - *The 2nd Hidden Space Project* **"URBAN ROOFTOPS"**

FHHS2012 - *The 3rd Hidden Space Project* **"HIGH STREET"**

FHHS2013 - *The 4th Hidden Space Project* **"THE RISE OF THE CREATIVE DISTRICT"**

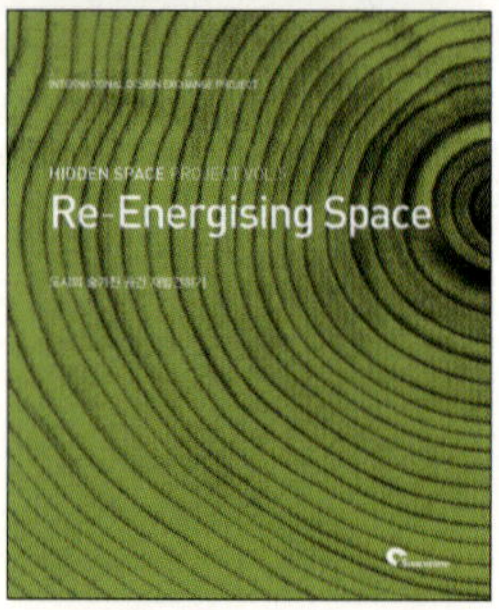

FHHS2014 - *The 5th Hidden Space Project* **"RE-ENERGISING SPACE"**

*** FHHS** (Future Horizon for Hidden Space)

2010년 [LIVING BRIDGES]주제로 영국 런던 메트로폴리탄대학과 한국의 가천대학교, 한국의 동양미래대학교가 공동으로 런던 아키텍 페스티벌 참가하면서 시작된 FHHS는 우리시대 숨겨진 공간들을 10년간 10개의 주제로 교류전을 기획, 현재 5회까지 진행했으며 이후 주제를 현재 기획중이다.
FHHS는 이후 계속해서 우리주변의 숨겨진 공간들을 찾아 열정과 지혜, 그리고 성찰의 노정(路程) 을 계속해 나갈 예정이다.

2014

INTERNATIONAL DESIGN EXCHANGE PROJECT

The 5th Hidden Space Project

RE-ENERGISING SPACE

Published By 2014 RE-ENERGISING SPACE Project
(Kaye Newman / Janette Harris / Park, Young Tae / Jun Zhang / Ahn, Sung Hee)

Total Design
Park, Young-Tae | DONGYANG MIRAE University
만든이 | 동양미래대학교 박영태
편집, 기획, 표지 디자인 | 박영태

펴낸날 | 2014. 6.20. 1쇄 발행
발행인 | 김기현
발행처 | spacetime · 시공문화사
등 록 | 1993년 3월12일 1-1493
주 소 | 120-796 서울시 서대문구 독립문공원길 13 극동프라자 5층
전 화 | 02-3147-1212
팩 스 | 02-3147-2626
ISBN | 978-89-5592-295-0

정가 | 23,000원